otentials

VI	VII	VIII		
		$1s^2$ 4.00260 $_2$**He** $^1S_0$ 24.59; 54.42 Helium		
$2s^22p^4$ 15.999$_4$ $^3P_2$ $_8$**O** 13.62; 35.1; 54.9 Oxygen	$2s^22p^5$ 18.998403 $^2P_{3/2}$ $_9$**F** 17.42; 35.0; 62.7 Fluorine	$2s^22p^6$ 20.179 $^1S_0$ $_{10}$**Ne** 21.56; 41.0; 63.5 Neon		
$3s^23p^4$ 32.06 $^3P_2$ $_{16}$**S** 10.36; 23.3; 34.8 Sulfur	$3s^23p^5$ 35.453 $^2P_{3/2}$ $_{17}$**Cl** 12.97; 23.8; 39.6 Chlorine	$3s^23p^6$ 39.948 $^1S_0$ $_{18}$**Ar** 15.76; 27.6; 40.9 Argon		
51.996 $3d^54s$ $_{24}$**Cr** $^7S_3$ 6.77; 16.5; 31.0 Chromium	54.9380 $3d^54s^2$ $_{25}$**Mn** $^6S_{5/2}$ 7.43; 15.6; 33.7 Manganese	55.84$_7$ $3d^64s^2$ $_{26}$**Fe** $^5D_4$ 7.90; 16.2; 30.6 Iron	58.9332 $3d^74s^2$ $_{27}$**Co** $^4F_{9/2}$ 7.86; 17.1; 33.5 Cobalt	58.69 $3d^84s^2$ $_{28}$**Ni** $^3F_4$ 7.64; 18.2; 35.3 Nickel
$4s^24p^4$ 78.9$_6$ $^3P_2$ $_{34}$**Se** 9.75; 21.2; 30.8 Selenium	$4s^24p^5$ 79.904 $^2P_{3/2}$ $_{35}$**Br** 11.81; 21.8; 35.9 Bromine	$4s^24p^6$ 83.80 $^1S_0$ $_{36}$**Kr** 14.00; 24.4; 37.0 Krypton		
95.94 $4d^55s$ $_{42}$**Mo** $^7S_3$ 7.10; 16.2; 27.2 Molybdenum	[98] $4d^55s^2$ $_{43}$**Tc** $^6S_{5/2}$ 7.28; 15.3; 29.5 Technetium	101.0$_7$ $4d^75s$ $_{44}$**Ru** $^5F_5$ 7.37; 16.8; 28.5 Ruthenium	102.9055 $4d^85s$ $_{45}$**Rh** $^4F_{9/2}$ 7.46; 18.1; 31.1 Rhodium	106.42 $4d^{10}$ $_{46}$**Pd** $^1S_0$ 8.34; 19.4; 32.9 Palladium
$5s^25p^4$ 127.6$_0$ $^3P_2$ $_{52}$**Te** 9.01; 18.6; 28.0 Tellurium	$5s^25p^5$ 126.9045 $^2P_{3/2}$ $_{53}$**I** 10.45; 19.1; 33.0 Iodine	$5s^25p^6$ 131.2$_9$ $^1S_0$ $_{54}$**Xe** 12.13; 21.0; 31.0 Xenon		
183.8$_5$ $5d^46s^2$ $_{74}$**W** $^5D_0$ Tungsten 7.98	186.207 $5d^56s^2$ $_{75}$**Re** $^6S_{5/2}$ Rhenium 7.88	190.2 $5d^66s^2$ $_{76}$**Os** $^5D_4$ Osmium 8.73	192.2$_2$ $5d^76s^2$ $_{77}$**Ir** $^4F_{9/2}$ Iridium 9.05	195.0$_8$ $5d^96s$ $_{78}$**Pt** $^3D_3$ Platinum 8.96; 18.6
$6s^26p^4$ [209] $^3P_2$ $_{84}$**Po** 8.42 Polonium	$6s^26p^5$ [210] $^2P_{3/2}$ $_{85}$**At** 9.0 Astatine	$6s^26p^6$ [222] $^1S_0$ $_{86}$**Rn** 10.75 Radon		

Standard Atomic Weights 1981
The values of A_r(E) are considered reliable to ± 1 in the last digit, or ± 3 if that digit is subscript (except the hydrogen case, when the uncertainty is enclosed in parentheses)

Key	
6.94$_1$	2s — Valence electron configuration
Symbol of the element — $_3$**Li**	$^2S_{1/2}$ — Ground state term
Atomic number — 3	
Lithium	5.39; 75.6; 122.4 — Ionization potential (eV) of Li I, Li II, Li III, respectively

4 $4f^96s^2$ $^6H_{15/2}$ n 5.86; 11.5; 21.9	162.5$_0$ $4f^{10}6s^2$ $_{66}$**Dy** $^5I_8$ Dysprosium 5.94; 11.7; 22.8	164.9304 $4f^{11}6s^2$ $_{67}$**Ho** $^4I_{15/2}$ Holmium 6.02; 11.8; 22.8	167.2$_6$ $4f^{12}6s^2$ $_{68}$**Er** $^3H_6$ Erbium 6.11; 11.9; 22.7	168.9342 $4f^{13}6s^2$ $_{69}$**Tm** $^2F_{7/2}$ Thulium 6.18; 12.1; 23.7	173.0$_4$ $4f^{14}6s^2$ $_{70}$**Yb** $^1S_0$ Ytterbium 6.25; 12.2; 25.0	174.967 $4f^{14}5d6s^2$ $_{71}$**Lu** $^2D_{3/2}$ Lutetium 5.43; 13.9; 21.0
$5f^97s^2$ $^6H_{15/2}$ um 6.23	[251] $5f^{10}7s^2$ $_{98}$**Cf** $^5I_8$ Californium 6.30	[252] $5f^{11}7s^2$ $_{99}$**Es** $^4I_{15/2}$ Einsteinium 6.42	[257] $5f^{12}7s^2$ $_{100}$**Fm** $^3H_6$ Fermium 6.5	[258] $5f^{13}7s^2$ $_{101}$**Md** $^2F_{7/2}$ Mendelevium 6.6	[259] $5f^{14}7s^2$ $_{102}$**No** $^1S_0$ Nobelium 6.6	[260] $5f^{14}7s^27p$ $_{103}$**Lr** $^2P_{1/2}$ Lawrencium

Feb/88

F.W. DAGBY

Spectroscopy

UBC.

31

Springer Series in Chemical Physics

Edited by J. P. Toennies

Springer Series in Chemical Physics
Editors: V. I. Goldanskii R. Gomer F. P. Schäfer J. P. Toennies

Volume 40 **High Resolution Spectroscopy of Transient Molecules**
Editor: E. Hirota

Volume 41 **High Resolution Spectral Atlas of Nitrogen Dioxide 559–597 nm**
By K. Uehara and H. Sasada

Volume 42 **Primary Processes of Bacterial Photosynthesis**
Editor: M. E. Michel-Beyerle

Volumes 1–39 are listed on the back inside cover

A. A. Radzig B. M. Smirnov

Reference Data on Atoms, Molecules, and Ions

With 114 Tables and 75 Figures

Springer-Verlag
Berlin Heidelberg New York Tokyo

Dr. Alexandre A. Radzig
Professor Boris M. Smirnov

I. V. Kurchatov Institute of Atomic Energy,
SU-Moscow, USSR

Title of the original Russian edition: *Spravochnik po atomnoi i molekularnoi fizike*

ISBN 3-540-12415-2 Springer-Verlag Berlin Heidelberg New York Tokyo
ISBN 0-387-12415-2 Springer-Verlag New York Heidelberg Berlin Tokyo

Library of Congress Cataloging in Publication Data. Radzig, A. A. (Alexandre Alexandrovitch), 1944- Reference data on atoms, molecules and ions. (Springer series in chemical physics ; 31) Translation of: Spravochnik po atomnoĭ i molekulîarnoĭ fizike. Includes index. 1. Atomic structure-Handbooks, manuals, etc. 2. Molecular structure-Handbooks, manuals, etc. 3. Ions-Handbooks, manuals etc. 4. Chemistry, Physical and theoretical-Handbooks, manuals, etc. I. Smirnov, B. M. (Boris Mikhaĭlovich) II. Title. III. Series: Springer series in chemical physics ; v. 31. QC173.R2613 1985 539'.02'02 84-23515

Typesetting: Schwetzinger Verlagsdruckerei, 6830 Schwetzingen
Offset printing: Beltz Offsetdruck, 6944 Hemsbach/Bergstr. Bookbinding: J. Schäffer OHG, 6718 Grünstadt
2153/3150-543210

To my dear *Catherine* (b. 78) –
World Patience Competition Prize Winner
(A. R.)

Preface

This reference book contains information about the structure and properties of atomic and molecular particles, as well as some of the nuclear parameters. It includes data which can be of use when studying atomic and molecular processes in the physics of gases, chemistry of gases and gas optics, in plasma physics and plasma chemistry, in physical chemistry and radiation chemistry, in geophysics, astrophysics, solid-state physics and a variety of cross-disciplinary fields of science and technology. Our aim was to collect carefully selected and estimated numerical values for a wide circle of microscopic parameters in a relatively "not thick" book. These values are of constant use in the work of practical investigators.

In essence, the book represents a substantially revised and extended edition of our reference book published in Russian in 1980. Two main reasons made it necessary to rework the material. On the one hand, a great deal of new high-quality data has appeared in the past few years and furthermore we have enlisted many sources of information previously inaccessible to us. On the other hand, we have tried to insert extensive information on new, rapidly progressing branches of physical research, such as multiply charged ions, Rydberg atoms, van der Waals and excimer molecules, complex ions, etc. All this brings us to the very edge of studies being carried out in the field.

Despite this, we have still paid great attention to presenting the material in a "compact" form, which is most convenient in practice, and have tried not to overload the book with unnecessary details, thus allowing the reader to form a clear view of the subject as a whole. However, the volume of the book has nearly doubled. One is left to think that this is a natural development of the situation which has been taking place during the last ten years.

Working on "Reference Data", we proceeded from the assumption that a practical manual of this kind allows one to deal with an enormous flow of new information when it is relevantly processed. We hope that our reference book will be of use to a wide circle of today's researchers, whether experimental or theoretical, who have been advancing for a long time or who are now following the paths of their own investigations.

We are very grateful to A. Volovick, who first translated the original text into English. It gives us great pleasure to acknowledge the help and fruitful cooperation we have received from "Springer-Verlag" during the time in which this book was being prepared for publication. Of course, many barriers along the way seemed insurmountable at first sight and only the attentive and patient attitude to the authors' intentions from the publishing-house staff helped the successful realization of the project. We would like to express here our sincere thanks to Ms. G.M. Hayes who copy-edited this manuscript and greatly contributed to its improvements. We are very indebted to Mr. C.-D. Bachem of the Book Production Department for his highly skilled work. And finally, we offer our sincere thanks to Dr. H.K.V. Lotsch without whose active participation this book would hardly have reached the reader.

Moscow, June 1985 *A.A. Radzig · B.M. Smirnov*

Contents*

* References are given just after the introductory text in each section

Part II **Molecules and Molecular Ions**

1. Introduction

In practice, a scientist who investigates atomic and molecular systems may face a situation when information about parameters of atoms and molecules is required. Then, one must turn to the relevant reference material, reviews, or original papers. It is good if one knows where to find these parameters and has an idea of reliability in determining the data involved. However, it may just as well be that the information required goes beyond the sphere of one's own scientific interests. In this case, the *values* of quantities to be determined are of primary significance, provided they are reliable, and *how* this information was obtained is of little importance or, sometimes, "interest". As is proved by experience, the aid of an expert is necessary to give guidelines in the search for such information and especially in an estimation of its reliability. Thus, offering the reader a "Reference Data on Atoms, Molecules and Ions" we presume to be such experts who have selected the most useful data on the parameters of atomic and molecular species, the error estimation of the data involved being given throughout.

In what way were the selection of data to be included in the reference book and their accuracy classification, which is of importance, carried out? The very idea of making our "Reference Data" as simple and informative as possible contributed much to the solution of the first task. Whenever ample information concerning a particular quantity was available, we restricted our consideration to the most simple and widely applied systems of species. This enabled us to compile in a rather compact form the quantitative data dealing with a wide range of individual properties of atomic and the simplest molecular species thoroughly investigated by physicists. Although the ways of determining average characteristics and the methods of estimating their accuracy and reliability are not referred to, processing of the information available as well as analysis of the methods used to gain information, have been performed in each case. Generally we tried to give an idea about the most reliable methods for determining the desired values in brief introductions at the beginning of nearly every section. We have introduced a classification scheme with several classes of accuracy and assigned most quantities to a particular class. The following arbitrary notation has been used throughout:

A ... for uncertainties within 1%
B ... for uncertainties within 3%
C ... for uncertainties within 10%

D ... for uncertainties within 30%
E ... for uncertainties larger than 30% .

When such a classification has not been made, the values presented have been truncated or rounded off to the point where the uncertainty was at most ±1 or 2 in the last digit quoted.

Let us touch upon the material of the "Reference Data". The principal information is contained in its tables and figures. We have mainly attempted to choose forms of data presentation which make values evident and informative. We did this by simplifying the phenomena involved. We were still able, however, to give a vivid description of a large variety of physical parameters which characterize atomic and molecular systems. As our desire is to make the "Reference Data" practical for wide use, a summary of fundamental physical constants and measurement units has been included as an introduction. It does not take much space in the text but, in our opinion, it is particularly useful for this book. Thereafter, apart from quantitative characteristics of atoms and molecules, we have included brief information about the structure of atoms and molecules, which helps to account for some of the parameters found in tables, figures and on diagrams. The introductory paragraphs may assist readers who have not dealt previously with these specific problems of atomic and molecular physics.

Literature lists supplied for each section include the most informative publications – reviews, reports, reference books, monographs, etc. If these sources of information were not available, original papers were referred to so that the reader might have a more detailed acquaintance with the particular problem. Although this approach towards the sources of the data cited is not a customary one, it is in good accord with our primary task, that is, to make the book simple, informative and not overloaded with details.

2. Units of Physical Quantities

In order to make this reference guide suitable for broader use in practical applications, we have included in this chapter a short list of measurement units of physical quantities, together with a brief report on fundamental physical constants and some natural systems of units, and tables of conversion factors for different measurement units in physics.

2.1 Systems of Units in Physics

A coherent system of units is a system based on a certain set of *base* units, which by convention are regarded as dimensionally independent. All other units in this system may be defined as *derived* units and can be expressed in terms of powers of the base quantities by algebraic relations. In practice, it is advisable to make up systems with three to six base units which should remain constant throughout, provide a possibility for check, and be reproducible and restorable if the standard is lost. By convention the following units have formed a set of base units [2.1.1, 2]:

Unit of Time. The *second* [s] is the duration of 9 192 631 770 periods of the radiation corresponding to the transition between the two hyperfine levels of the ground state of the caesium-133 atom $[6^2S_{1/2}(F=4) \to 6^2S_{1/2}(F=3)]$.

Unit of Length. The *metre* [m] is the length of the path travelled by light in vacuum during a time interval of 1/299 792 458 of a second (XVII Conférence Générale des Poids et Mesures, Paris 1983).

Unit of Mass. The *kilogram* [kg] is equal to the mass of the international prototype of the kilogram (this platinum-iridium prototype is kept in the Bureau International des Poids et Mesures, Sèvres, France).

Unit of Force. The *kilogram-force* [kgf] is equal to the weight of a 1-kilogram international prototype at a point on the earth's surface (the Bureau International des Poids et Mesures, Sèvres, France) where the value of the acceleration of gravity is fixed as $9.80665 \text{ m} \cdot \text{s}^{-2}$.

Unit of Electric Current. The *ampere* [A] is that constant current which, if maintained in two straight, parallel conductors of infinite length, of negligible circular cross section, placed 1 metre apart in vacuum, would produce between these conductors a force equal to $2 \cdot 10^{-7}$ newtons per metre of length [the numerical value of the permeability of vacuum (magnetic constant) μ_0 is fixed as $4\pi\,10^{-7}$ (exactly) and the permittivity of vacuum (electric constant) ε_0 is derived from $c = (\varepsilon_0\mu_0)^{-1/2}$, where c is the speed of light in vacuum].

Unit of Temperature. The *kelvin* [K], unit of thermodynamic temperature, is the fraction 1/273.16 of the thermodynamic temperature of the triple point of water [the unit "degree Celsius" [°C] is equal to the unit "kelvin" and by definition $t = T - 273.15\,\mathrm{K}$, where t is the Celsius temperature and T is the thermodynamic temperature].

Unit of Luminous Intensity. The *candela* [cd] is the luminous intensity, in the perpendicular direction, of a surface of 1/600 000 square metres of a black body at the temperature of freezing platinum under a pressure of 101 325 newtons per square metre (or 1 atm).

Unit of Amount of Substance. The *mole* [mol] is the amount of substance of a system which contains as many elementary entities (atoms, molecules, ions, electrons or other particles) as there are atoms in 0.012 kg of carbon-12.

The above units and their decimal multiples have served as a basis for the most important systems of units: the CGS [cm, g, s] unit system, most suitable for different fields of physics and the teaching of fundamental physics; the International System of Units (SI) [m, kg, s, A, K, mol, cd], especially recommended for usage in different fields of science and technology; the MKGFS unit system [m, kgf, s], adapted to mechanics but now almost discarded. Table 2.1 contains a list of symbols used to denote the units of physical quantities in the above systems; in Table 2.2 a list of prefixes for forming multiple and sub-multiple units is given [2.1.1, 2]. Conversion of units from one system to another is performed by means of dimensional formulas [2.1.3].

References

2.1.1 Document U.I.P. 20 (1978): "Symbols, units and nomenclature in physics", Physica **93 A**, 1–60 (1978)

2.1.2 D.N.Lapedes (ed.): *McGraw-Hill Dictionary of Physics and Mathematics* (McGraw-Hill, New York 1978)

2.1.3 P.W.Bridgman: *Dimensional Analysis*, 2nd ed. (Yale University Press, New Haven 1932)

Table 2.1. Name and symbol of units

Name	Symbol	Name	Symbol	Name	Symbol
ampere	A	gauss	G (Gs)	ohm	Ω
ångström (unified)	Å	gilbert	Gi	pascal	Pa
atomic mass unit	u	gram	g	poise	P
bar	bar	gray	Gy	radian	rad
barn	barn (b)	henry	H	röntgen	R
becquerel	Bq	hertz	Hz	second	s
calorie	cal	hour	h	siemens	S
candela	cd	joule	J	standard atmosphere	atm
centimetre	cm	kelvin	K	steradian	sr
coulomb	C	kilogram	kg	stokes	St
curie	Ci	litre	l	tesla	T
day	d	lumen	lm	torr	Torr
decibel	dB	lux	lx	volt	V
dyne	dyn	maxwell	Mx	watt	W
electronvolt	eV	metre	m	weber	Wb
erg	erg	minute	min	year	a
farad	F	mole	mol		
		newton	N		
		oersted	Oe		

Table 2.2. Prefixes for indicating decimal multiples or sub-multiples of a unit

Name	Multiple	Symbol	Name	Multiple	Symbol
exa	10^{18}	E	deci	10^{-1}	d
peta	10^{15}	P	centi	10^{-2}	c
tera	10^{12}	T	milli	10^{-3}	m
giga	10^{9}	G	micro	10^{-6}	μ
mega	10^{6}	M	nano	10^{-9}	n
kilo	10^{3}	k	pico	10^{-12}	p
hecto	10^{2}	h	femto	10^{-15}	f
deca	10	da	atto	10^{-18}	a

2.2 Fundamental Physical Constants

Table 2.3 presents numerical values of the most frequently used fundamental physical constants, expressed in units of the CGS and SI systems. These values have been obtained mainly on the basis of a CODATA bulletin [2.2.1], which contains a list of recommended consistent values of the fundamental constants for 1973. The procedure of statistical analysis of experimental and theoretical data, dealing with the determination of numerical values of fundamental physical constants, is described in [2.2.2]. Furthermore, some new results for the values of R_∞ [2.2.3], α^{-1} [2.2.4], N_A [2.2.5] and other physical constants [2.2.6] have also been taken into account in preparing Table 2.3.

The number of significant figures given for the numerical values of the fundamental physical constants was obtained as a result of rounding off. Further refinements may change only the last digit quoted, within the range ± 1. The last column of Table 2.3 gives the uncertainty of this determination of the fundamental constants.

References

2.2.1 Report of the CODATA Task Group on Fundamental Constants, August 1973: "Recommended Consistent Values of the Fundamental Constants, 1973", Codata Bull., No. 11 (December 1973)
2.2.2 E.R.Cohen, B.N.Taylor: J. Phys. Chem. Ref. Data **2**, 663–734 (1973)
2.2.3 S.R.Amin, C.D.Caldwell, W.Lichten: Phys. Rev. Lett. **47**, 1234 (1981)
2.2.4 D.C.Tsui, A.C.Gossard, B.F.Field, M.E.Cage, R.F.Dziuba: Phys. Rev. Lett. **48**, 3 (1982)
2.2.5 R.D.Deslattes: "The Avogadro Constant", Annu. Rev. Phys. Chem. **31**, 435–461 (1980)
2.2.6 E.R.Cohen: "Status of the Fundamental Constants", in *Atomic Masses and Fundamental Constants 6*, ed. by J. A. Nolen, Jr., W. Benenson (Plenum, New York 1980) pp. 525–540

Table 2.3. Fundamental physical constants

Quantity	Symbol and definition	Value Mantissa	Decimal multiple CGS unit	SI unit	Uncertainty (10^6 or ppm)
1	2	3	4	5	6
Speed of light in vacuum	c	2.99792458	10^{10} cm · s^{-1}	10^8 m · s^{-1}	0.004
Elementary charge	e	1.60219	10^{-20} e.m.u.	10^{-19} C	3
		4.80324	10^{-10} e.s.u	–	3
Planck constant	h	6.6262	10^{-27} erg · s	10^{-34} J · s	5
	$\hbar = h/2\pi$	1.05459	10^{-27} erg · s	10^{-34} J · s	5
Rydberg constant	$R_\infty = m_e e^4/4\pi\hbar^3 c$	1.097373151	10^5 cm^{-1}	10^7 m^{-1}	0.001
Fine structure constant	$\alpha = e^2/\hbar c$	7.297354	10^{-3}	10^{-3}	0.1
	α^{-1}	1.3703597	10^2	10^2	0.1
Josephson frequency-voltage ratio	$2e/h$	4.83594	–	10^{14} s^{-1} · V^{-1}	3
		1.44978	10^{17} s^{-1} · erg^{-1} · e.s.u.	–	3
Magnetic flux quantum	$\Phi_0 = h/2e$	2.06785	10^{-7} Mx	10^{-15} Wb	3
Bohr radius	$a_0 = \alpha/4\pi R_\infty$	5.291773	10^{-9} cm	10^{-11} m	0.1
Electron Compton wavelength	$\lambda_C = \alpha^2/2R_\infty$	2.426311	10^{-10} cm	10^{-12} m	0.2
	$\bar{\lambda}_C = \lambda_C/2\pi = \alpha a_0$	3.861595	10^{-11} cm	10^{-13} m	0.2
Classical electron radius	$r_e = \alpha\bar{\lambda}_C$	2.817942	10^{-13} cm	10^{-15} m	0.3
Avogadro constant	N_A	6.02210	10^{23} mol^{-1}	10^{26} kmol^{-1}	1
(Unified) atomic mass unit	$1u = 10^{-3}\,N_A^{-1}$ kg · mol^{-1}	1.66055	10^{-24} g	10^{-27} kg	1
Energy equivalent of atomic mass unit	$E(u) = m_u c^2$	1.49243	10^{-3} erg	10^{-10} J	1
		931.50 MeV	–	–	3

Table 2.3 (continued)

Quantity	Symbol and definition	Value Mantissa	Decimal multiple CGS unit	SI unit	Uncertainty (10^6 or ppm)
1	2	3	4	5	6
Electron rest mass	m_e	9.1095 5.48580	10^{-28} g 10^{-4} a.m.u.	10^{-31} kg 10^{-4} a.m.u.	5 0.4
Ratio, atomic mass unit to electron mass	m_u/m_e	1.822887	10^3	10^3	0.4
Proton rest mass	m_p	1.67265 1.0072765	10^{-24} g a.m.u.	10^{-27} kg a.m.u.	5 0.01
Ratio, proton mass to electron mass	m_p/m_e	1.8361525	10^3	10^3	0.04
Neutron rest mass	m_n	1.67495 1.0086650	10^{-24} g a.m.u.	10^{-27} kg a.m.u.	5 0.04
Specific electron charge	e/m_e	1.75880 5.2728	– 10^{17} e.s.u. · g^{-1}	10^{11} C · kg^{-1} –	3 3
Faraday constant	$F = eN_A$	9.6485 2.89253	– 10^{14} e.s.u. · mol^{-1}	10^4 C · mol^{-1} –	3 3
Quantum of circulation	$h/2m_e$	3.63695	dyn · s · g^{-1}	10^{-4} J · s · kg^{-1}	2
Permeability of vacuum	μ_0	1.2566371	–	10^{-6} H · m^{-1}	–
Permittivity of vacuum	$\varepsilon_0 = 1/\mu_0 c^2$	8.8541878	–	10^{-12} F · m^{-1}	0.008
Electron g factor	$g_e/2 = \mu_e/\mu_B$	1.00115966	–	–	0.004
Bohr magneton[a]	$\mu_B = e\hbar/2(c)m_e$	9.2741	10^{-21} erg · G^{-1}	10^{-24} J · T^{-1}	4
Nuclear magneton[a]	$\mu_N = e\hbar/2(c)m_p$	5.0508	10^{-24} erg · G^{-1}	10^{-27} J · T^{-1}	4

Electron magnetic moment	μ_e	9.2848	10^{-21} erg · G^{-1}	10^{-24} J · T^{-1}	4
Proton magnetic moment	μ_p	1.41062	10^{-23} erg · G^{-1}	10^{-26} J · T^{-1}	4
Ratio, electron to proton magnetic moments	μ_e/μ_p	6.5821069	10^2	10^2	0.010
Gravitational constant	G	6.67	10^{-5} dyn · cm^2 · g^{-2}	10^{-11} N · m^2 · kg^{-2}	600
Molar gas constant	R	8.3145	–	J · mol^{-1} · K^{-1}	20
		1.9859	–	cal · mol^{-1} · K^{-1}	20
Molar volume, ideal gas ($T_0 = 0\,°C = 273.15$ K; $p_0 = 1$ atm)	$V_m = RT_0/p_0$	2.2414	10^4 cm^3 · mol^{-1}	10^{-2} m^3 · mol^{-1}	20
Loschmidt constant	$L_0 = N_A/V_m$	2.6867	10^{19} cm^{-3}	10^{25} m^{-3}	20
Boltzmann constant	$k = R/N_A$	1.3807	10^{-16} erg · K^{-1}	10^{-23} J · K^{-1}	20
Stefan-Boltzmann constant	$\sigma = (\pi^2/60)\,k^4/\hbar^3c^2$	5.670	10^{-5} erg · s^{-1} · cm^{-2} · K^{-4}	10^{-8} W · m^{-2} · K^{-4}	100
First radiation constant	$c_1 = 2\pi hc^2$	3.7418	10^{-5} erg · cm^2 · s^{-1}	10^{-16} W · m^2	5
Second radiation constant	$c_2 = hc/k$	1.4388	cm · K	10^{-2} m · K	20
Acceleration of free fall:					
Standard (Sèvres, France)	g_n	9.80665	10^2 cm · s^{-2}	m · s^{-2}	–
Local – US datum	g (CB)[b]	9.801043	10^2 cm · s^{-2}	m · s^{-2}	0.02
Local – UK datum	g (BFS)[c]	9.811818	10^2 cm · s^{-2}	m · s^{-2}	0.02

[a] The formula contains the quantity in parentheses only for the CGS system of units
[b] CB Building, NBS, Washington
[c] BFS Building, NPL, England

2.3 Systems of Units Based on "Natural Standards"

To avoid an excessively large number of numerical multiples in the formulas of physical laws and definitions occurring in the field of atomic physics, these systems are introduced to decrease the number of universal constants. This can be done by reducing the number of base units so that many universal constants become equal to unity or a dimensionless constant number. Systems of units which have the maximum possible number of universal physical constants equal to one, are called "natural systems of units".

One of the most important natural systems – the system of Hartree units, often called the "system of atomic units" – puts equal to one (1) the charge (e) and mass (m_e) of the electron and the Planck constant ($\hbar$). Conversion to Hartree atomic units in formulas is made formally by putting $e = m_e = \hbar = 1$. These values may be combined to produce only one combination with a given dimension; this is then a unit for that particular quantity. This system makes use of the unit of length $\hbar^2/m_e e^2 \simeq 0.529 \cdot 10^{-8}$ cm and the unit of energy $m_e e^4/\hbar^2 \simeq 27.212$ eV and so on.

It is useful in quantum electrodynamics to work with a system of units in which the constants c, m_e and $\hbar$ are put equal to one. The unit of energy in such a system, also called the "system of relativistic units", coincides with the energy corresponding to the electron rest mass $m_e c^2 \simeq 8.2 \cdot 10^{-7}$ erg, the unit of length coincides with the electron Compton wavelength $\lambda\!\!\!^{-}_C \simeq 3.9 \cdot 10^{-11}$ cm, etc.

Systems of units so introduced are often thought to be convenient because the atomic and molecular parameters do not differ greatly in order of magnitude from 1. Tables 2.4, 5 include the corresponding Hartree atomic and relativistic units and their numerical values in the standard units of the CGS and SI systems.

Table 2.4. Hartree atomic units: $e = m_e = \hbar = 1$

Unit	Symbol	Value in CGS or SI units
Electric charge (equal to charge of electron)	e	$4.8032 \cdot 10^{-10}$ e.s.u = $1.6022 \cdot 10^{-20}$ e.m.u.
Mass (equal to electron mass)	m_e	$9.1095 \cdot 10^{-28}$ g
Angular momentum (Planck constant/2π)	$\hbar = h/2\pi$	$1.0546 \cdot 10^{-27}$ erg · s
Length (Bohr radius)	$a_0 = \hbar^2/m_e e^2$	$5.2918 \cdot 10^{-9}$ cm = 0.52918 Å
Velocity	$v_0 = e^2/\hbar$	$2.1877 \cdot 10^{8}$ cm · s^{-1}
Momentum	$p_0 = m_e e^2/\hbar$	$1.9929 \cdot 10^{-19}$ g · cm · s^{-1}
Energy	$\varepsilon_0 = m_e e^4/\hbar^2$	$4.3598 \cdot 10^{-11}$ erg = 27.212 eV
Time	$\tau_0 = \hbar^3/m_e e^4$	$2.4189 \cdot 10^{-17}$ s
Frequency	$\nu_0 = m_e e^4/\hbar^3$	$4.1341 \cdot 10^{16}$ s^{-1}
Electric field strength	$E_0 = m_e^2 e^5/\hbar^4$	$1.7153 \cdot 10^{7}$ e.s.u. · cm^{-2} = $5.1422 \cdot 10^{9}$ V · cm^{-1}
Potential	$\varphi_0 = m_e e^3/\hbar^2 = e/a_0$	$9.0767 \cdot 10^{-2}$ e.s.u. · cm^{-1} = 27.212 V
Magnetic moment	$p_M = \hbar^2/m_e e = 2\mu_B/\alpha$	$2.5418 \cdot 10^{-18}$ erg · G^{-1} = $2.5418 \cdot 10^{-21}$ J · T^{-1}
Dipole moment	$\mu_0 = e a_0$	$2.5418 \cdot 10^{-18}$ e.s.u. · cm = 2.5418 D (debye)
Cross section	$\sigma_0 = a_0^2$	$2.8003 \cdot 10^{-17}$ cm^2 = 0.28003 Å^2
Volume, polarizability	$V_0(\alpha_0) = a_0^3$	$1.4818 \cdot 10^{-25}$ cm^3 = $1.4818 \cdot 10^{-1}$ Å^3
Number density of particles	$n_0 = a_0^{-3}$	$6.7483 \cdot 10^{24}$ cm^{-3}
Two-body rate coefficient	$k_0 = e^2 a_0^2/\hbar = \hbar^3/m_e^2 e^2$	$6.126 \cdot 10^{-9}$ cm^3 · s^{-1}
Three-body rate coefficient	$\mathcal{K}_0 = e^2 a_0^5/\hbar = \hbar^9/m_e^5 e^8$	$9.078 \cdot 10^{-34}$ cm^6 · s^{-1}
Electric current density	$j_0 = e n_0 v_0 = m_e^3 e^9/\hbar^7$	$7.0911 \cdot 10^{23}$ e.s.u. · cm^{-2} · s^{-1} = $2.3653 \cdot 10^{14}$ A · cm^{-2}

Table 2.5. System of relativistic units: $c = m_e = \hbar = 1$

Unit	Symbol	Value in CGS or SI units
Electric charge	$Q = e/\alpha^{1/2} = (\hbar c)^{1/2}$	$5.6228 \cdot 10^{-9}$ e.s.u. = $1.8756 \cdot 10^{-18}$ C
Mass (equal to electron mass)	$m = m_e$	$9.1095 \cdot 10^{-28}$ g
Angular momentum (Planck constant/2π)	$\hbar = h/2\pi$	$1.0546 \cdot 10^{-27}$ erg · s
Length	$l = \hbar/m_e c = \lambda\!\!\!^{-}_C = \alpha a_0$	$3.8616 \cdot 10^{-11}$ cm
Time	$t = \hbar/m_e c^2$	$1.2881 \cdot 10^{-21}$ s
Velocity (equal to velocity of light in vacuum)	$v = c$	$2.9979 \cdot 10^{10}$ cm · s^{-1} = $2.9979 \cdot 10^{8}$ m · s^{-1}
Momentum	$p = m_e c$	$2.7310 \cdot 10^{-17}$ g · cm · s^{-1}
Energy	$\varepsilon = m_e c^2$	$8.1872 \cdot 10^{-7}$ erg = $8.1872 \cdot 10^{-14}$ J = $5.1100 \cdot 10^{5}$ eV
Acceleration	$a = m_e c^3/\hbar$	$2.3274 \cdot 10^{31}$ cm · s^{-2}
Frequency	$\nu = m_e c^2/\hbar$	$7.7634 \cdot 10^{20}$ s^{-1}
Force	$F = m_e^2 c^3/\hbar$	$2.1202 \cdot 10^{4}$ dyn = 0.21202 N
Dipole moment	$\mu = \hbar^{3/2}/m_e c^{1/2}$	$2.1713 \cdot 10^{-19}$ e.s.u. · cm = 0.21713 D (debye)
Electric field strength	$E = m_e^2 c^{5/2}/\hbar^{3/2}$	$3.7706 \cdot 10^{12}$ e.s.u. · cm^{-2} = $1.1304 \cdot 10^{15}$ V · cm^{-1}
Potential	$\varphi = m_e c^{3/2}/\hbar^{1/2} = e/\lambda\!\!\!^{-}_C \alpha^{1/2}$	$1.4561 \cdot 10^{2}$ e.s.u. · cm^{-1} = $4.3652 \cdot 10^{4}$ V
Magnetic moment	$p_M = \hbar^{3/2}/m_e c^{1/2} = 2\mu_B/\alpha^{1/2}$	$2.1713 \cdot 10^{-19}$ erg · G^{-1} = $2.1713 \cdot 10^{-22}$ J · T^{-1}
Cross section	$\sigma = \hbar^2/m_e^2 c^2 = \lambda\!\!\!^{-}{}_C^2 = \alpha^2 a_0^2$	$1.4912 \cdot 10^{-21}$ cm^2 = $1.4912 \cdot 10^{-5}$ Å^2
Volume, polarizability	$V(\alpha_d) = \hbar^3/m_e^3 c^3 = \lambda\!\!\!^{-}{}_C^3 = \alpha^3 a_0^3$	$5.7584 \cdot 10^{-32}$ cm^3 = $5.7584 \cdot 10^{-8}$ Å^3
Number density of particles	$n = m_e^3 c^3/\hbar^3 = \lambda\!\!\!^{-}{}_C^{-3}$	$1.7366 \cdot 10^{31}$ cm^{-3}
Electric current density	$j = m_e^3 c^{9/2}/\hbar^{5/2}$	$2.9273 \cdot 10^{33}$ e.s.u. · cm^{-2} · s^{-1} = $9.7644 \cdot 10^{23}$ A · cm^{-2}

2.4 Tables of Conversion Factors

To facilitate practical calculations Table 2.6, which lists some common units defined in terms of SI units, and tables for conversion between units (Tables 2.7–12) are useful [2.4.1]. To convert a quantity expressed in a unit in the left-hand column to the equivalent in a unit in the top row, multiply the quantity by the factor common to both units. When composing tables of conversion factors we used relationships based directly or indirectly on experimental data, the relevant definitions (partly from Table 2.6) and the calculated results founded on dimensional formulae [2.4.2]. The number of digits given in the numerical values of the conversion factors is such that any change on further refinement should be no more than one (1) in the last significant figure. All values are given in systems of units used in atomic and molecular physics.

References

2.4.1 D.N.Lapedes (ed.): *McGraw-Hill Dictionary of Physics and Mathematics* (McGraw-Hill, New York 1978)
2.4.2 P.W.Bridgman: *Dimensional Analysis*, 2nd ed. (Yale University Press, New Haven 1932)

Table 2.6. Some common units defined in terms of SI units

Quantity	Name of unit	Unit symbol	Definition of unit
Length	inch	in	0.0254 m
	foot	ft	0.3048 m
Mass	pound (avoirdupois)	lb	0.45359237 kg
Force	kilogram-force	kgf	9.80665 N
Energy	international steam calorie	cal_{IT}	4.1868 J
	thermochemical calorie	cal_{th}	4.184 J
	British thermal unit	Btu_{IT}	1055.05585262 J
Thermodynamic temperature (T)	degree Rankine	°R	(9/5) K
Customary temperature (t)	degree Celsius	°C	t [°C] = T [K] − 273.15
	degree Fahrenheit	°F	t [°F] = T [°R] − 459.67
Density	standard density of fluid mercury	ϱ_{Hg}	$13595.1\ kg \cdot m^{-3}$
Pressure	standard atmosphere	atm	101325 Pa
	torr	Torr (mm Hg)	(101325/760) Pa = 133.322368 Pa
	conventional millimeter of mercury	mmHg	(13.5951 · 9.80665) Pa = 133.322387 Pa

Table 2.7. Conversion factors for units of energy

Measurement unit	eV	J	erg	$kcal_{IT}$/mol	cm^{-1}	K	a.u.	Ry	MHz
1 eV	1	$1.6022 \cdot 10^{-19}$	$1.6022 \cdot 10^{-12}$	23.045	8065.48	11604	$3.6749 \cdot 10^{-2}$	$7.3498 \cdot 10^{-2}$	$2.4180 \cdot 10^{8}$
1 J	$6.2415 \cdot 10^{18}$	1	10^{7}	$1.4384 \cdot 10^{20}$	$5.0340 \cdot 10^{22}$	$7.2430 \cdot 10^{22}$	$2.2937 \cdot 10^{17}$	$4.5873 \cdot 10^{17}$	$1.5092 \cdot 10^{27}$
1 erg	$6.2415 \cdot 10^{11}$	10^{-7}	1	$1.4384 \cdot 10^{13}$	$5.0340 \cdot 10^{15}$	$7.2430 \cdot 10^{15}$	$2.2937 \cdot 10^{10}$	$4.5873 \cdot 10^{10}$	$1.5092 \cdot 10^{20}$
1 $kcal_{IT}$/mol	$4.3393 \cdot 10^{-2}$	$6.9524 \cdot 10^{-21}$	$6.9524 \cdot 10^{-14}$	1	$3.4999 \cdot 10^{2}$	$5.0356 \cdot 10^{2}$	$1.5946 \cdot 10^{-3}$	$3.1893 \cdot 10^{-3}$	$1.0492 \cdot 10^{7}$
1 cm^{-1}	$1.23985 \cdot 10^{-4}$	$1.9865 \cdot 10^{-23}$	$1.9865 \cdot 10^{-16}$	$2.8573 \cdot 10^{-3}$	1	1.4388	$4.5563 \cdot 10^{-6}$	$9.1127 \cdot 10^{-6}$	$2.9979 \cdot 10^{4}$
1 K	$8.617 \cdot 10^{-5}$	$1.3807 \cdot 10^{-23}$	$1.3807 \cdot 10^{-16}$	$1.9859 \cdot 10^{-3}$	$6.9502 \cdot 10^{-1}$	1	$3.1668 \cdot 10^{-6}$	$6.3335 \cdot 10^{-6}$	$2.0836 \cdot 10^{4}$
1 a.u.	27.2116	$4.3598 \cdot 10^{-18}$	$4.3598 \cdot 10^{-11}$	$6.2709 \cdot 10^{2}$	$2.1947 \cdot 10^{5}$	$3.1578 \cdot 10^{5}$	1	2	$6.5797 \cdot 10^{9}$
1 Ry	13.606	$2.1799 \cdot 10^{-18}$	$2.1799 \cdot 10^{-11}$	$3.1355 \cdot 10^{2}$	$1.0974 \cdot 10^{5}$	$1.5789 \cdot 10^{5}$	0.5	1	$3.2898 \cdot 10^{9}$
1 MHz	$4.1357 \cdot 10^{-9}$	$6.6262 \cdot 10^{-28}$	$6.6262 \cdot 10^{-21}$	$9.5308 \cdot 10^{-8}$	$3.3356 \cdot 10^{-5}$	$4.7993 \cdot 10^{-5}$	$1.5198 \cdot 10^{-10}$	$3.0397 \cdot 10^{-10}$	1

Table 2.8. Conversion factors for units of power

Measurement unit	$erg \cdot s^{-1}$	W	kW	$kgf \cdot m \cdot s^{-1}$	$cal_{IT} \cdot s^{-1}$	$kcal_{IT} \cdot h^{-1}$	hp	$Btu \cdot s^{-1}$
1 $erg \cdot s^{-1}$	1	10^{-7}	10^{-10}	$1.0197 \cdot 10^{-8}$	$2.3885 \cdot 10^{-8}$	$8.5985 \cdot 10^{-8}$	$1.3596 \cdot 10^{-10}$	$9.4782 \cdot 10^{-11}$
1 W	10^{7}	1	10^{-3}	0.1020	0.2388	0.8598	$1.3596 \cdot 10^{-3}$	$9.4782 \cdot 10^{-4}$
1 kW	10^{10}	10^{3}	1	$1.0197 \cdot 10^{2}$	$2.3885 \cdot 10^{2}$	$8.5985 \cdot 10^{2}$	1.3596	0.9478
1 $kgf \cdot m \cdot s^{-1}$	$9.80665 \cdot 10^{7}$	9.80665	$9.80665 \cdot 10^{-3}$	1	2.3423	8.4322	$1.3333 \cdot 10^{-2}$	$9.2949 \cdot 10^{-3}$
1 $cal_{IT} \cdot s^{-1}$	$4.1868 \cdot 10^{7}$	4.1868	$4.1868 \cdot 10^{-3}$	0.4269	1	3.6	$5.6925 \cdot 10^{-3}$	$3.9683 \cdot 10^{-3}$
1 $kcal_{IT} \cdot h^{-1}$	$1.1630 \cdot 10^{7}$	1.1630	$1.1630 \cdot 10^{-3}$	0.1186	0.2778	1	$1.5812 \cdot 10^{-3}$	$1.1023 \cdot 10^{-3}$
1 hp	$7.3550 \cdot 10^{9}$	735.50	$7.3550 \cdot 10^{-1}$	75	175.67	632.415	1	0.6971
1 $Btu_{IT} \cdot s^{-1}$	$1.0551 \cdot 10^{10}$	$1.0551 \cdot 10^{3}$	1.0551	107.59	252.00	907.185	1.4345	1

Table 2.9. Conversion factors for units of pressure

Measurement unit	$dyn \cdot cm^{-2}$	$N \cdot m^{-2}$	$kgf \cdot m^{-2}$	$kgf \cdot cm^{-2}$	atm	Torr
$1\ dyn \cdot cm^{-2}$	1	0.1	$1.0197 \cdot 10^{-2}$	$1.0197 \cdot 10^{-6}$	$9.8693 \cdot 10^{-7}$	$7.5007 \cdot 10^{-4}$
$1\ N \cdot m^{-2}$	10	1	$1.0197 \cdot 10^{-1}$	$1.0197 \cdot 10^{-5}$	$9.8693 \cdot 10^{-6}$	$7.5007 \cdot 10^{-3}$
$1\ kgf \cdot m^{-2}$	$9.807 \cdot 10^{1}$	9.807	1	10^{-4}	$9.6785 \cdot 10^{-5}$	$7.3556 \cdot 10^{-2}$
$1\ kgf \cdot cm^{-2}$	$9.807 \cdot 10^{5}$	$9.807 \cdot 10^{4}$	10^{4}	1	$9.6785 \cdot 10^{-1}$	$7.3556 \cdot 10^{2}$
1 atm	$1.01325 \cdot 10^{6}$	$1.01325 \cdot 10^{5}$	$1.0332 \cdot 10^{4}$	1.0332	1	760
1 Torr	$1.3332 \cdot 10^{3}$	$1.3332 \cdot 10^{2}$	13.595	$1.3595 \cdot 10^{-3}$	$1.3158 \cdot 10^{-3}$	1

Table 2.10. Conversion factors for units of cross section

Measurement unit	cm^2	$Å^2$	a_0^2	πa_0^2	barn
$1\ cm^2$	1	10^{16}	$3.5711 \cdot 10^{16}$	$1.1367 \cdot 10^{16}$	10^{24}
$1\ Å^2$	10^{-16}	1	3.5711	1.1367	10^{8}
$1\ a_0^2$	$2.8003 \cdot 10^{-17}$	$2.8003 \cdot 10^{-1}$	1	$3.1831 \cdot 10^{-1}$	$2.8003 \cdot 10^{7}$
$1\ \pi a_0^2$	$8.7973 \cdot 10^{-17}$	$8.7973 \cdot 10^{-1}$	3.1416	1	$8.7973 \cdot 10^{7}$
1 barn	10^{-24}	10^{-8}	$3.5711 \cdot 10^{-8}$	$1.1367 \cdot 10^{-8}$	1

Table 2.11. Relation between various temperature scales

	Kelvin	Celsius	Fahrenheit	Rankine	Reaumur
T[K]	1	C + 273.15	(5/9)(F + 459.67)	(5/9) R	(5/4) Re + 273.15
t[°C]	K – 273.15	1	(5/9)(F – 32)	(5/9)(R – 491.67)	(5/4) Re
t[°F]	(9/5)(K – 255.37)	(9/5) C + 32	1	R – 459.67	(9/4) Re + 32
T[°R]	(9/5) K	(9/5) C + 491.67	F + 459.67	1	(9/4) Re + 491.67
t[°Re]	(4/5)(K – 273.15)	(4/5) C	(4/9)(F – 32)	(4/9)(R – 491.67)	1

Table 2.12. Historically established nominal regions of electromagnetic radiation

Type	Wavelength interval [10^{-10} m or Å]	Energy interval [eV]	Effective temperature interval ΔT[K] for black body radiation: $h\nu = 2.82\,kT$
Radiowave, super-long	$10^{14} - 10^{18}$	$1.2 \cdot 10^{-10} - 1.2 \cdot 10^{-14}$	$5.1 \cdot 10^{-7} - 5.1 \cdot 10^{-11}$
long	$10^{13} - 10^{14}$	$1.2 \cdot 10^{-9} - 1.2 \cdot 10^{-10}$	$5.1 \cdot 10^{-6} - 5.1 \cdot 10^{-7}$
medium frequency	$10^{12} - 10^{13}$	$1.2 \cdot 10^{-8} - 1.2 \cdot 10^{-9}$	$5.1 \cdot 10^{-5} - 5.1 \cdot 10^{-6}$
short	$10^{11} - 10^{12}$	$1.2 \cdot 10^{-7} - 1.2 \cdot 10^{-8}$	$5.1 \cdot 10^{-4} - 5.1 \cdot 10^{-5}$
ultra-short	$10^{6} - 10^{11}$	$0.0124 - 1.2 \cdot 10^{-7}$	$51 - 5.1 \cdot 10^{-4}$
Infrared radiation, far	$10^{5} - 10^{7}$	$0.124 - 1.24 \cdot 10^{-3}$	510 – 5
near	$7500 - 10^{5}$	1.65 – 0.124	6803 – 510
Visible light, red	6500 – 7500	1.91 – 1.65	7849 – 6803
orange	5900 – 6500	2.10 – 1.91	8647 – 7849
yellow	5300 – 5900	2.34 – 2.10	9626 – 8647
green	4900 – 5300	2.53 – 2.34	10 412 – 9626
blue	4200 – 4900	2.95 – 2.53	12 148 – 10 412
violet	4000 – 4200	3.10 – 2.95	12 755 – 12 148
Ultraviolet radiation, near	2000 – 4000	6.20 – 3.10	25 510 – 12 755
far (vacuum)	100 – 2000	124 – 6.20	$5.1 \cdot 10^{5} - 25\,510$
X-ray radiation, soft	1 – 20	$1.24 \cdot 10^{4} - 620$	$5.1 \cdot 10^{7} - 2.6 \cdot 10^{6}$
hard	0.06 – 1	$2.07 \cdot 10^{5} - 1.24 \cdot 10^{4}$	$8.5 \cdot 10^{8} - 5.1 \cdot 10^{7}$
γ-ray radiation	$5 \cdot 10^{-4} - 2$	$2.48 \cdot 10^{7} - 6.2 \cdot 10^{3}$	$1.0 \cdot 10^{11} - 2.5 \cdot 10^{7}$

Part I **Atoms and Atomic Ions**

Part I provides quantitative information about the structural, energetic and spectroscopic parameters of neutral atoms, positive and negative atomic ions, and the simplest multiply charged ions.

3. Isotopic Composition, Atomic Mass Table and Atomic Weights of the Elements

Numerical parameters of stable and long-lived radioactive isotopes whose half-lives are longer than that of the neutron, are examined. Results are collected for the isotopic composition and atomic weights of the elements, the spins and magnetic moments of nuclei, their half-lives and modes of decay and the nuclidic masses.

3.1 Parameters of Stable and Long-Lived Isotopes

The next two tables list nuclear parameters for elements with different isotopic composition. Depending on the structure of the nuclei, all isotopes can be divided into three groups [3.1.1].

1) Even-even nuclei, which consist of an even number of protons Z and even number of neutrons N (keep in mind that the accepted symbol for the isotopes of the element X is: ${}^{A}_{Z}\mathrm{X}_{N}$, where the mass number is $A = Z + N$). Mechanical, magnetic and quadrupole moments of these nuclei are equal to zero.

2) Even-odd nuclei, having even Z and odd N and also odd-even nuclei with odd Z and even N. Their spins have half-integer values from 1/2 to 9/2 and their magnetic and quadrupole moments lie within a range which includes both positive and negative values. If the nuclei have spin $I = 1/2$, their quadrupole moments are equal to zero.

3) Odd-odd nuclei with odd values for both Z and N. Their spins have integer values from 1 to 7, and magnetic and quadrupole moments are different from zero, the quadrupole moments being rather small, as a rule.

Table 3.1 lists data on the stable isotopes – their representative isotopic composition, nuclear spin and magnetic moment, which is expressed in nuclear magnetons, $\mu_N = 5.0508 \cdot 10^{-24}\ \mathrm{erg\,G^{-1}}$.

Table 3.2 comprises ground-state characteristics of radioactive nuclei whose half-lives, $T_{1/2}$, are longer than that of a neutron ($\sim$ 10.6 min). Decay modes of radioactive nuclei are marked in the following manner: β^- – beta decay involving emission of an electron; β^+ – beta decay involving emission of a positron; α – alpha decay involving emission of an α particle (i.e. nucleus of the helium ion He^{2+}); EC – capture of an electron by a nucleus from the inner electronic shell; sf – spontaneous fission of a nucleus. If two decay channels are available, the more efficient one is mentioned first.

The information was obtained from [3.1.2–7] and supplemented by data obtained from scarce newer publications. The error in determining the values was taken into account when rounding off significant figures, and is within the range ± 1 for the last digit.

References

3.1.1 H.Kopfermann: *Nuclear Moments*, 3rd ed. (Academic, New York 1962)
3.1.2 International Union of Pure and Applied Chemistry, Commission on Atomic Weights and Isotopic Abundances: "Isotopic compositions of the elements – 1981", Pure Appl. Chem. **55**, 1120 (1983)
3.1.3 C.M.Lederer, V.S.Shirley (eds.): *Table of Isotopes*, 7th ed. (Wiley, New York 1978)
3.1.4 G.H.Fuller: J. Phys. Chem. Ref. Data **5**, 835 (1976)
3.1.5 G.H.Fuller, V. W. Cohen: Nucl. Data Tables **A 5**, 433 (1969)
3.1.6 J.Blachot, Ch.Fiche: "Tableau des isotopes radioactifs et des principaux rayonnements émis", Ann. Phys. (Paris), Suppl. to Vol. 6, 3–218 (1981)
3.1.7 R.L.Heath: "Table of the Isotopes", in *CRC Handbook of Chemistry and Physics*, ed. by R. C. Weast, 62nd ed. (CRC, Boca Raton 1981) pp. B 255–B 339

Table 3.1. Parameters of stable nuclides

Atomic number Z	Isotope	Representative isotopic composition [%]	Nuclear spin I	Magnetic moment in nuclear magnetons μ/μ_N
1	2	3	4	5
1	^{1}H	99.985	1/2	2.79285
	^{2}H	0.015	1	0.85744
2	^{3}He	$1.4\cdot10^{-4}$	1/2	−2.12762
	^{4}He	≈100	0	0
3	^{6}Li	7.5	1	0.82205
	^{7}Li	92.5	3/2	3.2564
4	^{9}Be	100	3/2	−1.178
5	^{10}B	19.9	3	1.8006
	^{11}B	80.1	3/2	2.6886
6	^{12}C	98.9	0	0
	^{13}C	1.1	1/2	0.7024
7	^{14}N	99.63	1	0.40376
	^{15}N	0.37	1/2	−0.2832
8	^{16}O	99.76	0	0
	^{17}O	0.04	5/2	−1.8938
	^{18}O	0.20	0	0
9	^{19}F	100	1/2	2.6289
10	^{20}Ne	90.5	0	0
	^{21}Ne	0.27	3/2	−0.6618
	^{22}Ne	9.2	0	0
11	^{23}Na	100	3/2	2.2175
12	^{24}Mg	79.0	0	0
	^{25}Mg	10.00	5/2	−0.8554
	^{26}Mg	11.01	0	0
13	^{27}Al	100	5/2	3.6415
14	^{28}Si	92.23	0	0
	^{29}Si	4.67	1/2	−0.5553
	^{30}Si	3.10	0	0
15	^{31}P	100	1/2	1.1316
16	^{32}S	95.0	0	0
	^{33}S	0.75	3/2	0.644
	^{34}S	4.2	0	0
	^{36}S	0.02	0	0
17	^{35}Cl	75.8	3/2	0.82187
	^{37}Cl	24.2	3/2	0.68412
18	^{36}Ar	0.34	0	0
	^{38}Ar	0.063	0	0
	^{40}Ar	99.60	0	0
19	^{39}K	93.26	3/2	0.3915
	^{40}K	0.0117	4	−1.298
	^{41}K	6.73	3/2	0.215
20	^{40}Ca	96.94	0	0
	^{42}Ca	0.647	0	0
	^{43}Ca	0.135	7/2	−1.317
	^{44}Ca	2.09	0	0
	^{46}Ca	$4\cdot10^{-3}$	0	0
	^{48}Ca	0.187	0	0
21	^{45}Sc	100	7/2	4.756
22	^{46}Ti	8.0	0	0
	^{47}Ti	7.3	5/2	−0.7885
	^{48}Ti	73.8	0	0
	^{49}Ti	5.5	7/2	−1.1042
	^{50}Ti	5.4	0	0
23	^{50}V	0.250	6	3.3475
	^{51}V	99.750	7/2	5.151
24	^{50}Cr	4.35	0	0
	^{52}Cr	83.79	0	0
	^{53}Cr	9.50	3/2	−0.47
	^{54}Cr	2.36	0	0
25	^{55}Mn	100	5/2	3.47
26	^{54}Fe	5.8	0	0
	^{56}Fe	92	0	0
	^{57}Fe	2.2	1/2	0.0907
	^{58}Fe	0.28	0	0
27	^{59}Co	100	7/2	4.6
28	^{58}Ni	68.27	0	0
	^{60}Ni	26.10	0	0
	^{61}Ni	1.13	3/2	−0.750
	^{62}Ni	3.59	0	0
	^{64}Ni	0.91	0	0
29	^{63}Cu	69.17	3/2	2.22
	^{65}Cu	30.83	3/2	2.38
30	^{64}Zn	48.6	0	0
	^{66}Zn	27.9	0	0
	^{67}Zn	4.1	5/2	0.876
	^{68}Zn	19	0	0
	^{70}Zn	0.6	0	0
31	^{69}Ga	60.1	3/2	2.017
	^{71}Ga	39.9	3/2	2.562
32	^{70}Ge	20	0	0
	^{72}Ge	27	0	0

Table 3.1 (continued)

Atomic number Z	Isotope	Representative isotopic composition [%]	Nuclear spin I	Magnetic moment in nuclear magnetons μ/μ_N
1	2	3	4	5
	^{73}Ge	7.8	9/2	−0.8795
	^{74}Ge	36	0	0
	^{76}Ge	7.8	0	0
33	^{75}As	100	3/2	1.439
34	^{74}Se	0.9	0	0
	^{76}Se	9.0	0	0
	^{77}Se	7.6	1/2	0.534
	^{78}Se	24	0	0
	^{80}Se	50	0	0
	^{82}Se	9	0	0
35	^{79}Br	50.7	3/2	2.1064
	^{81}Br	49.3	3/2	2.271
36	^{78}Kr	0.35	0	0
	^{80}Kr	2.25	0	0
	^{82}Kr	11.6	0	0
	^{83}Kr	11.5	9/2	−0.9707
	^{84}Kr	57	0	0
	^{86}Kr	17.3	0	0
37	^{85}Rb	72.17	5/2	1.353
	^{87}Rb	27.83	3/2	2.751
38	^{84}Sr	0.56	0	0
	^{86}Sr	9.86	0	0
	^{87}Sr	7.00	9/2	−1.093
	^{88}Sr	82.58	0	0
39	^{89}Y	100	1/2	−0.1374
40	^{90}Zr	51.45	0	0
	^{91}Zr	11.22	5/2	−1.304
	^{92}Zr	17.15	0	0
	^{94}Zr	17.38	0	0
	^{96}Zr	2.80	0	0
41	^{93}Nb	100	9/2	6.17
42	^{92}Mo	14.8	0	0
	^{94}Mo	9.25	0	0
	^{95}Mo	15.9	5/2	−0.914
	^{96}Mo	16.7	0	0
	^{97}Mo	9.55	5/2	−0.933
	^{98}Mo	24.1	0	0
	^{100}Mo	9.63	0	0
44	^{96}Ru	5.5	0	0
	^{98}Ru	1.9	0	0
	^{99}Ru	12.7	5/2	−0.64
	^{100}Ru	12.6	0	0
	^{101}Ru	17.0	5/2	−0.72
	^{102}Ru	31.6	0	0
	^{104}Ru	18.7	0	0
45	^{103}Rh	100	1/2	−0.088
46	^{102}Pd	1.02	0	0
	^{104}Pd	11.1	0	0
	^{105}Pd	22.3	5/2	−0.642
	^{106}Pd	27.3	0	0
	^{108}Pd	26.5	0	0
	^{110}Pd	11.7	0	0
47	^{107}Ag	51.84	1/2	−0.1137
	^{109}Ag	48.16	1/2	−0.1307
48	^{106}Cd	1.25	0	0
	^{108}Cd	0.89	0	0
	^{110}Cd	12.5	0	0
	^{111}Cd	12.8	1/2	−0.595
	^{112}Cd	24.1	0	0
	^{113}Cd	12.2	1/2	0.622
	^{114}Cd	28.7	0	0
	^{116}Cd	7.5	0	0
49	^{113}In	4.3	9/2	5.53
	^{115}In	95.7	9/2	5.54
50	^{112}Sn	0.97	0	0
	^{114}Sn	0.65	0	0
	^{115}Sn	0.36	1/2	−0.919
	^{116}Sn	14.5	0	0
	^{117}Sn	7.7	1/2	−1.001
	^{118}Sn	24.2	0	0
	^{119}Sn	8.6	1/2	−1.047
	^{120}Sn	32.6	0	0
	^{122}Sn	4.6	0	0
	^{124}Sn	5.8	0	0
51	^{121}Sb	57	5/2	3.36
	^{123}Sb	43	7/2	2.55
52	^{120}Te	0.096	0	0
	^{122}Te	2.60	0	0
	^{123}Te	0.91	1/2	−0.736
	^{124}Te	4.82	0	0
	^{125}Te	7.14	1/2	−0.887
	^{126}Te	18.95	0	0
	^{128}Te	31.69	0	0

Table 3.1 (continued)

Atomic number Z	Isotope	Representative isotopic composition [%]	Nuclear spin I	Magnetic moment in nuclear magnetons μ/μ_N
1	2	3	4	5
	^{130}Te	33.80	0	0
53	^{127}I	100	5/2	2.81
54	^{124}Xe	0.10	0	0
	^{126}Xe	0.09	0	0
	^{128}Xe	1.9	0	0
	^{129}Xe	26	1/2	−0.778
	^{130}Xe	4.1	0	0
	^{131}Xe	21	3/2	0.692
	^{132}Xe	27	0	0
	^{134}Xe	10.4	0	0
	^{136}Xe	8.9	0	0
55	^{133}Cs	100	7/2	2.582
56	^{130}Ba	0.106	0	0
	^{132}Ba	0.101	0	0
	^{134}Ba	2.42	0	0
	^{135}Ba	6.59	3/2	0.838
	^{136}Ba	7.9	0	0
	^{137}Ba	11.2	3/2	0.937
	^{138}Ba	71.7	0	0
57	^{138}La	0.09	5	3.70
	^{139}La	99.91	7/2	2.78
58	^{136}Ce	0.19	0	0
	^{138}Ce	0.25	0	0
	^{140}Ce	88.5	0	0
	^{142}Ce	11.1	0	0
59	^{141}Pr	100	5/2	4.1
60	^{142}Nd	27.1	0	0
	^{143}Nd	12.2	7/2	−1.06
	^{144}Nd	23.8	0	0
	^{145}Nd	8.3	7/2	−0.66
	^{146}Nd	17.2	0	0
	^{148}Nd	5.76	0	0
	^{150}Nd	5.64	0	0
62	^{144}Sm	3.1	0	0
	^{147}Sm	15.0	7/2	−0.81
	^{148}Sm	11.3	0	0
	^{149}Sm	13.8	7/2	−0.67
	^{150}Sm	7.4	0	0
	^{152}Sm	26.7	0	0
	^{154}Sm	22.7	0	0
63	^{151}Eu	48	5/2	3.47
	^{153}Eu	52	5/2	1.53
64	^{152}Gd	0.20	0	0
	^{154}Gd	2.18	0	0
	^{155}Gd	14.8	3/2	−0.26
	^{156}Gd	20.5	0	0
	^{157}Gd	15.6	3/2	−0.34
	^{158}Gd	24.8	0	0
	^{160}Gd	21.9	0	0
65	^{159}Tb	100	3/2	2.01
66	^{156}Dy	0.06	0	0
	^{158}Dy	0.10	0	0
	^{160}Dy	2.3	0	0
	^{161}Dy	18.9	5/2	−0.48
	^{162}Dy	25.5	0	0
	^{163}Dy	24.9	5/2	0.67
	^{164}Dy	28.2	0	0
67	^{165}Ho	100	7/2	4.17
68	^{162}Er	0.14	0	0
	^{164}Er	1.61	0	0
	^{166}Er	33.6	0	0
	^{167}Er	22.9	7/2	−0.57
	^{168}Er	26.8	0	0
	^{170}Er	14.9	0	0
69	^{169}Tm	100	1/2	−0.23
70	^{168}Yb	0.13	0	0
	^{170}Yb	3.1	0	0
	^{171}Yb	14.3	1/2	0.49
	^{172}Yb	21.9	0	0
	^{173}Yb	16.1	5/2	−0.68
	^{174}Yb	32	0	0
	^{176}Yb	12.7	0	0
71	^{175}Lu	97.41	7/2	2.23
	^{176}Lu	2.59	7	3.2
72	^{174}Hf	0.162	0	0
	^{176}Hf	5.21	0	0
	^{177}Hf	18.61	7/2	0.784
	^{178}Hf	27.30	0	0
	^{179}Hf	13.63	9/2	−0.633
	^{180}Hf	35.10	0	0
73	^{180}Ta	0.012	8	0
	^{181}Ta	99.988	7/2	2.4
74	^{180}W	0.1	0	0
	^{182}W	26.3	0	0

Table 3.1 (continued)

Atomic number Z	Isotope	Representative isotopic composition [%]	Nuclear spin I	Magnetic moment in nuclear magnetons μ/μ_N	Atomic number Z	Isotope	Representative isotopic composition [%]	Nuclear spin I	Magnetic moment in nuclear magnetons μ/μ_N
1	2	3	4	5	1	2	3	4	5
	^{183}W	14.3	1/2	0.118	79	^{197}Au	100	3/2	0.146
	^{184}W	30.7	0	0	80	^{196}Hg	0.1	0	0
	^{186}W	28.6	0	0		^{198}Hg	10.0	0	0
75	^{185}Re	37.40	5/2	3.19		^{199}Hg	16.8	1/2	0.506
	^{187}Re	62.60	5/2	3.22		^{200}Hg	23.1	0	0
76	^{184}Os	0.02	0	0		^{201}Hg	13.2	3/2	−0.560
	^{186}Os	1.6	0	0		^{202}Hg	29.8	0	0
	^{187}Os	1.6	1/2	0.065		^{204}Hg	6.8	0	0
	^{188}Os	13.3	0	0	81	^{203}Tl	29.52	1/2	1.62
	^{189}Os	16	3/2	0.7		^{205}Tl	70.48	1/2	1.64
	^{190}Os	26	0	0	82	^{204}Pb	1.4	0	0
	^{192}Os	41	0	0		^{206}Pb	24.1	0	0
77	^{191}Ir	37	3/2	0.15		^{207}Pb	22.1	1/2	0.59
	^{193}Ir	63	3/2	0.16		^{208}Pb	52.4	0	0
78	^{190}Pt	0.01	0	0	83	^{209}Bi	100	9/2	4.1
	^{192}Pt	0.8	0	0	90	^{232}Th[a]	100	0	0
	^{194}Pt	33	0	0	91	^{231}Pa[b]	100	3/2	2.0
	^{195}Pt	34	1/2	0.61	92	^{234}U[c]	0.005	0	0
	^{196}Pt	25	0	0		^{235}U[d]	0.720	7/2	−0.35
	^{198}Pt	7.2	0	0		^{238}U[e]	99.275	0	0

[a] $T_{1/2} = 1.4 \cdot 10^{10}$ a
[b] $T_{1/2} = 3.3 \cdot 10^{4}$ a
[c] $T_{1/2} = 2.4 \cdot 10^{5}$ a
[d] $T_{1/2} = 7.0 \cdot 10^{8}$ a
[e] $T_{1/2} = 4.5 \cdot 10^{9}$ a

Table 3.2. Parameters of long-lived radionuclides

Atomic number Z	Symbol and mass number	Nuclear spin I	Relative magnetic moment μ/μ_N	Half-life $T_{1/2}$	Decay mode
1	2	3	4	5	6
0	^{1}n	1/2	−1.9130	10.6 min	β^-
1	^{3}H	1/2	2.979	12.3 a	β^-
4	^{7}Be	3/2	–	53.4 d	EC
	^{10}Be	0	0	$1.6 \cdot 10^6$ a	β^-
6	^{11}C	3/2	1.0	0.34 h	β^+, EC
	^{14}C	0	0	5730 a	β^-
7	^{13}N	1/2	±0.322	0.166 h	β^+
9	^{18}F	1	0.8	1.83 h	β^+, EC
11	^{22}Na	3	1.75	2.6 a	β^+, EC
	^{24}Na	4	1.69	15.0 h	β^-
12	^{28}Mg	0	0	21.1 h	β^-
13	^{26}Al	5	2.8	$7.2 \cdot 10^5$ a	β^+, EC
14	^{31}Si	3/2	–	2.6 h	β^-
	^{32}Si	0	0	330 a	β^-
15	^{32}P	1	−0.252	14.4 d	β^-
	^{33}P	1/2	–	25.3 d	β^-
16	^{35}S	3/2	1.0	87.2 d	β^-
	^{38}S	0	0	2.84 h	β^-
17	^{36}Cl	2	1.285	$3.0 \cdot 10^5$ a	β^-, EC, β^+
	^{38}Cl	2	2.0	0.62 h	β^-
	^{39}Cl	3/2	–	0.93 h	β^-
18	^{37}Ar	3/2	0.9	35 d	EC
	^{39}Ar	7/2	−1.3	269 a	β^-
	^{41}Ar	7/2	–	1.83 h	β^-
	^{42}Ar	0	0	33 a	β^-
	^{44}Ar	0	0	0.20 h	β^-
19	^{40}K	4	−1.30	$1.3 \cdot 10^9$ a	β^-, EC, β^+
	^{42}K	2	−1.14	12.4 h	β^-
	^{43}K	3/2	±0.16	22.3 h	β^-
	^{44}K	2	–	0.37 h	β^-
	^{45}K	3/2	±0.17	0.3 h	β^-
20	^{41}Ca	7/2	−1.595	$1.4 \cdot 10^5$ a	EC
	^{45}Ca	7/2	–	164 d	β^-
	^{47}Ca	7/2	–	4.5 d	β^-
	^{48}Ca	0	0	$2.0 \cdot 10^{16}$ a	β^-
21	^{43}Sc	7/2	4.6	3.9 h	β^+, EC
	^{44}Sc	2	2.6	3.9 h	β^+, EC
	^{46}Sc	4	3.0	84 d	β^-
	^{47}Sc	7/2	5.3	3.4 d	β^-
	^{48}Sc	6	–	44 h	β^-
	^{49}Sc	7/2	–	0.96 h	β^-

Table 3.2 (continued)

Atomic number Z	Symbol and mass number	Nuclear spin I	Relative magnetic moment μ/μ_N	Half-life $T_{1/2}$	Decay mode
1	2	3	4	5	6
22	^{44}Ti	0	0	47 a	EC
	^{45}Ti	7/2	0.09	3.1 h	β^+, EC
23	^{47}V	3/2	–	0.54 h	β^+, EC
	^{48}V	4	±1.6	16 d	EC, β^+
	^{49}V	7/2	±4.5	0.90 a	EC
	^{50}V	6	3.347	$4 \cdot 10^{16}$ a	–
24	^{48}Cr	0	0	23.0 h	EC
	^{49}Cr	5/2	±0.48	0.70 h	β^+, EC
	^{51}Cr	7/2	±0.93	27.7 d	EC
25	^{51}Mn	5/2	±3.57	0.77 h	β^+, EC
	^{52}Mn	6	3.06	5.6 d	EC, β^+
	^{53}Mn	7/2	±5.02	$3.7 \cdot 10^6$ a	EC
	^{54}Mn	3	–	312 d	EC
	^{56}Mn	3	3.23	2.58 h	β^-
26	^{52}Fe	0	0	8.3 h	β^+, EC
	^{55}Fe	3/2	–	2.7 a	EC
	^{59}Fe	3/2	±1.1	44.5 d	β^-
	^{60}Fe	0	0	$3.0 \cdot 10^5$ a	β^-
27	^{55}Co	7/2	±4.5	17.5 h	β^+, EC
	^{56}Co	4	3.83	79 d	EC, β^+
	^{57}Co	7/2	4.72	271 d	EC
	^{58}Co	2	4.04	71 d	EC, β^+
	^{60}Co	5	3.80	5.3 a	β^-
	^{61}Co	7/2	–	1.7 h	β^-
28	^{56}Ni	0	0	6.1 d	EC
	^{57}Ni	3/2	–	1.5 d	EC, β^+
	^{59}Ni	3/2	–	$7.5 \cdot 10^4$ a	EC, β^+
	^{63}Ni	1/2	–	100 a	β^-
	^{65}Ni	5/2	–	2.52 h	β^-
	^{66}Ni	0	0	2.3 d	β^-
29	^{60}Cu	2	1.22	0.39 h	β^+, EC
	^{61}Cu	3/2	2.13	3.4 h	β^+, EC
	^{64}Cu	1	−0.22	12.7 h	EC, β^-, β^+
	^{67}Cu	3/2	–	62 h	β^-
30	^{62}Zn	0	0	9.3 h	EC, β^+
	^{63}Zn	3/2	−0.282	0.64 h	β^+, EC
	^{65}Zn	5/2	0.77	244 d	EC, β^+
	^{69}Zn	1/2	–	0.93 h	β^-
	^{72}Zn	0	0	46 h	β^-
31	^{65}Ga	3/2	–	0.25 h	β^+, EC
	^{66}Ga	0	–	9.5 h	β^+, EC
	^{67}Ga	3/2	1.85	3.3 d	EC
	^{68}Ga	1	±0.012	1.1 h	β^+, EC

Table 3.2 (continued)

Atomic number Z	Symbol and mass number	Nuclear spin I	Relative magnetic moment μ/μ_N	Half-life $T_{1/2}$	Decay mode
1	2	3	4	5	6
	^{70}Ga	1	–	0.35 h	β^-, EC
	^{72}Ga	3	−0.132	14.1 h	β^-
	^{73}Ga	3/2	–	4.9 h	β^-
32	^{66}Ge	0	0	2.3 h	EC, β^+
	^{67}Ge	1/2	–	0.3 h	β^+, EC
	^{68}Ge	0	0	287 d	EC
	^{69}Ge	5/2	–	39 h	EC, β^+
	^{71}Ge	1/2	0.55	11.8 d	EC
	^{75}Ge	1/2	–	1.38 h	β^-
	^{77}Ge	7/2	–	11.3 h	β^-
	^{78}Ge	0	0	1.5 h	β^-
33	^{69}As	5/2	–	0.25 h	β^+, EC
	^{70}As	4	–	0.9 h	β^+, EC
	^{71}As	5/2	–	2.7 d	EC, β^+
	^{72}As	2	±2.2	1.1 d	β^+, EC
	^{73}As	3/2	–	80 d	EC
	^{74}As	2	–	18 d	EC, β^-, β^+
	^{76}As	2	−0.91	1.1 d	β^-
	^{77}As	3/2	–	1.6 d	β^-
	^{78}As	2	–	1.5 h	β^-
34	^{70}Se	0	0	0.68 h	β^+, EC
	^{72}Se	0	0	8.4 d	EC
	^{73}Se	9/2	–	7.1 h	β^+, EC
	^{75}Se	5/2	±0.7	120 d	EC
	^{79}Se	7/2	−1.02	$6.5 \cdot 10^4$ a	β^-
	^{81}Se	1/2	–	0.3 h	β^-
	^{83}Se	9/2	–	0.38 h	β^-
35	^{74}Br	(0,1)	–	0.42 h	β^+, EC
	^{75}Br	3/2	–	1.6 h	β^+, EC
	^{76}Br	1	±0.55	16 h	β^+, EC
	^{77}Br	3/2	–	2.4 d	EC, β^+
	^{80}Br	1	±0.514	0.3 h	β^-, EC, β^+
	^{82}Br	5	1.63	1.47 d	β^-
	^{83}Br	3/2	–	2.4 h	β^-
	^{84}Br	2	–	0.53 h	β^-
36	^{74}Kr	0	0	0.2 h	β^+, EC
	^{76}Kr	0	0	15 h	EC
	^{77}Kr	5/2	–	1.2 h	β^+, EC
	^{79}Kr	1/2	–	35 h	EC, β^+
	^{81}Kr	7/2	–	$2 \cdot 10^5$ a	EC
	^{85}Kr	9/2	±1.00	10.7 a	β^-
	^{87}Kr	5/2	–	1.3 h	β^-
	^{88}Kr	0	0	2.8 h	β^-

Table 3.2 (continued)

Atomic number Z	Symbol and mass number	Nuclear spin I	Relative magnetic moment μ/μ_N	Half-life $T_{1/2}$	Decay mode
1	2	3	4	5	6
37	^{78}Rb	0	–	17.7 min	β^+, EC
	^{79}Rb	5/2	–	23 min	β^+, EC
	^{81}Rb	3/2	2.06	4.6 h	EC, β^+
	^{83}Rb	5/2	1.42	86 d	EC
	^{84}Rb	2	−1.32	33 d	EC, β^+, β^-
	^{86}Rb	2	−1.698	18.7 d	β^-, EC
	^{87}Rb	3/2	2.75	$4.8 \cdot 10^{10}$ a	β^-
	^{88}Rb	2	+0.51	0.30 h	β^-
	^{89}Rb	3/2	2.38	0.25 h	β^-
38	^{78}Sr	0	0	0.51 h	β^+, EC
	^{80}Sr	0	0	1.8 h	EC, β^+
	^{81}Sr	1/2	–	0.43 h	β^+, EC
	^{82}Sr	0	0	25 d	EC
	^{83}Sr	7/2	–	1.4 d	EC, β^+
	^{85}Sr	9/2	–	65 d	EC
	^{89}Sr	5/2	–	51 d	β^-
	^{90}Sr	0	0	29 a	β^-
	^{91}Sr	5/2	–	9.5 h	β^-
	^{92}Sr	0	0	2.7 h	β^-
39	^{84}Y	5	–	0.7 h	β^+, EC
	^{85}Y	1/2	–	2.7 h	β^+, EC
	^{86}Y	4	−1.1	15 h	EC, β^+
	^{87}Y	1/2	–	3.3 d	EC, β^+
	^{88}Y	4	–	107 d	EC, β^+
	^{90}Y	2	−1.63	2.7 d	β^-
	^{91}Y	1/2	±0.164	59 d	β^-
	^{92}Y	2	–	3.5 h	β^-
	^{93}Y	1/2	–	10 h	β^-
	^{94}Y	2	–	0.3 h	β^-
	^{95}Y	1/2	–	0.17 h	β^-
40	^{86}Zr	0	0	16 h	EC
	^{87}Zr	9/2	–	1.7 h	β^+, EC
	^{88}Zr	0	0	83 d	EC
	^{89}Zr	9/2	–	3.3 d	EC, β^+
	^{93}Zr	5/2	–	$1.5 \cdot 10^{6}$ a	β^-
	^{95}Zr	5/2	–	64 d	β^-
	^{97}Zr	1/2	–	17.0 h	β^-
41	^{88}Nb	8	–	0.24 h	β^+, EC
	^{89}Nb	1/2	–	1.1 h	EC, β^+
	^{90}Nb	8	–	14.6 h	β^+, EC
	^{91}Nb	9/2	–	$1.0 \cdot 10^{4}$ a	EC
	^{92}Nb	7	–	$3.5 \cdot 10^{7}$ a	EC
	^{94}Nb	6	–	$2 \cdot 10^{4}$ a	β^-
	^{95}Nb	9/2	–	35.0 d	β^-

Table 3.2 (continued)

Atomic number Z	Symbol and mass number	Nuclear spin I	Relative magnetic moment μ/μ_N	Half-life $T_{1/2}$	Decay mode
1	2	3	4	5	6
	^{96}Nb	6	–	0.97 d	β^-
	^{97}Nb	9/2	7	1.2 h	β^-
42	^{90}Mo	0	0	5.7 h	EC, β^+
	^{91}Mo	9/2	0	0.26 h	β^+, EC
	^{93}Mo	5/2	–	$3.5 \cdot 10^3$ a	EC
	^{99}Mo	1/2	±0.37	2.8 d	β^-
	^{101}Mo	1/2	–	0.24 h	β^-
	^{102}Mo	0	0	0.19 h	β^-
43	^{93}Tc	9/2	–	2.8 h	EC, β^+
	^{94}Tc	7	–	4.9 h	EC, β^+
	^{95}Tc	9/2	–	20 h	EC
	^{96}Tc	7	±5.4	4.3 d	EC
	^{97}Tc	9/2	–	$2.6 \cdot 10^6$ a	EC
	^{98}Tc	6	–	$4.2 \cdot 10^6$ a	β^-
	^{99}Tc	9/2	5.68	$2.1 \cdot 10^5$ a	β^-
	^{101}Tc	9/2	–	0.24 h	β^-
	^{104}Tc	–	–	0.3 h	β^-
44	^{94}Ru	0	0	0.86 h	EC
	^{95}Ru	5/2	–	1.6 h	EC, β^+
	^{97}Ru	5/2	–	2.9 d	EC
	^{103}Ru	3/2	0.18	39.3 d	β^-
	^{105}Ru	3/2	–	4.4 h	β^-
	^{106}Ru	0	0	1.0 a	β^-
45	^{97}Rh	9/2	–	0.5 h	β^+, EC
	^{99}Rh	1/2	–	16 d	EC, β^+
	^{100}Rh	1	–	21 h	EC, β^+
	^{101}Rh	1/2	–	3.3 a	EC
	^{102}Rh	6	4.1	2.9 a	EC
	^{105}Rh	7/2	–	1.5 d	β^-
	^{107}Rh	7/2	–	0.36 h	β^-
46	^{98}Pd	0	0	0.3 h	EC, β^+
	^{99}Pd	5/2	–	0.36 h	β^+, EC
	^{100}Pd	0	0	3.6 d	EC
	^{101}Pd	5/2	–	8.5 h	EC, β^+
	^{103}Pd	5/2	–	17 d	EC
	^{107}Pd	5/2	–	$6.5 \cdot 10^6$ a	β^-
	^{109}Pd	5/2	–	13.5 h	β^-
	^{111}Pd	5/2	–	0.39 h	β^-
	^{112}Pd	0	0	21 h	β^-
47	^{101}Ag	9/2	–	0.18 h	β^+, EC
	^{102}Ag	5	–	0.21 h	β^+, EC
	^{103}Ag	7/2	4.5	1.1 h	EC, β^+
	^{104}Ag	5	4.0	1.2 h	β^+, EC

Table 3.2 (continued)

Atomic number Z	Symbol and mass number	Nuclear spin I	Relative magnetic moment μ/μ_N	Half-life $T_{1/2}$	Decay mode
1	2	3	4	5	6
	^{105}Ag	1/2	±0.10	41 d	EC, β^+
	^{106}Ag	1	2.9	0.40 h	EC, β^+
	^{111}Ag	1/2	−0.15	7.5 d	β^-
	^{112}Ag	2	±0.055	3.1 h	β^-
	^{113}Ag	1/2	±0.16	5.4 h	β^-
	^{115}Ag	1/2	–	0.33 h	β^-
48	^{104}Cd	0	0	0.96 h	EC, β^+
	^{105}Cd	5/2	−0.74	0.93 h	EC, β^+
	^{107}Cd	5/2	−0.615	6.5 h	EC, β^+
	^{109}Cd	5/2	−0.83	1.3 a	EC
	^{113}Cd	1/2	0.622	$9 \cdot 10^{15}$ a	β^-
	^{115}Cd	1/2	−0.65	2.2 d	β^-
	^{117}Cd	1/2	–	2.5 h	β^-
	^{118}Cd	0	0	0.84 h	β^-
49	^{104}In	–	–	0.42 h	β^+, EC
	^{107}In	9/2	–	0.54 h	EC, β^+
	^{108}In	3	–	40 min	EC, β^+
	^{109}In	9/2	5.5	4.2 h	EC, β^+
	^{110}In	2	4.36	1.2 h	β^+, EC
	^{111}In	9/2	5.50	2.8 d	EC
	^{112}In	1	2.8	0.24 h	β^-, EC, β^+
	^{115}In	9/2	5.54	$4.4 \cdot 10^{14}$ a	β^-
	^{117}In	9/2	–	0.73 h	β^-
50	^{109}Sn	7/2	–	0.30 h	β^+, EC
	^{110}Sn	0	0	4 h	EC
	^{111}Sn	7/2	–	0.6 h	EC, β^+
	^{113}Sn	1/2	±0.88	115 d	EC
	^{121}Sn	3/2	±0.70	1.1 d	β^-
	^{123}Sn	11/2	–	129 d	β^-
	^{125}Sn	11/2	–	9.6 d	β^-
	^{126}Sn	0	0	10^5 a	β^-
	^{127}Sn	11/2	–	2.1 h	β^-
	^{128}Sn	0	0	0.99 h	β^-
51	^{115}Sb	5/2	3.46	0.54 h	EC, β^+
	^{116}Sb	3	–	0.26 h	EC, β^+
	^{117}Sb	5/2	3.4	2.8 h	EC, β^+
	^{119}Sb	5/2	3.45	38 h	EC
	^{120}Sb	8	±2.3	5.8 d	EC
	^{122}Sb	2	−1.90	2.7 d	β^-, EC, β^+
	^{124}Sb	3	±1.2	60 d	β^-
	^{125}Sb	7/2	±2.6	2.7 a	β^-
	^{126}Sb	8	±1.3	12 d	β^-
	^{127}Sb	7/2	±2.6	3.9 d	β^-
	^{128}Sb	8	±1.3	9.0 h	β^-

Table 3.2 (continued)

Atomic number Z	Symbol and mass number	Nuclear spin I	Relative magnetic moment μ/μ_N	Half-life $T_{1/2}$	Decay mode
1	2	3	4	5	6
	^{129}Sb	7/2	–	4.4 h	β^-
	^{130}Sb	8	–	40 min	β^-
	^{131}Sb	7/2	–	0.38 h	β^-
52	^{114}Te	0	0	0.25 h	EC, β^+
	^{116}Te	0	0	2.5 h	EC, β^+
	^{117}Te	1/2	–	1.0 h	EC, β^+
	^{118}Te	0	0	6 d	EC
	^{119}Te	1/2	±0.25	16 h	EC, β^+
	^{121}Te	1/2	–	17 d	EC
	^{123}Te	1/2	−0.736	10^{13} a	EC
	^{127}Te	3/2	±0.63	9.4 h	β^-
	^{129}Te	3/2	±0.70	1.2 h	β^-
	^{131}Te	3/2	±0.70	0.42 h	β^-
	^{132}Te	0	0	3.26 d	β^-
	^{133}Te	3/2	–	0.21 h	β^-
	^{134}Te	0	0	0.70 h	β^-
53	^{118}I	2	–	0.23 h	β^+, EC
	^{119}I	5/2	–	0.32 h	β^+, EC
	^{120}I	2	–	1.4 h	EC, β^+
	^{121}I	5/2	–	2.1 h	EC, β^+
	^{123}I	5/2	–	13.2 h	EC
	^{124}I	2	–	4.2 d	EC, β^+
	^{125}I	5/2	3	60 d	EC
	^{126}I	2	–	13 d	EC, β^-, β^+
	^{128}I	1	–	0.42 h	β^-, EC, β^+
	^{129}I	7/2	2.62	$1.6 \cdot 10^7$ a	β^-
	^{130}I	5	–	12.4 h	β^-
	^{131}I	7/2	2.74	8.0 d	β^-
	^{132}I	4	±3.09	2.3 h	β^-
	^{133}I	7/2	2.86	21 h	β^-
	^{134}I	4	–	0.88 h	β^-
	^{135}I	7/2	–	6.6 h	β^-
54	^{120}Xe	0	0	0.67 h	EC, β^+
	^{121}Xe	5/2	–	0.67 h	EC, β^+
	^{122}Xe	0	0	20 h	EC
	^{123}Xe	1/2	–	2.1 h	EC, β^+
	^{125}Xe	1/2	–	17 h	EC, β^+
	^{127}Xe	1/2	–	36.4 d	EC
	^{133}Xe	3/2	–	5.29 d	β^-
	^{135}Xe	3/2	–	9.1 h	β^-
	^{138}Xe	0	0	0.24 h	β^-
55	^{125}Cs	1/2	1.41	0.75 h	EC, β^+
	^{127}Cs	1/2	1.46	6.2 h	EC, β^+
	^{129}Cs	1/2	1.49	32 h	EC, β^+

Table 3.2 (continued)

Atomic number Z	Symbol and mass number	Nuclear spin I	Relative magnetic moment μ/μ_N	Half-life $T_{1/2}$	Decay mode
1	2	3	4	5	6
	^{130}Cs	1	1.46	0.5 h	EC, β^+, β^-
	^{131}Cs	5/2	3.53	9.7 d	EC
	^{132}Cs	2	2.23	6.5 d	EC, β^-, β^+
	^{134}Cs	4	2.99	2.06 a	β^-, EC
	^{135}Cs	7/2	2.73	$2.3 \cdot 10^6$ a	β^-
	^{136}Cs	5	3.7	13 d	β^-
	^{137}Cs	7/2	2.84	30 a	β^-
	^{138}Cs	3	0.70	0.54 h	β^-
56	^{124}Ba	0	0	0.20 h	EC, β^+
	^{126}Ba	0	0	1.67 h	EC, β^+
	^{127}Ba	1/2	–	0.22 h	β^+, EC
	^{128}Ba	0	0	2.4 d	EC
	^{129}Ba	1/2	–	2.2 h	EC, β^+
	^{131}Ba	1/2	–	12 d	EC
	^{133}Ba	1/2	–	10.5 a	EC
	^{139}Ba	7/2	–	1.4 h	β^-
	^{140}Ba	0	0	12.8 d	β^-
	^{141}Ba	3/2	–	0.3 h	β^-
	^{142}Ba	0	0	10.6 min	β^-
57	^{129}La	3/2	–	11.6 min	β^+, EC
	^{131}La	3/2	–	1.0 h	EC, β^+
	^{132}La	2	–	4.8 h	β^+, EC
	^{133}La	5/2	–	3.9 h	EC, β^+
	^{135}La	5/2	–	19 h	EC, β^+
	^{137}La	7/2	2.69	$6 \cdot 10^4$ a	EC
	^{138}La	5	3.70	$1.3 \cdot 10^{11}$ a	EC, β^-
	^{140}La	3	0.73	40.3 h	β^-
	^{141}La	–	–	3.9 h	β^-
	^{142}La	2	–	1.54 h	β^-
	^{143}La	–	–	0.24 h	β^-
58	^{130}Ce	0	0	0.42 h	EC, β^+
	^{132}Ce	0	0	4.2 h	EC
	^{133}Ce	9/2	–	5.4 h	EC, β^+
	^{134}Ce	0	0	76 h	EC
	^{135}Ce	1/2	–	17.6 h	EC, β^+
	^{137}Ce	3/2	±0.9	9 h	EC, β^+
	^{139}Ce	3/2	0.9	138 d	EC
	^{141}Ce	7/2	1.0	33 d	β^-
	^{142}Ce	0	0	$5 \cdot 10^{16}$ a	–
	^{143}Ce	3/2	–	33 h	β^-
	^{144}Ce	0	0	285 d	β^-
	^{146}Ce	0	0	0.24 h	β^-
59	^{134}Pr	2	–	0.28 h	β^+, EC
	^{135}Pr	3/2	–	0.37 h	EC, β^+

Table 3.2 (continued)

Atomic number Z	Symbol and mass number	Nuclear spin I	Relative magnetic moment μ/μ_N	Half-life $T_{1/2}$	Decay mode
1	2	3	4	5	6
	^{136}Pr	2	–	0.22 h	β^+, EC
	^{137}Pr	5/2	–	1.3 h	EC, β^+
	^{139}Pr	5/2	–	4.4 h	EC, β^+
	^{142}Pr	2	0.23	19 h	β^-, EC
	^{143}Pr	7/2	–	13.6 d	β^-
	^{144}Pr	0	–	0.29 h	β^-
	^{145}Pr	7/2	–	6.0 h	β^-
	^{146}Pr	–	–	0.40 h	β^-
	^{147}Pr	5/2	–	0.23 h	β^-
60	^{135}Nd	9/2	–	0.2 h	β^+, EC
	^{136}Nd	0	0	0.84 h	EC, β^+
	^{137}Nd	1/2	–	0.64 h	β^+, EC
	^{138}Nd	0	0	5.0 h	EC
	^{139}Nd	3/2	–	0.5 h	EC, β^+
	^{140}Nd	0	0	3.4 d	EC
	^{141}Nd	3/2	–	2.5 h	EC, β^+
	^{144}Nd	0	0	$2.1 \cdot 10^{15}$ a	α
	^{147}Nd	5/2	−0.55	11 d	β^-
	^{149}Nd	5/2	−0.35	1.7 h	β^-
	^{151}Nd	3/2	–	0.21 h	β^-
	^{152}Nd	0	0	0.19 h	β^-
61	^{141}Pm	5/2	–	0.35 h	β^+, EC
	^{143}Pm	5/2	±3.8	265 d	EC
	^{144}Pm	5	±1.7	363 d	EC
	^{145}Pm	5/2	–	18 a	EC, α
	^{146}Pm	(2, 3, 4)	–	5.5 a	EC, β^-
	^{147}Pm	7/2	2.7	2.62 a	β^-
	^{148}Pm	1	2.0	5.4 d	β^-
	^{149}Pm	7/2	±3.3	53 h	β^-
	^{150}Pm	1	–	2.7 h	β^-
	^{151}Pm	5/2	±1.6	28.4 h	β^-
62	^{140}Sm	0	0	0.25 h	EC, β^+
	^{142}Sm	0	0	1.2 h	EC, β^+
	^{145}Sm	7/2	±0.9	340 d	EC
	^{146}Sm	0	0	$1.0 \cdot 10^{8}$ a	α
	^{147}Sm	7/2	−0.81	$1.1 \cdot 10^{11}$ a	α
	^{148}Sm	0	0	$8 \cdot 10^{15}$ a	α
	^{149}Sm	7/2	−0.67	$1 \cdot 10^{16}$ a	α
	^{151}Sm	5/2	±0.35	90 a	β^-
	^{153}Sm	3/2	−0.022	47 h	β^-
	^{155}Sm	3/2	–	0.37 h	β^-
	^{156}Sm	0	0	9.4 h	β^-
63	^{145}Eu	5/2	–	6 d	EC, β^+
	^{146}Eu	4	–	4.6 d	EC, β^+

Table 3.2 (continued)

Atomic number Z	Symbol and mass number	Nuclear spin I	Relative magnetic moment μ/μ_N	Half-life $T_{1/2}$	Decay mode
1	2	3	4	5	6
	^{147}Eu	5/2	–	24 d	EC, β^+, α
	^{148}Eu	5	–	54 d	EC, β^+, α
	^{149}Eu	5/2	–	93 d	EC
	^{150}Eu	0	–	12.6 h	β^-, EC, β^+
	^{152}Eu	3	−1.94	13 a	EC, β^-, β^+
	^{154}Eu	3	±2.0	8.8 a	β^-, EC
	^{155}Eu	5/2	±1.9	5 a	β^-
	^{156}Eu	0	1.97	15.2 d	β^-
	^{157}Eu	5/2	–	15 h	β^-
	^{158}Eu	1	–	0.77 h	β^-
	^{159}Eu	5/2	–	0.31 h	β^-
64	^{145}Gd	1/2	–	0.40 h	β^+, EC
	^{146}Gd	0	0	48.3 d	EC, β^+
	^{147}Gd	7/2	–	1.59 d	EC, β^+
	^{148}Gd	0	0	93 a	α
	^{149}Gd	7/2	–	9.4 d	EC, α
	^{150}Gd	0	0	$1.8 \cdot 10^6$ a	α
	^{151}Gd	7/2	–	120 d	EC, α
	^{152}Gd	0	0	$1.1 \cdot 10^{14}$ a	α
	^{153}Gd	3/2	–	242 d	EC
	^{159}Gd	3/2	±0.4	18.6 h	β^-
65	^{147}Tb	5/2	–	1.7 h	EC, β^+
	^{148}Tb	2	–	1.0 h	EC, β^+
	^{149}Tb	(3/2, 5/2)	–	4.1 h	EC, α, β^+
	^{150}Tb	2	–	3.3 h	EC, β^+, α
	^{151}Tb	1/2	–	18 h	EC, β^+, α
	^{152}Tb	2	–	17 h	EC, β^+
	^{153}Tb	5/2	–	2.3 d	EC, β^+
	^{154}Tb	0	–	21 h	EC, β^+
	^{155}Tb	3/2	–	5.3 d	EC
	^{156}Tb	3	±1.4	5.3 d	EC
	^{157}Tb	3/2	±2.0	150 a	EC
	^{158}Tb	3	±1.7	150 a	EC, β^-
	^{160}Tb	3	±1.7	72 d	β^-
	^{161}Tb	3/2	–	6.9 d	β^-
	^{163}Tb	3/2	–	0.32 h	β^-
66	^{151}Dy	7/2	–	0.28 h	EC, β^+, α
	^{152}Dy	0	0	2.4 h	EC, α
	^{153}Dy	7/2	−0.71	6.4 h	EC, β^+, α
	^{154}Dy	0	0	10^7 a	α
	^{155}Dy	3/2	−0.34	10 h	EC, β^+
	^{157}Dy	3/2	−0.30	8.1 h	EC
	^{159}Dy	3/2	–	144 d	EC
	^{165}Dy	7/2	−0.52	2.3 h	β^-
	^{166}Dy	0	0	3.4 d	β^-

Table 3.2 (continued)

Atomic number Z	Symbol and mass number	Nuclear spin I	Relative magnetic moment μ/μ_N	Half-life $T_{1/2}$	Decay mode
1	2	3	4	5	6
67	^{154}Ho	1	–	0.20 h	β^+, EC, α
	^{155}Ho	5/2	–	0.8 h	EC, β^+, α
	^{156}Ho	1	–	0.93 h	β^+, EC
	^{157}Ho	7/2	–	0.21 h	β^+, EC
	^{158}Ho	5	–	0.19 h	EC, β^+
	^{159}Ho	7/2	–	0.55 h	EC
	^{160}Ho	5	–	0.43 h	EC, β^+
	^{161}Ho	7/2	–	2.5 h	EC
	^{162}Ho	1	–	0.25 h	EC, β^+
	^{163}Ho	7/2	–	33 a	EC
	^{164}Ho	1	–	0.48 h	EC, β^-
	^{166}Ho	0	–	1.1 d	β^-
	^{167}Ho	7/2	–	3.1 h	β^-
68	^{157}Er	3/2	–	0.4 h	β^+, EC
	^{158}Er	0	0	2.3 h	EC, β^+
	^{159}Er	3/2	–	0.6 h	EC, β^+
	^{160}Er	0	0	1.2 d	EC
	^{161}Er	3/2	−0.37	3.2 h	EC, β^+
	^{163}Er	5/2	0.56	1.25 h	EC, β^+
	^{165}Er	5/2	±0.65	10.4 h	EC
	^{169}Er	1/2	0.49	9.4 d	β^-
	^{171}Er	5/2	±0.7	7.5 h	β^-
	^{172}Er	0	0	2.1 d	β^-
69	^{161}Tm	7/2	–	0.63 h	EC, β^+
	^{162}Tm	1	–	0.36 h	EC, β^+
	^{163}Tm	1/2	−0.08	1.8 h	EC, β^+
	^{165}Tm	1/2	−0.14	30 h	EC, β^+
	^{166}Tm	2	±0.09	7.7 h	EC, β^+
	^{167}Tm	1/2	−0.20	9.2 d	EC
	^{168}Tm	3	–	93 d	EC, β^-
	^{170}Tm	1	±0.25	129 d	β^-, EC
	^{171}Tm	1/2	−0.23	1.9 a	β^-
	^{172}Tm	2	–	64 h	β^-
	^{173}Tm	1/2	–	8.2 h	β^-
	^{175}Tm	1/2	–	0.25 h	β^-
70	^{162}Yb	0	0	0.32 h	EC, β^+
	^{163}Yb	3/2	–	0.18 h	EC, β^+
	^{164}Yb	0	0	1.3 h	EC
	^{166}Yb	0	0	2.4 d	EC
	^{167}Yb	5/2	–	0.29 h	EC, β^+
	^{169}Yb	7/2	−0.6	32 d	EC
	^{175}Yb	7/2	±0.3	4.2 d	β^-
	^{177}Yb	9/2	–	1.9 h	β^-
	^{178}Yb	0	0	1.23 h	β^-

Table 3.2 (continued)

Atomic number Z	Symbol and mass number	Nuclear spin I	Relative magnetic moment μ/μ_N	Half-life $T_{1/2}$	Decay mode
1	2	3	4	5	6
71	^{165}Lu	–	–	0.2 h	EC, β^+
	^{167}Lu	7/2	–	0.86 h	EC, β^+
	^{169}Lu	7/2	–	1.4 d	EC, β^+
	^{170}Lu	0	–	2 d	EC, β^+
	^{171}Lu	7/2	–	8.2 d	EC, β^+
	^{172}Lu	4	–	6.7 d	EC
	^{173}Lu	7/2	2.3	1.4 a	EC
	^{174}Lu	1	1.9	3.3 a	EC, β^+
	^{176}Lu	7	3.2	$3.6 \cdot 10^{10}$ a	β^-
	^{177}Lu	7/2	2.24	6.7 d	β^-
	^{178}Lu	1	–	0.47 h	β^-
	^{179}Lu	7/2	–	4.6 h	β^-
72	^{168}Hf	0	0	0.43 h	EC
	^{170}Hf	0	0	16 h	EC
	^{171}Hf	7/2	–	12.1 h	EC, β^+
	^{172}Hf	0	0	1.9 a	EC
	^{173}Hf	1/2	–	1.0 d	EC
	^{174}Hf	0	0	$2 \cdot 10^{15}$ a	α
	^{175}Hf	5/2	–	70 d	EC
	^{181}Hf	1/2	–	42 d	β^-
	^{182}Hf	0	0	$9 \cdot 10^{6}$ a	β^-
	^{183}Hf	3/2	–	1.1 h	β^-
	^{184}Hf	0	0	4.1 h	β^-
73	^{171}Ta	7/2	–	0.39 h	EC, β^+
	^{172}Ta	3	–	0.61 h	EC, β^+
	^{173}Ta	5/2	–	3.7 h	EC, β^+
	^{174}Ta	4	–	1.2 h	EC, β^+
	^{175}Ta	7/2	–	10 h	EC, β^+
	^{176}Ta	1	–	8.1 h	EC, β^+
	^{177}Ta	7/2	–	2.4 d	EC, β^+
	^{178}Ta	7	–	2.2 h	EC
	^{179}Ta	7/2	–	1.82 a	EC
	^{182}Ta	3	±2.6	115 d	β^-
	^{183}Ta	7/2	–	5 d	β^-
	^{184}Ta	5	–	8.7 h	β^-
	^{185}Ta	7/2	–	0.8 h	β^-
74	^{173}W	–	–	0.28 h	EC, β^+
	^{174}W	0	0	0.48 h	EC
	^{175}W	1/2	–	0.57 h	EC, β^+
	^{176}W	0	0	2.3 h	EC
	^{177}W	1/2	–	2.25 h	EC, β^+
	^{178}W	0	0	22 d	EC
	^{179}W	7/2	–	0.63 h	EC
	^{181}W	9/2	–	121 d	EC

Table 3.2 (continued)

Atomic number Z	Symbol and mass number	Nuclear spin I	Relative magnetic moment μ/μ_N	Half-life $T_{1/2}$	Decay mode
1	2	3	4	5	6
	^{185}W	3/2	–	75 d	β^-
	^{187}W	3/2	–	1.0 d	β^-
	^{188}W	0	0	69 d	β^-
	^{189}W	3/2	–	0.19 h	β^-
	^{190}W	0	0	0.5 h	β^-
75	^{177}Re	5/2	–	0.23 h	EC, β^+
	^{178}Re	3	–	0.22 h	EC, β^+
	^{179}Re	5/2	–	0.33 h	EC, β^+
	^{181}Re	5/2	–	20 h	EC
	^{182}Re	7	–	64 h	EC
	^{183}Re	5/2	±3.1	70 d	EC
	^{184}Re	3	±2.5	38 d	EC
	^{186}Re	1	1.74	3.8 d	β^-, EC
	^{187}Re	5/2	3.22	$5 \cdot 10^{10}$ a	β^-
	^{188}Re	1	1.79	17 h	β^-
	^{189}Re	5/2	–	1.0 d	β^-
76	^{180}Os	0	0	0.37 h	EC
	^{181}Os	1/2	–	1.8 h	EC, β^+
	^{182}Os	0	0	22 h	EC
	^{183}Os	9/2	–	13 h	EC, β^+
	^{185}Os	1/2	–	94 d	EC
	^{186}Os	0	0	$2 \cdot 10^{15}$ a	α
	^{191}Os	9/2	–	15 d	β^-
	^{193}Os	3/2	–	1.27 d	β^-
	^{194}Os	0	0	6 a	β^-
	^{196}Os	0	0	0.58 h	β^-
77	^{182}Ir	(3)	–	0.25 h	EC, β^+
	^{183}Ir	(1/2, 3/2)	–	0.95 h	EC, β^+
	^{184}Ir	5	–	3.0 h	EC, β^+
	^{185}Ir	5/2	–	14 h	EC, β^+
	^{186}Ir	2	–	1.7 h	EC, β^+
	^{187}Ir	3/2	–	10.5 h	EC
	^{188}Ir	2	–	1.7 d	EC, β^+
	^{189}Ir	3/2	–	13.2 d	EC
	^{190}Ir	4	–	12 d	EC
	^{192}Ir	4	1.9	74 d	β^-, EC
	^{194}Ir	1	±0.4	19 h	β^-
	^{195}Ir	3/2	–	2.5 h	β^-
78	^{184}Pt	0	0	0.29 h	EC, β^+, α
	^{185}Pt	9/2	–	1.2 h	EC, β^+
	^{186}Pt	0	0	2.0 h	EC, α
	^{187}Pt	3/2	–	2.3 h	EC, β^+
	^{188}Pt	0	0	10.2 d	EC, α
	^{189}Pt	3/2	–	11 h	EC, β^+

Table 3.2 (continued)

Atomic number Z	Symbol and mass number	Nuclear spin I	Relative magnetic moment μ/μ_N	Half-life $T_{1/2}$	Decay mode
1	2	3	4	5	6
	^{190}Pt	0	0	$6 \cdot 10^{11}$ a	α
	^{191}Pt	3/2	–	2.9 d	EC
	^{193}Pt	1/2	–	50 a	EC
	^{197}Pt	1/2	±0.5	18 h	β^-
	^{199}Pt	5/2	–	0.5 h	β^-
	^{200}Pt	0	0	12.5 h	β^-
79	^{186}Au	–	–	0.18 h	EC, β^+
	^{189}Au	1/2	–	0.48 h	EC, β^+
	^{190}Au	1	±0.07	0.7 h	EC, β^+
	^{191}Au	3/2	±0.14	3.2 h	EC
	^{192}Au	1	±0.008	5 h	EC, β^+
	^{193}Au	3/2	±0.14	17.7 h	EC
	^{194}Au	1	±0.07	1.65 d	EC, β^+
	^{195}Au	3/2	±0.15	183 d	EC
	^{196}Au	2	±0.6	6.2 d	EC, β^-, β^+
	^{198}Au	2	0.59	2.7 d	β^-
	^{199}Au	3/2	0.27	3.1 d	β^-
	^{200}Au	1	–	0.81 h	β^-
	^{201}Au	3/2	–	0.43 h	β^-
80	^{190}Hg	0	0	0.33 h	EC
	^{191}Hg	3/2	–	0.82 h	EC, β^+
	^{192}Hg	0	0	4.9 h	EC
	^{193}Hg	3/2	−0.63	3.8 h	EC
	^{194}Hg	0	0	260 a	EC
	^{195}Hg	1/2	0.54	10 h	EC
	^{197}Hg	1/2	0.527	2.7 d	EC
	^{203}Hg	5/2	0.85	46.6 d	β^-
81	^{193}Tl	1/2	–	0.36 h	EC, β^+
	^{194}Tl	2	–	0.55 h	EC, β^+
	^{195}Tl	1/2	1.6	1.2 h	EC, β^+
	^{196}Tl	2	±0.07	1.8 h	EC, β^+
	^{197}Tl	1/2	1.6	2.8 h	EC, β^+
	^{198}Tl	2	±0.001	5.3 h	EC, β^+
	^{199}Tl	1/2	1.6	7.4 h	EC
	^{200}Tl	2	±0.04	1.1 d	EC, β^+
	^{201}Tl	1/2	1.6	3.0 d	EC
	^{202}Tl	2	±0.06	12.2 d	EC
	^{204}Tl	2	±0.09	3.8 a	β^-, EC
82	^{194}Pb	0	0	0.18 h	EC, β^+
	^{196}Pb	0	0	0.62 h	EC
	^{197}Pb	3/2	–	10 min	EC, β^+
	^{198}Pb	0	0	2.4 h	EC
	^{199}Pb	5/2	–	1.5 h	EC, β^+
	^{200}Pb	0	0	21.5 h	EC

Table 3.2 (continued)

Atomic number Z	Symbol and mass number	Nuclear spin I	Relative magnetic moment μ/μ_N	Half-life $T_{1/2}$	Decay mode
1	2	3	4	5	6
	^{201}Pb	5/2	–	9.4 h	EC, β^+
	^{202}Pb	0	0	$3 \cdot 10^5$ a	EC
	^{203}Pb	5/2	–	52 h	EC
	^{205}Pb	5/2	–	$1.4 \cdot 10^7$ a	EC
	^{209}Pb	9/2	–	3.3 h	β^-
	^{210}Pb	0	0	22 a	β^-, α
	^{211}Pb	9/2	–	0.6 h	β^-
	^{212}Pb	0	0	10.6 h	β^-
	^{214}Pb	0	0	0.45 h	β^-
83	^{198}Bi	7	–	0.20 h	EC, β^+
	^{199}Bi	9/2	–	0.45 h	EC
	^{200}Bi	7	–	0.61 h	EC, β^+
	^{201}Bi	9/2	–	1.8 h	EC, β^+
	^{202}Bi	5	–	1.7 h	EC, β^+
	^{203}Bi	9/2	4.6	11.8 h	EC, β^+
	^{204}Bi	6	4.3	11.2 h	EC
	^{205}Bi	9/2	4.2	15.3 d	EC, β^+
	^{206}Bi	6	4.6	6.2 d	EC
	^{207}Bi	9/2	–	38 a	EC, β^+
	^{208}Bi	5	–	$3.7 \cdot 10^5$ a	EC
	^{210}Bi	1	−0.045	5 d	β^-, α
	^{212}Bi	1	–	1.0 h	β^-, α
	^{213}Bi	9/2	–	0.76 h	β^-, α
	^{214}Bi	1	–	0.33 h	β^-, α
84	^{200}Po	0	0	0.19 h	EC, β^+, α
	^{201}Po	3/2	–	0.25 h	EC, β^+, α
	^{202}Po	0	0	0.75 h	EC, β^+, α
	^{203}Po	5/2	–	0.61 h	EC, β^+, α
	^{204}Po	0	0	3.5 h	EC, α
	^{205}Po	5/2	0.3	1.8 h	EC, β^+, α
	^{206}Po	0	0	8.8 d	EC, α
	^{207}Po	5/2	0.3	5.8 h	EC, β^+, α
	^{208}Po	0	0	2.90 a	α, EC
	^{209}Po	1/2	0.8	102 a	α, EC
	^{210}Po	0	0	138.4 d	α
85	^{205}At	9/2	–	0.44 h	EC, α, β^+
	^{206}At	5	–	0.49 h	EC, β^+, α
	^{207}At	9/2	–	1.8 h	EC, β^+, α
	^{208}At	0	–	1.6 h	EC, β^+, α
	^{209}At	9/2	–	5.4 h	EC, α
	^{210}At	5	–	8.1 h	EC, β^+, α
	^{211}At	9/2	–	7.21 h	EC, α
86	^{208}Rn	0	0	0.41 h	α, EC, β^+
	^{209}Rn	5/2	–	0.48 h	EC, α, β^+

Table 3.2 (continued)

Atomic number Z	Symbol and mass number	Nuclear spin I	Relative magnetic moment μ/μ_N	Half-life $T_{1/2}$	Decay mode
1	2	3	4	5	6
	^{210}Rn	0	0	2.4 h	α, EC
	^{211}Rn	1/2	–	14.6 h	EC, β^+, α
	^{212}Rn	0	0	0.4 h	α
	^{221}Rn	(7/2, 9/2)	–	0.42 h	β^-, α
	^{222}Rn	0	0	3.82 d	α
	^{223}Rn	–	–	43 min	β^-
	^{224}Rn	0	0	1.8 h	β^-
87	^{212}Fr	5	–	0.33 h	EC, β^+, α
	^{222}Fr	2	–	0.24 h	β^-, α
	^{223}Fr	–	–	0.36 h	β^-, α
88	^{223}Ra	1/2	–	11.4 d	α
	^{224}Ra	0	0	3.66 d	α
	^{225}Ra	3/2	–	14.8 d	β^-
	^{226}Ra	0	0	1600 a	α
	^{227}Ra	3/2	–	0.7 h	β^-
	^{228}Ra	0	0	5.8 a	β^-
	^{230}Ra	0	0	1.5 h	β^-
89	^{224}Ac	(0,1)	–	2.9 h	EC, α
	^{225}Ac	3/2	–	10 d	α
	^{226}Ac	1	–	1.2 d	β^-, EC, α
	^{227}Ac	3/2	1.1	21.8 a	β^-, α
	^{228}Ac	3	–	6.1 h	β^-
	^{229}Ac	3/2	–	1.05 h	β^-
90	^{226}Th	0	0	0.52 h	α
	^{227}Th	3/2	–	18.7 d	α
	^{228}Th	0	0	1.9 a	α
	^{229}Th	5/2	0.5	7300 a	α
	^{230}Th	0	0	$7.5 \cdot 10^4$ a	α
	^{231}Th	5/2	–	25.5 h	β^-
	^{232}Th	0	0	$1.4 \cdot 10^{10}$ a	α
	^{233}Th	1/2	–	0.37 h	β^-
	^{234}Th	0	0	24 d	β^-
	^{236}Th	0	0	0.62 h	β^-
91	^{227}Pa	–	–	0.64 h	α, EC
	^{228}Pa	3	–	22 h	EC, α
	^{229}Pa	5/2	–	1.4 d	EC, α
	^{230}Pa	2	–	17.4 d	EC, β^-, α
	^{231}Pa	3/2	±2.0	$3.28 \cdot 10^4$ a	α
	^{232}Pa	2	–	1.3 d	β^-
	^{233}Pa	3/2	3.5	27.0 d	β^-
	^{234}Pa	4	–	6.7 h	β^-
	^{235}Pa	3/2	–	0.4 h	β^-

Table 3.2 (continued)

Atomic number Z	Symbol and mass number	Nuclear spin I	Relative magnetic moment μ/μ_N	Half-life $T_{1/2}$	Decay mode
1	2	3	4	5	6
92	^{229}U	3/2	–	0.97 h	EC, α
	^{230}U	0	0	20.8 d	α
	^{231}U	5/2	–	4.2 d	EC, α
	^{232}U	0	0	69 a	α
	^{233}U	5/2	0.6	$1.59 \cdot 10^5$ a	α
	^{234}U	0	0	$2.45 \cdot 10^5$ a	α
	^{235}U	7/2	−0.3	$7.04 \cdot 10^8$ a	α
	^{236}U	0	0	$2.34 \cdot 10^7$ a	α
	^{237}U	1/2	–	6.7 d	β^-
	^{238}U	0	0	$4.47 \cdot 10^9$ a	α
	^{239}U	5/2	–	0.39 h	β^-
	^{240}U	0	0	14.1 h	β^-
93	^{231}Np	5/2	–	0.81 h	EC, α
	^{232}Np	4	–	0.25 h	EC
	^{233}Np	5/2	–	0.6 h	EC, α
	^{234}Np	0	–	4.4 d	EC, β^+
	^{235}Np	5/2	–	1.1 a	EC, α
	^{236}Np	6	–	$1.1 \cdot 10^5$ a	EC, β^-
	^{237}Np	5/2	3.1	$2.14 \cdot 10^6$ a	α
	^{238}Np	2	–	2.1 d	β^-
	^{239}Np	5/2	0.3	2.35 d	β^-
	^{240}Np	5	–	1.1 h	β^-
	^{241}Np	5/2	–	0.27 h	β^-
94	^{232}Pu	0	0	0.57 h	EC, α
	^{233}Pu	–	–	0.35 h	EC, α
	^{234}Pu	0	0	8.8 h	EC, α
	^{235}Pu	5/2	–	0.42 h	EC, α
	^{236}Pu	0	0	2.85 a	α
	^{237}Pu	7/2	–	45.3 d	EC, α
	^{238}Pu	0	0	88 a	α
	^{239}Pu	1/2	0.20	$2.41 \cdot 10^4$ a	α
	^{240}Pu	0	0	6540 a	α
	^{241}Pu	5/2	−0.71	14.4 a	β^-, α
	^{242}Pu	0	0	$3.8 \cdot 10^5$ a	α
	^{243}Pu	7/2	–	5.0 h	β^-
	^{244}Pu	0	0	$8.1 \cdot 10^7$ a	α
	^{245}Pu	9/2	–	10.5 h	β^-
	^{246}Pu	0	0	10.8 d	β^-
95	^{237}Am	5/2	–	1.2 h	EC, α
	^{238}Am	1	–	1.6 h	EC, α
	^{239}Am	5/2	–	12 h	EC, α
	^{240}Am	3	–	2.1 d	EC, α
	^{241}Am	5/2	1.6	432 a	α
	^{242}Am	1	0.39	16 h	β^-, EC

Table 3.2 (continued)

Atomic number Z	Symbol and mass number	Nuclear spin I	Relative magnetic moment μ/μ_N	Half-life $T_{1/2}$	Decay mode
1	2	3	4	5	6
	^{243}Am	5/2	1.6	$7.4 \cdot 10^3$ a	α
	^{244}Am	6	–	10.1 h	β^-
	^{245}Am	5/2	–	2 h	β^-
	^{246}Am	7	–	0.65 h	β^-
	^{247}Am	5/2	–	0.37 h	β^-
96	^{238}Cm	0	0	2.4 h	EC, α
	^{239}Cm	7/2	–	3 h	EC
	^{240}Cm	0	0	27 d	α
	^{241}Cm	1/2	–	33 d	EC, α
	^{242}Cm	0	0	163 d	α
	^{243}Cm	5/2	±0.4	28.5 a	α, EC
	^{244}Cm	0	0	18.1 a	α
	^{245}Cm	7/2	±0.5	$8.5 \cdot 10^3$ a	α
	^{246}Cm	0	0	$4.7 \cdot 10^3$ a	α
	^{247}Cm	9/2	±0.4	$1.56 \cdot 10^7$ a	α
	^{248}Cm	0	0	$3.4 \cdot 10^5$ a	α, sf
	^{249}Cm	1/2	–	1.1 h	β^-
	^{250}Cm	0	0	7400 a	sf
	^{251}Cm	1/2	–	17 min	β^-
97	^{243}Bk	3/2	–	4.5 h	EC, α
	^{244}Bk	4	–	4.4 h	EC, α
	^{245}Bk	3/2	–	4.9 d	EC, α
	^{246}Bk	2	–	1.8 d	EC
	^{247}Bk	3/2	–	1400 a	α
	^{248}Bk	1	–	24 h	β^-, EC
	^{249}Bk	7/2	±2	320 d	β^-, α
	^{250}Bk	2	–	3.2 h	β^-
	^{251}Bk	3/2	–	0.93 h	β^-
98	^{243}Cf	1/2	–	0.18 h	EC, α
	^{244}Cf	0	0	0.32 h	α
	^{245}Cf	–	–	0.73 h	EC, α
	^{246}Cf	0	0	35.7 h	α
	^{247}Cf	7/2	–	3.1 h	EC, α
	^{248}Cf	0	0	330 d	α
	^{249}Cf	9/2	–	351 a	α
	^{250}Cf	0	0	13.1 a	α
	^{251}Cf	1/2	–	900 a	α
	^{252}Cf	0	0	2.64 a	α, sf
	^{253}Cf	7/2	–	18 d	β^-, α
	^{254}Cf	0	0	60 d	sf, α
	^{255}Cf	–	–	2 h	β^-
	^{256}Cf	0	0	12 min	sf
99	^{248}Es	(2,0)	–	0.45 h	EC, α
	^{249}Es	7/2	–	1.7 h	EC, α

Table 3.2 (continued)

Atomic number Z	Symbol and mass number	Nuclear spin I	Relative magnetic moment μ/μ_N	Half-life $T_{1/2}$	Decay mode
1	2	3	4	5	6
	^{250}Es	1	–	2.2 h	EC
	^{251}Es	3/2	–	1.4 d	EC, α
	^{252}Es	(5,4)	–	472 d	α, EC
	^{253}Es	7/2	4.1	20.5 d	α
	^{254}Es	7	–	276 d	α
	^{255}Es	7/2	–	40 d	β^-, α, sf
	^{256}Es	1	–	22 min	β^-
100	^{250}Fm	0	0	0.5 h	α
	^{251}Fm	9/2	–	5.3 h	EC, α
	^{252}Fm	0	0	1.06 d	α
	^{253}Fm	1/2	–	3 d	EC, α
	^{254}Fm	0	0	3.2 h	α, sf
	^{255}Fm	7/2	–	20 h	α
	^{256}Fm	0	0	2.6 h	sf, α
	^{257}Fm	9/2	–	100 d	α, sf
101	^{255}Md	7/2	–	0.45 h	EC, α
	^{256}Md	–	–	1.3 h	EC, α
	^{257}Md	7/2	–	5.2 h	EC, α
	^{258}Md	8	–	55 d	α
	^{259}Md	7/2	–	1.6 h	sf
102	^{259}No	9/2	–	1.0 h	α, EC

3.2 Atomic Weights of the Elements and Atomic Mass Table

The data given below refer to the atomic weights of the elements and the nuclidic masses of the most abundant, long-lived isotopes. The term now adopted, "atomic weight of an element $A_r(E)$" (or "mean relative atomic mass"), implies the ratio of the average mass per atom of the element from the specified source to 1/12 of the mass of an atom of carbon-12.

Since most elements are made up of isotopes, the atomic weight of an element cannot be regarded as an invariant in nature but depends on the origin and treatment of the sample containing the element. Thus, it is important to establish consistent set of atomic weight values, expecially since the level and scope of scientific and technological applications of chemical elements is increasing all the time. This task was undertaken by the Commission on Atomic Weights working under the auspices of the International Union of Pure and Applied Chemistry. Data on the atomic weights $A_r(E)$ of the elements, which are cited below, are in accord with the official bulletin of this organization [3.2.1], issued in 1983. The values of $A_r(E)$ apply to the elements as they exist in materials of terrestrial origin and to certain artificial elements.

The accuracy with which the atomic weight of an element can be determined depends on the accuracy with which the isotope masses and isotopic abundance of the element can be measured for the specified sample. These total uncertainties are included in Table 3.3. The accuracy is shown by the form of the last significant figure. The values of the atomic weights are estimated to be accurate to within an error of not more than one ($\leqslant \pm 1$) or three ($\leqslant \pm 3$) in the last digit quoted if the latter is of normal size or is a subscript, respectively. Only in the case of hydrogen is the uncertainty reproduced directly. Values in brackets are used for radioactive elements whose atomic weights cannot be quoted precisely without knowing the origin of the elements; the value given characterizes the atomic mass number of the longest-lived isotope of that element.

Following the recommendations of the above-mentioned commission, the values of the atomic weights for about half of the elements are marked with superscripts, explained below, which assist the users in obtaining additional information [3.2.1]:

[g] geologically exceptional specimens are known in which the element has an isotopic composition outside the limits for normal material. The difference between the atomic weight of the element in such specimens and that given in Table 3.3 may exceed considerably the implied uncertainty.

[m] modified isotopic compositions may be found in commercially available material, which may have been subjected to undisclosed or inadvertent isotopic separation. Substantial deviations in the atomic weight of the element from that given in Table 3.3 can occur.

[r] range in isotopic composition of normal terrestrial material prevents a more precise atomic weight being given; the tabulated $A_r(E)$ value should be applicable to any normal material.

[L] Longest half-live isotope mass is chosen for the tabulated $A_r(E)$ value.

In the last column of Table 3.3 values of the relative atomic mass of some selected nuclides most widely used in modern technology are given [3.2.2, 3]. Symbols of the longest-lived radionuclides of an element are followed by an asterisk *. The numerical values of the nuclidic masses are expected to be accurate to within one unit in the last digit given.

References

3.2.1 International Union of Pure and Applied Chemistry, Commission on Atomic Weights and Isotopic Abundances: "Atomic weights of the elements – 1981": Pure Appl. Chem. **55**, 1102 (1983)

3.2.2 A.H. Wapstra, K. Bos: At. Data Nucl. Data Tables **19**, 177 (1977)

3.2.3 A.H. Wapstra, N.B. Gove: Nucl. Data Tables **9**, 265 (1971)

Table 3.3. Atomic weights A_r(E) of the elements (1981) and masses of the most abundant or most stable isotopes (1977) [$A_r(^{12}C) = 12$]

Atomic number Z	Element	Name	Atomic weight A_r(E) [a.m.u.]	Symbol and mass number of nuclide ^AX	Nuclidic mass [a.m.u.]
1	2	3	4	5	6
0	n	Neutron	–	^{1}n	1.00866
1	H	Hydrogen	1.00794(7)^{g,m,r}	^{1}H	1.007825
				^{2}H (D)	2.01410
				^{3}H (T)	3.01605
2	He	Helium	4.00260^g	^{3}He	3.01603
				^{4}He	4.00260
3	Li	Lithium	6.94_1^{g,m,r}	^{6}Li	6.01512
				^{7}Li	7.01600
4	Be	Beryllium	9.01218	^{9}Be	9.01218
5	B	Boron	10.81^{m,r}	^{10}B	10.0129
				^{11}B	11.0093
6	C	Carbon	12.011^r	^{12}C	12
				^{13}C	13.00335
7	N	Nitrogen	14.0067	^{14}N	14.00307
				^{15}N	15.00011
8	O	Oxygen	15.999_4^{g,r}	^{16}O	15.99491
				^{17}O	16.99913
				^{18}O	17.99916
9	F	Fluorine	18.998403	^{19}F	18.99840
10	Ne	Neon	20.179^{g,m}	^{20}Ne	19.99244
				^{22}Ne	21.99138
11	Na	Sodium	22.98977	^{23}Na	22.98977
12	Mg	Magnesium	24.305^g	^{24}Mg	23.9850
				^{25}Mg	24.9858
				^{26}Mg	25.9826
13	Al	Aluminium	26.98154	^{27}Al	26.9815
14	Si	Silicon	28.085_5	^{28}Si	27.9769
				^{29}Si	28.9765
15	P	Phosphorus	30.97376	^{31}P	30.97376
16	S	Sulfur	32.06^r	^{32}S	31.9721
				^{34}S	33.9679
17	Cl	Chlorine	35.453	^{35}Cl	34.9689
				^{37}Cl	36.9659
18	Ar	Argon	39.948^{g,r}	^{36}Ar	35.9675
				^{40}Ar	39.9624
19	K	Potassium	39.0983	^{39}K	38.9637
				^{41}K	40.9618
20	Ca	Calcium	40.08^g	^{40}Ca	39.9626
				^{42}Ca	41.9586
				^{44}Ca	43.9555

Table 3.3 (continued)

Atomic number Z	Element	Name	Atomic weight $A_r(E)$ [a.m.u.]	Symbol and mass number of nuclide ^{A}X	Nuclidic mass [a.m.u.]
1	2	3	4	5	6
21	Sc	Scandium	44.9559	^{45}Sc	44.95591
22	Ti	Titanium	47.8_8	^{48}Ti	47.94795
23	V	Vanadium	50.9415	^{51}V	50.94396
24	Cr	Chromium	51.996	^{52}Cr	51.9405
				^{53}Cr	52.9407
25	Mn	Manganese	54.9380	^{55}Mn	54.93805
26	Fe	Iron	55.84_7	^{54}Fe	53.9396
				^{56}Fe	55.9349
				^{57}Fe	56.9354
27	Co	Cobalt	58.9332	^{59}Co	58.93320
28	Ni	Nickel	58.69	^{58}Ni	57.9353
				^{60}Ni	59.9308
29	Cu	Copper	63.54^{r}_{6}	^{63}Cu	62.9296
				^{65}Cu	64.9278
30	Zn	Zinc	65.38	^{64}Zn	63.9291
				^{66}Zn	65.9260
				^{68}Zn	67.9248
31	Ga	Gallium	69.72	^{69}Ga	68.9256
				^{71}Ga	70.9247
32	Ge	Germanium	72.5_9	^{70}Ge	69.9242
				^{72}Ge	71.9221
				^{74}Ge	73.9212
33	As	Arsenic	74.9216	^{75}As	74.92160
34	Se	Selenium	78.9_6	^{78}Se	77.9173
				^{80}Se	79.9165
35	Br	Bromine	79.904	^{79}Br	78.9183
				^{81}Br	80.9163
36	Kr	Krypton	$83.80^{g,m}$	^{82}Kr	81.9135
				^{83}Kr	82.9141
				^{84}Kr	83.9115
				^{86}Kr	85.9106
37	Rb	Rubidium	85.467_8^{g}	^{85}Rb	84.9118
				^{87}Rb	86.9092
38	Sr	Strontium	87.62^{g}	^{86}Sr	85.9093
				^{87}Sr	86.9089
				^{88}Sr	87.9056
39	Y	Yttrium	88.9059	^{89}Y	88.90586
40	Zr	Zirconium	91.22^{g}	^{90}Zr	89.9047
				^{91}Zr	90.9056
				^{92}Zr	91.9050
				^{94}Zr	93.9063

Table 3.3 (continued)

Atomic number Z	Element	Name	Atomic weight A_r(E) [a.m.u.]	Symbol and mass number of nuclide ^{A}X	Nuclidic mass [a.m.u.]
1	2	3	4	5	6
41	Nb	Niobium	92.9064	^{93}Nb	92.90638
42	Mo	Molybdenum	95.94^{g}	^{92}Mo	91.9068
				^{95}Mo	94.9058
				^{96}Mo	95.9047
				^{98}Mo	97.9054
43	Tc	Technetium	[98]	^{97}Tc*	96.9063
				^{98}Tc*	97.9073
				^{99}Tc*	98.9063
44	Ru	Ruthenium	101.0_{7}^{g}	^{101}Ru	100.9056
				^{102}Ru	101.9043
				^{104}Ru	103.9054
45	Rh	Rhodium	102.9055	^{103}Rh	102.90550
46	Pd	Palladium	106.42^{g}	^{104}Pd	103.9040
				^{105}Pd	104.9051
				^{106}Pd	105.9035
				^{108}Pd	107.9039
				^{110}Pd	109.9052
47	Ag	Silver	107.868_{2}^{g}	^{107}Ag	106.9051
				^{109}Ag	108.9048
48	Cd	Cadmium	112.41^{g}	^{112}Cd	111.9028
				^{114}Cd	113.9034
49	In	Indium	114.82^{g}	^{113}In	112.9041
				^{115}In	114.9039
50	Sn	Tin	118.6_{9}	^{116}Sn	115.9017
				^{118}Sn	117.9016
				^{120}Sn	119.9022
51	Sb	Antimony	121.7_{5}	^{121}Sb	120.9038
				^{123}Sb	122.9042
52	Te	Tellurium	127.6_{0}^{g}	^{126}Te	125.9033
				^{128}Te	127.9045
				^{130}Te	129.9062
53	I	Iodine	126.9045	^{127}I	126.90448
54	Xe	Xenon	$131.2_{9}^{g,m}$	^{129}Xe	128.9048
				^{131}Xe	130.9051
				^{132}Xe	131.9041
55	Cs	Caesium	132.9054	^{133}Cs	132.90543
56	Ba	Barium	137.33^{g}	^{137}Ba	136.9058
				^{138}Ba	137.9052
57	La	Lanthanum	138.905_{5}^{g}	^{139}La	138.90636
58	Ce	Cerium	140.12^{g}	^{140}Ce	139.9054
				^{142}Ce	141.9092

Table 3.3 (continued)

Atomic number Z	Element	Name	Atomic weight $A_r(E)$ [a.m.u.]	Symbol and mass number of nuclide ^{A}X	Nuclidic mass [a.m.u.]
1	2	3	4	5	6
59	Pr	Praseodymium	140.9077	^{141}Pr	140.90766
60	Nd	Neodymium	144.2_4^{g}	^{142}Nd	141.9077
				^{143}Nd	142.9098
				^{144}Nd	143.9101
				^{146}Nd	145.9131
61	Pm	Promethium	[145]	^{145}Pm*	144.9128
				^{146}Pm*	145.9147
				^{147}Pm*	146.9151
62	Sm	Samarium	150.3_6^{g}	^{147}Sm	146.9149
				^{148}Sm	147.9148
				^{149}Sm	148.9172
				^{152}Sm	151.9197
				^{154}Sm	153.9222
63	Eu	Europium	151.96^{g}	^{151}Eu	150.9199
				^{153}Eu	152.9212
64	Gd	Gadolinium	157.2_5^{g}	^{155}Gd	154.9226
				^{156}Gd	155.9221
				^{157}Gd	156.9240
				^{158}Gd	157.9241
				^{160}Gd	159.9271
65	Tb	Terbium	158.9254	^{159}Tb	158.92535
66	Dy	Dysprosium	162.5_0	^{161}Dy	160.9269
				^{162}Dy	161.9268
				^{163}Dy	162.9287
				^{164}Dy	163.9292
67	Ho	Holmium	164.9304	^{165}Ho	164.93033
68	Er	Erbium	167.2_6	^{166}Er	165.9303
				^{167}Er	166.9321
				^{168}Er	167.9324
				^{170}Er	169.9355
69	Tm	Thulium	168.9342	^{169}Tm	168.93423
70	Yb	Ytterbium	173.0_4	^{171}Yb	170.9363
				^{172}Yb	171.9364
				^{173}Yb	172.9382
				^{174}Yb	173.9389
				^{176}Yb	175.9426
71	Lu	Lutetium	174.967	^{175}Lu	174.9408
				^{176}Lu	175.9427
72	Hf	Hafnium	178.4_9	^{177}Hf	176.9432
				^{178}Hf	177.9437
				^{179}Hf	178.9458
				^{180}Hf	179.9466

Table 3.3 (continued)

Atomic number Z	Element	Name	Atomic weight A_r(E) [a.m.u.]	Symbol and mass number of nuclide ^{A}X	Nuclidic mass [a.m.u.]
1	2	3	4	5	6
73	Ta	Tantalum	180.9479	^{181}Ta	180.94801
74	W	Tungsten	183.8_5	^{182}W	181.9482
				^{183}W	182.9502
				^{184}W	183.9510
				^{186}W	185.9544
75	Re	Rhenium	186.207	^{185}Re	184.9530
				^{187}Re	186.9558
76	Os	Osmium	190.2^{g}	^{188}Os	187.9559
				^{189}Os	188.9582
				^{190}Os	189.9585
				^{192}Os	191.9615
77	Ir	Iridium	192.2_2	^{191}Ir	190.9606
				^{193}Ir	192.9629
78	Pt	Platinum	195.0_8	^{194}Pt	193.9627
				^{195}Pt	194.9648
				^{196}Pt	195.9649
				^{198}Pt	197.9679
79	Au	Gold	196.9665	^{197}Au	196.96656
80	Hg	Mercury	200.5_9	^{198}Hg	197.9668
				^{199}Hg	198.9683
				^{200}Hg	199.9683
				^{201}Hg	200.9703
				^{202}Hg	201.9706
81	Tl	Thallium	204.383	^{203}Tl	202.9723
				^{205}Tl	204.9744
82	Pb	Lead	$207.2^{g,r}$	^{206}Pb	205.9745
				^{207}Pb	206.9759
				^{208}Pb	207.9766
83	Bi	Bismuth	208.9804	^{209}Bi	208.98039
84	Po	Polonium	[209]	^{208}Po*	207.9812
				^{209}Po*	208.9824
				^{210}Po*	209.9829
85	At	Astatine	[210]	^{209}At*	208.9862
				^{210}At*	209.9871
				^{211}At*	210.9875
86	Rn	Radon	[222]	^{211}Rn*	210.9906
				^{222}Rn*	222.01757
87	Fr	Francium	[223]	^{212}Fr*	211.9960
				^{222}Fr*	222.0175
				^{223}Fr*	223.0197

Table 3.3 (continued)

Atomic number Z	Element	Name	Atomic weight $A_r(E)$ [a.m.u.]	Symbol and mass number of nuclide ^AX	Nuclidic mass [a.m.u.]
1	2	3	4	5	6
88	Ra	Radium	$226.0254^{g,L}$	$^{226}Ra^*$	226.02541
				$^{228}Ra^*$	228.0311
89	Ac	Actinium	227.0278^{L}	$^{225}Ac^*$	225.0232
				$^{227}Ac^*$	227.02775
90	Th	Thorium	$232.0381^{g,L}$	$^{229}Th^*$	229.03176
				$^{230}Th^*$	230.03313
				$^{232}Th^*$	232.03805
91	Pa	Protactinium	231.0359^{L}	$^{230}Pa^*$	230.03453
				$^{231}Pa^*$	231.03588
				$^{233}Pa^*$	233.04024
92	U	Uranium	$238.0289^{g,m}$	$^{232}U^*$	232.0371
				$^{233}U^*$	233.03963
				$^{234}U^*$	234.04095
				$^{235}U^*$	235.04393
				$^{236}U^*$	236.04556
				$^{238}U^*$	238.05079
93	Np	Neptunium	237.0482^{L}	$^{235}Np^*$	235.04406
				$^{236}Np^*$	236.0466
				$^{237}Np^*$	237.04817
94	Pu	Plutonium	[244]	$^{238}Pu^*$	238.04956
				$^{239}Pu^*$	239.05216
				$^{240}Pu^*$	240.05381
				$^{241}Pu^*$	241.05685
				$^{242}Pu^*$	242.05874
				$^{244}Pu^*$	244.0642
95	Am	Americium	[243]	$^{241}Am^*$	241.05682
				$^{243}Am^*$	243.06137
96	Cm	Curium	[247]	$^{242}Cm^*$	242.05883
				$^{243}Cm^*$	243.06138
				$^{244}Cm^*$	244.06275
				$^{245}Cm^*$	245.06549
				$^{246}Cm^*$	246.06722
				$^{247}Cm^*$	247.0703
				$^{248}Cm^*$	248.0723
				$^{250}Cm^*$	250.0784
97	Bk	Berkelium	[247]	$^{247}Bk^*$	247.0703
				$^{249}Bk^*$	249.07498
98	Cf	Californium	[251]	$^{248}Cf^*$	248.0722
				$^{249}Cf^*$	249.07485
				$^{250}Cf^*$	250.07640
				$^{251}Cf^*$	251.0796
				$^{252}Cf^*$	252.0816
				$^{254}Cf^*$	254.0873

Table 3.3 (continued)

Atomic number Z	Element	Name	Atomic weight A_r(E) [a.m.u.]	Symbol and mass number of nuclide ^{A}X	Nuclidic mass [a.m.u.]
1	2	3	4	5	6
99	Es	Einsteinium	[252]	^{252}Es*	252.0828
				^{253}Es*	253.08482
				^{254}Es*	254.0880
100	Fm	Fermium	[257]	^{252}Fm*	252.0825
				^{253}Fm*	253.0852
				^{255}Fm*	255.0900
				^{257}Fm*	257.0951
101	Md	Mendelevium	[258]	^{258}Md*	258.0986
102	No	Nobelium	[259]	^{252}No*	252.0890
				^{256}No*	256.0943
				^{259}No*	259.1009
103	Lr	Lawrencium	[260]	^{260}Lr*	260.1054
104	(Unq)	(Unnilquadium)	[261]	^{261}(Unq)*	261.1087
105	(Unp)	(Unnilpentium)	[262]	^{262}(Unp)*	262.1138
106	(Unh)	(Unnilhexium)	[263]	^{263}(Unh)*	263.1184

4. Structure of Atomic Electron Shells

The systematics of electron quantum states in atoms is briefly reviewed along with the systematics of atomic terms and the filling-order of the electronic subshells. The normal electronic configurations and terms of atomic particles are presented, together with the Hartree-Fock and asymptotic parameters of valence electron wavefunctions and radial expectation values.

4.1 Electron Configurations and Ground-State Terms

States of Electrons in the Atom. Strictly speaking states of the combined system of electrons and nucleus, which form an atomic particle, can be regarded only as a whole. According to the approximate and most common model of an atomic particle, each electron moves in some effective centrally symmetric field produced by the nucleus and all the other electrons (or, as they put it, in a self-consistent field).

In the non-relativistic approximation the motion of an individual electron in the central field is completely characterized by the values of its orbital angular momentum $l = 0, 1, 2, \ldots$, by the projection of this momentum on some direction $l_z = m = l, l-1, \ldots, -l$, and also by the energy (all momenta in quantum mechanics are measured in units of $\hbar$ = Planck constant/2π, which is omitted for brevity). There are commonly accepted symbols to denote electron states with different l-values, namely letters of the Latin alphabet:

Value of l (in units of $\hbar$)	0	1	2	3	4	5	6
Designation	s	p	d	f	g	h	i

The states of electrons with a given l are numbered in order of increasing energy by the principal quantum number n, which takes the values $l + 1$, $l + 2 \ldots$. The states of individual electrons with different n and l are, usually, denoted by a symbol, which includes a digit indicating the value of n, and a letter indicating the value of l. If some of the electrons have the same values of n and l, their number is marked as a corresponding exponent. Thus, for example, the electron assembly in the ground-state lithium atom is denoted as follows:

$$\mathrm{Li}\,(1s^2 2s)\ ,$$

two electrons being in the state with the principal quantum number 1 and zero orbital momentum while one electron is in the state with $n = 2$ and $l = 0$. Similarly, outer electrons of the ground-state iron atom occupy the states

$$\mathrm{Fe}\,(3d^6 4s^2)\ ,$$

where six electrons are in the state with $n = 3$ and $l = 2$, and the other two are in the state with $n = 4$ and $l = 0$. The distribution of electrons in an atom according to states with different values of n and l is spoken of as the electron configuration.

Furthermore, each electron state in the central field is characterized by a definite parity, completely dependent on the orbital momentum l. Even-parity states correspond to even l and the respective electron wavefunction does not change on inversion transformation, and vice versa for odd l, i.e. the parity of a state with a given l is $P = (-1)^l$. Consequently, the state of a system of non-interacting electrons is even if the sum of the angular momenta of the electrons takes even values, so that $P = (-1)^{\Sigma l_i}$. The classification of electron states by their parity is necessary to fix the selection rules for radiative transitions of atomic particles.

Finally, a particle should be ascribed an intrinsic angular momentum ("spin") s with a projection s_z on the chosen direction, which is not connected with its motion in space. For the electron, $s = 1/2$ and $s_z = \pm 1/2$ (in units of $\hbar$). As a result, a complete description of the electron state is provided by the four quantum numbers n, l, l_z and s_z. When n and l are specified, there are $2(2l + 1)$ states of the electron with different values of the l_z and s_z quantum numbers (degeneracy of the nl level). According to the Pauli exclusion principle only one electron can occupy the state with the given quantum numbers n, l, l_z, s_z, and therefore not more than $2\,(2l + 1)$ electrons in an atomic particle can have equal values of n and l. The electrons with the same values of n and l are called equivalent and together they form either a filled (closed) or an unfilled (unclosed) shell of a given type. Thus, a closed s shell has 2 electrons, p – 6 electrons, d – 10 electrons, f – 14 electrons and so on. One can also use a terminology, according to which electrons with the principal quantum numbers $n = 1, 2, 3, 4, 5, 6\ldots$ are spoken of as electrons of the K, L, M, N, O, $P\ldots$ shells, respectively, so that a K shell may contain 2 electrons $(1s^2)$, L – 8 electrons $(2s^2 2p^6)$, M – 18 electrons $(3s^2 3p^6 3d^{10})$, N – 32 electrons $(4s^2 4p^6 4d^{10} 4f^{14})$, etc.

Consideration of relativistic effects results in the electron energy also depending on the mutual orientation of its angular momenta $\boldsymbol{l}$ and $\boldsymbol{s}$, which are not separately conserved. As a consequence, each energy level of an nl

electron must be characterized by the value of the total electronic angular momentum $j = l \pm 1/2$ (fine-structure splitting of levels), which is marked as a subscript to the right of the spectroscopic designation of l, e.g. $np_{1/2}$ or $nd_{5/2}$ and so on. The $(2j + 1)$ states are referred to the electron energy level nlj and differ from one another in the z component $m_j = l_z + s_z$ of the total angular momentum. The selection rule for the quantum number j has the representation $\Delta j = 0, \pm 1$. The ensemble of lines connected with transitions between the fine-structure components of the levels nl and $n'l'$ is called a multiplet (radiative transitions $nlj \to n'l'j'$).

Systematics of Energy Levels in Atomic Particles. Let us discuss the symbols adopted in spectroscopy, which denote the electron states of atoms and atomic ions. In the limit of the central field approximation, the energy level of an atomic particle is completely defined by stating the electron configuration $\{nl\}$. If non-central electrostatic interactions between electrons and the spin-orbit interaction are considered, the energy level $n_1l_1, n_2l_2 \ldots$ will split into a number of sublevels. Practically, it is sufficient to systematize these sublevels on the basis of two limiting approaches; in this case, one of the above interactions is regarded as small compared to the other one. The experimental data show that passing from the beginning of the periodic table to the heavy elements is accompanied by a more or less continuous transition from the case when the electrostatic interaction between electrons exceeds the spin-orbit interaction (LS-coupling) to the reverse case (jj-coupling). However, the pure case of jj-coupling occurs rather seldom and intermediate coupling is of greater importance.

Taking into account only the Coulomb interaction of the electrons results in the splitting of a level belonging to a given electron configuration into a number of sublevels (terms). The latter are characterized by the quantum numbers of the total orbital angular momentum of the electrons L and the total spin of the electrons S. Possible values for L and S are determined using the general rules of addition of separate electron momenta l_i and s_i and the Pauli principle. Values of the total orbital angular momentum L take the following symbols:

Values of L (in units of $\hbar$)	0	1	2	3	4	5	6	7	8
Designation	S	P	D	F	G	H	I	K	L

The parity of the term LS is marked by the right superscript added to the L symbol (the superscript o indicates odd states; even states have no marking). If relativistic effects are taken into consideration, the LS term will split into a number of components (fine-structure splitting) corresponding to different values of the total angular momentum of the electrons $\boldsymbol{J} = \boldsymbol{L} + \boldsymbol{S}$ within the range

$$L + S, L + S - 1, \ldots, |L - S| \quad \text{for given } L \text{ and } S.$$

To sum up, the energy levels of an atomic particle in this approximation are characterized by the values of L, S, J and parity, so that the full designation of a term has the form

$$^{2S+1}L_J^{\mathrm{o}}\,,$$

where the multiplicity of a term $(2S + 1)$ is given by the left superscript (it marks the number of levels with different J for the LS term when $L > S$); the value of the total angular momentum J of the electrons is indicated as a right subscript. For example, the symbol $^2P^{\mathrm{o}}_{3/2}$ denotes an odd energy level, with $L = 1$, $S = 1/2$, $J = 3/2$; the quantum numbers $L = 2$, $S = 2$, $J = 4$ and positive parity describe the $^5D_4$ level. Finally, it should be mentioned that in the framework of the LS-coupling scheme, each level with given values of L, S, J will remain degenerate with respect to the direction of the $\boldsymbol{J}$ vector; the multiplicity of this degeneracy is $(2J + 1)$. Only $(2L + 1)(2S + 1)$ states make up the LS term. They have different values of momentum projection on a given direction, and the sum $\Sigma_J(2J + 1) = (2L + 1)(2S + 1)$, if we take all possible J values for particular LS. We may, thus, conclude that the spin-orbit interaction does not change the number of states for the LS term.

For the LS-coupling scheme the typical energy level grouping is when the distance between various terms of one electron configuration is much less than that between similar terms of different configurations. Also, the distance between adjacent terms exceeds considerably the difference in energy between fine-structure components of each of the terms (in fact, this provides for the condition that relativistic effects be comparatively small). It has been found empirically that for a given electron configuration, the term having the greatest possible value of S and the largest value of L (compatible with this S) possesses the lowest energy (Hund's rule).

In the jj-coupling scheme the spin-orbit interaction of the electrons exceeds considerably the electrostatic interaction between electrons (or more precisely, that part of the electrostatic energy which is dependent on the quantum numbers L and S) and one may speak of conservation of total electron momentum j, as the momenta l and s are not conserved separately. In addition, the state of each electron is set by four quantum numbers $nljm_j$, where a given j has $l = j \pm 1/2$. If we neglect the electrostatic interaction between electrons, the energy of each electron will depend only on the nlj values and each j-state will have $(2j + 1)$-fold degeneracy; the atomic energy level is defined by the set of quantum numbers $n_il_ij_i$ for each electron. If we now take into account the electrostatic interaction between electrons, the energy level $\{n_il_ij_i\}$ will split into a number of sublevels characterized by values of the total angular momentum J of the electron subsystem. Possible values of J can be found by addition of the momenta j_i to give the total momentum of an electron configuration, taking into consideration the Pauli principle. The states with given values of the momenta j_1, j_2 and J are denoted as

$$(j_1 j_2)_J \ ,$$

so that $|j_1 - j_2| \leqslant J \leqslant j_1 + j_2$. For a given electron configuration the total number of levels with a definite value of J will be the same for the LS-coupling and for the jj-coupling schemes.

Highly excited states of some atomic particles (particularly noble gas atoms, atomic chlorine, etc.) follow the intermediate jl-coupling scheme, when the spin-orbit interaction of the electrons in the atomic core exceeds the electrostatic interaction between a highly excited valence electron and the electrons of the atomic core. The atomic core may be characterized by the values of its orbital and spin momenta, and its coupling with the highly excited electron can be described in the framework of the jj-coupling scheme. Let the quantum numbers L, S and j characterize the momenta of the electrons in the atomic core and let l characterize the angular momentum of the excited electron. If we consider the electrostatic interaction between the electron involved and the atomic core electrons, the state $LSjl$ will also split into a number of sublevels; the value of the angular momentum quantum number K ($j + l = K$) is related to each of these sublevels. If the spin-orbit interaction of the excited electron is taken into consideration, each of the $LSjlK$ levels splits into two components, which are ascribed the values of the total angular momentum J of the atomic electrons, $J = K \pm 1/2$.

As a result, the energy level of the atomic particle is characterized by a set of quantum numbers $LSjlKJ$, which are written

$$^{2S+1}Ljnl[K]_J \ ,$$

where n is the principal quantum number of the valence electron. For the jl-coupling scheme the typical energy level grouping is when the distance between the levels $LSjlK$ and $LSjlK'$ is much less than that between the $LSjlK$ and $L'S'j'lK$ levels relating to different states of atomic core, and much greater than the doublet splitting of the $LSjlK$ level according to the quantum number J.

Electron Configurations and Terms of the Atomic Ground States. The ground state of an atomic particle is the state with the lowest possible energy. When we pass from one unexcited atom to the next, the charge of the nucleus Z increases by 1 and one more electron is added to the electron shell, which then takes the lowest of the unoccupied states allowed by the Pauli principle. As a whole the binding energy of the added electron changes non-monotonically as the atomic number of the element (Z) increases and the successive filling of the electron shells is disturbed by a peculiar competition between ns, nd and nf states. This is due to the characteristic features of d and f states, when the outer electron is located much closer to the nucleus than in s and p

states. As a result, the groups of electron states filled in succession can be shown as follows:

$$\left.\begin{array}{r} 1s \ldots 2 \\ 2s2p \ldots 8 \\ 3s3p \ldots 8 \end{array}\right\} \text{electrons} \qquad \left.\begin{array}{r} 4s3d4p \ldots 18 \\ 5s4d5p \ldots 18 \\ 6s4f5d6p \ldots 32 \end{array}\right\} \text{electrons}$$
$$7s6d5f \ldots (?) \text{ electrons} .$$

The elements with closed *d* and *f* shells (or not containing these shells at all) are called elements of the principal groups and all the elements whose *d* and *f* shells are being filled are called elements of the intermediate groups. In Tables 4.1, 2 the electron configurations and terms of the ground states of these elements are given. Table 4.2 presents the configurations of the outer electrons around the closed electron shells. The left subscript added to the symbol of an element specifies its atomic number in the periodic table, i.e. the nuclear charge Z.

The above information was taken from general manuals on non-relativistic quantum mechanics of atoms (e.g. [4.1.1, 2]) and theories of atomic spectra [4.1.3].

References

4.1.1 L.D.Landau, E.M.Lifshitz: *Quantum Mechanics (Non-Relativistic Theory)*, 3rd ed. (Pergamon, London 1977)

4.1.2 E.U.Condon, H.Odabasi: *Atomic Structure* (Cambridge University Press, Cambridge 1980)

4.1.3 I.I.Sobelman: *Atomic Spectra and Radiative Transitions*, Springer Ser. Chem. Phys., Vol. 1 (Springer, Berlin, Heidelberg, New York 1979)

Table 4.1. Normal electron configurations and terms of principal group elements

Principal quantum number of electron	Electron shells being filled								Closed electron shells
	ns	ns^2	s^2np	s^2np^2	s^2np^3	s^2np^4	s^2np^5	s^2np^6	
2	$_3$Li	$_4$Be	$_5$B	$_6$C	$_7$N	$_8$O	$_9$F	$_{10}$Ne	$1s^2$
3	$_{11}$Na	$_{12}$Mg	$_{13}$Al	$_{14}$Si	$_{15}$P	$_{16}$S	$_{17}$Cl	$_{18}$Ar	$2s^2 2p^6$
4	$_{19}$K	$_{20}$Ca	–	–	–	–	–	–	$3s^2 3p^6$
4	$_{29}$Cu	$_{30}$Zn	$_{31}$Ga	$_{32}$Ge	$_{33}$As	$_{34}$Se	$_{35}$Br	$_{36}$Kr	$3d^{10}$
5	$_{37}$Rb	$_{38}$Sr	–	–	–	–	–	–	$4s^2 4p^6$
5	$_{47}$Ag	$_{48}$Cd	$_{49}$In	$_{50}$Sn	$_{51}$Sb	$_{52}$Te	$_{53}$I	$_{54}$Xe	$4d^{10}$
6	$_{55}$Cs	$_{56}$Ba	–	–	–	–	–	–	$5s^2 5p^6$
6	$_{79}$Au	$_{80}$Hg	$_{81}$Tl	$_{82}$Pb	$_{83}$Bi	$_{84}$Po	$_{85}$At	$_{86}$Rn	$4f^{14} 5d^{10}$
7	$_{87}$Fr	$_{88}$Ra							$6s^2 6p^6$
Ground-state term	$^2S_{1/2}$	$^1S_0$	$^2P^o_{1/2}$	$^3P_0$	$^4S^o_{3/2}$	$^3P_2$	$^2P^o_{3/2}$	$^1S_0$	

Table 4.2. Normal electron configurations and terms of intermediate group elements

I) Iron group elements (closed shells of $_{18}$Ar + 4s, 3d shells being filled)

Atom	$_{21}$Sc	$_{22}$Ti	$_{23}$V	$_{24}$Cr	$_{25}$Mn	$_{26}$Fe	$_{27}$Co	$_{28}$Ni
Outer electron configuration	$3d\,4s^2$	$3d^2\,4s^2$	$3d^3\,4s^2$	$3d^5\,4s$	$3d^5\,4s^2$	$3d^6\,4s^2$	$3d^7\,4s^2$	$3d^8\,4s^2$
Ground-state term	$^2D_{3/2}$	$^3F_2$	$^4F_{3/2}$	$^7S_3$	$^6S_{5/2}$	$^5D_4$	$^4F_{9/2}$	$^3F_4$

II) Palladium group elements (closed shells of $_{36}$Kr + 5s, 4d shells being filled)

Atom	$_{39}$Y	$_{40}$Zr	$_{41}$Nb	$_{42}$Mo	$_{43}$Tc	$_{44}$Ru	$_{45}$Rh	$_{46}$Pd
Outer electron configuration	$4d\,5s^2$	$4d^2\,5s^2$	$4d^4\,5s$	$4d^5\,5s$	$4d^5\,5s^2$	$4d^7\,5s$	$4d^8\,5s$	$4d^{10}$
Ground-state term	$^2D_{3/2}$	$^3F_2$	$^6D_{1/2}$	$^7S_3$	$^6S_{5/2}$	$^5F_5$	$^4F_{9/2}$	$^1S_0$

III) Platinum group elements ($_{57}$La contains closed shells of $_{54}$Xe + $6s^2\,5d$ shell; all other elements contain closed shells of $_{54}$Xe + closed 4f shell + 6s, 5d shells being filled)

Atom	$_{57}$La	$_{71}$Lu	$_{72}$Hf	$_{73}$Ta	$_{74}$W	$_{75}$Re	$_{76}$Os	$_{77}$Ir	$_{78}$Pt
Outer electron configuration	$5d\,6s^2$	$5d\,6s^2$	$5d^2\,6s^2$	$5d^3\,6s^2$	$5d^4\,6s^2$	$5d^5\,6s^2$	$5d^6\,6s^2$	$5d^7\,6s^2$	$5d^9\,6s$
Ground-state term	$^2D_{3/2}$	$^2D_{3/2}$	$^3F_2$	$^4F_{3/2}$	$^5D_0$	$^6S_{5/2}$	$^5D_4$	$^4F_{9/2}$	$^3D_3$

IV) Rare earth elements (closed shells of ${}_{54}$Xe + $6s$, $4f$ and $5d$ shells being filled)

Atom	${}_{58}$Ce	${}_{59}$Pr	${}_{60}$Nd	${}_{61}$Pm	${}_{62}$Sm	${}_{63}$Eu	${}_{64}$Gd	${}_{65}$Tb	${}_{66}$Dy	${}_{67}$Ho	${}_{68}$Er	${}_{69}$Tm	${}_{70}$Yb
Outer electron configuration	$4f5d6s^2$	$4f^3 6s^2$	$4f^4 6s^2$	$4f^5 6s^2$	$4f^6 6s^2$	$4f^7 6s^2$	$4f^7 5d 6s^2$	$4f^9 6s^2$	$4f^{10} 6s^2$	$4f^{11} 6s^2$	$4f^{12} 6s^2$	$4f^{13} 6s^2$	$4f^{14} 6s^2$
Ground-state term	${}^1G_4^o$	${}^4I_{9/2}^o$	${}^5I_4$	${}^6H_{5/2}^o$	${}^7F_0$	${}^8S_{7/2}^o$	${}^9D_2^o$	${}^6H_{15/2}^o$	${}^5I_8$	${}^4I_{15/2}^o$	${}^3H_6$	${}^2F_{7/2}^o$	${}^1S_0$

V) Actinide elements (closed shells of ${}_{86}$Rn + $7s$, $5f$ and $6d$ shells being filled)

Atom	${}_{89}$Ac	${}_{90}$Th	${}_{91}$Pa	${}_{92}$U	${}_{93}$Np	${}_{94}$Pu	${}_{95}$Am	${}_{96}$Cm
Outer electron configuration	$6d7s^2$	$6d^2 7s^2$	$5f^2({}^3H_4)6d7s^2$	$5f^3({}^4I_{9/2}^o)6d7s^2$	$5f^4({}^5I_4)6d7s^2$	$5f^6 7s^2$	$5f^7 7s^2$	$5f^7({}^8S_{7/2}^o)6d7s^2$
Ground-state term	${}^2D_{3/2}$	${}^3F_2$	${}^4K_{11/2}$	${}^5L_6^o$	${}^6L_{11/2}$	${}^7F_0$	${}^8S_{7/2}^o$	${}^9D_2^o$
Atom	${}_{97}$Bk	${}_{98}$Cf	${}_{99}$Es	${}_{100}$Fm	${}_{101}$Md	${}_{102}$No	${}_{103}$Lr	
Outer electron configuration	$5f^9 7s^2$	$5f^{10} 7s^2$	$5f^{11} 7s^2$	$5f^{12} 7s^2$	$5f^{13} 7s^2$	$5f^{14} 7s^2$	$7s^2 7p$	
Ground-state term	${}^6H_{15/2}^o$	${}^5I_8$	${}^4I_{15/2}^o$	${}^3H_6$	${}^2F_{7/2}^o$	${}^1S_0$	${}^2P_{1/2}^o$	

4.2 The Periodic Table

The features of electron structure and the physico-chemical properties of atoms related to them can be found in the periodic system of elements by D. I. Mendeleev. According to the method accepted at present all the elements occurring in nature are tabulated by horizontal rows into seven periods and by vertical columns into eight groups, some of the elements not fitting into this classification ($_{27}$Co, $_{28}$Ni, $_{45}$Rh, $_{46}$Pd, $_{77}$Ir, $_{78}$Pt, lanthanides, actinides). On the inside front cover one can see the resulting table of the elements, which specifies the symbol and the name of the element, its atomic number (Z) and atomic weight (or mass number of the most stable isotope), the designation of the outer electrons, their configuration and the first three ionization potentials of the atom in electronvolts [eV].

4.3 Parameters of Wavefunctions for Valence Electrons in Atoms, Positive and Negative Ions

Tables 4.3–5 consist of the parameters of the radial wavefunctions of the outer (valence) electrons in atomic particles (atoms, positive and negative ions) having a nuclear charge $Z \lesssim 90$. These functions are given in an analytical form and were obtained in the non-relativistic approximation by a series expansion on the basis set of atomic Slater orbitals [4.3.1, 2],

$$\varphi(nlm) = \sum_i C_i \chi_{n_i lm}(r, \theta, \phi) ,$$

$$\chi_{n_i lm}(r, \theta, \phi) = R_{n_i l}(r) \cdot Y_{lm}(\theta, \phi) ,$$

$$R_{n_i l}(r) = [(2n_i)!]^{-1/2}(2\zeta_i)^{n_i + 1/2} r^{n_i - 1} \exp(-\zeta_i r)$$

$$= N_i r^{n_i - 1} \exp(-\zeta_i r) .$$

Here, nlm are the quantum numbers of the electron, $Y_{lm}(\theta, \phi)$ is a normalized spherical function, and the radial parts of the φ and χ functions are normalized to unity with the weight function r^2. The coefficients C_i, ζ_i, N_i characterize the radial distribution of electron density. Tables 4.3–5 also include the calculated values of the means of powers of the distance between the valence electron and the nucleus (or radial expectation values):

$$\langle r^\alpha \rangle = \int_0^\infty r^\alpha |\varphi|^2 r^2 \, dr ,$$

where $\alpha = -1, 1, 2$, as well as the asymptotic parameters A and γ of the valence electron wavefunctions. These two parameters define the wavefunction amplitude in the range of distances r from the nucleus which are relatively large compared to the average size of atomic particles. The specified parameters characterize the solution of the one-electron Schrödinger equation in the asymptotic region at $r\gamma \gg 1$, where $(-\gamma^2/2)$ is the ionization potential of an atomic particle (in a.u.):

$$-\tfrac{1}{2}\nabla_r^2\varphi + [l(l+1)/r^2 - Z_c/r]\,\varphi = -(\gamma^2/2)\,\varphi(nl/r), \qquad r \to \infty\,,$$

where Z_c is the charge of the atomic core, equal to 0, 1 and 2 for the negative ion, atom and singly charged positive ion, respectively. The solution of this equation has the form

$$\lim_{r\gamma \gg 1} \varphi(nl/r) \simeq A r^{Z_c\gamma^{-1}-1} \exp(-r\gamma)\left[1 - \frac{Z_c}{2r\gamma^2}\left(\frac{Z_c}{\gamma} - 1\right) + \frac{l(l+1)}{2r\gamma} + 0\left(\frac{1}{r^2}\right)\right],$$

where the asymptotic parameter A is found by matching the above solution with the self-consistent field wavefunction of the valence electron for asymptotic distances [4.3.3].

The errors in the values of electronic wavefunction parameters in atomic particles included in Tables 4.3–5 are either specified or have been taken into consideration when rounding off the significant figures, so that they are within the range ± 1 for the last digit given.

References

4.3.1 E.Clementi, C.Roetti: At. Data Nucl. Data Tables **14**, 177 (1974)
4.3.2 A.D.McLean, R.S.McLean: At. Data Nucl. Data Tables **26**, 197 (1981)
4.3.3 A.V.Evseev, A.A.Radzig, B.M.Smirnov: Opt. Spectrosc. **44**, 833 (1978)

Table 4.3. Hartree-Fock and asymptotic parameters of valence electron wavefunctions for neutral atoms [a. u.]

$Z = 2$, Helium – He ($^1S_0$), $\varphi(1s)$

χ-basis set $n_i l$	ς_i	C_i	N_i
$1s$	1.4171	0.7684	3.374
$1s$	2.3768	0.2235	7.329
$1s$	4.3963	0.0408	18.44
$1s$	6.5270	−0.0099	33.35
$1s$	7.9425	0.0023	44.77

$\langle r\rangle = 0.927$; $\langle r^2\rangle = 1.185$;
$A = 2.87$ (A); $\gamma = 1.344$

$Z = 3$, Lithium – Li ($^2S_{1/2}$), $\varphi(2s)$

χ-basis set $n_i l$	ς_i	C_i	N_i
$1s$	2.4767	−0.1463	7.796
$1s$	4.6987	−0.0152	20.37
$2s$	0.3835	0.0038	0.1052
$2s$	0.6606	0.9805	0.4095
$2s$	1.07	0.1097	1.3675
$2s$	1.632	−0.1102	3.929

$\langle r\rangle = 3.874$; $\langle r^2\rangle = 17.74$;
$A = 0.82$ (B); $\gamma = 0.630$

$Z = 4$, Beryllium – Be ($^1S_0$), $\varphi(2s)$

χ-basis set $n_i l$	ς_i	C_i	N_i
$1s$	3.4712	−0.1709	12.93
$1s$	6.3686	−0.0146	32.14
$2s$	0.778	0.2119	0.6169
$2s$	0.9407	0.6250	0.9910
$2s$	1.4873	0.2666	3.115
$2s$	2.718	−0.0992	14.07

$\langle r\rangle = 2.649$: $\langle r^2\rangle = 8.426$;
$A = 1.62$ (B); $\gamma = 0.828$

$Z = 5$, Boron – B ($^2P^o_{1/2}$), $\varphi(2p)$

χ-basis set $n_i l$	ς_i	C_i	N_i
$2p$	0.8748	0.5362	0.8265
$2p$	1.3699	0.4034	2.536
$2p$	2.3226	0.1165	9.493
$2p$	5.5948	0.0082	85.49

$\langle r\rangle = 2.205$; $\langle r^2\rangle = 6.146$;
$A = 0.88$ (C); $\gamma = 0.781$

$Z = 6$, Carbon – C ($^3P_0$), $\varphi(2p)$

χ-basis set $n_i l$	ς_i	C_i	N_i
$2p$	0.9807	0.2824	1.100
$2p$	1.4436	0.5470	2.891
$2p$	2.6005	0.2320	12.59
$2p$	6.5100	0.0103	124.9

$\langle r\rangle = 1.743$; $\langle r^2\rangle = 3.890$;
$A = 1.30$ (C); $\gamma = 0.910$

$Z = 7$, Nitrogen – N ($^4S^o_{3/2}$), $\varphi(2p)$

χ-basis set $n_i l$	ς_i	C_i	N_i
$2p$	1.1607	0.2664	1.676
$2p$	1.7047	0.5232	4.381
$2p$	3.0394	0.2735	18.60
$2p$	7.1748	0.0129	159.2

$\langle r\rangle = 1.447$; $\langle r^2\rangle = 2.707$;
$A = 1.5$ (C); $\gamma = 1.034$

Table 4.3 (continued)

$Z = 8$, Oxygen – O ($^3P_2$), $\varphi(2p)$

χ-basis set $n_i l$	ς_i	C_i	N_i
2p	1.1439	0.1692	1.616
2p	1.8173	0.5797	5.141
2p	3.4499	0.3235	25.53
2p	7.5648	0.0166	181.7

$\langle r \rangle = 1.239$; $\langle r^2 \rangle = 2.001$;
$A = 1.3$ (*C*); $\gamma = 1.00$

$Z = 11$, Sodium – Na ($^2S_{1/2}$), $\varphi(3s)$

χ-basis set $n_i l$	ς_i	C_i	N_i
2s	3.8593	−0.1183	33.79
2s	2.3943	−0.0662	10.24
3s	1.2528	0.2794	0.9279
3s	0.7461	0.7814	0.1512

$\langle r \rangle = 4.209$; $\langle r^2 \rangle = 20.70$;
$A = 0.74$ (*B*); $\gamma = 0.615$

$Z = 9$, Fluorine – F ($^2P^o_{3/2}$), $\varphi(2p)$

χ-basis set $n_i l$	ς_i	C_i	N_i
2p	1.2657	0.1783	2.081
2p	2.0580	0.5619	7.016
2p	3.9285	0.3366	35.32
2p	8.2041	0.0190	222.6

$\langle r \rangle = 1.085$; $\langle r^2 \rangle = 1.544$;
$A = 1.59$ (*C*); $\gamma = 1.132$

$Z = 12$, Magnesium – Mg ($^1S_0$), $\varphi(3s)$

χ-basis set $n_i l$	ς_i	C_i	N_i
2s	4.4051	−0.1323	47.03
2s	2.9954	−0.1127	17.93
3s	1.4723	0.4724	1.633
3s	0.8917	0.6101	0.2823

$\langle r \rangle = 3.253$; $\langle r^2 \rangle = 12.42$;
$A = 1.32$ (*B*); $\gamma = 0.750$

$Z = 10$, Neon – Ne ($^1S_0$), $\varphi(2p)$

χ-basis set $n_i l$	ς_i	C_i	N_i
2p	1.4521	0.2180	2.934
2p	2.3817	0.5334	10.11
2p	4.4849	0.3293	49.19
2p	9.1346	0.0187	291.2

$\langle 1/r \rangle = 1.435$; $\langle r \rangle = 0.965$; $\langle r^2 \rangle = 1.228$;
$A = 1.75$ (*C*); $\gamma = 1.259$

$Z = 13$, Aluminium – Al ($^2P^o_{1/2}$), $\varphi(3p)$

χ-basis set $n_i l$	ς_i	C_i	N_i
2p	7.2078	−0.0448	161.1
2p	3.6541	−0.1498	29.47
3p	1.6828	0.2679	2.606
3p	0.9138	0.8038	0.3076

$\langle r \rangle = 3.434$; $\langle r^2 \rangle = 14.01$;
$A = 0.61$ (*C*); $\gamma = 0.663$

Table 4.3 (continued)

$Z = 14$, Silicon – Si ($^3P_0$), $\varphi(3p)$

χ-basis set $n_i l$	ς_i	C_i	N_i
$2p$	7.9691	0.0508	207.0
$2p$	4.1376	0.1774	40.21
$3p$	1.8190	−0.4216	3.422
$3p$	1.0646	−0.6577	0.5249

$\langle r \rangle = 2.788$; $\langle r^2 \rangle = 9.254$;
$A = 1.10$ (C); $\gamma = 0.774$

$Z = 15$, Phosphorus – P ($^4S^{\circ}_{3/2}$), $\varphi(3p)$

χ-basis set $n_i l$	ς_i	C_i	N_i
$2p$	8.7449	0.0526	261.1
$2p$	4.6304	0.2012	53.28
$3p$	2.0645	−0.4908	5.331
$3p$	1.2267	−0.5940	0.8620

$\langle r \rangle = 2.369$; $\langle r^2 \rangle = 6.694$;
$A = 1.65$ (C); $\gamma = 0.878$

$Z = 16$, Sulfur – S($^3P_2$), $\varphi(3p)$

χ-basis set $n_i l$	ς_i	C_i	N_i
$2p$	9.5125	−0.0524	322.3
$2p$	5.1205	−0.2201	68.51
$3p$	2.3379	0.5377	8.239
$3p$	1.3333	0.5615	1.154

$\langle r \rangle = 2.069$; $\langle r^2 \rangle = 5.116$;
$A = 1.11$ (C); $\gamma = 0.873$

$Z = 17$, Chlorine – Cl ($^2P^{\circ}_{3/2}$), $\varphi(3p)$

χ-basis set $n_i l$	ς_i	C_i	N_i
$2p$	10.29	−0.0514	392.0
$2p$	5.6130	−0.2377	86.19
$3p$	2.6242	0.5554	12.34
$3p$	1.4746	0.5519	1.642

$\langle r \rangle = 1.842$; $\langle r^2 \rangle = 4.059$;
$A = 1.78$ (C); $\gamma = 0.976$

$Z = 18$, Argon – Ar ($^1S_0$), $\varphi(3p)$

χ-basis set $n_i l$	ς_i	C_i	N_i
$2p$	11.07	−0.0494	470.8
$2p$	6.1066	−0.2514	106.4
$3p$	2.9034	0.5696	17.58
$3p$	1.6226	0.5431	2.294

$\langle 1/r \rangle = 0.814$; $\langle r \rangle = 1.663$; $\langle r^2 \rangle = 3.311$;
$A = 2.11$ (B); $\gamma = 1.076$

$Z = 19$, Potassium – K ($^2S_{1/2}$), $\varphi(4s)$

χ-basis set $n_i l$	ς_i	C_i	N_i
$2s$	8.5026	−0.0140	243.4
$2s$	6.7454	0.1097	136.5
$3s$	3.4800	−0.1555	33.15
$3s$	2.3268	−0.0788	8.102
$4s$	1.2053	0.4058	0.2611
$4s$	0.7277	0.6757	0.0270

$\langle r \rangle = 5.244$; $\langle r^2 \rangle = 31.54$;
$A = 0.52$ (C); $\gamma = 0.565$

Table 4.3 (continued)

$Z = 20$, Calcium – Ca ($^1S_0$), $\varphi(4s)$

χ-basis set $n_i l$	ς_i	C_i	N_i
2s	9.3003	−0.0212	304.6
2s	7.2676	0.1465	164.4
3s	3.6997	−0.2091	41.07
3s	2.5701	−0.0965	11.48
4s	1.4341	0.5179	0.5709
4s	0.8667	0.5836	0.0592

$\langle r \rangle = 4.218$; $\langle r^2 \rangle = 20.45$;
$A = 0.95$ (*C*); $\gamma = 0.670$

$Z = 21$, Scandium – Sc ($^2D_{3/2}$), $\varphi(4s)$

χ-basis set $n_i l$	ς_i	C_i	N_i
2s	7.7646	0.1508	194.0
3s	3.9271	−0.2234	50.61
3s	2.7006	−0.0792	13.65
4s	1.5384	0.5172	0.7829
4s	0.9141	0.5870	0.0752

$\langle r \rangle = 3.96$; $\langle r^2 \rangle = 18.07$;
$A = 1.11$ (*C*); $\gamma = 0.693$

$Z = 22$, Titanium – Ti ($^3F_2$), $\varphi(4s)$

χ-basis set $n_i l$	ς_i	C_i	N_i
2s	8.2522	0.1495	225.9
3s	4.1466	−0.2320	61.22
3s	2.7991	−0.0615	15.47
4s	1.6185	0.5172	0.9838
4s	0.9489	0.5874	0.0890

$\langle r \rangle = 3.766$; $\langle r^2 \rangle = 16.39$;
$A = 1.16$ (*C*); $\gamma = 0.708$

$Z = 23$, Vanadium – V ($^4F_{3/2}$), $\varphi(4s)$

χ-basis set $n_i l$	ς_i	C_i	N_i
2s	8.7282	0.1487	259.9
3s	4.3956	−0.2309	75.08
3s	2.9401	−0.054	18.37
4s	1.6973	0.5133	1.219
4s	0.9828	0.5919	0.1042

$\langle r \rangle = 3.607$; $\langle r^2 \rangle = 15.07$;
$A = 1.18$ (*B*); $\gamma = 0.704$

$Z = 24$, Chromium – Cr ($^7S_3$), $\varphi(4s)$

χ-basis set $n_i l$	ς_i	C_i	N_i
2s	9.1541	0.1458	292.8
3s	4.7273	−0.2023	96.84
3s	3.1374	−0.0524	23.07
4s	1.7444	0.4536	1.378
4s	0.9623	0.6545	0.0948

$\langle 1/r \rangle = 0.329$; $\langle r \rangle = 3.843$; $\langle r^2 \rangle = 17.21$;
$A = 1.13$ (*B*); $\gamma = 0.705$

$Z = 25$, Manganese – Mn ($^6S_{5/2}$), $\varphi(4s)$

χ-basis set $n_i l$	ς_i	C_i	N_i
2s	9.6693	0.1498	335.7
3s	4.9405	−0.2181	113.0
3s	3.2909	−0.0513	27.26
4s	1.8449	0.5046	1.773
4s	1.0441	0.6024	0.1368

$\langle r \rangle = 3.349$; $\langle r^2 \rangle = 13.05$;
$A = 1.31$ (*C*); $\gamma = 0.739$

Table 4.3 (continued)

$Z = 26$, Iron – Fe ($^5D_4$), $\varphi(4s)$

χ-basis set $n_i l$	ς_i	C_i	N_i
2s	10.131	0.1509	377.2
3s	5.2166	−0.2138	136.7
3s	3.4762	−0.0510	33.02
4s	1.9252	0.5016	2.148
4s	1.0774	0.6071	0.1576

$\langle r \rangle = 3.242$; $\langle r^2 \rangle = 12.25$;
$A = 1.40$ (C); $\gamma = 0.762$

$Z = 27$, Cobalt – Co ($^4F_{9/2}$), $\varphi(4s)$

χ-basis set $n_i l$	ς_i	C_i	N_i
2s	10.589	0.1518	421.3
3s	5.5025	−0.2074	164.8
3s	3.6714	−0.0521	39.98
4s	2.0014	0.4976	2.558
4s	1.1084	0.6125	0.1791

$\langle r \rangle = 3.144$; $\langle r^2 \rangle = 11.55$;
$A = 1.42$ (B); $\gamma = 0.760$

$Z = 28$, Nickel – Ni ($^3F_4$), $\varphi(4s)$

χ-basis set $n_i l$	ς_i	C_i	N_i
2s	11.05	0.1529	468.3
3s	5.7963	−0.2005	197.7
3s	3.8721	−0.0542	48.17
4s	2.0771	0.4929	3.023
4s	1.1389	0.6188	0.2023

$\langle r \rangle = 3.055$; $\langle r^2 \rangle = 10.93$;
$A = 1.42$ (A); $\gamma = 0.749$

$Z = 29$, Copper – Cu ($^2S_{1/2}$), $\varphi(4s)$

χ-basis set $n_i l$	ς_i	C_i	N_i
2s	11.45	0.1308	512.2
3s	6.1933	−0.1533	249.3
3s	4.0847	−0.0422	58.08
4s	2.0076	0.4143	2.594
4s	1.0368	0.6983	0.1326

$\langle 1/r \rangle = 0.382$; $\langle r \rangle = 3.331$; $\langle r^2 \rangle = 13.08$;
$A = 1.29$ (A); $\gamma = 0.754$

$Z = 30$, Zinc – Zn ($^1S_0$), $\varphi(4s)$

χ-basis set $n_i l$	ς_i	C_i	N_i
2s	11.91	0.1550	564.9
3s	6.4259	−0.1848	283.6
3s	4.2954	−0.0608	69.26
4s	2.2212	0.4843	4.088
4s	1.1951	0.6304	0.2513

$\langle r \rangle = 2.898$; $\langle r^2 \rangle = 9.869$;
$A = 1.69$ (C); $\gamma = 0.831$

$Z = 31$, Gallium – Ga ($^2P^o_{1/2}$), $\varphi(4p)$

χ-basis set $n_i l$	ς_i	C_i	N_i
3p	6.5222	−0.1261	298.8
3p	4.0381	−0.0571	55.79
4p	2.0596	0.4061	2.910
4p	1.0709	0.7055	0.1534

$\langle r \rangle = 3.424$; $\langle r^2 \rangle = 13.90$;
$A = 0.60$ (C); $\gamma = 0.664$

Table 4.3 (continued)

$Z = 32$, Germanium – Ge ($^3P_0$), $\varphi(4p)$

χ-basis set $n_i l$	ς_i	C_i	N_i
$3p$	6.7820	−0.1557	342.5
$3p$	4.2606	−0.0736	67.31
$4p$	2.2886	0.4669	4.677
$4p$	1.2516	0.6437	0.3094

$\langle r \rangle = 2.904$; $\langle r^2 \rangle = 9.951$;
$A = 1.29$ (*C*); $\gamma = 0.762$

$Z = 33$, Arsenic – As ($^4S^o_{3/2}$), $\varphi(4p)$

χ-basis set $n_i l$	ς_i	C_i	N_i
$3p$	7.0563	−0.1785	393.5
$3p$	4.5083	−0.0871	82.03
$4p$	2.4974	0.5156	6.928
$4p$	1.4078	0.5960	0.5252

$\langle r \rangle = 2.561$; $\langle r^2 \rangle = 7.716$;
$A = 1.58$ (*C*); $\gamma = 0.850$

$Z = 34$, Selenium – Se ($^3P_2$), $\varphi(4p)$

χ-basis set $n_i l$	ς_i	C_i	N_i
$3p$	7.2781	−0.2015	438.5
$3p$	4.6810	−0.0934	93.57
$4p$	2.7150	0.5509	10.09
$4p$	1.5114	0.5721	0.7229

$\langle r \rangle = 2.309$; $\langle r^2 \rangle = 6.256$;
$A = 1.52$ (*C*); $\gamma = 0.847$

$Z = 35$, Bromine – Br ($^2P^o_{3/2}$), $\varphi(4p)$

χ-basis set $n_i l$	ς_i	C_i	N_i
$3p$	7.6018	−0.2134	510.7
$3p$	5.0178	−0.1076	119.3
$4p$	2.9199	0.5821	14.00
$4p$	1.6241	0.5472	0.9993

$\langle r \rangle = 2.112$; $\langle r^2 \rangle = 5.224$;
$A = 1.83$ (*B*); $\gamma = 0.932$

$Z = 36$, Krypton – Kr ($^1S_0$), $\varphi(4p)$

χ-basis set $n_i l$	ς_i	C_i	N_i
$3p$	7.9359	−0.2192	593.7
$3p$	5.3714	−0.1237	151.4
$4p$	3.1274	0.6039	19.06
$4p$	1.7460	0.5302	1.384

$\langle 1/r \rangle = 0.669$; $\langle r \rangle = 1.952$;
$\langle r^2 \rangle = 4.455$; $A = 2.22$ (*B*); $\gamma = 1.014$

$Z = 37$, Rubidium – Rb ($^2S_{1/2}$), $\varphi(5s)$

χ-basis set $n_i l$	ς_i	C_i	N_i
$3s$	9.3132	0.0245	1039
$3s$	6.7684	0.0884	340.1
$4s$	3.8861	−0.1667	50.66
$4s$	2.5250	−0.0739	7.278
$5s$	1.3802	0.4481	0.1398
$5s$	0.804	0.6557	0.0072

$\langle r \rangle = 5.632$; $\langle r^2 \rangle = 36.18$;
$A = 0,48$ (*C*); $\gamma = 0.554$

Table 4.3 (continued)

Z = 38, Strontium – Sr ($^1S_0$), φ (5s)

χ-basis set $n_i l$	ς_i	C_i	N_i
3s	9.603	0.0228	1157
3s	7.1809	0.1242	418.4
4s	4.1612	−0.1986	68.92
4s	2.8205	−0.1154	11.98
5s	1.6304	0.5087	0.3494
5s	0.9613	0.6091	0.0191

$\langle r \rangle = 4.633$; $\langle r^2 \rangle = 24.50$;
$A = 0.86$ (C); $\gamma = 0.647$

Z = 39, Yttrium – Y ($^2D_{3/2}$), φ (5s)

χ-basis set $n_i l$	ς_i	C_i	N_i
3s	10.28	0.0196	1468
3s	7.4893	0.1383	484.7
4s	4.2802	−0.2364	78.25
4s	2.9141	−0.0806	13.87
5s	1.7052	0.5500	0.4473
5s	0.9973	0.5686	0.0234

$\langle r \rangle = 4.300$; $\langle r^2 \rangle = 21.14$;
$A = 1.02$ (C); $\gamma = 0.685$

Z = 40, Zirconium – Zr ($^3F_2$), φ (5s)

χ-basis set $n_i l$	ς_i	C_i	N_i
3s	10.81	0.0144	1754
3s	7.8217	0.1502	564.3
4s	4.4526	−0.2554	93.47
4s	3.0309	−0.0634	16.55
5s	1.7957	0.5559	0.5945
5s	1.0428	0.5633	0.0299

$\langle r \rangle = 4.078$; $\langle r^2 \rangle = 19.05$;
$A = 1.15$ (C); $\gamma = 0.709$

Z = 41, Niobium – Nb ($^6D_{1/2}$), φ (5s)

χ-basis set $n_i l$	ς_i	C_i	N_i
3s	11.31	0.0148	2049
3s	7.9335	0.1531	593.0
4s	4.6835	−0.2611	117.3
4s	3.1068	−0.0502	18.51
5s	1.8474	0.5507	0.6948
5s	1.0654	0.5668	0.0337

$\langle 1/r \rangle = 0.298$; $\langle r \rangle = 4.207$; $\langle r^2 \rangle = 20.37$;
$A = 1.16$ (C); $\gamma = 0.711$

Z = 42, Molybdenum – Mo ($^7S_3$), φ (5s)

χ-basis set $n_i l$	ς_i	C_i	N_i
3s	8.2694	0.1644	685.6
4s	4.9249	−0.2615	147.1
4s	3.2627	−0.053	23.06
5s	1.9412	0.5478	0.9123
5s	1.1033	0.5749	0.0408

$\langle 1/r \rangle = 0.307$; $\langle r \rangle = 4.079$; $\langle r^2 \rangle = 19.19$;
$A = 1.23$ (C); $\gamma = 0.722$

Z = 43, Technetium – Tc ($^6S_{5/2}$), φ (5s)

χ-basis set $n_i l$	ς_i	C_i	N_i
3s	8.8204	0.1750	859.3
4s	5.0392	−0.2743	163.1
4s	3.4213	−0.0350	28.56
5s	2.0196	0.5546	1.134
5s	1.1471	0.5655	0.0505

$\langle r \rangle = 3.650$; $\langle r^2 \rangle = 15.34$;
$A = 1.28$ (C); $\gamma = 0.736$

Table 4.3 (continued)

Z = 44, Ruthenium – Ru ($^5F_5$), φ (5s)

χ-basis set n_il	ς_i	C_i	N_i
3s	8.9592	0.1704	907.6
4s	5.3203	−0.2554	208.3
4s	3.5534	−0.0328	33.87
5s	2.0443	0.5238	1.213
5s	1.1310	0.5997	0.0468

$\langle 1/r\rangle = 0.324$; $\langle r\rangle = 3.877$; $\langle r^2\rangle = 17.42$;
$A = 1.22$ (C); $\gamma = 0.736$

Z = 45, Rhodium – Rh ($^4F_{9/2}$), φ (5s)

χ-basis set n_il	ς_i	C_i	N_i
3s	9.313	0.1722	1039
4s	5.5376	−0.2468	249.4
4s	3.7113	−0.0282	41.18
5s	2.0886	0.5119	1.365
5s	1.1416	0.6127	0.0492

$\langle 1/r\rangle = 0.331$; $\langle r\rangle = 3.795$; $\langle r^2\rangle = 16.73$;
$A = 1.19$ (C); $\gamma = 0.741$

Z = 46, Palladium – Pd ($^1S_0$), φ (4d)

χ-basis set n_il	ς_i	C_i	N_i
3d	16.12	−0.0685	7085
3d	9.1000	−0.2381	958.5
4d	5.79	0.2391	304.7
4d	3.4769	0.5889	30.71
4d	1.7379	0.3837	1.355

$\langle 1/r\rangle = 0.893$; $\langle r\rangle = 1.533$; $\langle r^2\rangle = 2.951$;
$A = 0.26$ (C); $\gamma = 0.783$

Z = 47, Silver – Ag ($^2S_{1/2}$), φ (5s)

χ-basis set n_il	ς_i	C_i	N_i
3s	10.14	0.1654	1399
4s	5.8718	−0.2176	324.6
4s	3.9877	−0.0557	56.90
5s	2.6640	0.2528	5.203
5s	1.6501	0.5102	0.3733
5s	1.0419	0.3948	0.0298

$\langle 1/r\rangle = 0.344$; $\langle r\rangle = 3.656$; $\langle r^2\rangle = 15.59$;
$A = 1.18$ (C); $\gamma = 0.746$

Z = 48, Cadmium – Cd ($^1S_0$), φ (5s)

χ-basis set n_il	ς_i	C_i	N_i
3s	10.48	0.1974	1568
4s	6.0667	−0.2692	376.0
4s	4.1340	−0.0222	66.92
5s	2.3520	0.5436	2.622
5s	1.2920	0.5819	0.0972

$\langle r\rangle = 3.237$; $\langle r^2\rangle = 12.17$;
$A = 1.6$ (C); $\gamma = 0.813$

Z = 49, Indium – In ($^2P^{\circ}_{1/2}$), φ (5p)

χ-basis set n_il	ς_i	C_i	N_i
3p	9.8279	0.1052	1255
4p	6.2086	−0.1849	417.2
4p	4.0263	−0.0553	59.42
5p	2.1421	0.4874	1.568
5p	1.1494	0.6415	0.0511

$\langle r\rangle = 3.778$; $\langle r^2\rangle = 16.65$;
$A = 0.58$ (C); $\gamma = 0.652$

Table 4.3 (continued)

$Z = 50$, Tin – Sn ($^3P_0$), $\varphi\,(5p)$

χ-basis set $n_i l$	ς_i	C_i	N_i
$3p$	10.18	0.1295	1419
$4p$	6.4839	−0.2121	507.1
$4p$	4.2737	−0.0805	77.71
$5p$	2.3587	0.5310	2.664
$5p$	1.3201	0.5938	0.1094

$\langle r \rangle = 3.286$; $\langle r^2 \rangle = 12.53$;
$A = 1.02$ (C); $\gamma = 0.735$

$Z = 51$, Antimony – Sb ($^4S^o_{3/2}$), $\varphi\,(5p)$

χ-basis set $n_i l$	ς_i	C_i	N_i
$3p$	10.59	0.1571	1631
$4p$	6.738	−0.2319	602.9
$4p$	4.5108	−0.1026	99.09
$5p$	2.5594	0.5610	4.174
$5p$	1.4736	0.5633	0.2004

$\langle r \rangle = 2.952$; $\langle r^2 \rangle = 10.07$;
$A = 1.67$ (C); $\gamma = 0.797$

$Z = 52$, Tellurium – Te ($^3P_2$), $\varphi\,(5p)$

χ-basis set $n_i l$	ς_i	C_i	N_i
$3p$	11.14	0.2002	1944
$4p$	6.9546	−0.2499	695.2
$4p$	4.7319	−0.1156	122.9
$5p$	2.7367	0.5934	6.033
$5p$	1.5618	0.5402	0.2759

$\langle r \rangle = 2.701$; $\langle r^2 \rangle = 8.411$;
$A = 1.65$ (C); $\gamma = 0.814$

$Z = 53$, Iodine – I ($^2P^o_{3/2}$), $\varphi\,(5p)$

χ-basis set $n_i l$	ς_i	C_i	N_i
$3p$	11.34	0.2190	2070
$4p$	7.2450	−0.2639	835.7
$4p$	4.9772	−0.1361	154.3
$5p$	2.9205	0.6140	8.625
$5p$	1.6711	0.5258	0.4002

$\langle r \rangle = 2.502$; $\langle r^2 \rangle = 7.201$;
$A = 1.94$ (C); $\gamma = 0.876$

$Z = 54$, Xenon – Xe ($^1S_0$), $\varphi\,(5p)$

χ-basis set $n_i l$	ς_i	C_i	N_i
$3p$	10.97	0.1679	1844
$4p$	7.6011	−0.2787	1037
$4p$	5.2292	−0.1577	192.7
$5p$	3.0947	0.6344	11.86
$5p$	1.7808	0.5103	0.5677

$\langle 1/r \rangle = 0.547$; $\langle r \rangle = 2.338$; $\langle r^2 \rangle = 6.277$;
$A = 2.4$ (C); $\gamma = 0.944$

$Z = 55$, Caesium – Cs ($^2S_{1/2}$), $\varphi\,(6s)$

χ-basis set $n_i l$	ς_i	C_i	N_i
$4s$	7.4567	0.1010	951.3
$4s$	5.3771	0.0352	218.4
$5s$	3.7211	−0.2060	32.69
$5s$	2.4237	−0.0357	3.093
$6s$	1.3934	0.5029	0.0357
$6s$	0.8237	0.6068	0.0012

$\langle 1/r \rangle = 0.192$; $\langle r \rangle = 6.30$; $\langle r^2 \rangle = 44.8$;
$A = 0.42$ (B); $\gamma = 0.535$

Table 4.3 (continued)

$Z = 56$, Barium – Ba ($^1S_0$), $\varphi(6s)$

χ-basis set $n_i l$	ς_i	C_i	N_i
4s	7.4982	0.1473	975.4
4s	5.2707	0.0279	199.7
5s	3.8574	−0.2729	39.85
5s	2.5976	−0.0307	4.528
6s	1.5886	0.5824	0.0838
6s	0.9525	0.5294	0.0030

$\langle 1/r \rangle = 0.234$; $\langle r \rangle = 5.25$; $\langle r^2 \rangle = 31.2$;
$A = 0.78$ (B); $\gamma = 0.619$

$Z = 57$, Lanthanum – La ($^2D_{3/2}$), $\varphi(6s)$

χ-basis set $n_i l$	ς_i	C_i	N_i
4s	7.8105	0.1457	1172
4s	5.7547	0.0474	296.5
5s	4.0678	−0.2912	53.37
5s	2.7554	−0.0282	6.263
6s	1.6938	0.5881	0.1271
6s	1.0087	0.5267	0.0044

$\langle 1/r \rangle = 0.250$; $\langle r \rangle = 4.93$; $\langle r^2 \rangle = 27.5$;
$A = 0.90$ (B); $\gamma = 0.640$

$Z = 58$, Cerium – Ce (^3H), $\varphi(6s)$

χ-basis set $n_i l$	ς_i	C_i	N_i
4s	8.1438	0.1310	1414
4s	6.1130	0.0600	389.1
5s	4.1855	−0.2800	62.43
5s	2.8240	−0.0363	7.170
6s	1.7238	0.5830	0.1425
6s	1.0217	0.5344	0.0048

$\langle 1/r \rangle = 0.252$; $\langle r \rangle = 4.88$; $\langle r^2 \rangle = 27.0$;
$A = 0.88$ (B); $\gamma = 0.638$

$Z = 59$, Praseodymium – Pr ($^4I^o_{9/2}$), $\varphi(6s)$

χ-basis set $n_i l$	ς_i	C_i	N_i
4s	8.3469	0.1209	1580
4s	6.1782	0.0506	408.1
5s	4.2105	−0.2464	64.51
5s	2.8012	−0.0511	6.858
6s	1.6796	0.5681	0.1203
6s	0.9939	0.5503	0.0040

$\langle 1/r \rangle = 0.242$; $\langle r \rangle = 5.06$; $\langle r^2 \rangle = 29.0$;
$A = 0.84$ (B); $\gamma = 0.634$

$Z = 60$, Neodymium – Nd ($^5I_4$), $\varphi(6s)$

χ-basis set $n_i l$	ς_i	C_i	N_i
4s	8.4569	0.1262	1676
4s	6.1672	0.0417	404.8
5s	4.2794	−0.2440	70.53
5s	2.8401	−0.0482	7.398
6s	1.7001	0.5671	0.1302
6s	1.0036	0.5512	0.0042

$\langle 1/r \rangle = 0.245$; $\langle r \rangle = 5.01$; $\langle r^2 \rangle = 28.4$;
$A = 0.85$ (B); $\gamma = 0.637$

$Z = 61$, Promethium – Pm ($^6H^o_{5/2}$), $\varphi(6s)$

χ-basis set $n_i l$	ς_i	C_i	N_i
4s	8.6393	0.1260	1845
4s	6.2631	0.0390	433.9
5s	4.3549	−0.2408	77.65
5s	2.8814	−0.0471	8.009
6s	1.7216	0.5655	0.1413
6s	1.0137	0.5531	0.0045

$\langle 1/r \rangle = 0.247$; $\langle r \rangle = 4.96$; $\langle r^2 \rangle = 27.8$;
$A = 0.86$ (B); $\gamma = 0.640$

Table 4.3 (continued)

$Z = 62$, Samarium – Sm ($^7F_0$), $\varphi(6s)$

χ-basis set n_il	ς_i	C_i	N_i
4s	8.7759	0.1287	1980
4s	6.2751	0.0326	437.7
5s	4.4125	−0.2393	83.47
5s	2.9124	−0.0431	8.495
6s	1.7395	0.5653	0.1511
6s	1.0222	0.5530	0.0048

$\langle 1/r \rangle = 0.250$; $\langle r \rangle = 4.91$; $\langle r^2 \rangle = 27.3$;
$A = 0.88$ (B); $\gamma = 0.644$

$Z = 63$, Europium – Eu ($^8S^o_{7/2}$), $\varphi(6s)$

χ-basis set n_il	ς_i	C_i	N_i
4s	8.9826	0.1257	2199
4s	6.4318	0.0336	489.1
5s	4.5009	−0.2344	93.10
5s	2.9622	−0.0459	9.325
6s	1.7632	0.5623	0.1650
6s	1.0330	0.5573	0.0051

$\langle 1/r \rangle = 0.252$; $\langle r \rangle = 4.86$; $\langle r^2 \rangle = 26.8$;
$A = 0.89$ (B); $\gamma = 0.646$

$Z = 64$, Gadolinium – Gd ($^9D^o_2$), $\varphi(6s)$

χ-basis set n_il	ς_i	C_i	N_i
4s	9.1369	0.1369	2374
4s	6.5080	0.0349	515.7
5s	4.6338	−0.2575	109.3
5s	3.0817	−0.0332	11.59
6s	1.8570	0.5742	0.2311
6s	1.0847	0.5453	0.0070

$\langle 1/r \rangle = 0.268$; $\langle r \rangle = 4.58$; $\langle r^2 \rangle = 23.9$;
$A = 1.01$ (B); $\gamma = 0.672$

$Z = 65$, Terbium – Tb ($^6H^o_{15/2}$), $\varphi(6s)$

χ-basis set n_il	ς_i	C_i	N_i
4s	9.3905	0.1207	2685
4s	6.7539	0.0341	609.4
5s	4.6649	−0.2272	113.3
5s	3.0505	−0.0469	10.96
6s	1.8070	0.5584	0.1935
6s	1.0530	0.5625	0.0058

$\langle 1/r \rangle = 0.257$; $\langle r \rangle = 4.77$; $\langle r^2 \rangle = 25.8$;
$A = 0.93$ (B); $\gamma = 0.657$

$Z = 66$, Dysprosium – Dy ($^5I_8$), $\varphi(6s)$

χ-basis set n_il	ς_i	C_i	N_i
4s	9.6221	0.1175	2996
4s	6.9324	0.0356	685.3
5s	4.7490	−0.2232	125.1
5s	3.0953	−0.0488	11.88
6s	1.8310	0.5548	0.2109
6s	1.0639	0.5672	0.0062

$\langle 1/r \rangle = 0.259$; $\langle r \rangle = 4.73$; $\langle r^2 \rangle = 25.3$;
$A = 0.94$ (B); $\gamma = 0.661$

$Z = 67$, Holmium – Ho ($^4I^o_{15/2}$), $\varphi(6s)$

χ-basis set n_il	ς_i	C_i	N_i
4s	9.8303	0.1146	3299
4s	7.0869	0.0370	756.7
5s	4.8430	−0.2189	139.3
5s	3.1492	−0.0517	13.06
6s	1.8533	0.5526	0.2281
6s	1.0739	0.5704	0.0066

$\langle 1/r \rangle = 0.262$; $\langle r \rangle = 4.69$; $\langle r^2 \rangle = 24.9$;
$A = 0.96$ (B); $\gamma = 0.665$

Table 4.3 (continued)

$Z = 68$, Erbium – Er ($^3H_6$), $\varphi(6s)$

χ-basis set $n_i l$	ς_i	C_i	N_i
4s	9.9785	0.1154	3529
4s	7.1382	0.0337	781.7
5s	4.9111	−0.2168	150.4
5s	3.1869	−0.0505	13.94
6s	1.8721	0.5516	0.2436
6s	1.0825	0.5717	0.0069

$\langle 1/r\rangle = 0.264$; $\langle r\rangle = 4.65$; $\langle r^2\rangle = 24.5$;
$A = 0.98$ (B); $\gamma = 0.670$

$Z = 69$, Thulium – Tm($^2F^{\circ}_{7/2}$), $\varphi(6s)$

χ-basis set $n_i l$	ς_i	C_i	N_i
4s	10.181	0.1132	3863
4s	7.2839	0.0343	856.1
5s	4.9964	−0.2134	165.3
5s	3.2341	−0.0522	15.12
6s	1.8938	0.5493	0.2625
6s	1.0923	0.5749	0.0073

$\langle 1/r\rangle = 0.266$; $\langle r\rangle = 4.61$; $\langle r^2\rangle = 24.1$;
$A = 0.99$ (B); $\gamma = 0.674$

$Z = 70$, Ytterbium – Yb ($^1S_0$), $\varphi(6s)$

χ-basis set $n_i l$	ς_i	C_i	N_i
4s	10.243	0.1185	3970
4s	7.1670	0.0250	796.0
5s	5.0197	−0.2146	169.6
5s	3.2463	−0.0451	15.43
6s	1.9054	0.5509	0.2732
6s	1.0981	0.5723	0.0076

$\langle 1/r\rangle = 0.268$; $\langle r\rangle = 4.57$; $\langle r^2\rangle = 23.7$;
$A = 1.01$ (B); $\gamma = 0.678$

$Z = 71$, Lutetium – Lu ($^2D_{3/2}$), $\varphi(6s)$

χ-basis set $n_i l$	ς_i	C_i	N_i
4s	11.412	0.0819	6457
4s	8.4433	0.0825	1664
5s	5.3283	−0.2196	235.5
5s	3.4966	−0.0674	23.22
6s	2.0541	0.5533	0.4452
6s	1.1768	0.5769	0.0119

$\langle 1/r\rangle = 0.289$; $\langle r\rangle = 4.27$; $\langle r^2\rangle = 20.7$;
$A = 0.92$ (C); $\gamma = 0.632$

$Z = 72$, Hafnium – Hf ($^3F_2$), $\varphi(6s)$

χ-basis set $n_i l$	ς_i	C_i	N_i
4s	11.462	0.0854	6585
4s	8.5797	0.0877	1789
5s	5.4990	−0.2324	280.1
5s	3.6349	−0.0632	28.74
6s	2.1449	0.5602	0.5898
6s	1.2234	0.5717	0.0153

$\langle 1/r\rangle = 0.303$; $\langle r\rangle = 4.08$; $\langle r^2\rangle = 19.0$;
$A = 1.31$ (B); $\gamma = 0.74$

$Z = 73$, Tantalum – Ta ($^4F_{3/2}$), $\varphi(6s)$

χ-basis set $n_i l$	ς_i	C_i	N_i
4s	11.941	0.0779	7917
4s	8.8985	0.1031	2108
5s	5.6269	−0.2465	317.9
5s	3.7320	−0.0513	33.22
6s	2.2190	0.5655	0.7354
6s	1.2608	0.5664	0.0187

$\langle 1/r\rangle = 0.314$; $\langle r\rangle = 3.94$; $\langle r^2\rangle = 17.7$;
$A = 1.41$ (B); $\gamma = 0.762$

Table 4.3 (continued)

$Z = 74$, Tungsten – W ($^5D_0$), $\varphi(6s)$

χ-basis set $n_i l$	ς_i	C_i	N_i
4s	12.372	0.0735	9287
4s	9.1741	0.1139	2418
5s	5.7436	−0.2599	355.9
5s	3.8144	−0.0365	37.47
6s	2.2807	0.5705	0.8790
6s	1.2912	0.5604	0.0218

$\langle 1/r\rangle = 0.325$; $\langle r\rangle = 3.82$; $\langle r^2\rangle = 16.7$; $A = 1.43$ (C); $\gamma = 0.766$

$Z = 75$, Rhenium – Re ($^6S_{5/2}$), $\varphi(6s)$

χ-basis set $n_i l$	ς_i	C_i	N_i
4s	12.407	0.0738	9405
4s	9.3385	0.1163	2619
5s	5.8936	−0.2673	410.1
5s	3.9277	−0.0286	44.01
6s	2.3463	0.5689	1.057
6s	1.3224	0.5620	0.0254

$\langle 1/r\rangle = 0.334$; $\langle r\rangle = 3.72$; $\langle r^2\rangle = 15.8$; $A = 1.42$ (B); $\gamma = 0.761$

$Z = 76$, Osmium – Os ($^5D_4$), $\varphi(6s)$

χ-basis set $n_i l$	ς_i	C_i	N_i
4s	12.644	0.0797	10240
4s	9.4703	0.1175	2789
5s	6.0015	−0.2821	453.1
5s	3.9932	−0.0119	48.20
6s	2.4005	0.5750	1.226
6s	1.3507	0.5533	0.0292

$\langle 1/r\rangle = 0.344$; $\langle r\rangle = 3.62$; $\langle r^2\rangle = 15.0$; $A = 1.66$ (B); $\gamma = 0.801$

$Z = 77$, Iridium – Ir ($^4F_{9/2}$), $\varphi(6s)$

χ-basis set $n_i l$	ς_i	C_i	N_i
4s	12.996	0.0757	11590
4s	9.7095	0.1262	3121
5s	6.1434	−0.2899	515.3
5s	4.0920	−0.0027	55.14
6s	2.4586	0.5765	1.432
6s	1.3786	0.5512	0.0333

$\langle 1/r\rangle = 0.353$; $\langle r\rangle = 3.53$; $\langle r^2\rangle = 14.3$; $A = 1.67$ (B); $\gamma = 0.816$

$Z = 78$, Platinum – Pt ($^3D_3$), $\varphi(6s)$

χ-basis set $n_i l$	ς_i	C_i	N_i
4s	12.671	0.0686	10340
4s	9.8174	0.1142	3280
5s	6.3379	−0.2654	611.6
5s	4.1982	0.0035	63.48
6s	2.4290	0.5436	1.324
6s	1.3248	0.5906	0.0257

$\langle 1/r\rangle = 0.335$; $\langle r\rangle = 3.72$; $\langle r^2\rangle = 15.9$; $A = 1.53$ (B); $\gamma = 0.812$

$Z = 79$, Gold – Au ($^2S_{1/2}$), $\varphi(6s)$

χ-basis set $n_i l$	ς_i	C_i	N_i
4s	10.264	0.1246	4007
5s	6.7013	−0.2183	831.1
5s	4.5321	−0.0697	96.70
6s	2.9804	0.3135	5.004
6s	1.7985	0.5632	0.1877
6s	1.1109	0.3106	0.0082

$\langle 1/r\rangle = 0.337$; $\langle r\rangle = 3.70$; $\langle r^2\rangle = 15.9$; $A = 1.58$ (A); $\gamma = 0.823$

Table 4.3 (continued)

Z = 80, Mercury – Hg ($^1S_0$), φ (6s)

χ-basis set $n_i l$	ς_i	C_i	N_i
4s	10.616	0.1573	4663
5s	6.9239	−0.2396	994.7
5s	4.7445	−0.0945	124.4
6s	3.2060	0.3268	8.041
6s	1.9846	0.5658	0.3560
6s	1.2299	0.2946	0.0159

$\langle 1/r\rangle = 0.376$; $\langle r\rangle = 3.33$; $\langle r^2\rangle = 12.8$;
$A = 1.96$ (A); $\gamma = 0.876$

Z = 81, Thallium – Tl ($^2P^o_{1/2}$), φ (6p)

χ-basis set $n_i l$	ς_i	C_i	N_i
4p	9.9954	0.0688	3556
5p	7.1101	−0.1213	1151
5p	4.8956	−0.1590	147.8
6p	3.5306	0.1538	15.05
6p	2.0206	0.5791	0.4001
6p	1.1626	0.4663	0.0110

$\langle 1/r\rangle = 0.317$; $\langle r\rangle = 3.92$; $\langle r^2\rangle = 17.7$;
$A = 0.55$ (C); $\gamma = 0.670$

Z = 82, Lead – Pb ($^3P_0$), φ (6p)

χ-basis set $n_i l$	ς_i	C_i	N_i
4p	10.121	0.0807	3762
5p	7.2008	−0.1586	1234
5p	4.9840	−0.1722	163.1
6p	3.6385	0.1793	18.30
6p	2.1973	0.6018	0.6899
6p	1.3144	0.4073	0.0244

$\langle 1/r\rangle = 0.363$; $\langle r\rangle = 3.42$; $\langle r^2\rangle = 13.4$;
$A = 1.09$ (C); $\gamma = 0.738$

Z = 83, Bismuth – Bi ($^4S^o_{3/2}$), φ (6p)

χ-basis set $n_i l$	ς_i	C_i	N_i
4p	9.7163	0.0704	3131
5p	7.0413	−0.2435	1091
5p	4.7856	−0.1091	130.4
6p	3.1649	0.3713	7.394
6p	2.0295	0.5668	0.4117
6p	1.3128	0.2253	0.0243

$\langle 1/r\rangle = 0.403$; $\langle r\rangle = 3.08$; $\langle r^2\rangle = 10.9$;
$A = 1.43$ (C); $\gamma = 0.732$

Z = 90, Thorium – Th ($^3F_2$), φ (7s)

χ-basis set $n_i l$	ς_i	C_i	N_i
5s	8.8021	0.1368	3724
5s	6.6025	0.0834	765.9
6s	4.6307	−0.2604	87.75
6s	3.1953	−0.0981	7.868
7s	2.2206	0.3728	0.2432
7s	1.4490	0.5617	0.0099
7s	0.9549	0.2371	0.0004

$\langle 1/r\rangle = 0.246$; $\langle r\rangle = 5.00$; $\langle r^2\rangle = 28.2$;
$A = 1.01$ (A); $\gamma = 0.67$

Z = 91, Protactinium – Pa ($5f^3 7s$ – ^4I), φ (7s)

χ-basis set $n_i l$	ς_i	C_i	N_i
5s	8.9073	0.1266	3975
5s	6.6099	0.0594	770.6
6s	4.6416	−0.2130	89.10
6s	3.1334	−0.1034	6.9283
7s	2.1303	0.3358	0.1782
7s	1.3842	0.5727	0.0070
7s	0.9090	0.2599	0.0003

$\langle 1/r\rangle = 0.227$; $\langle r\rangle = 5.35$; $\langle r^2\rangle = 32.3$;
$A = 0.90$ (A); $\gamma = 0.66$

Table 4.3 (continued)

$Z = 92$, Uranium – U ($5f^3 6d 7s^2$ – ^5K), $\varphi(7s)$

χ-basis set $n_i l$	ς_i	C_i	N_i
5s	9.0435	0.1392	4321
5s	6.7277	0.0598	849.3
6s	4.7475	−0.2333	103.2
6s	3.2296	−0.0939	8.433
7s	2.2190	0.3540	0.2419
7s	1.4398	0.5681	0.0094
7s	0.9435	0.2484	0.0004

$\langle 1/r \rangle = 0.240$; $\langle r \rangle = 5.08$; $\langle r^2 \rangle = 29.2$; $A = 0.99$ (A); $\gamma = 0.674$

Table 4.4. Hartree-Fock and asymptotic parameters of valence electron wavefunctions for positive atomic ions [a.u.]

$Z = 3$, Lithium – Li^+ ($^1S_0$), $\varphi(1s)$

χ-basis set $n_i l$	ς_i	C_i	N_i
1s	2.4638	0.8946	7.734
1s	4.7036	0.1184	20.40
1s	6.4669	−0.0023	32.89
1s	1.3579	0.0044	3.165

$A = 6.5$ (B); $\gamma = 2.358$

$Z = 4$, Beryllium – Be^+ ($^2S_{1/2}$), $\varphi(2s)$

χ-basis set $n_i l$	ς_i	C_i	N_i
1s	3.4977	−0.2046	13.08
1s	6.5023	−0.0157	33.16
2s	1.1838	1.1080	1.761
2s	2.6277	−0.1329	12.92

$A = 2.67$ (A); $\gamma = 1.157$

$Z = 5$, Boron – B^+ ($^1S_0$), $\varphi(2s)$

χ-basis set $n_i l$	ς_i	C_i	N_i
1s	4.4299	−0.2029	18.65
1s	7.8634	−0.0194	44.10
2s	1.5924	0.7349	3.695
2s	4.0102	−0.0922	37.19
2s	1.2502	0.3453	2.018

$A = 6.7$ (C); $\gamma = 1.360$

$Z = 6$, Carbon – C^+ ($^2P^o_{1/2}$), $\varphi(2p)$

χ-basis set $n_i l$	ς_i	C_i	N_i
2p	1.8778	0.3998	5.580
2p	3.078	0.1230	19.19
2p	1.3927	0.5119	2.643
2p	7.0112	0.0078	150.3

$A = 2.53$ (C); $\gamma = 1.339$

Table 4.4 (continued)

$Z = 7$, Nitrogen – N^+ ($^3P_0$), $\varphi(2p)$

χ-basis set $n_i l$	ς_i	C_i	N_i
2p	1.9759	0.5629	6.337
2p	3.3858	0.2026	24.36
2p	1.4789	0.2761	3.071
2p	8.0019	0.0088	209.1

$A = 2.9$ (C); $\gamma = 1.475$

$Z = 8$, Oxygen – O^+ ($^4S^o_{3/2}$), $\varphi(2p)$

χ-basis set $n_i l$	ς_i	C_i	N_i
2p	2.2445	0.5537	8.715
2p	3.8417	0.2331	33.40
2p	1.6413	0.2588	3.985
2p	8.5940	0.0107	250.0

$A = 3.3$ (C); $\gamma = 1.607$

$Z = 9$, Fluorine – F^+ ($^3P_2$), $\varphi(2p)$

χ-basis set $n_i l$	ς_i	C_i	N_i
2p	2.3317	0.6033	9.586
2p	4.2480	0.2913	42.95
2p	1.6422	0.1609	3.991
2p	9.5323	0.0111	323.9

$A = 3.1$ (C); $\gamma = 1.603$

$Z = 10$, Neon – Ne^+ ($^2P^o_{3/2}$), $\varphi(2p)$

χ-basis set $n_i l$	ς_i	C_i	N_i
2p	2.5710	0.5872	12.24
2p	4.7236	0.3052	56.00
2p	1.7524	0.1672	4.694
2p	9.811	0.0138	348.1

$A = 3.4$ (C); $\gamma = 1.735$

$Z = 11$, Sodium – Na^+ ($^1S_0$), $\varphi(2p)$

χ-basis set $n_i l$	ς_i	C_i	N_i
2p	2.8948	0.5428	16.46
2p	5.2651	0.3048	73.45
2p	1.9589	0.2163	6.201
2p	10.69	0.0141	431.8

$A = 3.7$ (C); $\gamma = 1.864$

$Z = 12$, Magnesium – Mg^+ ($^2S_{1/2}$), $\varphi(3s)$

χ-basis set $n_i l$	ς_i	C_i	N_i
2s	4.5253	−0.2058	50.30
3s	1.4135	0.65	1.416
3s	1.0632	0.4065	0.523
3s	3.5536	−0.0958	35.67

$A = 2.31$ (A); $\gamma = 1.051$

$Z = 13$, Aluminium – Al^+ ($^1S_0$), $\varphi(3s)$

χ-basis set $n_i l$	ς_i	C_i	N_i
2s	4.8362	−0.2429	59.39
3s	1.7751	0.5546	3.142
3s	1.2538	0.5263	0.9306
3s	3.8880	−0.0981	48.86

$A = 3.1$ (C); $\gamma = 1.176$

$Z = 14$, Silicon – Si^+ ($^2P^o_{1/2}$), $\varphi(3p)$

χ-basis set $n_i l$	ς_i	C_i	N_i
2p	5.7813	−0.1658	92.80
3p	1.7078	0.6127	2.745
3p	1.1477	0.4500	0.6830
3p	4.2737	−0.0895	68.04

$A = 1.8$ (C); $\gamma = 1.096$

Table 4.4 (continued)

Z = 15, Phosphorus – P^+ ($^3P_0$), φ (3p)

χ-basis set n_il	ς_i	C_i	N_i
2p	6.4620	−0.1775	122.6
3p	2.0224	0.5888	4.960
3p	1.3437	0.4831	1.186
3p	4.8429	−0.1044	105.4

A = 2.5 (C); γ = 1.204

Z = 16, Sulfur – S^+ ($^4S^o_{3/2}$), φ (3p)

χ-basis set n_il	ς_i	C_i	N_i
2p	6.8792	−0.1940	143.3
3p	2.3323	0.5724	8.169
3p	1.5361	0.5079	1.894
3p	5.2639	−0.1054	141.1

A = 3.2 (C); γ = 1.309

Z = 17, Chlorine – Cl^+ ($^3P_2$), φ (3p)

χ-basis set n_il	ς_i	C_i	N_i
2p	7.4953	−0.1995	177.6
3p	2.6483	0.5730	12.74
3p	1.6571	0.5213	2.470
3p	5.8297	−0.1145	201.7

A = 3.1 (C); γ = 1.323

Z = 18, Argon – Ar^+ ($^2P^o_{3/2}$), φ (3p)

χ-basis set n_il	ς_i	C_i	N_i
2p	7.9504	−0.2094	205.8
3p	2.9565	0.5752	18.73
3p	1.7999	0.5296	3.298
3p	6.3001	−0.1166	264.6

A = 3.4 (C); γ = 1.425

Z = 19, Potassium – K^+ ($^1S_0$), φ (3p)

χ-basis set n_il	ς_i	C_i	N_i
2p	8.6173	−0.2100	251.7
3p	3.2482	0.5822	26.04
3p	1.9456	0.5298	4.332
3p	6.8838	−0.1264	360.9

A = 3.9 (C); γ = 1.525

Z = 20, Calcium – Ca^+ ($^2S_{1/2}$), φ (4s)

χ-basis set n_il	ς_i	C_i	N_i
3s	3.9588	−0.2037	52.05
3s	3.0161	−0.1088	20.09
4s	3.091	−0.0560	18.07
4s	1.4260	0.6038	0.5564
4s	1.0026	0.5348	0.1140
4s	0.8683	−0.0699	0.0597

A = 1.62 (B); γ = 0.934

Table 4.4 (continued)

$Z = 22$, Titanium – Ti^+ ($^4F_{3/2}$), φ (4s)

χ-basis set $n_i l$	ς_i	C_i	N_i
3s	4.8903	−0.2565	109.0
3s	0.3421	0.0007	0.0099
4s	3.7168	−0.1538	41.46
4s	1.7424	0.4723	1.371
4s	1.1924	0.6000	0.2487
4s	0.9598	0.0113	0.0937

$A = 2.06$ (B); $\gamma = 0.999$

$Z = 30$, Zinc – Zn^+ ($^2S_{1/2}$), φ (4s)

χ-basis set $n_i l$	ς_i	C_i	N_i
3s	6.9485	−0.0448	372.9
3s	5.7019	−0.2575	186.6
4s	5.0410	0.0237	163.4
4s	2.1523	0.6092	3.548
4s	1.3285	0.4852	0.4046
4s	0.9337	−0.0296	0.0827

$A = 3.01$ (B); $\gamma = 1.149$

$Z = 26$, Iron – Fe^+ ($^6D_{9/2}$), φ (4s)

χ-basis set $n_i l$	ς_i	C_i	N_i
3s	7.0523	−0.0591	392.7
3s	4.7126	−0.3075	95.80
4s	3.8393	0.0340	47.97
4s	1.8653	0.6795	1.863
4s	1.1933	0.4029	0.2496
4s	0.8315	−0.0279	0.0491

$A = 2.68$ (B); $\gamma = 1.091$

$Z = 35$, Bromine – Br^+ ($^3P_2$), φ (4p)

χ-basis set $n_i l$	ς_i	C_i	N_i
3p	8.6400	−0.1204	799.4
3p	5.8999	−0.2374	210.3
4p	4.7239	0.0360	122.0
4p	2.8909	0.5699	13.38
4p	1.8189	0.4931	1.663
4p	1.3175	0.0191	0.3898

$A = 2.5$ (D); $\gamma = 1.266$

$Z = 29$, Copper – Cu^+ ($^1S_0$), φ (3d)

χ-basis set $n_i l$	ς_i	C_i	N_i
3d	4.7056	0.3357	95.30
3d	7.3863	0.2380	461.8
3d	2.9366	0.3617	18.30
3d	1.6928	0.2292	2.661

$A = 0.46$ (D); $\gamma = 1.221$

$Z = 36$, Krypton – Kr^+ ($^2P^o_{3/2}$), φ (4p)

χ-basis set $n_i l$	ς_i	C_i	N_i
3p	9.3140	−0.1004	1040
3p	6.5114	−0.2558	297.0
4p	5.1030	−0.0293	172.6
4p	3.3385	0.5171	25.58
4p	2.0615	0.5726	2.922
4p	1.3318	0.0409	0.4091

$A = 3.7$ (C); $\gamma = 1.338$

Table 4.4 (continued)

$Z = 37$, Rubidium – Rb^+ ($^1S_0$), $\varphi(4p)$

χ-basis set $n_i l$	ς_i	C_i	N_i
$3p$	8.2772	−0.2204	687.9
$3p$	5.7561	−0.1811	192.9
$4p$	3.8396	0.3530	47.99
$4p$	2.6913	0.4829	9.699
$4p$	1.8758	0.3079	1.911

$A = 3.82$ (B); $\gamma = 1.416$

$Z = 48$, Cadmium – Cd^+ ($^2S_{1/2}$), $\varphi(5s)$

χ-basis set $n_i l$	ς_i	C_i	N_i
$4s$	5.9977	−0.289	357.1
$4s$	4.0741	−0.0834	62.66
$5s$	3.3311	0.1504	17.78
$5s$	2.1883	0.5485	1.764
$5s$	1.4477	0.4366	0.1818

$A = 2.8$ (C); $\gamma = 1.115$

$Z = 38$, Strontium – Sr^+ ($^2S_{1/2}$), $\varphi(4p)$

χ-basis set $n_i l$	ς_i	C_i	N_i
$3p$	8.5295	−0.2413	764.1
$3p$	6.0050	−0.1957	223.7
$4p$	4.0192	0.4121	58.95
$4p$	2.7693	0.5428	11.03
$4p$	1.9765	0.1899	2.418

$A = 1.39$ (B); $\gamma = 0.900$

$Z = 53$, Iodine – I^+ ($^3P_2$), $\varphi(5p)$

χ-basis set $n_i l$	ς_i	C_i	N_i
$3p$	11.65	0.2860	2279
$4p$	7.3691	−0.2397	902.1
$4p$	5.1336	−0.2002	177.3
$5p$	3.3611	0.3261	18.68
$5p$	2.4710	0.4539	3.440
$5p$	1.7320	0.3598	0.4873

$A = 2.9$ (C); $\gamma = 1.186$

$Z = 47$, Silver – Ag^+ ($^1S_0$), $\varphi(4d)$

χ-basis set $n_i l$	ς_i	C_i	N_i
$3d$	16.52	−0.0757	7731
$3d$	9.3647	−0.2533	1060
$4d$	5.7044	0.3214	285.0
$4d$	3.5287	0.5448	32.82
$4d$	2.0105	0.3180	2.610

$A = 0.7$ (D); $\gamma = 1.257$

$Z = 54$, Xenon – Xe^+ ($^2P^o_{3/2}$), $\varphi(5p)$

χ-basis set $n_i l$	ς_i	C_i	N_i
$3p$	13.25	−0.1174	3572
$3p$	12.23	0.3505	2696
$4p$	7.5294	−0.2548	993.8
$4p$	5.3351	−0.2065	210.9
$5p$	3.4263	0.4312	20.76
$5p$	2.4313	0.4318	3.147
$5p$	1.7727	0.2832	0.5538

$A = 3.2$ (C); $\gamma = 1.249$

Table 4.4 (continued)

$Z = 55$, Caesium – Cs^+ ($^1S_0$), $\varphi(5p)$

χ-basis set $n_i l$	ς_i	C_i	N_i
4*p*	8.0550	−0.2269	1346
4*p*	5.6750	−0.2904	278.4
5*p*	4.4297	0.1671	85.28
5*p*	3.0135	0.6225	10.25
5*p*	1.9471	0.3876	0.9277

$\langle 1/r \rangle = 0.608$; $\langle r \rangle = 2.10$; $\langle r^2 \rangle = 5.01$;
$A = 3.62$ (*C*); $\gamma = 1.358$

$Z = 56$, Barium – Ba^+ ($^2S_{1/2}$), $\varphi(6s)$

χ-basis set $n_i l$	ς_i	C_i	N_i
4*s*	7.7939	0.1468	1161
4*s*	5.8169	0.0644	311.2
5*s*	4.0202	−0.2665	50.02
5*s*	2.7249	−0.1338	5.891
6*s*	1.9612	0.2782	0.3295
6*s*	1.4065	0.5472	0.0380
6*s*	1.0525	0.2924	0.0058

$\langle 1/r \rangle = 0.261$; $\langle r \rangle = 4.69$; $\langle r^2 \rangle = 24.5$;
$A = 1.54$ (*D*); $\gamma = 0.857$

$Z = 71$, Lutetium – Lu^+ ($^1S_0$), $\varphi(6s)$

χ-basis set $n_i l$	ς_i	C_i	N_i
4*s*	11.326	0.1056	6240
4*s*	8.4320	0.0989	1654
5*s*	5.5825	−0.2217	304.3
5*s*	3.7421	−0.1613	33.72
6*s*	2.4780	0.3923	1.507
6*s*	1.6445	0.5975	0.1049
6*s*	1.1677	0.1436	0.0113

$\langle 1/r \rangle = 0.336$; $\langle r \rangle = 3.66$; $\langle r^2 \rangle = 15.0$;
$A = 2.63$ (*B*); $\gamma = 1.011$

$Z = 72$, Hafnium – Hf^+ ($^2D_{3/2}$), $\varphi(6s)$

χ-basis set $n_i l$	ς_i	C_i	N_i
4*s*	11.700	0.0961	7223
4*s*	8.7814	0.1181	1986
5*s*	5.8335	−0.2113	387.6
5*s*	3.9729	−0.1895	46.87
6*s*	2.7515	0.3034	2.977
6*s*	1.8496	0.5950	0.2252
6*s*	1.2954	0.2464	0.0222

$\langle 1/r \rangle = 0.351$; $\langle r \rangle = 3.51$; $\langle r^2 \rangle = 13.8$;
$A = 2.88$ (*C*); $\gamma = 1.046$

$Z = 79$, Gold – Au^+ ($^1S_0$), $\varphi(5d)$

χ-basis set $n_i l$	ς_i	C_i	N_i
4*d*	13.192	0.1792	12400
4*d*	8.9100	0.2154	2120
5*d*	5.6869	−0.4241	337.0
5*d*	3.4795	−0.5414	22.60
5*d*	2.0680	−0.2394	1.292

$\langle 1/r \rangle = 0.856$; $\langle r \rangle = 1.53$; $\langle r^2 \rangle = 2.75$;
$A = 0.56$ (*C*); $\gamma = 1.227$

$Z = 80$, Mercury – Hg^+ ($^2S_{1/2}$), $\varphi(6s)$

χ-basis set $n_i l$	ς_i	C_i	N_i
4*s*	13.504	0.0725	13770
4*s*	10.491	0.1651	4421
5*s*	6.9412	−0.2690	1008
5*s*	4.7650	−0.1123	127.4
6*s*	3.2739	0.3364	9.214
6*s*	2.1067	0.5876	0.5247
6*s*	1.4330	0.2266	0.0429

$\langle 1/r \rangle = 0.413$; $\langle r \rangle = 3.00$; $\langle r^2 \rangle = 10.2$;
$A = 3.71$ (*C*); $\gamma = 1.174$

Table 4.4 (continued)

$Z = 81$, Thallium – Tl^+ ($^1S_0$), $\varphi(6s)$

χ-basis set $n_i l$	ς_i	C_i	N_i
$4s$	13.972	0.0772	16050
$4s$	10.701	0.1813	4834
$5s$	6.9235	−0.3276	994.4
$5s$	4.7041	−0.0633	118.7
$6s$	3.1913	0.4388	7.804
$6s$	2.1003	0.5299	0.5145
$6s$	1.4712	0.1669	0.0509

$\langle 1/r \rangle = 0.443$; $\langle r \rangle = 2.81$; $\langle r^2 \rangle = 8.93$;
$A = 4.30$ (C); $\gamma = 1.225$

$Z = 82$, Lead – Pb^+ ($^2P^o_{1/2}$), $\varphi(6p)$

χ-basis set $n_i l$	ς_i	C_i	N_i
$4p$	12.536	0.1233	9853
$4p$	9.8439	0.0706	3320
$5p$	7.0393	−0.1865	1089
$5p$	5.0087	−0.1776	167.6
$6p$	4.0239	0.1177	35.22
$6p$	2.4688	0.5758	1.471
$6p$	1.5579	0.4654	0.0738

$\langle 1/r \rangle = 0.393$; $\langle r \rangle = 3.14$; $\langle r^2 \rangle = 11.1$;
$A = 1.73$ (C); $\gamma = 1.051$

$Z = 83$, Bismuth – Bi^+ ($^3P_0$), $\varphi(6p)$

χ-basis set $n_i l$	ς_i	C_i	N_i
$4p$	12.745	0.1265	10610
$4p$	10.107	0.0946	3738
$5p$	7.3046	−0.2197	1335
$5p$	5.0862	−0.1785	182.4
$6p$	3.6661	0.2055	19.22
$6p$	2.4650	0.5780	1.457
$6p$	1.6336	0.3679	0.1005

$\langle 1/r \rangle = 0.429$; $\langle r \rangle = 2.89$; $\langle r^2 \rangle = 9.41$;
$A = 2.24$ (C); $\gamma = 1.108$

Table 4.5. Hartree-Fock and asymptotic parameters of valence electron wavefunctions for negative atomic ions [a.u.]

$Z = 3$, Lithium – Li^- (^1S), $\varphi(2s)$

χ-basis set $n_i l$	ς_i	C_i	N_i
1s	2.4747	−0.1003	7.786
1s	4.6921	−0.0110	20.33
2s	0.2676	0.3977	0.0428
2s	0.5340	0.5609	0.2406
2s	1.0113	0.2048	1.188
2s	1.6628	−0.0791	4.117

$A = 1.0$ (D); $\gamma = 0.212$

$Z = 8$, Oxygen – O^- (^2P), $\varphi(2p)$

χ-basis set $n_i l$	ς_i	C_i	N_i
2p	1.7440	0.5275	4.638
2p	3.4297	0.3056	25.16
2p	0.8639	0.2889	0.8011
2p	0.4083	0.0089	0.1230

$A = 0.65$ (D); $\gamma = 0.328$

$Z = 5$, Boron – B^- (^3P), $\varphi(2p)$

χ-basis set $n_i l$	ς_i	C_i	N_i
2p	1.3068	0.1684	2.254
2p	2.0109	0.1587	6.621
2p	0.8673	0.4544	0.8089
2p	0.4205	0.3629	0.1324

$A = 0.39$ (D); $\gamma = 0.148$

$Z = 9$, Fluorine – F^- ($^1S_0$), $\varphi(2p)$

χ-basis set $n_i l$	ς_i	C_i	N_i
2p	2.0754	0.4704	7.165
2p	3.9334	0.3084	35.43
2p	1.4660	0.0988	3.005
2p	0.9568	0.2470	1.034

$A = 0.84$ (C); $\gamma = 0.500$

$Z = 6$, Carbon – C^- (^4S), $\varphi(2p)$

χ-basis set $n_i l$	ς_i	C_i	N_i
2p	1.4705	0.3871	3.028
2p	2.4996	0.2037	11.41
2p	0.8837	0.3691	0.8476
2p	0.5368	0.1583	0.2438
2p	5.7105	0.0134	89.98

$A = 0.74$ (D); $\gamma = 0.306$

$Z = 11$, Sodium – Na^- (^1S), $\varphi(3s)$

χ-basis set $n_i l$	ς_i	C_i	N_i
2s	4.1303	−0.0954	40.03
3s	0.9574	0.5117	0.3620
3s	0.4169	0.6262	0.0197
3s	3.1346	−0.0291	22.99

$A = 1.0$ (D); $\gamma = 0.201$

Table 4.5 (continued)

Z = 13, Aluminium – Al^- (^3P), $\varphi(3p)$

χ-basis set $n_i l$	ς_i	C_i	N_i
3p	1.7365	0.1501	2.909
3p	0.4698	0.4561	0.0300
3p	3.9148	−0.0627	50.05
3p	1.0069	0.5603	0.4319

A = 0.51 (D); γ = 0.192

Z = 16, Sulfur – S^- (^2P), $\varphi(3p)$

χ-basis set $n_i l$	ς_i	C_i	N_i
2p	7.0774	−0.1515	153.9
3p	2.6206	0.2974	12.28
3p	0.8997	0.3330	0.2913
3p	5.4249	−0.1043	156.8
3p	1.6580	0.5236	2.474

A = 0.10 (E); γ = 0.391

Z = 14, Silicon – Si^- (^4S), $\varphi(3p)$

χ-basis set $n_i l$	ς_i	C_i	N_i
2p	6.1438	−0.1147	108.0
3p	2.0551	0.1838	5.246
3p	0.6855	0.3953	0.1125
3p	4.4933	−0.0856	81.08
3p	1.2940	0.5570	1.039

A = 1.1 (D); γ = 0.320

Z = 17, Chlorine – Cl^- (^1S), $\varphi(3p)$

χ-basis set $n_i l$	ς_i	C_i	N_i
2p	7.6297	−0.1622	185.7
3p	2.9265	0.3296	18.08
3p	1.0156	0.3027	0.4452
3p	5.9400	−0.1134	215.4
3p	1.8319	0.5244	3.508

A = 1.34 (C); γ = 0.516

Z = 15, Phosphorus – P^- (^3P), $\varphi(3p)$

χ-basis set $n_i l$	ς_i	C_i	N_i
2p	6.5044	−0.1370	124.6
3p	2.2228	0.3228	6.904
3p	0.7470	0.3101	0.1519
3p	4.9018	−0.0918	109.9
3p	1.3676	0.5174	1.261

A = 0.9 (D); γ = 0.238

Z = 19, Potassium – K^- (^1S), $\varphi(4s)$

χ-basis set $n_i l$	ς_i	C_i	N_i
2s	7.585	0.0464	183.0
3s	3.9935	−0.0762	53.66
3s	2.6383	−0.0923	12.58
4s	0.3520	0.3923	0.0010
4s	1.2652	0.2439	0.3248
4s	0.7045	0.5658	0.0233

A = 0.9 (E); γ = 0.192

Table 4.5 (continued)

Z = 26, Iron – Fe^- (^4S), $\varphi(4s)$

χ-basis set $n_i l$	ς_i	C_i	N_i
3s	4.7306	−0.1188	97.08
4s	1.9328	0.2895	2.186
4s	0.9992	0.5668	0.1123
4s	0.4855	0.3736	0.0044

A = 0.93 (D); γ = 0.171

Z = 29, Copper – Cu^- (^1S), $\varphi(4s)$

χ-basis set $n_i l$	ς_i	C_i	N_i
3s	6.6632	0.2757	322.0
3s	6.3797	0.4153	276.5
4s	2.0688	0.3062	2.969
4s	1.0758	0.5415	0.1565
4s	0.5267	0.3820	0.0063

A = 1.2 (E); γ = 0.301

Z = 27, Cobalt – Co^- (^3F), $\varphi(4s)$

χ-basis set $n_i l$	ς_i	C_i	N_i
3s	5.2702	−0.1156	141.7
4s	2.0219	0.2713	2.678
4s	1.0777	0.5407	0.1578
4s	0.5259	0.4133	0.0062

A = 1.1 (D); γ = 0.257

Z = 34, Selenium – Se^- (^2P), $\varphi(4p)$

χ-basis set $n_i l$	ς_i	C_i	N_i
3p	8.7108	−0.0767	822.5
3p	5.8440	−0.2	203.4
4p	2.6621	0.4699	9.234
4p	1.5616	0.4889	0.8373
4p	0.8938	0.1964	0.0680

A = 1.3 (C); γ = 0.385

Z = 28, Nickel – Ni^- (^2D), $\varphi(4s)$

χ-basis set $n_i l$	ς_i	C_i	N_i
3s	6.0288	−0.1491	226.9
4s	2.198	0.2311	3.899
4s	1.2013	0.5154	0.2572
4s	0.5850	0.4742	0.0101

A = 1.3 (E); γ = 0.309

Z = 35, Bromine – Br^- (^1S), $\varphi(4p)$

χ-basis set $n_i l$	ς_i	C_i	N_i
3p	8.6078	−0.0915	789.0
3p	5.8779	−0.2293	207.6
4p	2.8155	0.5170	11.88
4p	1.6858	0.4350	1.182
4p	1.0414	0.1889	0.1353

A = 1.49 (B); γ = 0.498

Table 4.5 (continued)

Z = 37, Rubidium – Rb^- (^1S), φ (5s)

χ-basis set $n_i l$	ς_i	C_i	N_i
4s	3.8861	−0.1106	50.66
4s	2.5250	−0.0697	7.278
5s	1.5190	0.2348	0.2368
5s	0.9375	0.3586	0.0167
5s	0.5629	0.5857	0.0010

A = 0.8 (E); γ = 0.189

Z = 51, Antimony – Sb^- (^3P), φ (5p)

χ-basis set $n_i l$	ς_i	C_i	N_i
3p	10.63	0.1405	1650
4p	6.7509	−0.2009	608.1
4p	4.5451	−0.1007	102.5
5p	2.6459	0.4433	5.011
5p	1.5355	0.5471	0.2512
5p	0.8545	0.2089	0.0100

A = 1.2 (D); γ = 0.278

Z = 47, Silver – Ag^- (^1S), φ (5s)

χ-basis set $n_i l$	ς_i	C_i	N_i
4s	5.8596	−0.1751	321.6
4s	3.9820	−0.0186	56.53
5s	2.2709	0.3327	2.162
5s	1.2026	0.5534	0.0655
5s	0.5962	0.3754	0.0014

A = 1.3 (E); γ = 0.309

Z = 52, Tellurium – Te^- (^2P), φ (5p)

χ-basis set $n_i l$	ς_i	C_i	N_i
3p	11.03	0.1707	1881
4p	6.9706	−0.2238	702.4
4p	4.7572	−0.1156	125.9
5p	2.8219	0.4836	7.141
5p	1.6523	0.5322	0.3761
5p	0.9349	0.1790	0.0164

A = 1.5 (D); γ = 0.374

Z = 50, Tin – Sn^- (^4S), φ (5p)

χ-basis set $n_i l$	ς_i	C_i	N_i
3p	10.33	0.1177	1491
4p	6.4872	−0.1781	508.3
4p	4.2862	−0.0809	78.74
5p	2.4685	0.3899	3.421
5p	1.4281	0.5598	0.1686
5p	0.8017	0.2471	0.0070

A = 1.3 (D); γ = 0.303

Z = 53, Iodine – I^- (^1S), φ (5p)

χ-basis set $n_i l$	ς_i	C_i	N_i
3p	11.69	0.2207	2306
4p	7.1112	−0.2449	768.5
4p	4.9399	−0.1226	149.1
5p	3.0017	0.5122	10.03
5p	1.7759	0.5207	0.5592
5p	1.0267	0.1574	0.0275

A = 1.9 (D); γ = 0.475

5. Energetics of Neutral Atoms

Numerical data are presented for the ionization potentials of neutral atoms, the fine- and hyperfine-structure splitting of low-lying energy levels, the level and transition isotope shifts in atoms and the quantum defects of atomic Rydberg states. Values of the atomic scalar polarizability and the magnetic susceptibility are also given.

5.1 Ionization Potentials of Atoms

Table 5.1 gives the values of the ionization potentials (IP) of neutral atoms, i.e. the minimum energy "spent" in the transition of a valence electron into the continuous spectrum. The first ionization limit for a series of optical transitions with respect to the ground level of an atom is also presented. These values can be regarded as ionization potentials too, but expressed in the units cm^{-1} (conversion factor: $1\,eV = 8065.48\ cm^{-1}$). In a separate column of Table 5.1 the configuration of the valence electron shell and the electron term of the ground state of the atom are specified.

The experimental values of optical ionization potentials are determined either by assuming an appropriate function, such as the Ritz formula, for the energy levels of a long Rydberg series and extrapolating to the limit, or by considering the levels whose binding energies are so close to the hydrogenic value that the difference can be expressed entirely in terms of a polarization of the atomic core [5.1.1]. The data derived from optical spectra are the most accurate and reliable.

In addition, part of the numerical data was obtained by the methods of photoionization-threshold measurement, photoelectron spectroscopy and laser spectroscopy of atomic Rydberg states, electron-impact appearance potential measurement, surface ionization, and so on. Finally, for one-electron systems we used the theoretical values of Erickson's ionization potentials [5.1.2]; many QED effects were taken into account in the calculation of these values.

The basic information on ionization potentials of atomic species is contained in [5.1.3–9], we also made use of recent journal publications. The values listed in Table 5.1 are truncated to the point where the uncertainty is at most ± 1 or 2 in the last significant figure quoted.

References

5.1.1 B.Edlèn: "Atomic Spectra", in *Spectroscopy I*, ed. by S.Flügge, Encyclopedia of Physics, Vol. 27 (Springer, Berlin, Göttingen, Heidelberg, New York 1964) pp. 79–220

5.1.2 G.W.Erickson: J. Phys. Chem. Ref. Data **6**, 831 (1977)

5.1.3 C.E.Moore: *Ionization Potentials and Ionization Limits Derived From Optical Spectra*, Nat. Stand. Ref. Data Ser. Nat. Bur. Stand. **34** (1970)

5.1.4 W.C.Martin, L.Hagan, J.Reader, J.Sugar: "Ground levels and IP's for lanthanide and actinide atoms and ions", J. Phys. Chem. Ref. Data **3**, 771 (1974)

5.1.5 W.C.Martin, R.Zalubas, L.Hagan: *Atomic Energy Levels – The Rare-Earth Elements*, Nat. Stand. Ref. Data Ser. Nat. Bur. Stand. **60** (1978)

5.1.6 R.L.Kelly, D.E.Harrison: "IP's, experimental and theoretical, of the elements hydrogen to krypton", At. Data **3**, 177 (1971)

5.1.7 E.G.Rauh, R.J.Ackermann: "The first ionization potentials of the transition metals", J. Chem. Phys. **70**, 1004 (1979)

5.1.8 E.F.Worden, J.G.Conway: "Multistep Laser Photoionization of the Lanthanides and Actinides", in *Lanthanide and Actinide Chemistry and Spectroscopy*, American Chem. Soc. Symposium Series 131, ed. by N.M.Edelstein (American Chemical Society, Washington, D.C. 1980) Chap. 19, pp. 381–425

5.1.9 H.Odabasi: "Regularities in ionization potentials", Phys. Scr. **19**, 313 (1979)

Table 5.1. Ionization potentials of neutral atoms

Atomic number Z	Element X	XI Valence electron configuration and electronic term	Ionization limit [cm^{-1}] (*upper line*); Ionization potential IP [eV] (*lower line*)
1	2	3	4
1	H	$1s - {}^2S_{1/2}$	109678.774 13.5985
	D	$1s - {}^2S_{1/2}$	109708.617 13.6022
	T	$1s - {}^2S_{1/2}$	109718.546 13.6035
2	^{3}He	$1s^2 - {}^1S_0$	198300.3 24.5863
	^{4}He	$1s^2 - {}^1S_0$	198310.77 24.5876
3	Li	$2s - {}^2S_{1/2}$	43487.15 5.39176
4	Be	$2s^2 - {}^1S_0$	75192.5 9.3228
5	B	$2p - {}^2P^o_{1/2}$	66928.1 8.2981
6	C	$2p^2 - {}^3P_0$	90820.4 11.260
7	N	$2p^3 - {}^4S^o_{3/2}$	117225.7 14.534

Table 5.1 (continued)

Atomic number Z	Element X	XI Valence electron configuration and electronic term	Ionization limit [cm^{-1}] (*upper line*); Ionization potential IP [eV] (*lower line*)
1	2	3	4
8	O	$2p^4 - {}^3P_2$	109837.0 13.618
9	F	$2p^5 - {}^2P^o_{3/2}$	140524.5 17.423
10	Ne	$2p^6 - {}^1S_0$	173929.7 21.565
11	Na	$3s - {}^2S_{1/2}$	41449.4 5.13912
12	Mg	$3s^2 - {}^1S_0$	61671.0 7.6463
13	Al	$3p - {}^2P^o_{1/2}$	48278.42 5.9858
14	Si	$3p^2 - {}^3P_0$	65747.8 8.1517
15	P	$3p^3 - {}^4S^o_{3/2}$	84580.8 10.4868
16	S	$3p^4 - {}^3P_2$	83559.3 10.36012
17	Cl	$3p^5 - {}^2P^o_{3/2}$	104591 12.968
18	Ar	$3p^6 - {}^1S_0$	127109.8 15.760
19	K	$4s - {}^2S_{1/2}$	35009.81 4.34070
20	Ca	$4s^2 - {}^1S_0$	49306.0 6.1132
21	Sc	$3d\,4s^2 - {}^2D_{3/2}$	52922 6.5615
22	Ti	$3d^2\,4s^2 - {}^3F_2$	55000 6.82
23	V	$3d^3\,4s^2 - {}^4F_{3/2}$	54360 6.74
24	Cr	$3d^5\,4s - {}^7S_3$	54570 6.766
25	Mn	$3d^5\,4s^2 - {}^6S_{5/2}$	59959.4 7.43408
26	Fe	$3d^6\,4s^2 - {}^5D_4$	63740 7.9024
27	Co	$3d^7\,4s^2 - {}^4F_{9/2}$	63400 7.86

Table 5.1 (continued)

Atomic number Z	Element X	XI Valence electron configuration and electronic term	Ionization limit [cm^{-1}] (*upper line*); Ionization potential IP [eV] (*lower line*)
1	2	3	4
28	Ni	$3d^8 4s^2 - {}^3F_4$	61600 7.637
29	Cu	$4s - {}^2S_{1/2}$	62317.4 7.7264
30	Zn	$4s^2 - {}^1S_0$	75769.3 9.3943
31	Ga	$4p - {}^2P^o_{1/2}$	48387.63 5.99935
32	Ge	$4p^2 - {}^3P_0$	63713.2 7.8995
33	As	$4p^3 - {}^4S^o_{3/2}$	78950 9.789
34	Se	$4p^4 - {}^3P_2$	78658.2 9.752
35	Br	$4p^5 - {}^2P^o_{3/2}$	95284.8 11.814
36	Kr	$4p^6 - {}^1S_0$	112914.5 13.9997
37	Rb	$5s - {}^2S_{1/2}$	33690.88 4.17717
38	Sr	$5s^2 - {}^1S_0$	45932.1 5.69490
39	Y	$4d\,5s^2 - {}^2D_{3/2}$	50144 6.217
40	Zr	$4d^2 5s^2 - {}^3F_2$	55100 6.837
41	Nb	$4d^4 5s - {}^6D_{1/2}$	55500 6.88
42	Mo	$4d^5 5s - {}^7S_3$	57300 7.099
43	Tc	$4d^5 5s^2 - {}^6S_{5/2}$	58700 7.28
44	Ru	$4d^7 5s - {}^5F_5$	59410 7.366
45	Rh	$4d^8 5s - {}^4F_{9/2}$	60200 7.46
46	Pd	$4d^{10} - {}^1S_0$	67236 8.336
47	Ag	$5s - {}^2S_{1/2}$	61106.6 7.5763

Table 5.1 (continued)

Atomic number Z	Element X	XI Valence electron configuration and electronic term	Ionization limit [cm^{-1}] (*upper line*); Ionization potential IP [eV] (*lower line*)
1	2	3	4
48	Cd	$5s^2 - {}^1S_0$	72540.1 8.9939
49	In	$5s^2 5p - {}^2P^{\circ}_{1/2}$	46670.11 5.78640
50	Sn	$5p^2 - {}^3P_0$	59232.7 7.3440
51	Sb	$5p^3 - {}^4S^{\circ}_{3/2}$	69430 8.609
52	Te	$5p^4 - {}^3P_2$	72670 9.010
53	I	$5p^5 - {}^2P^{\circ}_{3/2}$	84295.0 10.451
54	Xe	$5p^6 - {}^1S_0$	97834,4 12.130
55	Cs	$6s - {}^2S_{1/2}$	31406.47 3.89394
56	Ba	$6s^2 - {}^1S_0$	42034.90 5.21171
57	La	$5d\,6s^2 - {}^2D_{3/2}$	44980 5.577
58	Ce	$4f\,5d\,6s^2 - {}^1G^{\circ}_4$	44670 5.539
59	Pr	$4f^3 6s^2 - {}^4I^{\circ}_{9/2}$	44100 5.47
60	Nd	$4f^4 6s^2 - {}^5I_4$	44560 5.525
61	Pm	$4f^5 6s^2 - {}^6H^{\circ}_{5/2}$	45000 5.58
62	Sm	$4f^6 6s^2 - {}^7F_0$	45520 5.644
63	Eu	$4f^7 6s^2 - {}^8S^{\circ}_{7/2}$	45735 5.670
64	Gd	$4f^7 5d\,6s^2 - {}^9D^{\circ}_2$	49603 6.150
65	Tb	$4f^9 6s^2 - {}^6H^{\circ}_{15/2}$	47300 5.864
66	Dy	$4f^{10} 6s^2 - {}^5I_8$	47900 5.939
67	Ho	$4f^{11} 6s^2 - {}^4I^{\circ}_{15/2}$	48570 6.022

Table 5.1 (continued)

Atomic number Z	Element X	XI Valence electron configuration and electronic term	Ionization limit [cm^{-1}] (*upper line*); Ionization potential IP [eV] (*lower line*)
1	2	3	4
68	Er	$4f^{12}6s^2 - {}^3H_6$	49260 6.108
69	Tm	$4f^{13}6s^2 - {}^2F^o_{7/2}$	49880 6.184
70	Yb	$4f^{14}6s^2 - {}^1S_0$	50441 6.254
71	Lu	$4f^{14}5d6s^2 - {}^2D_{3/2}$	43762.4 5.426
72	Hf	$5d^2 6s^2 - {}^3F_2$	55600 6.8
73	Ta	$5d^3 6s^2 - {}^4F_{3/2}$	63600 7.89
74	W	$5d^4 6s^2 - {}^5D_0$	64000 7.98
75	Re	$5d^5 6s^2 - {}^6S_{5/2}$	64000 7.88
76	Os	$5d^6 6s^2 - {}^5D_4$	70450 8.73
77	Ir	$5d^7 6s^2 - {}^4F_{9/2}$	73000 9.05
78	Pt	$5d^9 6s - {}^3D_3$	72300 8.96
79	Au	$5d^{10} 6s - {}^2S_{1/2}$	74409.0 9.2256
80	Hg	$6s^2 - {}^1S_0$	84184.1 10.4376
81	Tl	$6s^2 6p - {}^2P^o_{1/2}$	49266.7 6.1083
82	Pb	$6p^2 - {}^3P_0$	59819.6 7.4167
83	Bi	$6p^3 - {}^4S^o_{3/2}$	58762 7.2856
84	Po	$6p^4 - {}^3P_2$	67885.3 8.417
85	At	$6p^5 - {}^2P^o_{3/2}$	– 9.0
86	Rn	$6p^6 - {}^1S_0$	86692 10.75
87	Fr	$7s - {}^2S_{1/2}$	– 4.0

Table 5.1 (continued)

Atomic number Z	Element X	XI Valence electron configuration and electronic term	Ionization limit [cm^{-1}] (*upper line*); Ionization potential IP [eV] (*lower line*)
1	2	3	4
88	Ra	$7s^2 - {}^1S_0$	42573.4 5.2785
89	Ac	$6d7s^2 - {}^2D_{3/2}$	42000 5.2
90	Th	$6d^27s^2 - {}^3F_2$	49000 6.1
91	Pa	$5f^2({}^3H_4)6d7s^2(4, 3/2)_{11/2}$	47000 6.0
92	U	$5f^3({}^4I^{o}_{9/2})6d7s^2(9/2, 3/2)^{o}_{6}$	49960 6.194
93	Np	$5f^4({}^5I_4)6d7s^2(4, 3/2)_{11/2}$	50540 6.266
94	Pu	$5f^67s^2 - {}^7F_0$	49000 6.06
95	Am	$5f^77s^2 - {}^8S^{o}_{7/2}$	48300 6.0
96	Cm	$5f^7({}^8S^{o}_{7/2})6d7s^2(7/2, 3/2)^{o}_{2}$	48600 6.02
97	Bk	$5f^97s^2 - {}^6H^{o}_{15/2}$	50200 6.23
98	Cf	$5f^{10}7s^2 - {}^5I_8$	50800 6.30
99	Es	$5f^{11}7s^2 - {}^4I^{o}_{15/2}$	51800 6.42
100	Fm	$5f^{12}7s^2 - {}^3H_6$	52000 6.5
101	Md	$\{5f^{13}7s^2 - {}^2F^{o}_{7/2}\}$	53000 6.6
102	No	$\{5f^{14}7s^2 - {}^1S_0\}$	54000 6.6

5.2 Quantum Defects of Atomic Rydberg States

Table 5.2 gives numerical values of the quantum defects of highly excited (or Rydberg) levels of atomic particles. The magnitude of the quantum defect δ_l characterizes the contribution of the non-Coulomb interaction between an excited electron and the atomic core (mainly by means of core polarization and the exchange interaction between the excited electron and the inner-shell electrons) in a range of electron coordinates scaled to the size of the core. This interaction results in the negative shift of atomic energy levels towards the boundary of the continuous spectrum. The position of the electron energy level $T_{n,l}$ for an atomic Rydberg state is determined according to the modified Rydberg-Ritz formula [5.2.1–3]

$$T_{n,l} = T_\infty - \frac{R_M Z_c^2}{(n^*)^2} = T_\infty - \frac{R_M Z_c^2}{(n-\delta_l)^2} ,$$

where T_∞ is the ionization limit for a given electron configuration (n, l), R_M is the Rydberg constant for the nuclide with mass M, equal to $R_\infty/(1 + m_e/M)$, Z_c is the charge of the atomic core and $n^* = n - \delta_l$ is the effective principal quantum number of the excited electron. In the limit $n \to \infty$ the quantum defect of the level δ_l does not depend on the principal quantum number of the electron, but on its orbital angular momentum and on the total angular momentum of the atomic particle, J.

Table 5.2 presents limiting ($n \to \infty$) values of the quantum defect δ_l, derived by comparing the above expression for $T_{n,l}$ to the experimental energy-level values of highly excited atomic states, known from the processing of optical spectra. The information concerning the parameters of Rydberg series in optical spectra of atomic particles was taken from reviews [5.2.4–9] and numerous recent publications in journals. The last column of Table 5.2 shows the range of values investigated for the principal quantum number Δn of the excited electron.

References

5.2.1 M.J.Seaton: Proc. Phys. Soc. (London) **88**, 801 (1966); Comments At. Mol. Phys. **D 2**, 37 (1970); Rep. Prog. Phys. **46**, 167 (1983)
5.2.2 K.T.Lu, U.Fano: Phys. Rev. **A 2**, 81 (1970)
5.2.3 U.Fano: J. Opt. Soc. Am. **65**, 979 (1975)
5.2.4 W.C.Martin: J. Opt. Soc. Am. **70**, 784 (1980)
5.2.5 S.A.Edelstein, T.F.Gallagher: "Rydberg atoms", Adv. At. Mol. Phys. **14**, 365 (1978)
5.2.6 H.Walther, K.W.Rothe (eds.): *Laser Spectroscopy IV*, Proc. 4th Int. Conf., Rottach-Egern, Fed. Rep. Germany, June 11–15, 1979; Springer Ser. Opt. Sci., Vol. 21, Part III – Rydberg States (Springer, Berlin, Heidelberg, New York 1979) p. 235
5.2.7 D.Kleppner: "The spectroscopy of highly excited atoms", in *Progress in Atomic Spectroscopy*, Part B, ed. by W.Hanle, H.Kleinpoppen (Plenum, New York 1979) Chap. 16, p. 713
5.2.8 A.Lindgård, S.E.Nielsen: At. Data Nucl. Data Tables **19**, 533 (1977)
5.2.9 C.-J.Lorenzen, K.Niemax: Phys. Scripta **27**, 300 (1983)

Table 5.2. Quantum defects δ_l of atomic Rydberg states

Atomic number Z	Element (normal electron configuration and term)	Valence electron shell and electronic term	Quantum defect δ_l (accuracy)	Investigated range of principal quantum number Δn
1	2	3	4	5
2	He	$1sns-{}^3S_1$	0.297(*A*)	8–17
	$(1s^2-{}^1S_0)$	$1sns-{}^1S_0$	0.139(*A*)	8–15
		$1snp-{}^3P^o_{2,1,0}$	0.063(*C*)	8–22
		$1snp-{}^1P^o_1$	0.0123(*A*)	8–20
		$1snd-{}^{3,1}D$	0.0027(*D*)	8–21
3	Li	$ns-{}^2S_{1/2}$	0.400(*A*)	5–12
	$(1s^2 2s-{}^2S_{1/2})$	$np-{}^2P^o_{1/2,3/2}$	0.047(*A*)	5–12
		$nd-{}^2D_{3/2,5/2}$	0.0021(*A*)	5–12
		$nf-{}^2F^o_{5/2,7/2}$	−0.00008(*C*)	5–8
4	Be^+	$ns-{}^2S_{1/2}$	0.259(*A*)	5–12
	$(1s^2 2s-{}^2S_{1/2})$	$np-{}^2P^o_{1/2,3/2}$	0.050(*A*)	5–12
		$nd-{}^2D_{3/2,5/2}$	0.002(*A*)	5–12
5	B	$nd-{}^2D_{3/2,5/2}$	0.042(*D*)	7–30
	$(2s^2 2p-{}^2P^o_{1/2})$			
6	C	$2pnd-{}^3F^o_3$	0.06(*D*)	7–22
	$(2p^2-{}^3P_0)$	$2pnd-{}^1F^o_3$	0.04(*D*)	10–29
11	Na	$ns-{}^2S_{1/2}$	1.348(*A*)	30–40
	$(2p^6 3s-{}^2S_{1/2})$	$np-{}^2P^o_{1/2}$	0.855(*A*)	20–35
		$np-{}^2P^o_{3/2}$	0.855(*A*)	20–35
		$nd-{}^2D_{3/2,5/2}$	0.0155(*A*)	10–20
		$nf-{}^2F^o_{5/2,7/2}$	0.00145(*A*)	10–20
		$ng-{}^2G_{7/2,9/2}$	0.00044(*C*)	10–20
12	Mg	$3snp-{}^1P^o_1$	1.03(*B*)	5–40
	$(3s^2-{}^1S_0)$	$3snd-{}^1D_2$	0.57(*C*)	7–25
		$3snf-{}^3F^o,\ {}^1F^o$	0.049(*C*)	7–15
	Mg^+	$ns-{}^2S_{1/2}$	1.071(*A*)	5–10
	$(3s-{}^2S_{1/2})$	$np-{}^2P^o_{1/2,3/2}$	0.700(*A*)	5–10
		$nd-{}^2D_{3/2,5/2}$	0.043(*C*)	5–9
		$nf-{}^2F^o_{5/2,7/2}$	0.0030(*C*)	5–10
13	Al	$nd-{}^2D$	1.0(*D*)	8–35
	$(3s^2 3p-{}^2P^o_{1/2})$			
19	K	$ns-{}^2S_{1/2}$	2.180(*A*)	6–55
	$(3p^6 4s-{}^2S_{1/2})$	$np-{}^2P^o_{1/2}$	1.714(*A*)	10–20
		$np-{}^2P^o_{3/2}$	1.711(*A*)	10–20
		$nd-{}^2D_{3/2,5/2}$	0.277(*A*)	10–50
		$nf-{}^2F^o_{5/2,7/2}$	0.010(*C*)	7–14
20	Ca	$4sns-{}^3S_1$	2.46(*A*)	6–18
	$(4s^2-{}^1S_0)$	$4sns-{}^1S_0$	2.34(*A*)	7–30
		$4snp-{}^3P^o_2$	1.96(*B*)	6–60
		$4snp-{}^1P^o_1$	1.85(*B*)	10–80
		$4snd-{}^3D_3$	0.75(*C*)	10–17
		$4snf-{}^3F^o_4$	0.095(*C*)	6–13
		$4snf-{}^1F^o_3$	0.09(*D*)	7–28

Table 5.2 (continued)

Atomic number Z	Element (normal electron configuration and term)	Valence electron shell and electronic term	Quantum defect δ_l (accuracy)	Investigated range of principal quantum number Δn
1	2	3	4	5
		$3dnp - {}^1P_1^o$	$1.70(B)$	4–61
		$3dnp - {}^3P_1^o$	$2.2(C)$	4–33
		$3dnp - {}^3D_1^o$	$1.75(C)$	4–58
		$3dnf - {}^1P_1^o$	$0.20(D)$	4–39
	Ca^+ ($4s - {}^2S_{1/2}$)	$ns - {}^2S_{1/2}$	$1.805(A)$	5–12
		$np - {}^2P^o_{1/2,3/2}$	$1.437(A)$	5–12
		$nd - {}^2D_{3/2,5/2}$	$0.627(A)$	5–12
		$nf - {}^2F^o_{5/2,7/2}$	$0.028(A)$	5–12
25	Mn ($3d^5 4s^2 - {}^6S_{5/2}$)	$4s\,({}^7S)\,np - {}^8P^o$	$2.139(A)$	10–23
		$4s\,({}^7S)\,np - {}^6P^o$	$2.030(A)$	15–42
		$4s\,({}^5S)\,np - {}^6P^o$	$2.135(A)$	13–35
29	Cu ($3d^{10} 4s - {}^2S_{1/2}$)	$ns - {}^2S_{1/2}$	$2.60(A)$	5–10
30	Zn ($4s^2 - {}^1S_0$)	$4snp - {}^1P_1^o$	$2.10(A)$	10–66
		$4snd - {}^1D_2$	$1.23(A)$	10–20
		$4snp - {}^3P_1^o$	$2.20(B)$	6–12
37	Rb ($4p^6 5s - {}^2S_{1/2}$)	$ns - {}^2S_{1/2}$	$3.131(A)$	9–116
		$np - {}^2P^o_{1/2}$	$2.655(A)$	13–68
		$np - {}^2P^o_{3/2}$	$2.641(A)$	13–68
		$nd - {}^2D_{3/2,5/2}$	$1.347(A)$	7–124
		$nf - {}^2F^o_{5/2,7/2}$	$0.0163(A)$	10–12
38	Sr ($4p^6 5s^2 - {}^1S_0$)	$5sns - {}^1S_0$	$3.269(A)$	10–80
		$5snp - {}^1P_1^o$	$2.72(A)$	15–33
		$5snd - {}^1D_2$	$2.37(B)$	30–60
		$5snd - {}^3D_2$	$2.63(B)$	25–37
	Sr^+ ($4p^6 5s - {}^2S_{1/2}$)	$ns - {}^2S_{1/2}$	$2.710(A)$	5–12
		$np - {}^2P^o_{1/2,3/2}$	$2.320(A)$	5–12
		$nd - {}^2D_{3/2,5/2}$	$1.455(A)$	4–12
		$nf - {}^2F^o_{5/2,7/2}$	$0.062(A)$	4–12
47	Ag ($4d^{10} 5s - {}^2S_{1/2}$)	$np - {}^2P^o_{1/2,3/2}$	$3.03(A)$	10–60
48	Cd ($5s^2 - {}^1S_0$)	$5snp - {}^1P_1^o$	$3.07(A)$	5–50
		$5snd - {}^1D$	$2.23(A)$	5–26
		$5snp - {}^3P_1^o$	$3.14(A)$	5–31
55	Cs ($5p^6 6s - {}^2S_{1/2}$)	$ns - {}^2S_{1/2}$	$4.049(A)$	10–54
		$np - {}^2P^o_{1/2}$	$3.591(A)$	20–50
		$np - {}^2P^o_{3/2}$	$3.559(A)$	20–50
		$nd - {}^2D_{3/2}$	$2.475(A)$	10–60
		$nd - {}^2D_{5/2}$	$2.466(A)$	10–60
		$nf - {}^2F^o_{5/2}$	$0.0334(A)$	20–100
		$nf - {}^2F^o_{7/2}$	$0.0335(A)$	20–100
		$ng - {}^2G_{7/2,9/2}$	$0.007(C)$	25–35

Table 5.2 (continued)

Atomic number Z	Element (normal electron configuration and term)	Valence electron shell and electronic term	Quantum defect δ_l (accuracy)	Investigated range of principal quantum number Δn
1	2	3	4	5
56	Ba	$6sns - {}^1S_0$	4.20(*A*)	8–60
	$(6s^2 - {}^1S_0)$	$6snd - {}^3D_2$	2.80(*A*)	15–30
		$6snd - {}^1D_2$	2.70(*A*)	15–80
	Ba^+	$ns - {}^2S_{1/2}$	3.583(*A*)	7–12
	$(6s - {}^2S_{1/2})$	$np - {}^2P^o_{1/2,3/2}$	3.180(*A*)	6–12
		$nd - {}^2D_{3/2,5/2}$	2.380(*A*)	5–12
		$nf - {}^2F^o_{5/2,7/2}$	0.845(*A*)	5–12
70	Yb	$6sns - {}^1S_0$	4.28(*B*)	15–65
	$(4f^{14}6s^2 - {}^1S_0)$	$6snd - {}^1D_2$	2.70(*B*)	10–80
		$6snd - {}^3D_2$	2.80(*B*)	10–60
79	Au $(5d^{10}6s - {}^2S_{1/2})$	$np - {}^2P^o_{1/2}$	4.03(*A*)	10–42
82	Pb	$6pnp - [1/2, 3/2]_2$	4.21(*A*)	7–48
	$(6p^2 - {}^3P_0)$	$6pnp - [1/2, 3/2]_1$	4.23(*A*)	7–20
		$6pnp - [1/2, 1/2]_0$	4.30(*A*)	7–23
		$6pnf - 1/2\,[5/2]_2$	1.02(*B*)	7–26
83	Bi	$6p^2\,({}^3P_0)\,ns - {}^4P_{1/2}$	4.89(*A*)	7–18
	$(6p^3 - {}^4S^o_{3/2})$	$6p^2nd - {}^2D_{3/2}$	3.20(*B*)	6–42
		$6p^2nd - {}^2D_{5/2}$	3.23(*B*)	6–20
88	Ra $(7s^2 - {}^1S_0)$	$7snp - {}^1P^o_1$	4.63(*A*)	13–52
	Ra^+	$ns - {}^2S_{1/2}$	4.585(*A*)	8–12
	$(7s - {}^2S_{1/2})$	$np - {}^2P^o_{1/2,3/2}$	4.135(*A*)	7–12
		$nd - {}^2D_{3/2,5/2}$	3.270(*A*)	6–12
		$nf - {}^2F^o_{5/2,7/2}$	1.845(*A*)	6–12

5.3 Fine-Structure Splitting of Atomic Energy Levels

The interaction between the orbital and spin momenta of the electrons (called spin-orbit interaction) causes the dependence of the atomic energy levels (in the framework of the *LS*-coupling scheme) on the orbital ($\boldsymbol{L}$) and spin ($\boldsymbol{S}$) momenta of the atom, as well as on the total angular momentum of an atom $\boldsymbol{J} = \boldsymbol{L} + \boldsymbol{S}$, where J can have the values $L + S,\ L + S - 1, \ldots,$ $|L - S|$. The energy difference between adjacent components of the split multiplet term with given values of L and S is [5.3.1]

$$E_J - E_{J-1} = AJ\ ,$$

where A is the fine-structure-splitting constant, which depends on the electron configuration and the momenta L and S; the lower index to E is the value of the total atomic angular momentum.

This relationship is called the Landé interval rule. The constant A can be either positive or negative and the multiplet is called normal (minimum value $J = |L - S|$ corresponds to the lowest component) or inverted (maximum value $J = L + S$ corresponds to the lowest component), respectively.

The measured values of the splitting energy $\Delta E_{J,J-1}$ [cm^{-1}] for adjacent components of low-lying atomic multiplets are presented in Figs. 7.1–43.

In the case of highly excited atomic levels, the energy intervals between adjacent components of the split term with fixed values of L and S can be approximated by the expression

$$E_J - E_{J-1} = K/n^{*3} ,$$

where $n^* = n - \delta_l$ is the effective principal quantum number of the electron. Values of the coefficient K for some Rydberg states of alkali atoms are included in Table 5.3 [5.3.2–9].

The errors in measuring the fine-structure splitting of the atomic energy levels were allowed for in rounding off the significant figures, so that they are within the range ± 1 for the last digit given.

Table 5.3. Parameters of fine-structure doublet splitting of high-lying Rydberg states in alkalis: $\Delta_{fs} = K/n^{*3}$, $n^* = n - \delta_l$, δ_l is the quantum defect, Δn is the range of electron principal quantum number observed in fs experiments

Fine-structure splitting parameters		Li	Na	K	Rb	Cs
$^2P^o_{1/2,3/2}$	K[10^6 MHz]	–	5.37(A)	20.4(B)	86.0(A)	214(A)
	K[cm^{-1}]	–	179(A)	680(B)	2870(A)	7140(A)
	δ_l	0.047	0.855	1.712	2.647	3.5699
	Δn	7–10	10–40	9–21	13–68	10–80
$^2D_{3/2,5/2}$	K[10^6 MHz]	0.029(B)	−0.092(B)	−1.17(B)	10.6(B)	60.1(A)
	K[cm^{-1}]	0.97(B)	−3.07(B)	−39(B)	350(B)	2006(A)
	δ_l	0.002	0.0155	0.277	1.347	2.471
	Δn	7–10	7–30	10–36	10–40	10–50
$^2F^o_{5/2,7/2}$	K[10^6 MHz]	0.0144(B)	0.014(D)	–	–	−0.975(C)
	K[cm^{-1}]	0.48(B)	0.47(D)	–	–	−32.5(C)
	δ_l	−0.00008	0.00145	0.010	0.016	0.0335
	Δn	7–10	11–17	7–14	7–12	10–30
$^2G_{7/2,9/2}$	K[10^6 MHz]	0.009(C)	–	–	–	–
	K[cm^{-1}]	0.3(C)	–	–	–	–
	δ_l	~0	0.00044	–	–	0.0070
	Δn	8–9	10–20	–	–	25–35

References

5.3.1 E.U.Condon, G.H.Shortley: *The Theory of Atomic Spectra*, 4th ed. (Cambridge University Press, Cambridge 1957)
5.3.2 L.R.Pendrill: Phys. Scripta **27**, 371 (1983)
5.3.3 T.F.Gallagher, R.M.Hill, S.A.Edelstein: Phys. Rev. **A 14**, 744 (1976) (Na – n^2F, n^2G)
5.3.4 W.E.Cooke, T.F.Gallagher, R.M.Hill, S.A.Edelstein: Phys. Rev. **A 16**, 1141 (1977) (Li – n^2F, n^2G)
5.3.5 S.Liberman, J.Pinard: Phys. Rev. **A 20**, 507 (1979) (Rb – n^2P)
5.3.6 T.N.Chang, F.Larijani: J. Phys. **B 13**, 1307 (1980) (Li, Na, K, Rb, Cs – n^2D)
5.3.7 D.C.Thompson, M.S.O'Sullivan, B.P.Stoicheff, Gen-Xing Xu: Can. J. Phys. **61**, 949 (1983) (K – n^2D)
5.3.8 M.S.O'Sullivan, B.P.Stoicheff: Can. J. Phys. **61**, 940 (1983) (Cs – n^2D)
5.3.9 P.Goy, J.M.Raimond, G.Vitrant, S.Haroche: Phys. Rev. **A 26**, 2733 (1982) (Cs – n^2P, n^2D, n^2F)

5.4 Hyperfine Structure of Atomic Energy Levels

The positions of the atomic energy levels and the structure of atomic spectra are also influenced by the interaction of the electrons with the nucleus. Certain properties of a nucleus, such as the finite mass of the nucleons, the spin and the spatial distribution of charges, distinguish it from a fixed point charge and give rise to additional terms in the atomic Hamiltonian, responsible for particular interactions. Taking into account the interaction of the electrons with the angular momentum (spin) of a nucleus leads to the splitting of the atomic energy levels into a number of hyperfine components, each component being characterized by a set of four quantum numbers $\{J, I, F, m_F\}$, where J is the total angular momentum of the electrons, I is the spin of a nucleus, $\boldsymbol{F} = \boldsymbol{J} + \boldsymbol{I}$ is the total angular momentum of the atom, and m_F is the projection of the total angular momentum $\boldsymbol{F}$ on the quantization axis.

The interaction of the electrons with the nuclear multipole moments of the lowest order contributes much to the hyperfine splitting of the atomic energy levels. To a first approximation, I and J can be regarded as being conserved; the total energy of an atomic level is then presented as a sum [5.4.1, 2]

$$E_F = E_J + E_{M1} + E_{E2} = E_J + \frac{h}{2} AK + \frac{3h}{8} B \frac{K(K+1) - \frac{4}{3} I(I+1) J(J+1)}{I(2I-1) J(2J-1)}, \quad I, J \geqslant 1,$$

where E_J is the energy of the level unperturbed by the interaction of the electrons with the nuclear moments, E_{M1} is the interaction energy for the system electrons – nuclear dipole moment, and E_{E2} is the interaction energy for the system electrons – nuclear quadrupole moment. Furthermore, the value $K = F(F+1) - I(I+1) - J(J+1)$, and finally, A and B are the hyperfine splitting constants of the atomic energy levels, the magnetic dipole interaction being one and a half orders of magnitude larger than the electric quadrupole interaction. The quadrupole interaction constant B is zero for states where $J \leqslant 1/2$ owing to the spherical symmetry of the charge distribution.

The usual way to designate the hyperfine splitting of energy levels is as follows: $\Delta\nu(F, F') = \Delta E_{FF'}/h$, where $\Delta E_{FF'}$ is the energy separation between two hyperfine components with total angular momenta F and $F' = F - 1$, measured when an external magnetic field is absent. The dependence of the hyperfine splitting $\Delta\nu(F, F')$ on the constants A, B is given by

$$\Delta\nu(F, F') = AF + \frac{3BF(F^2 + 1/2)}{I(2I-1)J(2J-1)}, \quad I, J \geqslant 1 .$$

Table 5.4 presents measured values of $\Delta\nu(F, F')$ (in units of MHz and 10^{-3} cm^{-1}), as well as the hyperfine splitting constants A, B [MHz] for the low-lying atomic levels in the most stable nuclides. The errors in the determination of these values were allowed for when truncating the significant figures, so that they were within the range ± 1 for the last digit quoted. When selecting the material, we made use of the data offered in reviews [5.4.3–6], and we supplemented them by the results of later publications.

References

5.4.1 A.J.Freeman, R.H.Frankel: *Hyperfine Interactions* (Academic, New York 1967)
5.4.2 I.Lindgren, A.Rosen: Case Stud. At. Phys. **4**, 93 (1974)
5.4.3 G.H.Fuller: J. Phys. Chem. Ref. Data **5**, 835 (1976)
5.4.4 G.H.Fuller, V.W.Cohen: Nucl. Data Tables **A 5**, 433 (1969)
5.4.5 E.Arimondo, M.Inguscio, P.Violino: Rev. Mod. Phys. **49**, 31 (1977)
5.4.6 S.Penselin: "Recent Developments and Results of the Atomic-Beam Magnetic-Resonance Method", in *Progress in Atomic Spectroscopy*, Part A, ed. by W.Hanle, H.Kleinpoppen (Plenum, New York 1978) Chap. 10, pp. 463–490

Table 5.4. Hyperfine splitting of low-lying atomic energy levels (stable nuclides)

Atomic number Z	Isotope (ground-state term, nuclear spin I)	Electronic term	Quantum numbers of total angular momentum (F, F')	hf splitting $\Delta\nu(F, F')$ A, B [MHz]	hf splitting $\Delta E(F, F')$ $[10^{-3}\ cm^{-1}]$
1	2	3	4	5	6
1	$^1H\,(^2S_{1/2})$	$1\,^2S_{1/2}$	(1, 0)	1420.40575	47.3796
	$(I = 1/2)$	$2\,^2S_{1/2}$	(1, 0)	177.5568	5.92266
	$^2H\,(^2S_{1/2})$	$1\,^2S_{1/2}$	(3/2, 1/2)	327.38435	10.9204
	$(I = 1)$	$2\,^2S_{1/2}$	(3/2, 1/2)	40.9244	1.36509
	$^3H\,(^2S_{1/2})(I = 1/2)$	$1\,^2S_{1/2}$	(1, 0)	1516.70147	50.5917
2	$^3He\,(^1S_0)(I = 1/2)$	$2\,^3S_1$	(3/2, 1/2)	6739.701	224.812
	$^3He^+\,(^2S_{1/2})$	$1\,^2S_{1/2}$	(1, 0)	8665.6499	289.055
3	$^6Li\,(^2S_{1/2})$	$2\,^2S_{1/2}$	(3/2, 1/2)	228.20526	7.61211
	$(I = 1)$	$2\,^2P_{1/2}$	–	$A = 17.37$	–
		$2\,^2P_{3/2}$	–	$A = -1.16$	–
				$B = -0.1$	–
	$^7Li\,(^2S_{1/2})$	$2\,^2S_{1/2}$	(2, 1)	803.50409	26.80203
	$(I = 3/2)$	$2\,^2P_{1/2}$	–	$A = 45.9$	–
		$2\,^2P_{3/2}$	–	$A = -3.06$	–
				$B = -0.2$	–
4	$^9Be\,(^1S_0)$	$2\,^3P_1$	(5/2, 3/2)	354.44	11.823
	$(I = 3/2)$		(3/2, 1/2)	202.95	6.7697
		$2\,^3P_2$	(7/2, 5/2)	435.48	14.526
			(5/2, 3/2)	312.02	10.408
			(3/2, 1/2)	187.62	6.2583
5	$^{10}B\,(^2P_{1/2})(I = 3)$	$2\,^2P_{1/2}$	(7/2, 5/2)	429.05	14.312
	$^{11}B\,(^2P_{1/2})$	$2\,^2P_{1/2}$	(2, 1)	732.15	24.422
	$(I = 3/2)$	$2\,^2P_{3/2}$	(3, 2)	222.7	7.428
			(2, 1)	144.0	4.803
			(1, 0)	71	2.37
6	$^{13}C\,(^3P_0)$	$2\,^3P_1$	(3/2, 1/2)	4.3	0.14
	$(I = 1/2)$	$2\,^3P_2$	(5/2, 3/2)	372.6	12.43
7	$^{14}N\,(^4S_{3/2})$	$2\,^4S_{3/2}$	–	$A = 10.45093$	–
	$(I = 1)$			$B = 1.3$	–
8	$^{17}O\,(^3P_2)$	$2\,^3P_2$	–	$A = -219.6$	–
	$(I = 5/2)$	$2\,^3P_1$	–	$A = 4.7$	–
9	$^{19}F\,(^2P_{3/2})$	$2\,^2P_{3/2}$	(2, 1)	4020	134
	$(I = 1/2)$	$2\,^2P_{1/2}$	(1, 0)	10250	342
10	$^{21}Ne\,(^1S_0)$	$3\,^3P_2$	(7/2, 5/2)	1034.5	34.51
	$(I = 3/2)$		(5/2, 3/2)	599.4	19.99
			(3/2, 1/2)	303.9	10.14
11	$^{23}Na\,(^2S_{1/2})$	$3\,^2S_{1/2}$	(2, 1)	1771.62613	59.09513
	$(I = 3/2)$	$3\,^2P_{1/2}$	–	$A = 94.3$	–
		$3\,^2P_{3/2}$	–	$A = 18.7$	–
				$B = 2.9$	–

Table 5.4 (continued)

Atomic number Z	Isotope (ground-state term, nuclear spin I)	Electronic term	Quantum numbers of total angular momentum (F, F')	hf splitting $\Delta\nu(F, F')$ A, B [MHz]	$\Delta E(F, F')$ [10^{-3} cm^{-1}]
1	2	3	4	5	6
12	^{25}Mg ($^1S_0$)	$3\,^3P_1$	(7/2, 5/2)	516.1	17.22
	(I = 5/2)		(5/2, 3/2)	350.0	11.7
		$3\,^3P_2$	(9/2, 7/2)	567.3	18.92
			(7/2, 5/2)	452.3	15.09
			(5/2, 3/2)	329.0	10.97
			(3/2, 1/2)	199.8	6.66
13	^{27}Al ($^2P_{1/2}$)	$3\,^2P_{1/2}$	(3, 2)	1506.1	50.24
	(I = 5/2)	$3\,^2P_{3/2}$	(4, 3)	392	13.1
			(3, 2)	274	9.14
		$4\,^2S_{1/2}$	–	A = 420	–
15	^{31}P ($^4S_{3/2}$)(I = 1/2)	$3\,^4S_{3/2}$	–	A = 55.06	–
17	^{35}Cl ($^2P_{3/2}$)	$3\,^2P_{3/2}$	(3, 2)	670.0135	22.349
	(I = 3/2)		(2, 1)	355.2210	11.849
			(1, 0)	150.1736	5.009
		$3\,^2P_{1/2}$	(2, 1)	2074.38	69.19
	^{37}Cl ($^2P_{3/2}$)	$3\,^2P_{3/2}$	(3, 2)	555.3043	18.523
	(I = 3/2)		(2, 1)	298.1277	9.944
			(1, 0)	127.4408	4.251
		$3\,^2P_{1/2}$	(2, 1)	1726.7	57.60
19	^{39}K ($^2S_{1/2}$)	$4\,^2S_{1/2}$	(2, 1)	461.71972	15.40132
	(I = 3/2)	$4\,^2P_{1/2}$	–	A = 27.8	–
		$4\,^2P_{3/2}$	–	A = 6.1	–
				B = 2.8	–
	^{40}K ($^2S_{1/2}$)	$4\,^2S_{1/2}$	–	A = −285.73	–
	(I = 4)	$4\,^2P_{1/2}$	–	A = −34.5	–
		$4\,^2P_{3/2}$	–	A = −7.5	–
				B = −3	–
	^{41}K ($^2S_{1/2}$)	$4\,^2S_{1/2}$	(2, 1)	254.01387	8.47300
	(I = 3/2)	$4\,^2P_{1/2}$	–	A = 15.2	–
		$4\,^2P_{3/2}$	–	A = 3.4	–
				B = 3.3	
21	^{45}Sc ($^2D_{3/2}$)	$3\,^2D_{3/2}$	(5, 4)	1329	44.3
	(I = 7/2)		(4, 3)	1085.8	36.22
		$3\,^2D_{5/2}$	(6, 5)	635.0	21.18
			(5, 4)	543.8	18.14
			(4, 3)	444.7	14.83
		$4\,^4F_{3/2}$	–	A = 158.5	–
				B = −5.2	–
		$4\,^4F_{5/2}$	–	A = 154.0	–
				B = −6.5	–
		$4\,^4F_{7/2}$	–	A = 250.0	–
				B = −9.1	–

Table 5.4 (continued)

Atomic number Z	Isotope (ground-state term, nuclear spin I)	Electronic term	Quantum numbers of total angular momentum (F, F')	hf splitting: $\Delta\nu(F, F')$ A, B [MHz]	hf splitting: $\Delta E(F, F')$ [10^{-3} cm^{-1}]
1	2	3	4	5	6
		$4\,^4F_{9/2}$	–	$A = 286.0$	–
				$B = -15$	–
22	^{47}Ti ($^3F_2$)	$3\,^3F_2$	–	$A = -85.703$	–
	($I = 5/2$)			$B = 25.70$	–
	^{49}Ti ($^3F_2$)	$3\,^3F_2$	–	$A = -85.726$	–
	($I = 7/2$)			$B = 21.07$	–
23	^{51}V ($^4F_{3/2}$)	$3\,^4F_{3/2}$	–	$A = 560.07$	–
	($I = 7/2$)			$B = 3.98$	–
24	^{53}Cr ($^7S_3$)	$3\,^7S_3$	(9/2, 7/2)	371.7	12.40
	($I = 3/2$)		(7/2, 5/2)	289.09	9.643
			(5/2, 3/2)	206.50	6.888
25	^{55}Mn ($^6S_{5/2}$)	$3\,^6S_{5/2}$	–	$A = -72.4208$	–
	($I = 5/2$)			$B = -0.018$	–
		$4\,^6D_{9/2}$	–	$A = 510.3$	–
				$B = 132.2$	–
		$4\,^6D_{7/2}$	–	$A = 458.9$	–
				$B = 21.7$	–
		$4\,^6D_{5/2}$	–	$A = 436.7$	–
				$B = -46.8$	–
		$4\,^6D_{3/2}$	–	$A = 469.4$	–
				$B = -65.1$	–
		$4\,^6D_{1/2}$	–	$A = 882.1$	–
26	^{57}Fe ($^5D_4$)	$3\,^5D_4$	–	$A = 38.08$	–
	($I = 1/2$)	$4\,^5F_5$	–	$A = 87.25$	–
		$4\,^5F_4$	–	$A = 78.43$	–
		$4\,^5F_3$	–	$A = 69.63$	–
		$4\,^5F_2$	–	$A = 55.99$	–
27	^{59}Co ($^4F_{9/2}$)	$3\,^4F_{9/2}$	(8, 7)	3655	121.9
	($I = 7/2$)		(7, 6)	3169.4	105.7
			(6, 5)	2695	89.9
			(5, 4)	2230.6	74.40
			(4, 3)	1774.5	59.19
28	^{61}Ni ($^3F_4$)	$3\,^3F_4$	–	$A = -215.04$	–
	($I = 3/2$)			$B = -56.9$	–
29	^{63}Cu ($^2S_{1/2}$)($I = 3/2$)	$4\,^2S_{1/2}$	(2, 1)	11733.8174	391.398
	^{65}Cu ($^2S_{1/2}$)($I = 3/2$)	$4\,^2S_{1/2}$	(2, 1)	12568.780	419.250
30	^{67}Zn ($^1S_0$)	$4\,^3P_2$	(9/2, 7/2)	2418.1	80.66
	($I = 5/2$)		(7/2, 5/2)	1855.7	61.90
			(5/2, 3/2)	1312.1	43.77
			(3/2, 1/2)	781.9	26.08

Table 5.4 (continued)

Atomic number Z	Isotope (ground-state term, nuclear spin I)	Electronic term	Quantum numbers of total angular momentum (F, F')	hf splitting $\Delta\nu(F, F')$ A, B [MHz]	$\Delta E(F, F')$ [10^{-3} cm^{-1}]
1	2	3	4	5	6
31	$^{69}Ga\,(^2P_{1/2})$	$4\,^2P_{1/2}$	(2, 1)	2677.987	89.328
	$(I = 3/2)$	$4\,^2P_{3/2}$	–	$A = 190.794$	–
				$B = 62.522$	–
		$5\,^2S_{1/2}$	(2, 1)	2140	71.3
	$^{71}Ga\,(^2P_{1/2})$	$4\,^2P_{1/2}$	(2, 1)	3402.69	113.50
	$(I = 3/2)$	$4\,^2P_{3/2}$	(3, 2)	766.696	25.574
			(2, 1)	445.470	14.859
			(1, 0)	203.043	6.773
		$5\,^2S_{1/2}$	(2, 1)	2720	90.6
32	$^{73}Ge\,(^3P_0)$	$4\,^3P_1$	–	$A = 15.55$	–
	$(I = 9/2)$			$B = -54.57$	–
		$4\,^3P_2$	–	$A = -64.427$	–
				$B = 111.8$	–
33	$^{75}As\,(^4S_{3/2})$	$4\,^4S_{3/2}$	–	$A = -66.20$	–
	$(I = 3/2)$			$B = -0.53$	–
			(3, 2)	819.45	27.33
			(2, 1)	595.12	19.85
36	$^{83}Kr\,(^1S_0)$	$5p\,[1/2]_1$	–	$A = -143.0$	–
	$(I = 9/2)$				
37	$^{85}Rb\,(^2S_{1/2})$	$5\,^2S_{1/2}$	(3, 2)	3035.732	101.261
	$(I = 5/2)$	$5\,^2P_{1/2}$	–	$A = 120.7$	–
		$5\,^2P_{3/2}$	–	$A = 25.0$	–
				$B = 26.0$	–
		$4\,^2D_{5/2}$	–	$A = -5$	–
		$4\,^2D_{3/2}$	–	$A = 7$	–
		$6\,^2S_{1/2}$	–	$A = 239$	–
	$^{87}Rb\,(^2S_{1/2})$	$5\,^2S_{1/2}$	(2, 1)	6834.6826	227.98
	$(I = 3/2)$	$5\,^2P_{1/2}$	–	$A = 406$	–
		$5\,^2P_{3/2}$	–	$A = 84.9$	–
				$B = 12.6$	–
		$4\,^2D_{5/2}$	–	$A = -17$	–
		$4\,^2D_{3/2}$	–	$A = 25$	–
		$6\,^2S_{1/2}$	–	$A = 810$	–
39	$^{89}Y\,(^2D_{3/2})$	$4\,^2D_{3/2}$	–	$A = -57.2$	–
	$(I = 1/2)$	$4\,^2D_{5/2}$	–	$A = -28.7$	–
47	$^{107}Ag\,(^2S_{1/2})$	$5\,^2P_{3/2}$	–	$A = -32$	–
	$(I = 1/2)$				
49	$^{113}In\,(^2P_{1/2})$	$5\,^2P_{1/2}$	(5, 4)	11385	379.8
	$(I = 9/2)$	$6\,^2S_{1/2}$	(5, 4)	8410	281
	$^{115}In\,(^2P_{1/2})$	$5\,^2P_{1/2}$	(5, 4)	11410	380.6
	$(I = 9/2)$	$5\,^2P_{3/2}$	–	$A = 242.165$	–
		$6\,^2S_{1/2}$	(5, 4)	8430	281

Table 5.4 (continued)

Atomic number Z	Isotope (ground-state term, nuclear spin I)	Electronic term	Quantum numbers of total angular momentum (F, F')	hf splitting $\Delta\nu(F, F')$ A, B [MHz]	$\Delta E(F, F')$ [10^{-3} cm^{-1}]
1	2	3	4	5	6
51	^{123}Sb ($^4S_{3/2}$)	$5\,^4S_{3/2}$	(5, 4)	815.6	27.20
	($I = 7/2$)		(4, 3)	648.5	21.63
			(3, 2)	484.0	16.1
52	^{125}Te ($^3P_2$)	$5\,^3P_2$	–	$A = -1010.3$	–
	($I = 1/2$)	$5\,^3P_1$	–	$A = 782.5$	–
		$5\,^1D_2$	–	$A = -2887.0$	–
53	^{127}I ($^2P_{3/2}$)	$5\,^2P_{3/2}$	(4, 3)	4226.17	140.97
	($I = 5/2$)		(3, 2)	1965.9	65.58
			(2, 1)	737.49	24.60
54	^{129}Xe ($^1S_0$)($I = 1/2$)	$6\,^3P_2$	(5/2, 3/2)	5961.258	198.85
	^{131}Xe ($^1S_0$)	$6\,^3P_2$	(7/2, 5/2)	2693.623	89.850
	($I = 3/2$)		(5/2, 3/2)	1608.348	53.649
			(3/2, 1/2)	838.764	27.978
55	^{133}Cs ($^2S_{1/2}$)	$6\,^2S_{1/2}$	(4, 3)	9192.63177	306.63342
	($I = 7/2$)	$6\,^2P_{1/2}$	–	$A = 292$	–
		$6\,^2P_{3/2}$	–	$A = 50.3$	–
				$B = -0.4$	–
		$5\,^2D_{3/2}$	–	$A = 16.3$	–
		$5\,^2D_{5/2}$	–	$A = -22$	–
		$7\,^2S_{1/2}$	–	$A = 550$	–
56	^{135}Ba ($^1S_0$)	$5\,^3D_1$	–	$A = -470$	–
	($I = 3/2$)			$B = 12$	–
		$5\,^3D_2$	–	$A = 371$	–
				$B = 18$	–
		$5\,^3D_3$	–	$A = 408$	–
				$B = 20$	–
		$5\,^1D_2$	–	$A = -73.4$	–
				$B = 38.7$	
	^{137}Ba ($^1S_0$)	$5\,^3D_1$	–	$A = -520$	–
	($I = 3/2$)			$B = 17$	–
		$5\,^3D_2$	–	$A = 414$	–
				$B = 27$	–
		$5\,^3D_3$	–	$A = 455$	–
				$B = 40$	–
		$5\,^1D_2$	–	$A = -82.2$	–
				$B = 59.6$	–
57	^{139}La ($^2D_{3/2}$)	$5\,^2D_{3/2}$	(5, 4)	737.97	24.62
	($I = 7/2$)		(4, 3)	551.98	18.41
			(3, 2)	391.6	13.06
		$5\,^2D_{5/2}$	(6, 5)	1120.90	37.39
			(5, 4)	912.79	30.45
			(4, 3)	716.29	23.89
			(3, 2)	529.1	17.65

Table 5.4 (continued)

Atomic number Z	Isotope (ground-state term, nuclear spin I)	Electronic term	Quantum numbers of total angular momentum (F, F')	hf splitting $\Delta\nu(F, F')$ A, B [MHz]	$\Delta E(F, F')$ $[10^{-3}\ cm^{-1}]$
1	2	3	4	5	6
		$6\,^4F_{3/2}$	(5, 4)	2390.6	79.74
			(4, 3)	1925.5	64.23
		$6\,^4F_{5/2}$	(6, 5)	1808.9	60.34
			(5, 4)	1503.2	50.14
			(4, 3)	1199.8	40.02
		$6\,^4P_{1/2}$	(4, 3)	9840.6	328.2
		$6\,^4P_{3/2}$	(4, 3)	3707.8	123.68
		$6\,^4P_{5/2}$	(4, 3)	3216.5	107.3
59	$^{141}Pr\ (^4I_{9/2})$ $(I = 5/2)$	$4\,^4I_{9/2}$	–	$A = 926.209$	–
				$B = -11.88$	–
		$4\,^4I_{11/2}$	–	$A = 730.393$	–
				$B = -11.88$	–
		$4\,^4I_{13/2}$	–	$A = 613.240$	–
				$B = -12.85$	–
		$4\,^4I_{15/2}$	–	$A = 541.575$	–
				$B = -14.56$	–
60	$^{143}Nd\ (^5I_4)$ $(I = 7/2)$	$4\,^5I_4$	(15/2, 13/2)	1418	47.3
			(13/2, 11/2)	1257.5	41.95
			(11/2, 9/2)	1084.7	36.18
			(9/2, 7/2)	901.5	30.07
			(7/2, 5/2)	710	23.7
		$4\,^5I_5$	–	$A = -153.68$	–
				$B = 115.7$	–
	$^{145}Nd\ (^5I_4)$ $(I = 7/2)$	$4\,^5I_4$	–	$A = -121.63$	–
				$B = 64.6$	–
		$4\,^5I_5$	–	$A = -95.53$	–
				$B = 61.0$	–
62	$^{147}Sm\ (^7F_0)$ $(I = 7/2)$	$4\,^7F_1$	–	$A = -33.494$	–
				$B = -58.692$	–
		$4\,^7F_2$	–	$A = -41.184$	–
				$B = -62.23$	–
		$4\,^7F_3$	–	$A = -50.240$	–
				$B = -33.68$	–
	$^{149}Sm\ (^7F_0)$ $(I = 7/2)$	$4\,^7F_1$	–	$A = -27.611$	–
				$B = 16.962$	–
		$4\,^7F_2$	–	$A = -33.951$	–
				$B = 17.99$	–
		$4\,^7F_3$	–	$A = -41.418$	–
				$B = 9.75$	–
63	$^{151}Eu\ (^8S_{7/2})$ $(I = 5/2)$	$4\,^8S_{7/2}$	(6, 5)	120.67	4.025
			(5, 4)	100.29	3.345
			(4, 3)	80.05	2.67

Table 5.4 (continued)

Atomic number Z	Isotope (ground-state term, nuclear spin I)	Electronic term	Quantum numbers of total angular momentum (F, F')	hf splitting $\Delta\nu(F, F')$ A, B [MHz]	hf splitting $\Delta E(F, F')$ [10^{-3} cm^{-1}]
1	2	3	4	5	6
	^{153}Eu ($^8S_{7/2}$) ($I = 5/2$)	$4\,^8S_{7/2}$	(6, 5)	54.04	1.803
			(5, 4)	44.00	1.47
			(4, 3)	35.00	1.17
64	^{155}Gd ($^9D_2$) ($I = 3/2$)	$5\,^9D_2$	–	$A = 36.575$	–
				$B = 179.4$	–
		$5\,^9D_3$	–	$A = 4.92$	–
				$B = -406.67$	–
		$5\,^9D_4$	–	$A = -6.86$	–
				$B = -352.8$	–
	^{157}Gd ($^9D_2$) ($I = 3/2$)	$5\,^9D_2$	–	$A = 47.96$	–
				$B = 191.2$	–
		$5\,^9D_3$	–	$A = 6.45$	–
				$B = -433.2$	–
		$5\,^9D_4$	–	$A = -9.00$	–
				$B = -375.9$	–
65	^{159}Tb ($^6H_{15/2}$) ($I = 3/2$)	$4\,^6H_{15/2}$	–	$A = 673.75$	–
				$B = 1449.3$	–
		$4\,^6H_{13/2}$	–	$A = 682.91$	–
				$B = 1167.5$	–
		$5\,^8G_{13/2}$	–	$A = 532.20$	–
				$B = 928.9$	–
66	^{161}Dy ($^5I_8$) ($I = 5/2$)	$4\,^5I_8$	–	$A = -116.232$	–
				$B = 1091.57$	–
	^{163}Dy ($^5I_8$) ($I = 5/2$)	$4\,^5I_8$	–	$A = 162.7543$	–
				$B = 1152.86$	–
		$4\,^5I_7$	–	$A = 177.53$	–
				$B = 1066.4$	–
67	^{165}Ho ($^4I_{15/2}$) ($I = 7/2$)	$4\,^4I_{15/2}$	(9, 8)	7184.8	239.7
			(8, 7)	6540.8	218.2
			(7, 6)	5842.4	194.9
			(6, 5)	5096.3	170.0
			(5, 4)	4309.3	143.7
68	^{167}Er ($^3H_6$) ($I = 7/2$)	$4\,^3H_6$	–	$A = -120.486$	–
				$B = -4552.96$	–
69	^{169}Tm ($^2F_{7/2}$) ($I = 1/2$)	$4\,^2F_{7/2}$	(4, 3)	1496.5507	49.920
			–	$A = -374.13766$	–
71	^{175}Lu ($^2D_{3/2}$) ($I = 7/2$)	$5\,^2D_{3/2}$	(5, 4)	2051.2201	68.421
			(4, 3)	345.497	11.524
			(3, 2)	496.578	16.564
		$5\,^2D_{5/2}$	(6, 5)	1837.570	61.295
			(5, 4)	800.343	26.70
			(4, 3)	161.815	5.398

Table 5.4 (continued)

Atomic number Z	Isotope (ground-state term, nuclear spin I)	Electronic term	Quantum numbers of total angular momentum (F, F')	hf splitting $\Delta\nu(F, F')$ A, B [MHz]	$\Delta E(F, F')$ $[10^{-3}\ \mathrm{cm}^{-1}]$
1	2	3	4	5	6
			(3, 2)	157.73	5.26
			(2, 1)	238.058	7.941
	^{176}Lu ($^2D_{3/2}$) ($I = 7$)	$5\,^2D_{3/2}$	–	$A = 137.9$	–
				$B = 2131$	–
		$5\,^2D_{5/2}$	–	$A = 104.0$	–
				$B = 2624$	–
72	^{177}Hf ($^3F_2$) ($I = 7/2$)	$5\,^3F_2$	(11/2, 9/2)	991.792	33.08
			(9/2, 7/2)	477.008	15.91
			(7/2, 5/2)	162.887	5.433
			(5/2, 3/2)	4.864	0.16
	^{179}Hf ($^3F_2$) ($I = 9/2$)	$5\,^3F_2$	(13/2, 11/2)	82.132	2.74
			(11/2, 9/2)	392.848	13.104
			(9/2, 7/2)	541.9104	18.076
			(7/2, 5/2)	558.672	18.635
73	^{181}Ta ($^4F_{3/2}$) ($I = 7/2$)	$5\,^4F_{3/2}$	–	$A = 509.08$	–
				$B = -1012.24$	–
		$5\,^4F_{5/2}$	–	$A = 313.47$	–
				$B = -834.8$	–
		$5\,^4F_{7/2}$	–	$A = 264.41$	–
				$B = -787.5$	–
		$5\,^4F_{9/2}$	–	$A = 256.62$	–
				$B = -650.4$	–
		$5\,^4P_{1/2}$	–	$A = 884.17$	–
		$5\,^4P_{3/2}$	–	$A = 379$	–
				$B = -1350$	–
74	^{183}W ($^5D_0$) ($I = 1/2$)	$5\,^5D_1$	–	$A = 29.12$	–
		$6\,^7S_3$	–	$A = 505.6$	–
		$5\,^5D_2$	–	$A = 56.3$	–
		$5\,^5D_3$	–	$A = 78.0$	–
		$5\,^5D_4$	–	$A = 88.3$	–
75	^{185}Re ($^6S_{5/2}$) ($I = 5/2$)	$5\,^6S_{5/2}$	–	$A = -56.596$	–
				$B = 29.635$	–
		$5\,^4P_{5/2}$	–	$A = 880.44$	–
				$B = 1618.5$	–
	^{187}Re ($^6S_{5/2}$) ($I = 5/2$)	$5\,^6S_{5/2}$	–	$A = -57.149$	–
				$B = 28.05$	–
		$5\,^4P_{5/2}$	–	$A = 889.24$	–
				$B = 1531.7$	–
		$6\,^6D_{9/2}$	–	$A = 2600$	–
				$B = 2000$	–

Table 5.4 (continued)

Atomic number Z	Isotope (ground-state term, nuclear spin I)	Electronic term	Quantum numbers of total angular momentum (F, F')	hf splitting $\Delta\nu(F, F')$ A, B [MHz]	$\Delta E(F, F')$ $[10^{-3}\ \text{cm}^{-1}]$
1	2	3	4	5	6
77	$^{191}\text{Ir}\,(^4F_{9/2})$	$5\,^4F_{9/2}$	(6, 5)	659.265	21.991
	$(I = 3/2)$		(5, 4)	189.440	6.319
			(4, 3)	84.050	2.804
	$^{193}\text{Ir}\,(^4F_{9/2})$	$5\,^4F_{9/2}$	(6, 5)	660.090	22.018
	$(I = 3/2)$		(5, 4)	224.478	7.488
			(4, 3)	33.535	1.119
78	$^{195}\text{Pt}\,(^3D_3)$	$5\,^3D_3$	–	$A = 5702.6$	–
	$(I = 1/2)$	$5\,^3D_2$	–	$A = -2609.6$	–
		$6\,^3F_4$	(9/2, 7/2)	3820.56	127.4
79	$^{197}\text{Au}\,(^2S_{1/2})$	$6\,^2S_{1/2}$	(2, 1)	6099.320	203.452
	$(I = 3/2)$	$5\,^2D_{5/2}$	–	$A = 80.24$	–
				$B = 1049.8$	–
		$5\,^2D_{3/2}$	–	$A = 199.842$	–
				$B = 911.077$	–
80	$^{199}\text{Hg}\,(^1S_0)(I = 1/2)$	$6\,^3P_2$	–	$A = 9066.45$	–
	$^{201}\text{Hg}\,(^1S_0)$	$6\,^3P_2$	(7/2, 5/2)	11382.629	379.68
	$(I = 3/2)$		(5/2, 3/2)	8629.522	287.85
			(3/2, 1/2)	5377.49	179.37
		$6\,^3D_3$	–	$A = -2450$	–
				$B = 60$	–
81	$^{203}\text{Tl}\,(^2P_{1/2})$	$6\,^2P_{1/2}$	(1, 0)	21105.45	704.0026
	$(I = 1/2)$	$6\,^2P_{3/2}$	(2, 1)	524.0599	17.4808
	$^{205}\text{Tl}\,(^2P_{1/2})$	$6\,^2P_{1/2}$	(1, 0)	21310.83	710.8534
	$(I = 1/2)$	$6\,^2P_{3/2}$	(2, 1)	530.0765	17.6815
82	$^{207}\text{Pb}\,(^3P_0)(I = 1/2)$	$6\,^1D_2$	(5/2, 3/2)	1524.5	50.85
83	$^{209}\text{Bi}\,(^4S_{3/2})$	$6\,^4S_{3/2}$	(6, 5)	2884.67	96.22
	$(I = 9/2)$		(5, 4)	2171.42	72.43
			(4, 3)	1584.50	52.85
		$6\,^2P_{3/2}$	(6, 5)	3598.65	120.04
			(5, 4)	2251.04	75.09
			(4, 3)	1311.9	43.76
92	$^{235}\text{U}\,(^5L_6^o)$	$6\,^5L_6^o$	–	$A = -60.56$	–
	$(I = 7/2)$			$B = 4104.1$	–
		$6\,^5K_5^o$	–	$A = -68.35$	–
				$B = 40.1$	–
93	$^{237}\text{Np}\,(^6L_{11/2})$	$5\,^6L_{11/2}$	–	$A = 778$	–
	$(I = 5/2)$			$B = 645$	–

5.5 Isotope Shifts of Low-Lying Atomic Levels

In the spectra of elements which have several isotopes the splitting of lines into a number of components is observed. Each of the lines characterizes a definite isotope. This isotopic structure is determined by the influence of the electron-nucleus interaction on the position of the atomic energy levels. The relevant terms in the atomic Hamiltonian, recorded in the centre-of-mass system, allow for the motion of the nucleons about the atom's centre of mass (normal or Bohr mass effect), the electron exchange interaction dependent on the nuclear mass (specific mass effect) and the interaction of the valence electrons (usually *s* electrons and to a smaller extent *p* electrons) with the extended nuclear charge (field or volume effect) [5.5.1]. The observed transition isotope shift $\Delta\nu$ for two different isotopes 1 and 2 is given by the sum of three parts [5.5.2, 3].

$$\Delta\nu = \Delta\nu_{\mathrm{NMS}} + \Delta\nu_{\mathrm{SMS}} + \Delta\nu_{\mathrm{VS}} ,$$

where $\Delta\nu_{\mathrm{NMS}} = \nu m_e (M_2 - M_1)/M_1 M_2$ is the normal mass shift, which is of particular importance for the light elements ($Z \lesssim 30$). ν is the transition frequency for an infinite nuclear mass, M_1, M_2 are the nuclear masses and m_e is the electron mass. $\Delta\nu_{\mathrm{SMS}}$ is the specific mass shift and $\Delta\nu_{\mathrm{VS}}$ is the volume shift between the lines of two isotopes, prevailing in the spectra of heavy isotopes ($Z \gtrsim 50$). In the spectra of medium-weight elements ($Z \sim 20$–55) the absolute value of the observed isotope shift is small and difficult to measure accurately.

The transition isotope shift $\Delta\nu$ occurs as a result of level isotope shifts, and the relation between $\Delta\nu$ and the isotope shifts of the upper term ($\Delta T'$) and lower term (ΔT) can be expressed as follows:

$$\Delta\nu = \Delta T' - \Delta T \, .$$

By convention, the transition isotope shift is said to be positive if the lines of the heavier isotope are shifted to higher frequences. Similarly, the level isotope shift is ascribed a negative sign if the level of the heavier isotope lies deeper (its distance to the boundary with the continuous spectrum is greater) than that of the lighter one. When the nuclei have magnetic moments and, naturally, the hyperfine structure of the levels can be observed, the level isotope shift should be referred to the centres of gravity of the hyperfine structure components.

Table 5.5 incorporates the values of measured level isotope shifts $\Delta T(A_1 - A_2)$ for the low-lying terms of stable and long-lived isotopes with mass numbers A_1 and A_2 ($A = Z + N$). Table 5.6 gives the values of transition isotope shifts $\Delta\nu(A_1 - A_2)$ for the resonance lines of elements. To illustrate the relative contribution of isotope and hyperfine splittings of levels, Figs. 5.1–4 depict the term diagrams for hydrogen and some alkali atoms [5.5.4–7].

The above information is based on material in reviews [5.5.2, 3, 8–15], bibliographies [5.5.16, 17] and data of the latest publications. In accordance with the reported errors of the measurements, the listed numerical values were rounded off and are estimated to be accurate to within an error ± 1 in the last digit quoted.

References

5.5.1 H.Kopfermann: *Nuclear Moments*, 3rd ed. (Academic, New York 1962)
5.5.2 A.R.Striganov, Yu.P.Dontsov: Usp. Fiz. Nauk **55**, 315 (1955)
5.5.3 A.F.Golovin, A.R.Striganov: Usp. Fiz. Nauk **93**, 111 (1967)
5.5.4 G.W.Erickson: J. Phys. Chem. Ref. Data **6**, 831 (1977) (H, D, T)
5.5.5 C.-J.Lorenzen, K.Niemax: J. Phys. **B 15**, L 139 (1982) (^{6,7}Li)
5.5.6 L.R.Pendrill, K.Niemax: J. Phys. **B 15**, L 147 (1982) (^{39,41}K)
5.5.7 P.Grundevik, M.Gustavsson, A.Rosén, S.Svanberg: Z. Phys. **A 283**, 127 (1977) (^{85,87}Rb)
5.5.8 P.Brix, H.Kopfermann: "Hyperfeinstruktur der Atomterme und Atomlinien", in Landolt-Börnstein, *Zahlenwerte und Funktionen aus Physik, Chemie, Astronomie, Geophysik und Technik*, Vol. 1, Part 5, 6th ed. (Springer, Berlin, Göttingen, Heidelberg 1952) pp. 1–69
5.5.9 P.Brix, H.Kopfermann: Rev. Mod. Phys. **30**, 517 (1958)
5.5.10 D.A.Shirley: Rev. Mod. Phys. **36**, 339 (1964)
5.5.11 D.N.Stacey: Rep. Prog. Phys. **29**, 176 (1966)
5.5.12 K.Heilig, A.Steudel: At. Data Nucl. Data Tables **14**, 613 (1974)
5.5.13 K.Heilig, A.Steudel: "New Developments of Classical Optical Spectroscopy", in *Progress in Atomic Spectroscopy*, Part A, ed. by W.Hanle, H.Kleinpoppen (Plenum, New York 1978) pp. 263–328
5.5.14 J.Bauche, R.J.Champeau: Adv. At. Mol. Phys. **12**, 39 (1976)
5.5.15 S.Gerstenkorn: Comm. At. Mol. Phys. **9**, 1 (1979)
5.5.16 K.Heilig: "Bibliography on experimental optical isotope shifts, 1918 through October 1976", Spectrochim. Acta B **32**, pp. 1–57 (1977)
5.5.17 R.Zalubas, A.Albright: *Bibliography on Atomic Energy Levels and Spectra, July 1975 through June 1979*, Nat. Bur. Stand. (U.S.) Spec. Publ. **363**, Suppl. 2 (1980)

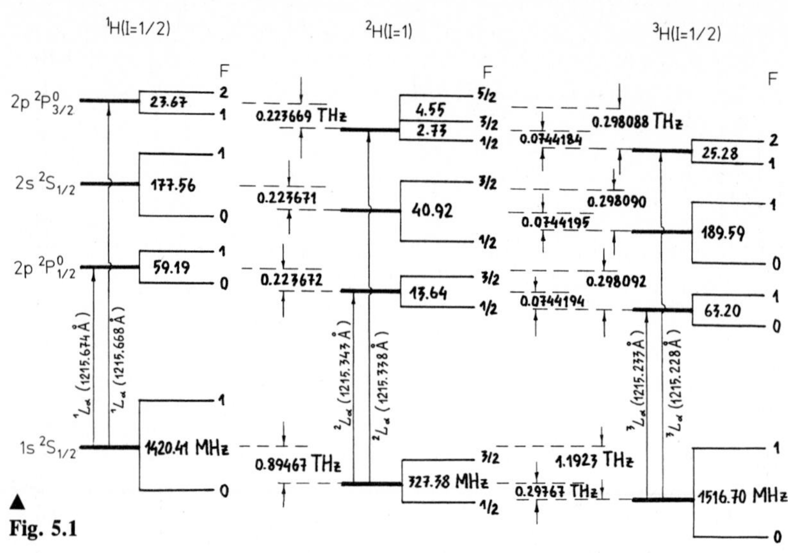

▲
Fig. 5.1

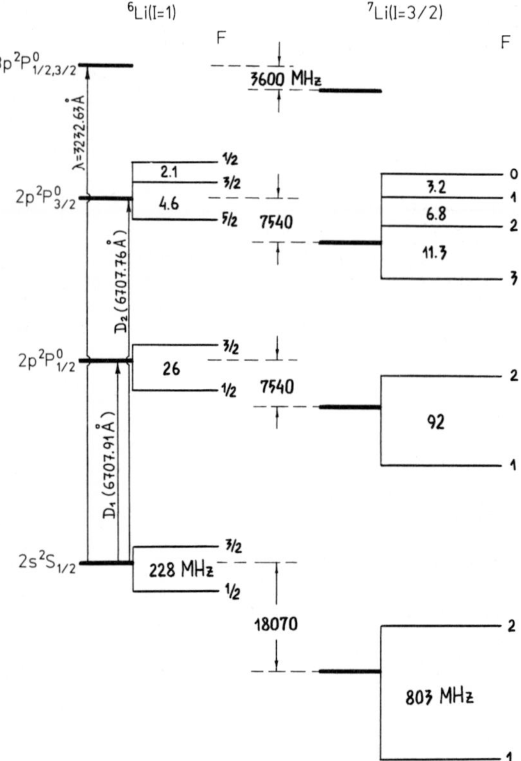

Fig. 5.1 Isotope shift and hyperfine splitting of hydrogen levels ($n \leq 2$)

Fig. 5.2 Isotope shift and hyperfine splitting of low-lying lithium levels

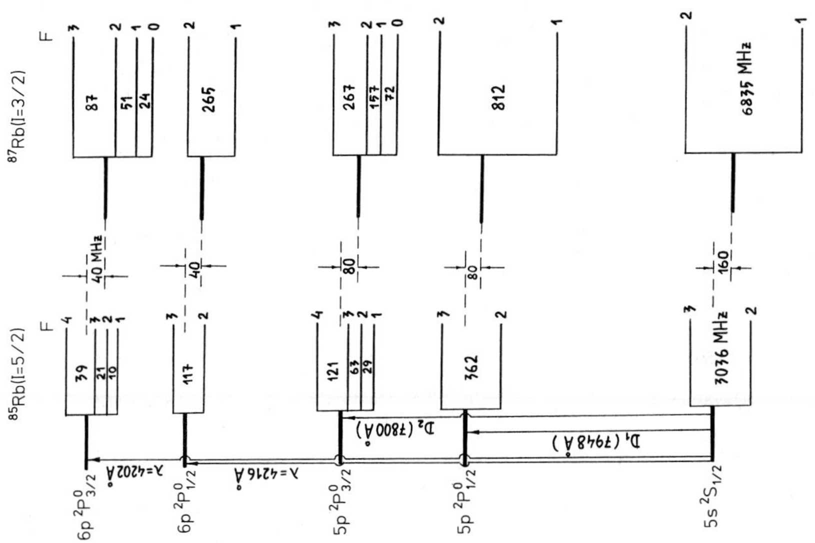

Fig. 5.4 Isotope shift and hyperfine splitting of low-lying rubidium levels

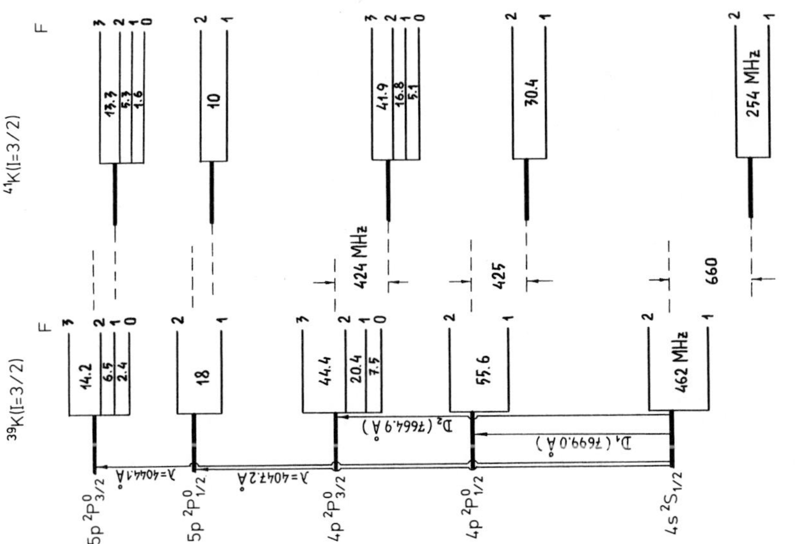

Fig. 5.3 Isotope shift and hyperfine splitting of low-lying potassium levels

Table 5.5. Isotope shifts of atomic energy levels

Atomic number Z	Element	Electronic term	Level isotope shift $\Delta T(A_1 - A_2)$	
			$A_1 - A_2$	ΔT [cm^{-1}]
1	2	3	4	5
1	H	$1\,S_{1/2}$	1 – 2	–29.84284
	($1s - {}^2S_{1/2}$)		1 – 3	–39.77218
		$2\,P_{1/2}$	1 – 2	– 7.46090
			1 – 3	– 9.94326
		$2\,S_{1/2}$	1 – 2	– 7.46085
			1 – 3	– 9.94322
		$2\,P_{3/2}$	1 – 2	– 7.46080
			1 – 3	– 9.94313
		$3\,P_{1/2}$	1 – 2	– 3.31595
			1 – 3	– 4.41922
		$3\,S_{1/2}$	1 – 2	– 3.31594
			1 – 3	– 4.41921
		$3\,P_{3/2}$	1 – 2	– 3.31592
			1 – 3	– 4.41918
		$3\,D_{3/2}$	1 – 2	– 3.31592
			1 – 3	– 4.41918
		$3\,D_{5/2}$	1 – 2	– 3.31591
			1 – 3	– 4.41917
2	He	$1\,{}^1S_0$	3 – 4	–10.5
	($1s^2 - {}^1S_0$)	$2\,{}^3S_1$	3 – 4	– 1.86
		$2\,{}^1S_0$	3 – 4	– 1.56
		$2\,{}^3P$	3 – 4	– 0.68
		$2\,{}^1P_1$	3 – 4	– 1.68
		$3\,{}^3S_1$	3 – 4	– 0.7
		$3\,{}^1S_0$	3 – 4	– 0.64
		$3\,{}^3P$	3 – 4	– 0.41
		$3\,{}^3D$	3 – 4	– 0.54
		$3\,{}^1D$	3 – 4	– 0.57
		$3\,{}^1P_1$	3 – 4	– 0.71
3	Li	$2\,{}^2S_{1/2}$	6 – 7	– 0.603
	($2s - {}^2S_{1/2}$)	$2\,{}^2P^o$	6 – 7	– 0.251
		$3\,{}^2P^o$	6 – 7	– 0.120
12	Mg	$3\,{}^1S_0$	24 – 25	– 0.115
	($3s^2 - {}^1S_0$)	$3\,{}^3P^o$	24 – 25	– 0.032
		$3\,{}^1P_1^o$	24 – 25	– 0.068
		$4\,{}^3S_1$	24 – 25	– 0.045
19	K	$4\,{}^2S_{1/2}$	39 – 41	– 0.0220
	($4s - {}^2S_{1/2}$)	$4\,{}^2P^o$	39 – 41	– 0.0142
29	Cu	$4\,{}^2S_{1/2}$	63 – 65	– 0.018
	($4s - {}^2S_{1/2}$)	$3d^9\,4s^2 - {}^2D_{5/2}$	63 – 65	– 0.085
		$3d^9\,4s^2 - {}^2D_{3/2}$	63 – 65	– 0.074
37	Rb	$5\,{}^2S_{1/2}$	85 – 87	– 0.0053
	($5s - {}^2S_{1/2}$)	$5\,{}^2P^o$	85 – 87	– 0.0027
		$6\,{}^2P^o$	85 – 87	– 0.0013

Table 5.6. Isotope shifts in atomic resonance lines

Atomic number Z	Element (ground-state term)	Transition array	Wave-length λ [Å]	Mass numbers $A_1 - A_2$	Isotope shift in line $\Delta\nu(A_1 - A_2)$ [10^{-3} cm^{-1}]
1	2	3	4	5	6
1	H ($1s - {}^2S_{1/2}$)	$1\,{}^2S_{1/2} - 2\,{}^2P^o$	1215.7	1 – 2	$2.238 \cdot 10^4$
				1 – 3	$2.983 \cdot 10^4$
				2 – 3	$7.447 \cdot 10^3$
2	He ($1s^2 - {}^1S_0$)	$1\,{}^1S_0 - 2\,{}^1P_1^o$	584.3	3 – 4	$8.8 \cdot 10^3$
3	Li ($2s - {}^2S_{1/2}$)	$2\,{}^2S_{1/2} - 2\,{}^2P^o$	6708	6 – 7	351.3
5	B ($2p - {}^2P_{1/2}^o$)	$2p\,{}^2P^o - 3s\,{}^2S_{1/2}$	2497	10 – 11	–170
6	C ($2p^2 - {}^3P_0$)	$2p^2\,{}^1S_0 - 2p3s\,{}^1P_1^o$	2478.6	12 – 13	–160
7	N ($2p^3 - {}^4S_{3/2}^o$)	$3s\,{}^2P_{3/2} - 3p\,{}^2P_{3/2}^o$	8629.2	14 – 15	70
		$3s\,{}^4P_{5/2} - 3p\,{}^4P_{5/2}^o$	8216.3	14 – 15	–60
8	O ($2p^4 - {}^3P_2$)	$3s\,{}^3S_1 - 3p\,{}^3P_1$	8446.8	16 – 18	140
10	Ne ($2p^6 - {}^1S_0$)	$3s'\,[1/2]_1^o - 3p\,[5/2]_1$	7173.9	20 – 22	70
		$3s\,[3/2]_2^o - 3p\,[1/2]_1$	7032.4	20 – 22	50
11	Na ($3s - {}^2S_{1/2}$)	$3s\,{}^2S_{1/2} - 4p\,{}^2P_{1/2}^o$	3303.0	23 – 24	24
12	Mg ($3s^2 - {}^1S_0$)	$3s^2\,{}^1S_0 - 3s3p\,{}^1P_1^o$	2852.1	24 – 25	50
				24 – 26	60
18	Ar ($3p^6 - {}^1S_0$)	$4s\,[3/2]_2^o - 4p'\,[3/2]_1$	7147.0	36 – 40	20
		$4s'\,[1/2]_1^o - 5p\,[1/2]_0$	4510.7	36 – 40	50
19	K ($4s - {}^2S_{1/2}$)	$4s\,{}^2S_{1/2} - 4p\,{}^2P_{1/2}^o$	7699.0	39 – 40	4.19
				39 – 41	7.85
		$4s\,{}^2S_{1/2} - 4p\,{}^2P_{3/2}^o$	7664.9	39 – 40	4.22
				39 – 41	7.88
29	Cu ($3d^{10}4s - {}^2S_{1/2}$)	$4s\,{}^2S_{1/2} - 4p\,{}^2P_{1/2}^o$	3274.0	63 – 65	20
30	Zn ($4s^2 - {}^1S_0$)	$4s^2\,{}^1S_0 - 4s4p\,{}^1P_1^o$	2138.6	$\Delta A = 2$ (mean value)	16
31	Ga ($4p - {}^2P_{1/2}^o$)	$4p\,{}^2P_{1/2}^o - 5s\,{}^2S_{1/2}$	4033.0	69 – 71	–1.1
		$4p\,{}^2P_{3/2}^o - 5s\,{}^2S_{1/2}$	4172.1	69 – 71	–1.3
36	Kr ($4p^6 - {}^1S_0$)	$5s\,[3/2]_1^o - 5p\,[5/2]_2$	8776.7	82 – 84	2
37	Rb ($5s - {}^2S_{1/2}$)	$5s\,{}^2S_{1/2} - 5p\,{}^2P_{1/2}^o$	7947.6	85 – 87	2.6
		$5s\,{}^2S_{1/2} - 5p\,{}^2P_{3/2}^o$	7800.2	85 – 87	2.6
38	Sr ($5s^2 - {}^1S_0$)	$5s^2\,{}^1S_0 - 5s5p\,{}^1P_1^o$	4607.3	84 – 88	9.0
				86 – 88	4.2
				87 – 88	1.5

Table 5.6 (continued)

Atomic number Z	Element (ground-state term)	Transition array	Wave-length λ [Å]	Mass numbers $A_1 - A_2$	Isotope shift in line $\Delta\nu(A_1 - A_2)$ $[10^{-3}\ \mathrm{cm}^{-1}]$
1	2	3	4	5	6
40	Zr $(4d^2 5s^2 - {}^3F_2)$	$4d^3 5s\,{}^5F_5 - 4d^3 5p\,{}^5G_6^{\circ}$	4687.8	90 – 92	−12
				92 – 94	−7
				94 – 96	−5
47	Ag $(4d^{10} 5s - {}^2S_{1/2})$	$4d^{10} 5s\,{}^2S_{1/2} - 5p\,{}^2P_{1/2}^{\circ}$	3382.9	107 – 109	−15
		$4d^{10} 5s\,{}^2S_{1/2} - 5p\,{}^2P_{3/2}^{\circ}$	3280.7	107 – 109	−15
48	Cd $(5s^2 - {}^1S_0)$	$5s^2\,{}^1S_0 - 5s5p\,{}^3P_1^{\circ}$	3261.0	$\Delta A = 2$ (mean value)	−15
49	In $(5p - {}^2P_{1/2}^{\circ})$	$5p\,{}^2P_{1/2}^{\circ} - 6s\,{}^2S_{1/2}$	4101.8	113 – 115	8.6
		$5p\,{}^2P_{3/2}^{\circ} - 6s\,{}^2S_{1/2}$	4511.3	113 – 115	8.5
54	Xe $(5p^6 - {}^1S_0)$	$6s\,[3/2]_2^{\circ} - 6p\,[3/2]_2$	8231.6	134 – 136	−3
55	Cs $(6s - {}^2S_{1/2})$	$6s\,{}^2S_{1/2} - 6p\,{}^2P_{3/2}^{\circ}$	8521.1	133 – 134	1.2
				133 – 135	1.2
70	Yb $(4f^{14} 6s^2 - {}^1S_0)$	$6s^2\,{}^1S_0 - 6s6p\,{}^1P_1^{\circ}$	3988.0	174 – 176	17
				174 – 172	18
71	Lu $(5d\,6s^2 - {}^2D_{3/2})$	$5d\,6s^2\,{}^2D_{3/2} - 6p\,{}^4F_{3/2}^{\circ}$	5736.5	175 – 176	−13.1
		$5d\,6s^2\,{}^2D_{5/2} - 6p\,{}^4F_{7/2}^{\circ}$	5421.9	175 – 176	−13.6
79	Au $(6s - {}^2S_{1/2})$	$6s\,{}^2S_{1/2} - 6p\,{}^2P_{1/2}^{\circ}$	2675.9	195 – 197	−97
80	Hg $(5d^{10} 6s^2 - {}^1S_0)$	$6s^2\,{}^1S_0 - 6s6p\,{}^3P_1^{\circ}$	2536.5	198 – 199	−9
				198 – 200	−160
				199 – 201	−210
				200 – 202	−180
				202 – 204	−160
82	Pb $(6p^2 - {}^3P_0)$	$6p^2\,{}^3P_0 - 6p7s\,{}^3P_1^{\circ}$	2833.1	207 – 208	−47
				206 – 208	−75
				204 – 208	−140
				202 – 208	−207

5.6 Atoms in Static Electric and Magnetic Fields. Atomic Polarizabilities and Magnetic Susceptibilities

Electric Field. The homogeneous electric field removes the degeneracy of atomic levels according to the direction of the electronic angular momentum J, leaving the levels with $M_J = \pm |M_J|$ unsplit. Here M_J is the projection of the momentum J on a chosen direction. This is known as the Stark effect. The shift ΔE_n of the energy level E_n in a rather weak static electric field of strength $\mathscr{E}$ is (in the lowest order of perturbation theory)

$$\Delta E_n = -\tfrac{1}{2}\alpha_{zz}^{(n)}\mathscr{E}^2 ,$$

where $\alpha_{zz}^{(n)}$ is the component of the atomic polarizability tensor in the direction of the external field $\mathscr{E} = \mathscr{E}_z$, which also defines the mean value of the atomic dipole moment induced by the field:

$$\overline{d_z^{(n)}} = \frac{\partial \Delta E_n}{\partial \mathscr{E}} = \alpha_{zz}^{(n)}\mathscr{E}_z .$$

The atomic polarizability depends on the unperturbed atomic state n, as well as on the quantum number M_J. This dependence can be expressed in general as follows [5.6.1]:

$$\alpha^{(n)} = \alpha_n + 2\beta_n[M_J^2 - \tfrac{1}{3}J(J+1)] , \tag{5.1}$$

where α_n, β_n are constants. The second term in (5.1) is equal to zero in a summation over all values of M_J, and it is also equal to zero when $J = 1/2$. Thus, the atomic polarizability averaged over all the M-states of the J-level is a scalar quantity.

The atomic polarizability is simply connected with the permittivity of a gas or vapour, and this relationship has the form

$$\varepsilon = 1 + 4\pi N\alpha , \tag{5.2}$$

where N is the number density of the gas particles. If the angular momentum of the atoms composing the gas is different from zero, then the value of the polarizability averaged over the projections M_J of the electron angular momentum enters (5.2).

Table 5.7 presents values of atomic scalar polarizabilities [5.6.2–5]. These values are grouped into accuracy classes according to the definition in the Introduction.

Magnetic Field. The magnetic field splits the atomic energy levels removing completely their degeneracy with respect to the direction of the total atomic angular momentum J. This is the Zeeman effect. The energy splitting of the level characterized by the quantum numbers J, L, S (when the LS-coupling scheme is valid) in a rather weak homogeneous magnetic field of strength H is [5.6.1]

$$\Delta E_J = \mu_B g M_J H , \tag{5.3}$$

where $\mu_B = e\hbar/2m_e c$ is the Bohr magneton, M_J is the projection of the total electronic angular momentum on the direction of the magnetic field, and the Landé g factor (gyromagnetic ratio) is given by

$$g = 1 + [J(J+1) - L(L+1) + S(S+1)]/[2J(J+1)] .$$

To derive (5.3), we proceeded from the assumption that the quadratic field effects could be neglected.

If the spin and orbital angular momenta of an atom are equal to zero ($L = S = 0$), the shift of the atomic energy level is [5.6.1]

$$\Delta E = -\chi H^2/2 ,$$

where the factor χ is called the magnetic susceptibility of the atom. This also characterizes the average magnetic moment of the atom induced by the field:

$$\bar{\mu}_z = \chi H , \quad H = H_z .$$

In Table 5.8 measured and calculated values of the magnetic susceptibility χ [10^{-6} cm^3/mol] are given [5.6.6, 7].

References

5.6.1 L.D.Landau, E.M.Lifshitz: *Quantum Mechanics (Non-Relativistic Theory)* 3rd ed. (Pergamon, London 1977)
5.6.2 R.R.Teachout, R.T.Pack: At. Data **3**, 195 (1971)
5.6.3 P.J.Leonard: At. Data Nucl. Data Tables **14**, 21 (1974)
5.6.4 T.M.Miller, B.Bederson: "Atomic and molecular polarizabilities – A review of recent advances", Adv. At. Mol. Phys. **13**, 1–55 (1977)
5.6.5 A.Dalgarno: Adv. Phys. **11**, 281 (1962)
5.6.6 E.A.Reinsch, W.Meyer: Phys. Rev. **A 14**, 915 (1976)
5.6.7 W.R.Johnson, F.D.Feiok: Phys. Rev. **168**, 22 (1968)

Table 5.7. Dipole polarizabilities of atomic species

Atomic number Z	Element (electronic term)	Dipole polarizability [Å^3]	[a_0^3]	Accuracy
1	2	3	4	5
1	H (^2L)	$\alpha^{(n)}(L) = n^6 + \frac{7}{4} n^4 (l^2 + l + 2) a_0^3$, where n, l are the principal and orbital quantum numbers of electron		Exact
	H ($1s - {}^2S$)	0.6668	4.5	Exact
	H ($2p - {}^2P^o$)	26.08	176	Exact
	H^- (^1S)	30.5	206	*B*
2	He ($1\,^1S$)	0.205	1.383	*A*
	He ($2\,^3S$)	46.8	316	*A*
	He ($2\,^1S$)	119	803	*A*
3	Li (^2S)	24	162	*B*
	Li^+ (^1S)	0.0285	0.1925	*A*
	Li^+ ($2\,^3S$)	6.965	47.0	*A*
	Li^+ ($2\,^1S$)	14.7	99	*B*
	Li^- (^1S)	96	650	*C*
4	Be (^1S)	5.6	38	*B*
5	B (^2P)	3.0	20.5	*B*
6	C (^3P)	1.75	11.8	*B*
7	N (^4S)	1.11	7.5	*B*
8	O (^3P)	0.80	5.41	*B*
9	F (^2P)	0.56	3.76	*B*
	F^- (^1S)	0.76	5.1	*C*
10	Ne (^1S)	0.397	2.68	*A*
	Ne* (^3P)	27.6	186	*B*
11	Na (^2S)	24	162	*B*
	Na^+ (^1S)	1.7	12	*C*
12	Mg (^1S)	11	72	*C*
13	Al (^2P)	8.7	59	*C*
14	Si (^2P)	5.5	37	*C*
15	P (^4P)	3.6	24	*C*
16	S (^3P)	2.7	18	*C*
17	Cl (^2P)	2.1	14	*C*
	Cl^- (^1S)	2.8	19	*C*
18	Ar (^1S)	1.64	11.08	*A*
	Ar* (^3P)	7.1	47.8	*B*
19	K (^2S)	43	290	*B*
	K^+ (^1S)	1.2	8.1	*C*
20	Ca (^1S)	25	170	*C*
21	Sc (^2D)	23	160	*C*
22	Ti (^3F)	22	150	*D*

Table 5.7 (continued)

Atomic number Z	Element (electronic term)	Dipole polarizability [Å^3]	[a_0^3]	Accuracy
1	2	3	4	5
23	V (^4F)	19	130	*D*
24	Cr (^7S)	11	74	*D*
25	Mn (^6S)	15	100	*D*
26	Fe (^5D)	13	90	*D*
27	Co (^4F)	11	74	*D*
28	Ni (^3F)	10	70	*D*
29	Cu (^2S)	6	40	*D*
30	Zn (^1S)	7	50	*D*
35	Br (^2P)	4	30	*D*
	Br^- (^1S)	4.1	28	*C*
36	Kr (^1S)	2.48	16.74	*A*
	Kr* (^3P)	50.6	341	*B*
37	Rb (^2S)	48	320	*B*
	Rb^+ (^1S)	1.8	12.1	*C*
38	Sr (^1S)	28	190	*C*
47	Ag (^2S)	10	67	*D*
48	Cd (^1S)	9	60	*D*
53	I (^2P)	4	27	*D*
54	Xe (^1S)	4.01	27.06	*A*
	Xe* (^3P)	63.2	426	*B*
55	Cs (^2S)	60	400	*B*
56	Ba (^1S)	40	270	*C*
74	W (^5D)	17	115	*D*
80	Hg (^1S)	5.1	34	*B*
82	Pb (^3P)	7.3	49	*B*

Table 5.8. Magnetic susceptibilities χ of atoms

Atom	χ [10^{-6} cm^3/mol]	Atom	χ [10^{-6} cm^3/mol]
He	1.884 (*A*)	Ar	19.3 (*B*)
Be	13 (*C*)	Ca	44 (*C*)
Ne	7.0 (*B*)	Kr	29 (*B*)
Mg	23 (*C*)	Xe	45 (*B*)

6. Energetics of Atomic Ions

Numerical data are presented for the ionization potentials of singly, doubly and triply charged atomic ions and also for the electron affinities of atoms. The energy levels of the lowest ($n = 2$) excited states in hydrogen- and heliumlike atomic ions are also assessed.

6.1 Ionization Potentials of Atomic Ions

Table 6.1 gives the ionization potentials (IP) of atomic ions with a net charge of +1, +2 and +3 (i.e. for ions X II, X III, X IV, if spectroscopic notation is used). Each value corresponds to the minimum energy which must be applied to the valence electron to transfer it into the continuum. There are also values of the ionization limit for a series of optical transitions, measured with respect to the ground state of the ion. These values are ionization potentials as well, expressed in units of cm^{-1} (conversion factor: 1 eV = 8065.48 cm^{-1}). In a separate column of Table 6.1 the configuration of the valence electron shell and the electronic term of the ionic ground state are specified. The designations of electron configurations and terms enclosed in braces are less reliable.

Key information concerning ionization potentials and ionization limits of positive atomic ions can be found in [5.1.1–9]. The numerical values listed are estimated to be accurate to within an error of not more than one, or occasionally two, in the last significant figure quoted.

Table 6.1. Ionization potentials of singly, doubly and triply ionized atoms

Atomic number Z	Element X	X II Valence electron configuration and electronic term	Ionization limit [cm^{-1}] (*upper line*); Ionization potential IP [eV] (*lower line*)	X III Valence electron configuration and electronic term	Ionization limit [cm^{-1}] (*upper line*); Ionization potential IP [eV] (*lower line*)	X IV Valence electron configuration and electronic term	Ionization limit [cm^{-1}] (*upper line*); Ionization potential IP [eV] (*lower line*)
1	2	3	4	5	6	7	8
2	He	$1s-{}^2S_{1/2}$	438908.89 54.418	–	–	–	–
3	Li	$1s^2-{}^1S_0$	610078 75.641	$1s-{}^2S_{1/2}$	987661.03 122.45	–	–
4	Be	$2s-{}^2S_{1/2}$	146882.9 18.211	$1s^2-{}^1S_0$	1241250 153.90	$1s-{}^2S_{1/2}$	1756018.8 217.72
5	B	$2s^2-{}^1S_0$	202887 25.155	$2s-{}^2S_{1/2}$	305931 37.931	$1s^2-{}^1S_0$	2092001.4 259.38
6	C	$2p-{}^2P^o_{1/2}$	196665 24.384	$2s^2-{}^1S_0$	386241 47.89	$2s-{}^2S_{1/2}$	520178 64.49
7	N	$2p^2-{}^3P_0$	238750 29.602	$2p-{}^2P^o_{1/2}$	382704 47.45	$2s^2-{}^1S_0$	624866 77.47
8	O	$2p^3-{}^4S^o_{3/2}$	283240 35.118	$2p^2-{}^3P_0$	443085 54.936	$2p-{}^2P^o_{1/2}$	624382 77.414
9	F	$2p^4-{}^3P_2$	282059 34.971	$2p^3-{}^4S^o_{3/2}$	505780 62.71	$2p^2-{}^3P_0$	702830 87.14
10	Ne	$2p^5-{}^2P^o_{3/2}$	330389 40.963	$2p^4-{}^3P_2$	512000 63.46	$2p^3-{}^4S^o_{3/2}$	783300 97.12
11	Na	$2p^6-{}^1S_0$	381390 47.287	$2p^5-{}^2P^o_{3/2}$	577650 71.620	$2p^4-{}^3P_2$	797800 98.92

12	Mg	$3s - {}^2S_{1/2}$	121267.6 15.0354	$2p^6 - {}^1S_0$	646400 80.144	$2p^5 - {}^2P^o_{3/2}$	881290 109.27
13	Al	$3s^2 - {}^1S_0$	151863 18.829	$3s - {}^2S_{1/2}$	229446 28.448	$2p^6 - {}^1S_0$	967800 119.99
14	Si	$3p - {}^2P^o_{1/2}$	131838 16.3460	$3s^2 - {}^1S_0$	270139 33.493	$3s - {}^2S_{1/2}$	364093 45.142
15	P	$3p^2 - {}^3P_0$	159451 19.770	$3p - {}^2P^o_{1/2}$	243601 30.203	$3s^2 - {}^1S_0$	414923 51.444
16	S	$3p^3 - {}^4S^o_{3/2}$	188233 23.338	$3p^2 - {}^3P_0$	280900 34.83	$3p - {}^2P^o_{1/2}$	381540 47.305
17	Cl	$3p^4 - {}^3P_2$	192070 23.814	$3p^3 - {}^4S^o_{3/2}$	319500 39.61	$3p^2 - {}^3P_0$	431230 53.47
18	Ar	$3p^5 - {}^2P^o_{3/2}$	222848.2 27.630	$3p^4 - {}^3P_2$	329966 40.911	$3p^3 - {}^4S^o_{3/2}$	482400 59.81
19	K	$3p^6 - {}^1S_0$	255100 31.63	$3p^5 - {}^2P^o_{3/2}$	369450 45.81	$3p^4 - {}^3P_2$	491300 60.91
20	Ca	$4s - {}^2S_{1/2}$	95751.9 11.872	$3p^6 - {}^1S_0$	410642 50.913	$3p^5 - {}^2P^o_{3/2}$	542600 67.3
21	Sc	$3d\,4s - {}^3D_1$	103237 12.800	$3d - {}^2D_{3/2}$	199677 24.757	$3p^6 - {}^1S_0$	592730 73.49
22	Ti	$3d^24s - {}^4F_{3/2}$	109490 13.58	$3d^2 - {}^3F_2$	221740 27.49	$3d - {}^2D_{3/2}$	348973 43.27
23	V	$3d^4 - {}^5D_0$	118200 14.66	$3d^3 - {}^4F_{3/2}$	236400 29.31	$3d^2 - {}^3F_2$	376700 46.71
24	Cr	$3d^5 - {}^6S_{5/2}$	133000 16.50	$3d^4 - {}^5D_0$	249700 31.0	$3d^3 - {}^4F_{3/2}$	396000 49.2
25	Mn	$3d^54s - {}^7S_3$	126145 15.640	$3d^5 - {}^6S_{5/2}$	271550 33.67	$3d^4 - {}^5D_0$	413000 51.2
26	Fe	$3d^64s - {}^6D_{9/2}$	130560 16.188	$3d^6 - {}^5D_4$	247200 30.65	$3d^5 - {}^6S_{5/2}$	442000 54.8

Table 6.1 (continued)

Atomic number Z	Element X	X II Valence electron configuration and electronic term	Ionization limit [cm^{-1}] (*upper line*); Ionization potential IP [eV] (*lower line*)	X III Valence electron configuration and electronic term	Ionization limit [cm^{-1}] (*upper line*); Ionization potential IP [eV] (*lower line*)	X IV Valence electron configuration and electronic term	Ionization limit [cm^{-1}] (*upper line*); Ionization potential IP [eV] (*lower line*)
1	2	3	4	5	6	7	8
27	Co	$3d^8 - {}^3F_4$	137790 17.084	$3d^7 - {}^4F_{9/2}$	270200 33.5	$3d^6 - {}^5D_4$	414000 51.3
28	Ni	$3d^9 - {}^2D_{5/2}$	146541.6 18.169	$3d^8 - {}^3F_4$	285000 35.3	$3d^7 - {}^4F_{9/2}$	443000 54.9
29	Cu	$3d^{10} - {}^1S_0$	163669 20.293	$3d^9 - {}^2D_{5/2}$	297140 36.84	$3d^8 - {}^3F_4$	463000 57.4
30	Zn	$4s - {}^2S_{1/2}$	144892 17.964	$3d^{10} - {}^1S_0$	320390 39.72	$3d^9 - {}^2D_{5/2}$	480500 59.57
31	Ga	$4s^2 - {}^1S_0$	165460 20.51	$4s - {}^2S_{1/2}$	247700 30.7	$3d^{10} - {}^1S_0$	517600 64.2
32	Ge	$4p - {}^2P^o_{1/2}$	128521.3 15.935	$4s^2 - {}^1S_0$	276036 34.2	$4s - {}^2S_{1/2}$	368700 45.7
33	As	$4p^2 - {}^3P_0$	149900 18.59	$4p - {}^2P^o_{1/2}$	228670 28.4	$4s^2 - {}^1S_0$	404370 50.1
34	Se	$4p^3 - {}^4S^o_{3/2}$	170700 21.16	$4p^2 - {}^3P_0$	248580 30.82	$4p - {}^2P^o_{1/2}$	346400 42.95
35	Br	$4p^4 - {}^3P_2$	175900 21.81	$4p^3 - {}^4S^o_{3/2}$	289529 35.90	$4p^2 - {}^3P_0$	381600 47.3
36	Kr	$4p^5 - {}^2P^o_{3/2}$	196475 24.360	$4p^4 - {}^3P_2$	298000 36.95	$4p^3 - {}^4S^o_{3/2}$	423600 52.5

37	Rb	$4p^6 - {}^1S_0$	220105 27.290	$4p^5 - {}^2P^o_{3/2}$	316600 39.2	$4p^4 - {}^3P_2$	424400 52.6
38	Sr	$5s - {}^2S_{1/2}$	88964 11.030	$4p^6 - {}^1S_0$	345880 42.88	$4p^5 - {}^2P^o_{3/2}$	453900 56.28
39	Y	$5s^2 - {}^1S_0$	98700 12.24	$5s - {}^2S_{1/2}$	165540 20.525	$4p^6 - {}^1S_0$	488830 60.61
40	Zr	$4d^25s - {}^4F_{3/2}$	105900 13.13	$4d^2 - {}^3F_2$	186000 23.1	$4d - {}^2D_{3/2}$	277606 34.419
41	Nb	$4d^4 - {}^5D_0$	115500 14.32	$4d^3 - {}^4F_{3/2}$	202000 25.0	$4d^2 - {}^3F_2$	304000 37.7
42	Mo	$4d^5 - {}^6S_{5/2}$	130300 16.16	$4d^4 - {}^5D_0$	219100 27.2	$4d^3 - {}^4F_{3/2}$	374180 46.4
43	Tc	$4d^55s - {}^7S_3$	123000 15.26	$4d^5 - {}^6S_{5/2}$	238300 29.5	$4d^4 - {}^5D_0$	–
44	Ru	$4d^7 - {}^4F_{9/2}$	135200 16.76	$4d^6 - {}^5D_4$	229600 28.5	$4d^5 - {}^6S_{5/2}$	–
45	Rh	$4d^8 - {}^3F_4$	145800 18.08	$4d^7 - {}^4F_{9/2}$	250500 31.1	$4d^6 - {}^5D_4$	–
46	Pd	$4d^9 - {}^2D_{5/2}$	156700 19.43	$4d^8 - {}^3F_4$	265600 32.9	$4d^7 - {}^4F_{9/2}$	–
47	Ag	$4d^{10} - {}^1S_0$	173300 21.49	$4d^9 - {}^2D_{5/2}$	280900 34.8	$4d^8 - {}^3F_4$	–
48	Cd	$5s - {}^2S_{1/2}$	136374.7 16.908	$4d^{10} - {}^1S_0$	302200 37.47	$4d^9 - {}^2D_{5/2}$	–
49	In	$5s^2 - {}^1S_0$	152195 18.87	$5s - {}^2S_{1/2}$	226100 28.0	$4d^{10} - {}^1S_0$	460000 57.0
50	Sn	$5s^25p - {}^2P^o_{1/2}$	118017 14.632	$5s^2 - {}^1S_0$	246020 30.50	$5s - {}^2S_{1/2}$	328550 40.74
51	Sb	$5p^2 - {}^3P_0$	133327.5 16.53	$5s^25p - {}^2P^o_{1/2}$	204248 25.32	$5s^2 - {}^1S_0$	356160 44.16

Table 6.1 (continued)

Atomic number Z	Element X	X II Valence electron configuration and electronic term	Ionization limit [cm^{-1}] (*upper line*); Ionization potential IP [eV] (*lower line*)	X III Valence electron configuration and electronic term	Ionization limit [cm^{-1}] (*upper line*); Ionization potential IP [eV] (*lower line*)	X IV Valence electron configuration and electronic term	Ionization limit [cm^{-1}] (*upper line*); Ionization potential IP [eV] (*lower line*)
1	2	3	4	5	6	7	8
52	Te	$5p^3 - {}^4S^o_{3/2}$	150000 18.6	$5p^2 - {}^3P_0$	225500 27.96	$5s^25p - {}^2P^o_{1/2}$	301776 37.42
53	I	$5p^4 - {}^3P_2$	154304 19.131	$5p^3 - {}^4S^o_{3/2}$	266000 33.0	$5p^2 - {}^3P_0$	–
54	Xe	$5p^5 - {}^2P^o_{3/2}$	169200 20.98	$5p^4 - {}^3P_2$	250000 31.0	$5p^3 - {}^4S^o_{3/2}$	360000 45
55	Cs	$5p^6 - {}^1S_0$	187000 23.15	$5p^5 - {}^2P^o_{3/2}$	270000 33.4	$5p^4 - {}^3P_2$	370000 46
56	Ba	$6s - {}^2S_{1/2}$	80686.9 10.004	$5p^6 - {}^1S_0$	289000 35.8	$5p^5 - {}^2P^o_{3/2}$	380000 47
57	La	$5d^2 - {}^3F_2$	89200 11.1	$5d - {}^2D_{3/2}$	154675 19.18	$5p^6 - {}^1S_0$	403000 49.9
58	Ce	$4f5d^2 - {}^4H^o_{7/2}$	87000 10.8	$4f^2 - {}^3H_4$	162900 20.20	$4f - {}^2F^o_{5/2}$	296500 36.76
59	Pr	$4f^3\,({}^4I^o_{9/2})\,6s\,(9/2,\,1/2)^o_4$	85000 10.6	$4f^3 - {}^4I^o_{9/2}$	174400 21.62	$4f^2 - {}^3H_4$	314400 38.98
60	Nd	$4f^46s - {}^6I_{7/2}$	86000 10.7	$4f^4 - {}^5I_4$	179000 22.1	$4f^3 - {}^4I^o_{9/2}$	326000 40.4
61	Pm	$4f^56s - {}^7H^o_2$	88000 10.9	$4f^5 - {}^6H^o_{5/2}$	180000 22.3	$4f^4 - {}^5I_4$	330000 41.0

62	Sm	$4f^66s-{}^8F_{1/2}$	89000 11.1	$4f^6-{}^7F_0$	189000 23.4	$4f^5-{}^6H^o_{5/2}$	330000 41.4
63	Eu	$4f^76s-{}^9S^o_4$	90660 11.24	$4f^7-{}^8S^o_{7/2}$	201000 24.9	$4f^6-{}^7F_0$	340000 42.7
64	Gd	$4f^75d6s-{}^{10}D^o_{5/2}$	97000 12.1	$4f^75d-{}^9D^o_2$	166000 20.6	$4f^7-{}^8S^o_{7/2}$	350000 44.0
65	Tb	$4f^9\,({}^6H^o_{15/2})\,6s\,(15/2,\ 1/2)^o_8$	93000 11.5	$4f^9-{}^6H^o_{15/2}$	177000 21.9	$4f^8-{}^7F_6$	317000 39.4
66	Dy	$4f^{10}\,({}^5I_8)\,6s\,(8,\ 1/2)_{17/2}$	94000 11.7	$\{4f^{10}-{}^5I_8\}$	184000 22.8	$4f^9-{}^6H^o_{15/2}$	330000 41.4
67	Ho	$4f^{11}\,({}^4I^o_{15/2})\,6s\,(15/2,\ 1/2)^o_8$	95000 11.8	$4f^{11}-{}^4I^o_{15/2}$	184000 22.8	$4f^{10}-{}^5I_8$	340000 42.5
68	Er	$4f^{12}\,({}^3H_6)\,6s\,(6,\ 1/2)_{13/2}$	96000 11.9	$4f^{12}-{}^3H_6$	183000 22.7	$4f^{11}-{}^4I^o_{15/2}$	340000 42.7
69	Tm	$4f^{13}\,({}^2F^o_{7/2})\,6s\,(7/2,\ 1/2)^o_4$	97000 12.1	$4f^{13}-{}^2F^o_{7/2}$	191000 23.7	$4f^{12}-{}^3H_6$	340000 42.7
70	Yb	$4f^{14}6s-{}^2S_{1/2}$	98300 12.18	$4f^{14}-{}^1S_0$	202100 25.05	$4f^{13}-{}^2F^o_{7/2}$	351000 43.6
71	Lu	$4f^{14}6s^2-{}^1S_0$	112000 13.9	$4f^{14}6s-{}^2S_{1/2}$	169050 20.96	$4f^{14}-{}^1S_0$	365000 45.25
72	Hf	$5d6s^2-{}^2D_{3/2}$	120000 14.9	$5d^2-{}^3F_2$	187800 23.3	$5d-{}^2D_{3/2}$	269000 33.4
73	Ta	$5d^36s-{}^5F_1$	–	–	–	–	–
74	W	$5d^46s-{}^6D_{1/2}$	–	–	–	–	–
75	Re	$5d^56s-{}^7S_3$	–	–	–	–	–
76	Os	$5d^66s-{}^6D_{9/2}$	–	–	–	–	–
77	Ir	$5d^76s-{}^5F_5$	–	–	–	–	–
78	Pt	$5d^9-{}^2D_{5/2}$	149720 18.56	–	–	–	–

Table 6.1 (continued)

Atomic number Z	Element X	XII Valence electron configuration and electronic term	Ionization limit [cm^{-1}] (*upper line*); Ionization potential IP [eV] (*lower line*)	XIII Valence electron configuration and electronic term	Ionization limit [cm^{-1}] (*upper line*); Ionization potential IP [eV] (*lower line*)	XIV Valence electron configuration and electronic term	Ionization limit [cm^{-1}] (*upper line*); Ionization potential IP [eV] (*lower line*)
1	2	3	4	5	6	7	8
79	Au	$5d^{10} - {}^1S_0$	165000 20.5	$5d^9 - {}^2D_{5/2}$	270000 34	$5d^8 - {}^3F_4$	350000 43
80	Hg	$5d^{10}6s - {}^2S_{1/2}$	151280 18.76	$5d^{10} - {}^1S_0$	276000 34.2	$5d^9 - {}^2D_{5/2}$	370000 46
81	Tl	$6s^2 - {}^1S_0$	164760 20.43	$6s - {}^2S_{1/2}$	240770 29.85	–	–
82	Pb	$6s^26p - {}^2P^o_{1/2}$	121245.1 15.033	$6s^2 - {}^1S_0$	257590 31.94	$6s - {}^2S_{1/2}$	341440 42.33
83	Bi	$6p^2 - {}^3P_0$	135000 16.7	$6s^26p - {}^2P^o_{1/2}$	206180 25.56	$6s^2 - {}^1S_0$	365500 45.3
84	Po	–	–	–	–	–	–
85	At	–	–	–	–	–	–
86	Rn	–	–	–	–	–	–
87	Fr	–	–	–	–	–	–
88	Ra	$7s - {}^2S_{1/2}$	81842.3 10.15	–	–	–	–
89	Ac	$7s^2 - {}^1S_0$	95000 11.75	$6p^67s - {}^2S_{1/2}$	– 20	–	–
90	Th	$6d^27s - {}^4F_{3/2}$	96000 11.9	$5f6d - {}^3H^o_4$	148000 18.3	$5f - {}^2F^o_{5/2}$	231900 28.7

91	Pa	$5f^27s^2-{}^3H_4$	–	–	–	$\{5f^2-{}^3H_4\}$	–
92	U	$5f^37s^2-{}^4I^o_{9/2}$	96000 11.9	–	160000 20	$5f^3-{}^4I^o_{9/2}$	300000 37
93	Np	$5f^36d7s^2-{}^5L^o_6$	–	–	–	$5f^4-{}^5I_4$	–
94	Pu	$5f^67s-{}^8F_{1/2}$	–	–	–	$5f^5-{}^6H^o_{5/2}$	–
95	Am	$5f^77s-{}^9S^o_4$	–	$5f^7-{}^8S^o_{7/2}$	–	$5f^6-{}^7F_6$	–
96	Cm	$5f^77s^2-{}^8S^o_{7/2}$	–	–	–	$5f^7-{}^8S^o_{7/2}$	–
97	Bk	$5f^97s-{}^7H^o_8$	–	–	–	$5f^8-{}^7F_6$	–
98	Cf	$5f^{10}({}^5I_8)\,7s\,(8,\,1/2)_{17/2}$	–	–	–	$5f^9-{}^6H^o_{15/2}$	–
99	Es	$5f^{11}({}^4I^o_{15/2})\,7s\,(15/2,\,1/2)^o_8$	–	$5f^{11}-{}^4I^o_{15/2}$	–	$\{5f^{10}-{}^5I_8\}$	–
100	Fm	–	–	–	–	$\{5f^{11}-{}^4I^o_{15/2}\}$	–
101	Md	–	–	–	–	$\{5f^{12}-{}^3H_6\}$	–
102	No	$\{5f^{14}7s-{}^2S_{1/2}\}$	–	$\{5f^{14}-{}^1S_0\}$	–	$\{5f^{13}-{}^2F^o_{7/2}\}$	–

6.2 Electron Affinities of Atoms

Table 6.2 gives the values of electron affinities (EA) of atomic species, i.e. the amount of energy which is needed to separate the valence electron from the negative ion. The table shows the spectroscopic symbols for the negative ion states, the designation of the outer electron shell and the binding energy (EA) of an electron. Data in the table are grouped according to the accuracy classes defined in the Introduction. The cases when the absence of a negative ion of a given sort is reported are divided into two classes of reliability: (a) a stable negative ion of a given element does not exist; (b) the existence of a stable negative ion is deemed unlikely, but it is necessary to investigate further.

Data for Table 6.2 were taken from a monograph [6.2.1] which, together with other monographs [6.2.2, 3] and reviews [6.2.4–9], describes the modern methods of determining the electron binding energy.

References

6.2.1 B.M.Smirnov: *Negative Ions* (McGraw-Hill, New York 1981)
6.2.2 H.S.W.Massey: *Negative Ions*, 3rd ed. (Cambridge University Press, Cambridge 1976)
6.2.3 F.M.Page, G.C.Goode: *Negative Ions and the Magnetron* (Wiley, London 1969)
6.2.4 B.H.Steiner: "Photodetachment Cross Sections and Electron Affinities", in *Case Studies in Atomic Collision Physics II*, ed. by E.W.McDaniel, M.R.C.McDowell (North-Holland, Amsterdam 1972) Chap. 7, pp. 485–545
6.2.5 H.Hotop, W.C.Lineberger: "Binding energies in atomic negative ions", J. Phys. Chem. Ref. Data **4**, 539–576 (1975)
6.2.6 H.Walther: "Atomic and Molecular Spectroscopy with Lasers", in *Laser Spectroscopy of Atoms and Molecules*, ed. by H.Walther, Topics Appl. Phys., Vol. 2 (Springer, Berlin, Heidelberg, New York 1976) pp. 1–124
6.2.7 B.L.Moiseiwitsch: "Negative Ions", in *Atomic Processes and Applications*, ed. by P.G.Burke, B.L.Moiseiwitsch (North-Holland, Amsterdam 1976) Chap. 9
6.2.8 R. R. Corderman, W. C. Lineberger: "Negative ion spectroscopy", Annu. Rev. Phys. Chem. **30**, 347–378 (1979)
6.2.9 R.D.Mead, A.E.Stevens, W.C.Lineberger: "Photodetachment in Negative Ion Beams", in *Gas Phase Ion Chemistry*, ed. by M.T.Bowers, Vol. 3 (Academic, New York 1984) Chap. 22, pp. 213–248

Table 6.2. Electron affinities EA of atoms

Atomic number Z	Negative ion and its term	Electron configuration	EA [eV]	Classes of accuracy and reliability
1	2	3	4	5
1	H^- (^1S)	$1s^2$	0.75421	*A*
2	He^- (^4P)	$1s\,2s\,2p$	0.077	*B*
3	Li^- (^1S)	$1s^2\,2s^2$	0.618	*A*
4	Be^- (^2S)	$2s^2\,3s$	doesn't exist	*a*
5	B^- (^3P)	$2s^2\,2p^2$	0.28	*C*
6	C^- (^4S)	$2s^2\,2p^3$	1.263	*A*
	C^- (^2D)	$2s^2\,2p^3$	0.035	*B*
7	N^-	$2s^2\,2p^4$	doesn't exist	*a*
8	O^- (^2P)	$2s^2\,2p^5$	1.46	*A*
9	F^- (^1S)	$2s^2\,2p^6$	3.40	*A*
10	Ne^-	$2p^6\,3s$	doesn't exist	*a*
11	Na^- (^1S)	$3s^2$	0.5479	*A*
12	Mg^-	–	doesn't exist	*a*
13	Al^- ($^3P_0$)	$3p^2$	0.44	*B*
	Al^- ($^1D_2$)	$3p^2$	0.33	*B*
14	Si^- (^4S)	$3p^3$	1.39	*A*
	Si^- (^2D)	$3p^3$	0.52	*B*
	Si^- (^2P)	$3p^3$	0.030	*B*
15	P^- (^3P)	$3p^4$	0.746	*A*
16	S^- (^2P)	$3p^5$	2.07712	*A*
17	Cl^- (^1S)	$3p^6$	3.62	*A*
18	Ar^-	–	doesn't exist	*a*
19	K^- (^1S)	$4s^2$	0.5015	*A*
20	Ca^- (^2D)	$3d\,4s^2$	doesn't exist	*a*
21	Sc^- (^1D)	$3d\,4s^2\,4p$	0.19	*C*
	Sc^- (^3D)	$3d\,4s^2\,4p$	0.04	*D*
22	Ti^- (^4F)	$3d^3\,4s^2$	0.08	*D*
23	V^- (^5D)	$3d^4\,4s^2$	0.53	*C*
24	Cr^- (^6S)	$3d^5\,4s^2$	0.67	*B*
25	Mn^- (^5D)	$3d^6\,4s^2$	doesn't exist	*b*
26	Fe^- (^4F)	$3d^7\,4s^2$	0.16	*D*
27	Co^- (^3F)	$3d^8\,4s^2$	0.66	*B*
28	Ni^- (^2D)	$3d^9\,4s^2$	1.16	*A*
29	Cu^- (^1S)	$3d^{10}\,4s^2$	1.23	*A*
30	Zn^-	–	doesn't exist	*b*
31	Ga^- (^3P)	$4p^2$	0.3	*D*
32	Ge^- (^4S)	$4p^3$	1.2	*C*
33	As^- (^3P)	$4p^4$	0.80	*C*
34	Se^- (^2P)	$4p^5$	2.021	*A*

Table 6.2 (continued)

Atomic number Z	Negative ion and its term	Electron configuration	EA [eV]	Classes of accuracy and reliability
1	2	3	4	5
35	Br^- (^1S)	$4p^6$	3.37	*A*
36	Kr^-	–	doesn't exist	*a*
37	Rb^- (^1S)	$5s^2$	0.4859	*A*
38	Sr^- (^2D)	$4d\,5s^2$	doesn't exist	*b*
39	Y^- (^1D)	$4d\,5s^2\,5p$	0.31	*C*
	Y^- (^3D)	$4d\,5s^2\,5p$	0.16	*D*
40	Zr^- (^4F)	$4d^3\,5s^2$	0.43	*B*
41	Nb^- (^5D)	$4d^4\,5s^2$	0.89	*B*
42	Mo^- (^6S)	$4d^5\,5s^2$	0.75	*B*
43	Tc^- (^5D)	$4d^6\,5s^2$	0.6	*D*
44	Ru^- (^4F)	$4d^7\,5s^2$	1.1	*D*
45	Rh^- (^3F)	$4d^8\,5s^2$	1.14	*A*
46	Pd^- (^2D)	$4d^9\,5s^2$	0.56	*B*
47	Ag^- (^1S)	$4d^{10}\,5s^2$	1.30	*A*
48	Cd^-	–	doesn't exist	*a*
49	In^- (^3P)	$5p^2$	0.3	*D*
50	Sn^- (^4S)	$5p^3$	1.2	*C*
51	Sb^- (^3P)	$5p^4$	1.1	*C*
52	Te^- (^2P)	$5p^5$	1.971	*A*
53	I^- (^1S)	$5p^6$	3.059	*A*
54	Xe^-	–	doesn't exist	*a*
55	Cs^- (^1S)	$6s^2$	0.4716	*A*
56	Ba^- (^2D)	$5d\,6s^2$	doesn't exist	*a*
57	La^- (^3F)	$5d^2\,6s^2$	0.5	*D*
72	Hf^- (^4F)	$5d^3\,6s^2$	doesn't exist	*b*
73	Ta^- (^5D)	$5d^4\,6s^2$	0.32	*C*
74	W^- (^6S)	$5d^5\,6s^2$	0.82	*A*
75	Re^- (^5D)	$5d^6\,6s^2$	0.15	*D*
76	Os^- (^4F)	$5d^7\,6s^2$	1.4	*D*
77	Ir^- (^3F)	$5d^8\,6s^2$	1.57	*A*
78	Pt^- (^2D)	$5d^9\,6s^2$	2.13	*A*
79	Au^- (^1S)	$5d^{10}\,6s^2$	2.3086	*A*
80	Hg^-	$6s^2\,6p$, $6s^2\,7s$	doesn't exist	*a*
81	Tl^- (^3P)	$6p^2$	0.3	*D*
82	Pb^- (^4S)	$6p^3$	0.37	*B*
83	Bi^- (^3P)	$6p^4$	0.95	*B*
84	Po^- (^2P)	$6p^5$	1.9	*D*
85	At^- (^1S)	$6p^6$	2.9	*D*

6.3 Energy Levels of Multiply Charged Atomic Ions

In Table 6.3 the theoretical values of the ionization potential and excitation energy for $2S$ and $2P$ states of hydrogen-like ions are given for a wide range of nuclear charge, $Z \lesssim 90$. These data were taken from [6.3.1, 2], which took into account many small effects of quantumelectrodynamic theory of one-electron systems. The accuracy of the numerical results for the energy levels of one-electron multiply charged atomic ions still surpasses essentially the accuracy with which they can be determined from the experimentally observed spectra in the wavelength range $\lambda \lesssim 10$ Å.

Information on ionization potentials and excited energy states with principal quantum number $n = 2$ in helium-like atomic ions ($Z < 100$) is presented in Table 6.4. To assist readers in identifying low-lying energy levels and in locating transitions, we present in Fig. 6.1 schematic diagrams of energy levels, with their labels, for some ions ($Z = 2$, 10, 40 and 80) in the helium isoelectronic sequence. As Z increases along the isoelectronic sequence, some of these levels are rearranged due to various contributions to the ion energy from the electron-electron interaction, relativistic effects and quantumelectrodynamic corrections.

Numerical data for energy levels are also shown in Table 6.4. For ions with $Z \leqslant 10$ we used the results of spectroscopic observations [6.3.3–5] and "exact" non-relativistic variational calculations from [6.3.6–8], for $10 < Z < 42$ our data are based on the results of theoretical work [6.3.9–13] which attempts to take into account all the effects needed for energy calculations in the framework of relativistic theory. Finally, for $Z > 43$ we used the calculated relativistic energy values from [6.3.14–16]. These energy data agree with the results derived from the observations of spectra of helium-like ions with $10 < Z \lesssim 30$.

The numerical values for energy levels included in Tables 6.3, 4 were rounded off in such a manner that further possible refinement might change only the last quoted significant figure within the range $\pm 1 - \pm 2$.

References

6.3.1 G.W.Erickson: J. Phys. Chem. Ref. Data **6**, 831 (1977)
6.3.2 P.J.Mohr: At. Data Nucl. Data Tables **29**, 453 (1983) ($10 \leq Z \leq 40$)
6.3.3 W.A.Davis, R.Marrus: Phys. Rev. **A 15**, 1963 (1977)
6.3.4 J.Hata, I.P.Grant: J. Phys. **B 14**, 2111 (1981)
6.3.5 M.F.Stamp, I.A.Armour, N.J.Peacock, J.D.Siever: J. Phys. **B 14**, 3551 (1981)
6.3.6 Y.Accad, C.L.Pekeris, B.Schiff: Phys. Rev. **A 4**, 516 (1971)
6.3.7 B.Schiff, Y.Accad, C.L.Pekeris: Phys. Rev. **A 8**, 2272 (1973)
6.3.8 K.Frankowski, C.L.Pekeris: Phys. Rev. **146**, 46 (1966)
6.3.9 A.M.Ermolaev: Phys. Rev. **A 8**, 1651 (1973)
6.3.10 A.M.Ermolaev, M.Jones: J. Phys. **B 7**, 199 (1974)
6.3.11 U.I.Safronova: Phys. Scr. **23**, 241 (1981)
6.3.12 W.C.Martin: Phys. Scr. **24**, 725 (1981)

6.3.13 R.DeSerio, H.G.Berry, R.L.Brooks, J.Hardis, A.E.Livingston, S.J.Hinterlong: Phys. Rev. **A 24**, 1872 (1981)
6.3.14 W.R.Johnson, C.D.Lin: Phys. Rev. **A 14**, 565 (1976)
6.3.15 L.N.Ivanov, E.P.Ivanova, U.I.Safronova: J. Quant. Spectrosc. Radiat. Transfer **15**, 553 (1975)
6.3.16 G.W.F.Drake: Phys. Rev. **A 19**, 1387 (1979)

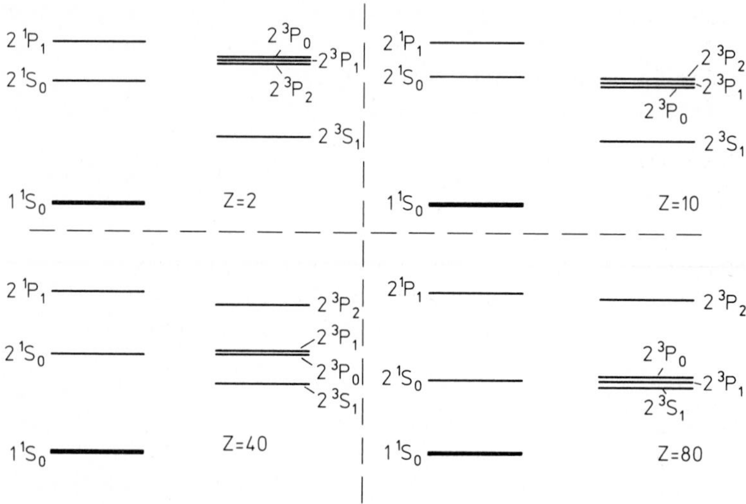

Fig. 6.1. Ordering of low-lying levels for helium isoelectronic sequence

Table 6.3. Low-lying terms for the H (nl)-like isoelectronic sequence: $n = 1, 2$; $2 \leqslant Z \leqslant 94$

Atomic number Z	Ground-state term; ionization potential IP $[10^6\ \mathrm{cm}^{-1}]$	Excited-state terms	Excitation energy $T_k\,[10^6\ \mathrm{cm}^{-1}]$
1	2	3	4
2	^{4}He II $(1s - {}^2S_{1/2})$ 0.43890889	$2p - {}^2P_{1/2}$ $2s - {}^2S_{1/2}$ $2p - {}^2P_{3/2}$	0.3291793 0.3291798 0.3291852
3	^{7}Li III $(1s - {}^2S_{1/2})$ 0.9876610	$2p - {}^2P_{1/2}$ $2s - {}^2S_{1/2}$ $2p - {}^2P_{3/2}$	0.7407344 0.7407364 0.7407640
4	^{9}Be IV $(1s - {}^2S_{1/2})$ 1.7560188	$2p - {}^2P_{1/2}$ $2s - {}^2S_{1/2}$ $2p - {}^2P_{3/2}$	1.316979 1.316985 1.317073
5	^{11}B V $(1s - {}^2S_{1/2})$ 2.744108	$2p - {}^2P_{1/2}$ $2s - {}^2S_{1/2}$ $2p - {}^2P_{3/2}$	2.057998 2.058012 2.058227
6	^{12}C VI $(1s - {}^2S_{1/2})$ 3.9520615	$2p - {}^2P_{1/2}$ $2s - {}^2S_{1/2}$ $2p - {}^2P_{3/2}$	2.963878 2.963904 2.964353
7	^{14}N VII $(1s - {}^2S_{1/2})$ 5.380089	$2p - {}^2P_{1/2}$ $2s - {}^2S_{1/2}$ $2p - {}^2P_{3/2}$	4.034761 4.034807 4.035642
8	^{16}O VIII $(1s - {}^2S_{1/2})$ 7.028394	$2p - {}^2P_{1/2}$ $2s - {}^2S_{1/2}$ $2p - {}^2P_{3/2}$	5.270782 5.270855 5.272285
9	^{19}F IX $(1s - {}^2S_{1/2})$ 8.897240	$2p - {}^2P_{1/2}$ $2s - {}^2S_{1/2}$ $2p - {}^2P_{3/2}$	6.672119 6.672231 6.674527
10	^{20}Ne X $(1s - {}^2S_{1/2})$ 10.986873	$2p - {}^2P_{1/2}$ $2s - {}^2S_{1/2}$ $2p - {}^2P_{3/2}$	8.238937 8.239100 8.242610
11	^{23}Na XI $(1s - {}^2S_{1/2})$ 13.29767	$2p - {}^2P_{1/2}$ $2s - {}^2S_{1/2}$ $2p - {}^2P_{3/2}$	9.971493 9.971720 9.976874
12	^{24}Mg XII $(1s - {}^2S_{1/2})$ 15.82994	$2p - {}^2P_{1/2}$ $2s - {}^2S_{1/2}$ $2p - {}^2P_{3/2}$	11.869985 11.870294 11.877612
13	^{27}Al XIII $(1s - {}^2S_{1/2})$ 18.58412	$2p - {}^2P_{1/2}$ $2s - {}^2S_{1/2}$ $2p - {}^2P_{3/2}$	13.93472 13.93513 13.94524
14	^{28}Si XIV $(1s - {}^2S_{1/2})$ 21.56060	$2p - {}^2P_{1/2}$ $2s - {}^2S_{1/2}$ $2p - {}^2P_{3/2}$	16.16596 16.16649 16.18011
15	^{31}P XV $(1s - {}^2S_{1/2})$ 24.7599	$2p - {}^2P_{1/2}$ $2s - {}^2S_{1/2}$ $2p - {}^2P_{3/2}$	18.56405 18.56473 18.58272

Table 6.3 (continued)

Atomic number Z	Ground-state term; ionization potential IP $[10^6\ cm^{-1}]$	Excited-state terms	Excitation energy $T_k\,[10^6\ cm^{-1}]$
1	2	3	4
16	^{32}S XVI $(1s - {}^2S_{1/2})$	$2p - {}^2P_{1/2}$	21.12930
	28.1825	$2s - {}^2S_{1/2}$	21.13015
		$2p - {}^2P_{3/2}$	21.15349
17	^{35}Cl XVII $(1s - {}^2S_{1/2})$	$2p - {}^2P_{1/2}$	23.86212
	31.8289	$2s - {}^2S_{1/2}$	23.86316
		$2p - {}^2P_{3/2}$	23.89298
18	^{40}Ar XVIII $(1s - {}^2S_{1/2})$	$2p - {}^2P_{1/2}$	26.76290
	35.6998	$2s - {}^2S_{1/2}$	26.76417
		$2p - {}^2P_{3/2}$	26.80174
19	^{39}K XIX $(1s - {}^2S_{1/2})$	$2p - {}^2P_{1/2}$	29.83198
	39.7956	$2s - {}^2S_{1/2}$	29.83352
		$2p - {}^2P_{3/2}$	29.88027
20	^{40}Ca XX $(1s - {}^2S_{1/2})$	$2p - {}^2P_{1/2}$	33.06990
	44.1172	$2s - {}^2S_{1/2}$	33.07174
		$2p - {}^2P_{3/2}$	33.12926
21	^{45}Sc XXI $(1s - {}^2S_{1/2})$	$2p - {}^2P_{1/2}$	36.47716
	48.6652	$2s - {}^2S_{1/2}$	36.47934
		$2p - {}^2P_{3/2}$	36.54940
22	^{48}Ti XXII $(1s - {}^2S_{1/2})$	$2p - {}^2P_{1/2}$	40.05419
	53.4404	$2s - {}^2S_{1/2}$	40.05675
		$2p - {}^2P_{3/2}$	40.14133
23	^{51}V XXIII $(1s - {}^2S_{1/2})$	$2p - {}^2P_{1/2}$	43.80155
	58.4435	$2s - {}^2S_{1/2}$	43.80453
		$2p - {}^2P_{3/2}$	43.90581
24	^{52}Cr XXIV $(1s - {}^2S_{1/2})$	$2p - {}^2P_{1/2}$	47.7198
	63.675	$2s - {}^2S_{1/2}$	47.7232
		$2p - {}^2P_{3/2}$	47.8436
25	^{55}Mn XXV $(1s - {}^2S_{1/2})$	$2p - {}^2P_{1/2}$	51.8095
	69.137	$2s - {}^2S_{1/2}$	51.8135
		$2p - {}^2P_{3/2}$	51.9555
26	^{56}Fe XXVI $(1s - {}^2S_{1/2})$	$2p - {}^2P_{1/2}$	56.0713
	74.829	$2s - {}^2S_{1/2}$	56.0759
		$2p - {}^2P_{3/2}$	56.2425
27	^{59}Co XXVII $(1s - {}^2S_{1/2})$	$2p - {}^2P_{1/2}$	60.5059
	80.752	$2s - {}^2S_{1/2}$	60.5111
		$2p - {}^2P_{3/2}$	60.7053
28	^{58}Ni XXVIII $(1s - {}^2S_{1/2})$	$2p - {}^2P_{1/2}$	65.1139
	86.908	$2s - {}^2S_{1/2}$	65.1198
		$2p - {}^2P_{3/2}$	65.3449
29	^{63}Cu XXIX $(1s - {}^2S_{1/2})$	$2p - {}^2P_{1/2}$	69.8961
	93.298	$2s - {}^2S_{1/2}$	69.9027
		$2p - {}^2P_{3/2}$	70.1624

Table 6.3 (continued)

Atomic number Z	Ground-state term; ionization potential IP $[10^6\ cm^{-1}]$	Excited-state terms	Excitation energy $T_k\ [10^6\ cm^{-1}]$
1	2	3	4
30	^{64}Zn XXX $(1s - {}^2S_{1/2})$	$2p - {}^2P_{1/2}$	74.8532
	99.922	$2s - {}^2S_{1/2}$	74.8606
		$2p - {}^2P_{3/2}$	75.1588
31	$^{69}Ga^{30+}$ $(1s - {}^2S_{1/2})$	$2p - {}^2P_{1/2}$	79.9854
	106.782	$2s - {}^2S_{1/2}$	79.9943
		$2p - {}^2P_{3/2}$	80.3351
32	$^{74}Ge^{31+}$ $(1s - {}^2S_{1/2})$	$2p - {}^2P_{1/2}$	85.2952
	113.879	$2s - {}^2S_{1/2}$	85.3045
		$2p - {}^2P_{3/2}$	85.6925
33	$^{75}As^{32+}$ $(1s - {}^2S_{1/2})$	$2p - {}^2P_{1/2}$	90.7818
	121.214	$2s - {}^2S_{1/2}$	90.7921
		$2p - {}^2P_{3/2}$	91.2322
34	$^{80}Se^{33+}$ $(1s - {}^2S_{1/2})$	$2p - {}^2P_{1/2}$	96.4466
	128.79	$2s - {}^2S_{1/2}$	96.4580
		$2p - {}^2P_{3/2}$	96.9552
35	$^{79}Br^{34+}$ $(1s - {}^2S_{1/2})$	$2p - {}^2P_{1/2}$	102.290
	136.60	$2s - {}^2S_{1/2}$	102.303
		$2p - {}^2P_{3/2}$	102.863
36	$^{84}Kr^{35+}$ $(1s - {}^2S_{1/2})$	$2p - {}^2P_{1/2}$	108.314
	144.66	$2s - {}^2S_{1/2}$	108.328
		$2p - {}^2P_{3/2}$	108.957
37	$^{85}Rb^{36+}$ $(1s - {}^2S_{1/2})$	$2p - {}^2P_{1/2}$	114.520
	152.96	$2s - {}^2S_{1/2}$	114.535
		$2p - {}^2P_{3/2}$	115.238
38	$^{88}Sr^{37+}$ $(1s - {}^2S_{1/2})$	$2p - {}^2P_{1/2}$	120.907
	161.51	$2s - {}^2S_{1/2}$	120.924
		$2p - {}^2P_{3/2}$	121.708
39	$^{89}Y^{38+}$ $(1s - {}^2S_{1/2})$	$2p - {}^2P_{1/2}$	127.477
	170.30	$2s - {}^2S_{1/2}$	127.496
		$2p - {}^2P_{3/2}$	128.369
40	$^{90}Zr^{39+}$ $(1s - {}^2S_{1/2})$	$2p - {}^2P_{1/2}$	134.232
	179.35	$2s - {}^2S_{1/2}$	134.252
		$2p - {}^2P_{3/2}$	135.222
41	$^{93}Nb^{40+}$ $(1s - {}^2S_{1/2})$	$2p - {}^2P_{1/2}$	141.16
	188.64	$2s - {}^2S_{1/2}$	141.19
		$2p - {}^2P_{3/2}$	142.26
42	$^{98}Mo^{41+}$ $(1s - {}^2S_{1/2})$	$2p - {}^2P_{1/2}$	148.29
	198.18	$2s - {}^2S_{1/2}$	148.32
		$2p - {}^2P_{3/2}$	149.50
47	$^{107}Ag^{46+}$ $(1s - {}^2S_{1/2})$	$2p - {}^2P_{1/2}$	186.76
	249.74	$2s - {}^2S_{1/2}$	186.80
		$2p - {}^2P_{3/2}$	188.69

Table 6.3 (continued)

Atomic number Z	Ground-state term; ionization potential IP [10^6 cm^{-1}]	Excited-state terms	Excitation energy T_k [10^6 cm^{-1}]
1	2	3	4
48	$^{114}Cd^{47+}$ ($1s - {}^2S_{1/2}$)	$2p - {}^2P_{1/2}$	195.04
	260.84	$2s - {}^2S_{1/2}$	195.08
		$2p - {}^2P_{3/2}$	197.14
49	$^{115}In^{48+}$ ($1s - {}^2S_{1/2}$)	$2p - {}^2P_{1/2}$	203.50
	272.20	$2s - {}^2S_{1/2}$	203.55
		$2p - {}^2P_{3/2}$	205.80
50	$^{120}Sn^{49+}$ ($1s - {}^2S_{1/2}$)	$2p - {}^2P_{1/2}$	212.17
	283.83	$2s - {}^2S_{1/2}$	212.22
		$2p - {}^2P_{3/2}$	214.67
51	$^{121}Sb^{50+}$ ($1s - {}^2S_{1/2}$)	$2p - {}^2P_{1/2}$	221.0
	295.7	$2s - {}^2S_{1/2}$	221.1
		$2p - {}^2P_{3/2}$	223.7
52	$^{130}Te^{51+}$ ($1s - {}^2S_{1/2}$)	$2p - {}^2P_{1/2}$	230.1
	307.9	$2s - {}^2S_{1/2}$	230.2
		$2p - {}^2P_{3/2}$	233.1
53	$^{127}I^{52+}$ ($1s - {}^2S_{1/2}$)	$2p - {}^2P_{1/2}$	239.4
	320.4	$2s - {}^2S_{1/2}$	239.4
		$2p - {}^2P_{3/2}$	242.6
54	$^{132}Xe^{53+}$ ($1s - {}^2S_{1/2}$)	$2p - {}^2P_{1/2}$	248.9
	333.1	$2s - {}^2S_{1/2}$	248.9
		$2p - {}^2P_{3/2}$	252.3
55	$^{133}Cs^{54+}$ ($1s - {}^2S_{1/2}$)	$2p - {}^2P_{1/2}$	258.5
	346.1	$2s - {}^2S_{1/2}$	258.6
		$2p - {}^2P_{3/2}$	262.3
56	$^{138}Ba^{55+}$ ($1s - {}^2S_{1/2}$)	$2p - {}^2P_{1/2}$	268.4
	359.4	$2s - {}^2S_{1/2}$	268.5
		$2p - {}^2P_{3/2}$	272.4
72	$^{180}Hf^{71+}$ ($1s - {}^2S_{1/2}$)	$2p - {}^2P_{1/2}$	457.0
	613.7	$2s - {}^2S_{1/2}$	457.2
		$2p - {}^2P_{3/2}$	469.0
73	$^{181}Ta^{72+}$ ($1s - {}^2S_{1/2}$)	$2p - {}^2P_{1/2}$	470.9
	632.4	$2s - {}^2S_{1/2}$	471.0
		$2p - {}^2P_{3/2}$	483.5
74	$^{184}W^{73+}$ ($1s - {}^2S_{1/2}$)	$2p - {}^2P_{1/2}$	484.9
	651.4	$2s - {}^2S_{1/2}$	485.1
		$2p - {}^2P_{3/2}$	498.3
75	$^{187}Re^{74+}$ ($1s - {}^2S_{1/2}$)	$2p - {}^2P_{1/2}$	499.2
	670.8	$2s - {}^2S_{1/2}$	499.4
		$2p - {}^2P_{3/2}$	513.5
76	$^{192}Os^{75+}$ ($1s - {}^2S_{1/2}$)	$2p - {}^2P_{1/2}$	514
	691	$2s - {}^2S_{1/2}$	514
		$2p - {}^2P_{3/2}$	529

Table 6.3 (continued)

Atomic number Z	Ground-state term; ionization potential IP $[10^6\ cm^{-1}]$	Excited-state terms	Excitation energy $T_k\,[10^6\ cm^{-1}]$
1	2	3	4
77	$^{193}Ir^{76+}$ $(1s-{}^2S_{1/2})$	$2p-{}^2P_{1/2}$	529
	711	$2s-{}^2S_{1/2}$	529
		$2p-{}^2P_{3/2}$	545
78	$^{195}Pt^{77+}$ $(1s-{}^2S_{1/2})$	$2p-{}^2P_{1/2}$	544
	731	$2s-{}^2S_{1/2}$	544
		$2p-{}^2P_{3/2}$	561
79	$^{197}Au^{78+}$ $(1s-{}^2S_{1/2})$	$2p-{}^2P_{1/2}$	559
	752	$2s-{}^2S_{1/2}$	560
		$2p-{}^2P_{3/2}$	577
80	$^{202}Hg^{79+}$ $(1s-{}^2S_{1/2})$	$2p-{}^2P_{1/2}$	575
	774	$2s-{}^2S_{1/2}$	575
		$2p-{}^2P_{3/2}$	594
81	$^{205}Tl^{80+}$ $(1s-{}^2S_{1/2})$	$2p-{}^2P_{1/2}$	591
	795	$2s-{}^2S_{1/2}$	591
		$2p-{}^2P_{3/2}$	611
82	$^{208}Pb^{81+}$ $(1s-{}^2S_{1/2})$	$2p-{}^2P_{1/2}$	607
	818	$2s-{}^2S_{1/2}$	608
		$2p-{}^2P_{3/2}$	629
83	$^{209}Bi^{82+}$ $(1s-{}^2S_{1/2})$	$2p-{}^2P_{1/2}$	624
	840	$2s-{}^2S_{1/2}$	624
		$2p-{}^2P_{3/2}$	647
90	$^{232}Th^{89+}$ $(1s-{}^2S_{1/2})$	$2p-{}^2P_{1/2}$	749
	1011	$2s-{}^2S_{1/2}$	750
		$2p-{}^2P_{3/2}$	783
92	$^{238}U^{91+}$ $(1s-{}^2S_{1/2})$	$2p-{}^2P_{1/2}$	788
	1064	$2s-{}^2S_{1/2}$	789
		$2p-{}^2P_{3/2}$	825
94	Pu^{93+} $(1s-{}^2S_{1/2})$	$2p-{}^2P_{1/2}$	829
	1120	$2s-{}^2S_{1/2}$	830
		$2p-{}^2P_{3/2}$	870

Table 6.4. Low-lying terms for the He (nl)-like isoelectronic sequence: $n = 1, 2$; $2 \leqslant Z < 100$

Atomic number Z	Ground-state term; ionization potential IP [10^6 cm^{-1}]	Excited-state terms	Excitation energy T_k [10^6 cm^{-1}]
1	2	3	4
2	He I ($1s^2 - {}^1S_0$)	$1s\,2s - {}^3S_1$	0.1598561
	0.1983108	${}^1S_0$	0.1662775
		$1s\,2p - {}^3P_2$	0.1690869
		${}^3P_1$	0.1690869
		${}^3P_0$	0.1690879
		${}^1P_1$	0.1711350
3	Li II ($1s^2 - {}^1S_0$)	$1s\,2s - {}^3S_1$	0.476035
	0.610078	${}^1S_0$	0.491375
		$1s\,2p - {}^3P_1$	0.494260
		${}^3P_2$	0.494262
		${}^3P_0$	0.494265
		${}^1P_1$	0.501811
4	Be III ($1s^2 - {}^1S_0$)	$1s\,2s - {}^3S_1$	0.956515
	1.241250	${}^1S_0$	0.981178
		$1s\,2p - {}^3P_1$	0.983368
		${}^3P_0$	0.983380
		${}^3P_2$	0.983383
		${}^1P_1$	0.997454
5	B IV ($1s^2 - {}^1S_0$)	$1s\,2s - {}^3S_1$	1.60116
	2.092001	${}^1S_0$	1.63572
		$1s\,2p - {}^3P_1$	1.63654
		${}^3P_0$	1.63655
		${}^3P_2$	1.63659
		${}^1P_1$	1.65798
6	C V ($1s^2 - {}^1S_0$)	$1s\,2s - {}^3S_1$	2.41127
	3.16240	${}^1S_0$	2.45501
		$1s\,2p - {}^3P_1$	2.45516
		${}^3P_0$	2.45517
		${}^3P_2$	2.45529
		${}^1P_1$	2.48337
7	N VI ($1s^2 - {}^1S_0$)	$1s\,2s - {}^3S_1$	3.38589
	4.45275	$1s\,2p - {}^3P_0$	3.43830
		${}^3P_1$	3.43832
		${}^3P_2$	3.43861
		$1s\,2s - {}^1S_0$	3.43930
		$1s\,2p - {}^1P_1$	3.47379
8	O VII ($1s^2 - {}^1S_0$)	$1s\,2s - {}^3S_1$	4.52464
	5.96311	$1s\,2p - {}^3P_0$	4.58562
		${}^3P_1$	4.58568
		${}^3P_2$	4.58624
		$1s\,2s - {}^1S_0$	4.58850
		$1s\,2p - {}^1P_1$	4.62920
9	F VIII ($1s^2 - {}^1S_0$)	$1s\,2s - {}^3S_1$	5.83031
	7.69381	$1s\,2p - {}^3P_0$	5.89990

Table 6.4 (continued)

Atomic number Z	Ground-state term; ionization potential IP [10^6 cm^{-1}]	Excited-state terms	Excitation energy T_k [10^6 cm^{-1}]
1	2	3	4
		$^3P_1$	5.90005
		$^3P_2$	5.90101
		$1s\,2s - {}^1S_0$	5.90290
		$1s\,2p - {}^1P_1$	5.94990
10	Ne IX ($1s^2 - {}^1S_0$) 9.64500	$1s\,2s - {}^3S_1$	7.30205
		$1s\,2p - {}^3P_0$	7.38032
		$^3P_1$	7.38062
		$^3P_2$	7.38217
		$1s\,2s - {}^1S_0$	7.38260
		$1s\,2p - {}^1P_1$	7.43660
11	Na X ($1s^2 - {}^1S_0$) 11.8170	$1s\,2s - {}^3S_1$	8.9353
		$1s\,2p - {}^3P_0$	9.0223
		$^3P_1$	9.0229
		$^3P_2$	9.0252
		$1s\,2s - {}^1S_0$	9.0280
		$1s\,2p - {}^1P_1$	9.0887
12	Mg XI ($1s^2 - {}^1S_0$) 14.2101	$1s\,2s - {}^3S_1$	10.7364
		$1s\,2p - {}^3P_0$	10.8325
		$^3P_1$	10.8331
		$^3P_2$	10.8366
		$1s\,2s - {}^1S_0$	10.8390
		$1s\,2p - {}^1P_1$	10.9069
13	Al XII ($1s^2 - {}^1S_0$) 16.8247	$1s\,2s - {}^3S_1$	12.7035
		$1s\,2p - {}^3P_0$	12.8082
		$^3P_1$	12.8095
		$^3P_2$	12.8146
		$1s\,2s - {}^1S_0$	12.8161
		$1s\,2p - {}^1P_1$	12.8915
14	Si XIII ($1s^2 - {}^1S_0$) 19.6612	$1s\,2s - {}^3S_1$	14.8367
		$1s\,2p - {}^3P_0$	14.9505
		$^3P_1$	14.9523
		$^3P_2$	14.9594
		$1s\,2s - {}^1S_0$	14.9595
		$1s\,2p - {}^1P_1$	15.0428
15	P XIV ($1s^2 - {}^1S_0$) 22.720	$1s\,2s - {}^3S_1$	17.136
		$1s\,2p - {}^3P_0$	17.259
		$^3P_1$	17.261
		$1s\,2s - {}^1S_0$	17.269
		$1s\,2p - {}^3P_2$	17.271
		$^1P_1$	17.360
16	S XV ($1s^2 - {}^1S_0$) 26.002	$1s\,2s - {}^3S_1$	19.602
		$1s\,2p - {}^3P_0$	19.734
		$^3P_1$	19.738
		$1s\,2s - {}^1S_0$	19.746

Table 6.4 (continued)

Atomic number Z	Ground-state term; ionization potential IP $[10^6\ \mathrm{cm}^{-1}]$	Excited-state terms	Excitation energy $T_k\,[10^6\ \mathrm{cm}^{-1}]$
1	2	3	4
		$1s\,2p - {}^3P_2$	19.751
		${}^1P_1$	19.846
17	Cl XVI $(1s^2 - {}^1S_0)$	$1s\,2s - {}^3S_1$	22.235
	29.507	$1s\,2p - {}^3P_0$	22.377
		${}^3P_1$	22.382
		$1s\,2s - {}^1S_0$	22.390
		$1s\,2p - {}^3P_2$	22.398
		${}^1P_1$	22.500
18	Ar XVII $(1s^2 - {}^1S_0)$	$1s\,2s - {}^3S_1$	25.036
	33.236	$1s\,2p - {}^3P_0$	25.187
		${}^3P_1$	25.193
		$1s\,2s - {}^1S_0$	25.201
		$1s\,2p - {}^3P_2$	25.215
		${}^1P_1$	25.322
19	K XVIII $(1s^2 - {}^1S_0)$	$1s\,2s - {}^3S_1$	28.005
	37.189	$1s\,2p - {}^3P_0$	28.166
		${}^3P_1$	28.173
		$1s\,2s - {}^1S_0$	28.181
		$1s\,2p - {}^3P_2$	28.201
		${}^1P_1$	28.313
20	Ca XIX $(1s^2 - {}^1S_0)$	$1s\,2s - {}^3S_1$	31.145
	41.367	$1s\,2p - {}^3P_0$	31.316
		${}^3P_1$	31.324
		$1s\,2s - {}^1S_0$	31.331
		$1s\,2p - {}^3P_2$	31.359
		${}^1P_1$	31.477
21	Sc XX $(1s^2 - {}^1S_0)$	$1s\,2s - {}^3S_1$	34.45
	45.77	$1s\,2p - {}^3P_0$	34.63
		${}^3P_1$	34.64
		$1s\,2s - {}^1S_0$	34.65
		$1s\,2p - {}^3P_2$	34.68
		${}^1P_1$	34.81
22	Ti XXI $(1s^2 - {}^1S_0)$	$1s\,2s - {}^3S_1$	37.93
	50.40	$1s\,2p - {}^3P_0$	38.12
		${}^3P_1$	38.13
		$1s\,2s - {}^1S_0$	38.14
		$1s\,2p - {}^3P_2$	38.18
		${}^1P_1$	38.31
23	V XXII $(1s^2 - {}^1S_0)$	$1s\,2s - {}^3S_1$	41.57
	55.26	$1s\,2p - {}^3P_0$	41.77
		${}^3P_1$	41.78
		$1s\,2s - {}^1S_0$	41.79
		$1s\,2p - {}^3P_2$	41.85
		${}^1P_1$	41.98

Table 6.4 (continued)

Atomic number Z	Ground-state term; ionization potential IP [10^6 cm^{-1}]	Excited-state terms	Excitation energy T_k [10^6 cm^{-1}]
1	2	3	4
24	Cr XXIII ($1s^2 - {}^1S_0$)	$1s\,2s - {}^3S_1$	45.38
	60.35	$1s\,2p - {}^3P_0$	45.60
		${}^3P_1$	45.61
		$1s\,2s - {}^1S_0$	45.61
		$1s\,2p - {}^3P_2$	45.69
		${}^1P_1$	45.83
25	Mn XXIV ($1s^2 - {}^1S_0$)	$1s\,2s - {}^3S_1$	49.37
	65.66	$1s\,2p - {}^3P_0$	49.59
		${}^3P_1$	49.61
		$1s\,2s - {}^1S_0$	49.61
		$1s\,2p - {}^3P_2$	49.71
		${}^1P_1$	49.85
26	Fe XXV ($1s^2 - {}^1S_0$)	$1s\,2s - {}^3S_1$	53.53
	71.20	$1s\,2p - {}^3P_0$	53.76
		${}^3P_1$	53.78
		$1s\,2s - {}^1S_0$	53.78
		$1s\,2p - {}^3P_2$	53.90
		${}^1P_1$	54.04
27	Co XXVI ($1s^2 - {}^1S_0$)	$1s\,2s - {}^3S_1$	57.86
	76.98	$1s\,2p - {}^3P_0$	58.10
		${}^3P_1$	58.12
		$1s\,2s - {}^1S_0$	58.12
		$1s\,2p - {}^3P_2$	58.26
		${}^1P_1$	58.41
28	Ni XXVII ($1s^2 - {}^1S_0$)	$1s\,2s - {}^3S_1$	62.36
	82.99	$1s\,2p - {}^3P_0$	62.62
		$1s\,2s - {}^1S_0$	62.64
		$1s\,2p - {}^3P_1$	62.64
		${}^3P_2$	62.80
		${}^1P_1$	62.95
29	Cu XXVIII ($1s^2 - {}^1S_0$)	$1s\,2s - {}^3S_1$	67.04
	89.22	$1s\,2p - {}^3P_0$	67.30
		$1s\,2s - {}^1S_0$	67.33
		$1s\,2p - {}^3P_1$	67.33
		${}^3P_2$	67.52
		${}^1P_1$	67.67
30	Zn XXIX ($1s^2 - {}^1S_0$)	$1s\,2s - {}^3S_1$	71.90
	95.70	$1s\,2p - {}^3P_0$	72.16
		$1s\,2s - {}^1S_0$	72.20
		$1s\,2p - {}^3P_1$	72.20
		${}^3P_2$	72.41
		${}^1P_1$	72.58
31	Ga XXX ($1s^2 - {}^1S_0$)	$1s\,2s - {}^3S_1$	76.9
	102.4	$1s\,2p - {}^3P_0$	77.2

Table 6.4 (continued)

Atomic number Z	Ground-state term; ionization potential IP $[10^6\ cm^{-1}]$	Excited-state terms	Excitation energy $T_k [10^6\ cm^{-1}]$
1	2	3	4
		$1s\,2s - {}^1S_0$	77.2
		$1s\,2p - {}^3P_1$	77.2
		${}^3P_2$	77.5
		${}^1P_1$	77.7
32	Ge^{30+} $(1s^2 - {}^1S_0)$ 109.3	$1s\,2s - {}^3S_1$	82.1
		$1s\,2p - {}^3P_0$	82.4
		$1s\,2s - {}^1S_0$	82.4
		$1s\,2p - {}^3P_1$	82.4
		${}^3P_2$	82.7
		${}^1P_1$	82.9
33	As^{31+} $(1s^2 - {}^1S_0)$ 116.5	$1s\,2s - {}^3S_1$	87.5
		$1s\,2p - {}^3P_0$	87.8
		$1s\,2s - {}^1S_0$	87.8
		$1s\,2p - {}^3P_1$	87.8
		${}^3P_2$	88.2
		${}^1P_1$	88.3
34	Se^{32+} $(1s^2 - {}^1S_0)$ 123.9	$1s\,2s - {}^3S_1$	93.0
		$1s\,2p - {}^3P_0$	93.4
		$1s\,2s - {}^1S_0$	93.4
		$1s\,2p - {}^3P_1$	93.4
		${}^3P_2$	93.8
		${}^1P_1$	94.0
35	Br^{33+} $(1s^2 - {}^1S_0)$ 131.6	$1s\,2s - {}^3S_1$	98.8
		$1s\,2p - {}^3P_0$	99.1
		$1s\,2s - {}^1S_0$	99.1
		$1s\,2p - {}^3P_1$	99.1
		${}^3P_2$	99.6
		${}^1P_1$	99.8
36	Kr^{34+} $(1s^2 - {}^1S_0)$ 139.5	$1s\,2s - {}^3S_1$	104.7
		$1s\,2p - {}^3P_0$	105.0
		$1s\,2s - {}^1S_0$	105.1
		$1s\,2p - {}^3P_1$	105.1
		${}^3P_2$	105.6
		${}^1P_1$	105.8
37	Rb^{35+} $(1s^2 - {}^1S_0)$ 147.6	$1s\,2s - {}^3S_1$	110.7
		$1s\,2p - {}^3P_0$	111.1
		$1s\,2s - {}^1S_0$	111.1
		$1s\,2p - {}^3P_1$	111.1
		${}^3P_2$	111.7
		${}^1P_1$	111.9
38	Sr^{36+} $(1s^2 - {}^1S_0)$ 156.0	$1s\,2s - {}^3S_1$	117.0
		$1s\,2p - {}^3P_0$	117.4
		$1s\,2s - {}^1S_0$	117.5
		$1s\,2p - {}^3P_1$	117.5

Table 6.4 (continued)

Atomic number Z	Ground-state term; ionization potential IP $[10^6\ cm^{-1}]$	Excited-state terms	Excitation energy $T_k\,[10^6\ cm^{-1}]$
1	2	3	4
		$^3P_2$	118.1
		$^1P_1$	118.3
39	$Y^{37+}\ (1s^2 - {}^1S_0)$ 164.7	$1s\,2s - {}^3S_1$	123.5
		$1s\,2p - {}^3P_0$	123.9
		$1s\,2s - {}^1S_0$	124.0
		$1s\,2p - {}^3P_1$	124.0
		$^3P_2$	124.7
		$^1P_1$	124.9
40	$Zr^{38+}\ (1s^2 - {}^1S_0)$ 173.5	$1s\,2s - {}^3S_1$	130.1
		$1s\,2p - {}^3P_0$	130.6
		$1s\,2s - {}^1S_0$	130.6
		$1s\,2p - {}^3P_1$	130.6
		$^3P_2$	131.4
		$^1P_1$	131.6
41	$Nb^{39+}\ (1s^2 - {}^1S_0)$ 182.6	$1s\,2s - {}^3S_1$	136.9
		$1s\,2p - {}^3P_0$	137.3
		$1s\,2s - {}^1S_0$	137.3
		$1s\,2p - {}^3P_1$	137.4
		$^3P_2$	138.3
		$^1P_1$	138.5
42	$Mo^{40+}\ (1s^2 - {}^1S_0)$ 192.0	$1s\,2s - {}^3S_1$	144.0
		$1s\,2p - {}^3P_0$	144.4
		$1s\,2s - {}^1S_0$	144.4
		$1s\,2p - {}^3P_1$	144.4
		$^3P_2$	145.5
		$^1P_1$	145.6
50	$Sn^{48+}\ (1s^2 - {}^1S_0)$ 276.9	$1s\,2s - {}^3S_1$	207.1
		$1s\,2p - {}^3P_1$	207.7
		$1s\,2s - {}^1S_0$	207.7
		$1s\,2p - {}^3P_0$	207.7
		$^3P_2$	209.9
		$^1P_1$	210.2
60	$Nd^{58+}\ (1s^2 - {}^1S_0)$ 407.1	$1s\,2s - {}^3S_1$	303.8
		$1s\,2p - {}^3P_1$	304.6
		$1s\,2s - {}^1S_0$	304.7
		$1s\,2p - {}^3P_0$	304.8
		$^3P_2$	309.6
		$^1P_1$	309.9
70	$Yb^{68+}\ (1s^2 - {}^1S_0)$ 567.5	$1s\,2s - {}^3S_1$	422.6
		$1s\,2p - {}^3P_1$	423.6
		$1s\,2s - {}^1S_0$	423.7
		$1s\,2p - {}^3P_0$	423.8
		$^3P_2$	433.4
		$^1P_1$	433.8

Table 6.4 (continued)

Atomic number Z	Ground-state term; ionization potential IP $[10^6\ \mathrm{cm}^{-1}]$	Excited-state terms	Excitation energy $T_k\,[10^6\ \mathrm{cm}^{-1}]$
1	2	3	4
80	Hg^{78+} $(1s^2-{}^1S_0)$	$1s\,2s-{}^3S_1$	566
	762	$1s\,2p-{}^3P_1$	567
		$1s\,2s-{}^1S_0$	567
		$1s\,2p-{}^3P_0$	568
		${}^3P_2$	585
		${}^1P_1$	586
90	Th^{88+} $(1s^2-{}^1S_0)$	$1s\,2s-{}^3S_1$	739
	998	$1s\,2p-{}^3P_1$	740
		$1s\,2s-{}^1S_0$	741
		$1s\,2p-{}^3P_0$	741
		${}^3P_2$	772
		${}^1P_1$	773
100	Fm^{98+} $(1s^2-{}^1S_0)$	$1s\,2s-{}^3S_1$	948
	1285	$1s\,2p-{}^3P_1$	950
		$1s\,2s-{}^1S_0$	950
		$1s\,2p-{}^3P_0$	951
		${}^3P_2$	1003
		${}^1P_1$	1004

7. Spectroscopic Characteristics of Neutral Atoms

Numerical data are presented for low-lying atomic terms, oscillator strengths of the most intense optical transitions, radiative lifetimes of low-lying resonant excited states and Rydberg levels of atoms with one valence electron and for typical parameters of atomic metastable states. The spectra of the most intense optical transitions are reproduced on atomic Grotrian diagrams.

7.1 Low-Lying Atomic Terms

Table 7.1 presents values of the excitation energy T [cm^{-1}] for low-lying levels of neutral atoms. These states characterize, as a rule, some of the first terms of the major electron configurations for a given atom. The energy levels were derived from the optical spectra of atoms. They are listed in order of increasing excitation energy with respect to the ground-state level ($T_0 = 0$); designation of the terms and configurations of the excited valence electrons is also provided. The ground-state term symbol and the value of the optical limit (ionization limit) for a series of transitions converging to the ground-state level of a singly ionized atom are given for convenience in a separate column of Table 7.1. If the distance between adjacent energy levels $T_{J1}, T_{J2}, \ldots$ of a given multiplet term is small, only the position of the multiplet centre of gravity $\overline{T}$, characterized by the relation $\overline{T} = \Sigma g_i T_i / \Sigma g_i$, is given in Table 7.1. Here $g_i = 2J_i + 1$ is the statistical weight of the level, J_i is the quantum number of the total electron angular momentum and T_i is the energy of the multiplet component.

The main information concerning atomic energy levels can be found in monographs and special issues from the National Bureau of Standards [7.1.1–4] and in the monograph [7.1.5]; we have also made use of the numerous publications of the last decade relating to particular elements (see bibliographies in [7.1.6–10]. The number of significant figures which we give in Table 7.1 was determined in each case according to the authors' estimated accuracy and the values are considered reliable to $\sim \pm 1$ in the last significant figure.

References

7.1.1 C.E.Moore: *Atomic Energy Levels*, Nat. Bur. Stand. (U.S.) Circ. **467**, Vol. 1 (1949); Vol. 2 (1952); Vol. 3 (1958)

7.1.2 W.C.Martin, R.Zalubas, L.Hagan: *Atomic Energy Levels – The Rare-Earth Elements*, Nat. Stand. Ref. Data Ser. Nat. Bur. Stand. **60** (1978)

7.1.3 C.E.Moore: *Selected Tables of Atomic Spectra*, Nat. Stand. Ref. Data Ser. Nat. Bur. Stand. **3**, Sects. 2, 3, 5, 6, 7 (1967–1976)

7.1.4 W.C.Martin, R.Zalubas: J. Phys. Chem. Ref. Data **12**, 323 (1983) (and references therein on Na- and K-row elements)

7.1.5 S.Bashkin, J.Stoner, Jr.: *Atomic Energy Levels and Grotrian Diagrams* Vol. 1, 1975; Addenda to Vol. 1 (1978); Vol. 2 (1978); Vol. 3 (1981); Vol. 4 (1982) (North-Holland, Amsterdam)

7.1.6 B.Edlén: "Term Analysis of Atomic Spectra", in *Beam-Foil Spectroscopy*, ed. by I.Sellin, C.Pegg, Vol. 1 (Plenum, New York 1976) pp. 1–9

7.1.7 J.Blaise, J.-F.Wyart, J.G.Conway, E.F.Worden: Phys. Scr. **22**, 224 (1980) (and references therein on actinide energy levels)

7.1.8 L.Hagan, W.C.Martin: *Bibliography on Atomic Energy Levels and Spectra, July 1968 through June 1971*, Nat. Bur. Stand. (U.S.) Spec. Publ. **363** (1972)

7.1.9 L.Hagan: *Bibliography on Atomic Energy Levels and Spectra, July 1971 through June 1975*, Nat. Bur. Stand. (U.S.) Spec. Publ. **363**, Suppl. 1 (1977)

7.1.10 R.Zalubas, A.Albright: *Bibliography on Atomic Energy Levels and Spectra, July 1975 through June 1979*, Nat. Bur. Stand. (U.S.) Spec. Publ. **363**, Suppl. 2 (1980)

Table 7.1. Low-lying energy levels of atoms

Atomic number Z	Element (ground-state term); ionization limit [cm^{-1}]	Excited state	Energy level [cm^{-1}]
1	2	3	4
1	^{1}H $(1s-{}^2S_{1/2})$	$2p\,(^2P^o_{1/2})$	82258.921
	109678.77	$2s\,(^2S_{1/2})$	82258.956
		$2p\,(^2P^o_{3/2})$	82259.286
		$3p\,(^2P^o_{1/2})$	97492.213
		$3s\,(^2S_{1/2})$	97492.223
		$3p\,(^2P^o_{3/2})$, $3d\,(^2D_{3/2})$	97492.321
		$3d\,(^2D_{5/2})$	97492.357
2	He $(1s^2-{}^1S_0)$	$2s\,(^3S_1)$	159856.08
	198310.77	$2s\,(^1S_0)$	166277.54
		$2p\,(^3P^o_2)$	169086.87
		$2p\,(^3P^o_1)$	169086.95
		$2p\,(^3P^o_0)$	169087.93
		$2p\,(^1P^o_1)$	171135.00
		$3s\,(^3S_1)$	183236.89
		$3s\,(^1S_0)$	184864.93
		$3p\,(^3P^o_2)$	185564.65
		$3p\,(^3P^o_1)$	185564.68
		$3p\,(^3P^o_0)$	185564.95
		$3d\,(^3D_3)$	186101.65
		$3d\,(^3D_2)$	186101.65
		$3d\,(^3D_1)$	186101.70
		$3d\,(^1D_2)$	186105.07
		$3p\,(^1P^o_1)$	186209.47
3	Li $(2s-{}^2S_{1/2})$	$2p\,(^2P^o_{1/2})$	14903.7
	43487.15	$2p\,(^2P^o_{3/2})$	14904.0
		$3s\,(^2S_{1/2})$	27206.1
		$3p\,(^2P^o_{1/2,\,3/2})$	30925.4
		$3d\,(^2D_{3/2})$	31283.1
		$3d\,(^2D_{5/2})$	31283.1
		$4s\,(^2S_{1/2})$	35012.1
		$4p\,(^2P^o_{1/2,\,3/2})$	36469.6
		$4d\,(^2D_{3/2,\,5/2})$	36623.4
		$4f\,(^2F^o_{5/2,\,7/2})$	36630.2
4	Be $(2s^2-{}^1S_0)$	$2p\,(^3P^o_0)$	21978.3
	75192.5	$2p\,(^3P^o_1)$	21978.9
		$2p\,(^3P^o_2)$	21981.3
		$2p\,(^1P^o_1)$	42565.4
		$3s\,(^3S_1)$	52080.9
		$3s\,(^1S_0)$	54677.3
		$2p^2\,(^1D_2)$	56882.4
		$3p\,(^3P^o_{0,\,1})$	58907.4
		$3p\,(^3P_2)$	58907.8
		$2p^2\,(^3P_0)$	59693.6
		$2p^2\,(^3P_1)$	59695.1
		$2p^2\,(^3P_2)$	59697.1
		$3p\,(^1P^o_1)$	60187.3

Table 7.1 (continued)

Atomic number Z	Element (ground-state term); ionization limit [cm^{-1}]	Excited state	Energy level [cm^{-1}]
1	2	3	4
5	B $(2p - {}^2P^o_{1/2})$ 66928.1	$2p\,({}^2P^o_{3/2})$	15.25
		$2s\,2p^2\,({}^4P_{1/2})$	28870
		$2s\,2p^2\,({}^4P_{3/2})$	28875
		$2s\,2p^2\,({}^4P_{5/2})$	28881
		$3s\,({}^2S_{1/2})$	40039.65
		$2s\,2p^2\,({}^2D_{3/2,\,5/2})$	47857.12
		$3p\,({}^2P^o_{1/2,\,3/2})$	48613.01
		$3d\,({}^2D_{3/2,\,5/2})$	54767.74
		$4s\,({}^2S_{1/2})$	55010.18
		$4p\,({}^2P^o_{1/2,\,3/2})$	57786.80
6	C $(2p^2 - {}^3P_0)$ 90820.4	$2p^2\,({}^3P_1)$	16.4
		$2p^2\,({}^3P_2)$	43.4
		$2p^2\,({}^1D_2)$	10192.6
		$2p^2\,({}^1S_0)$	21648.0
		$2s\,2p^3\,({}^5S^o_2)$	33735.2
		$2p\,3s\,({}^3P^o_0)$	60333.4
		$2p\,3s\,({}^3P^o_1)$	60352.6
		$2p\,3s\,({}^3P^o_2)$	60393.1
		$2p\,3s\,({}^1P^o_1)$	61981.8
		$2s\,2p^3\,({}^3D^o_3)$	64086.9
		$2s\,2p^3\,({}^3D^o_1)$	64089.8
		$2s\,2p^3\,({}^3D^o_2)$	64090.9
7	N $(2p^3 - {}^4S^o_{3/2})$ 117225.7	$2p\,({}^2D^o_{5/2})$	19224.46
		$2p\,({}^2D^o_{3/2})$	19233.18
		$2p\,({}^2P^o_{1/2})$	28838.92
		$2p\,({}^2P^o_{3/2})$	28839.31
		$3s\,({}^4P_{1/2})$	83284.07
		$3s\,({}^4P_{3/2})$	83317.83
		$3s\,({}^4P_{5/2})$	83364.62
		$3s\,({}^2P_{1/2})$	86137.35
		$3s\,({}^2P_{3/2})$	86220.51
		$2s\,2p^4\,({}^4P_{5/2})$	88107.26
		$2s\,2p^4\,({}^4P_{3/2})$	88151.17
		$2s\,2p^4\,({}^4P_{1/2})$	88170.57
8	O $(2p^4 - {}^3P_2)$ 109837.0	$2p\,({}^3P_1)$	158.26
		$2p\,({}^3P_0)$	226.98
		$2p\,({}^1D_2)$	15867.86
		$2p\,({}^1S_0)$	33792.58
		$3s\,({}^5S^o_2)$	73768.20
		$3s\,({}^3S^o_1)$	76794.98
		$3p\,({}^5P_1)$	86625.76
		$3p\,({}^5P_2)$	86627.78
		$3p\,({}^5P_3)$	86631.45
		$3p\,({}^3P_1)$	88630.59
		$3p\,({}^3P_2)$	88631.15

Table 7.1 (continued)

Atomic number Z	Element (ground-state term); ionization limit [cm^{-1}]	Excited state	Energy level [cm^{-1}]
1	2	3	4
		$3p\,(^3P_0)$	88631.30
		$4s\,(^5S_2^o)$	95476.73
		$4s\,(^3S_1^o)$	96225.05
9	F $(2p^5 - {}^2P_{3/2}^o)$ 140524.5	$2p\,(^2P_{1/2}^o)$	404.1
		$3s\,(^4P_{5/2})$	102405.7
		$3s\,(^4P_{3/2})$	102680.4
		$3s\,(^4P_{1/2})$	102840.4
		$3s\,(^2P_{3/2})$	104731.0
		$3s\,(^2P_{1/2})$	105056.3
		$3p\,(^4P_{5/2}^o)$	115917.9
		$3p\,(^4P_{3/2}^o)$	116040.9
		$3p\,(^4P_{1/2}^o)$	116143.6
		$3p\,(^4D_{7/2}^o)$	116987.4
		$3p\,(^4D_{5/2}^o)$	117164.0
		$3p\,(^4D_{3/2}^o)$	117308.6
		$3p\,(^4D_{1/2}^o)$	117392.0
		$3p\,(^2D_{5/2}^o)$	117622.9
		$3p\,(^2D_{3/2}^o)$	117872.9
		$3p\,(^2S_{1/2}^o)$	118405.3
		$3p\,(^4S_{3/2}^o)$	118427.8
		$3p\,(^2P_{3/2}^o)$	118936.8
		$3p\,(^2P_{1/2}^o)$	119081.8
10	Ne $(2p^6 - {}^1S_0)$ 173929.7	$2p^5\,(^2P_{3/2}^o)\,3s\,[3/2]_2^o$	134041.84
		$3s\,[3/2]_1^o$	134459.29
		$2p^5\,(^2P_{1/2}^o)\,3s'\,[1/2]_0^o$	134818.64
		$3s'\,[1/2]_1^o$	135888.72
		$3p\,[1/2]_1$	148257.79
		$3p\,[5/2]_3$	149657.04
		$3p\,[5/2]_2$	149824.22
		$3p\,[3/2]_1$	150121.59
		$3p\,[3/2]_2$	150315.86
		$3p'\,[3/2]_1$	150772.11
		$3p'\,[3/2]_2$	150858.51
		$3p\,[1/2]_0$	150917.43
		$3p'\,[1/2]_1$	151038.45
		$3p'\,[1/2]_0$	152970.73
11	Na $(3s - {}^2S_{1/2})$ 41449.4	$3p\,(^2P_{1/2}^o)$	16956.17
		$3p\,(^2P_{3/2}^o)$	16973.37
		$4s\,(^2S_{1/2})$	25739.99
		$3d\,(^2D_{5/2})$	29172.84
		$3d\,(^2D_{3/2})$	29172.89
		$4p\,(^2P_{1/2}^o)$	30266.99
		$4p\,(^2P_{3/2}^o)$	30272.58
		$5s\,(^2S_{1/2})$	33200.68
		$4d\,(^2D_{5/2})$	34548.73

Table 7.1 (continued)

Atomic number Z	Element (ground-state term); ionization limit [cm^{-1}]	Excited state	Energy level [cm^{-1}]
1	2	3	4
		$4d\,(^2D_{3/2})$	34548.77
		$4f\,(^2F_{5/2,\,7/2})$	34586.92
12	Mg $(3s^2 - {}^1S_0)$ 61671.0	$3p\,(^3P^o_0)$	21850.41
		$3p\,(^3P^o_1)$	21870.46
		$3p\,(^3P^o_2)$	21911.18
		$3p\,(^1P^o_1)$	35051.26
		$4s\,(^3S_1)$	41197.40
		$4s\,(^1S_0)$	43503.33
		$3d\,(^1D_2)$	46403.06
		$4p\,(^3P^o_0)$	47841.12
		$4p\,(^3P^o_1)$	47844.41
		$4p\,(^3P^o_2)$	47851.16
		$3d\,(^3D_2)$	47957.03
		$3d\,(^3D_3)$	47957.04
		$3d\,(^3D_1)$	47957.06
		$4p\,(^1P^o_1)$	49346.73
13	Al $(3p - {}^2P^o_{1/2})$ 48278.4	$3p\,(^2P^o_{3/2})$	112.06
		$4s\,(^2S_{1/2})$	25347.76
		$3s\,3p^2\,(^4P_{1/2})$	29020.4
		$3s\,3p^2\,(^4P_{3/2})$	29067.0
		$3s\,3p^2\,(^4P_{5/2})$	29142.8
		$3d\,(^2D_{3/2})$	32435.43
		$3d\,(^2D_{5/2})$	32436.78
		$4p\,(^2P^o_{1/2})$	32949.80
		$4p\,(^2P^o_{3/2})$	32965.64
		$5s\,(^2S_{1/2})$	37689.41
14	Si $(3p^2 - {}^3P_0)$ 65747.8	$3p^2\,(^3P_1)$	77.12
		$3p^2\,(^3P_2)$	223.16
		$3p^2\,(^1D_2)$	6298.85
		$3p^2\,(^1S_0)$	15394.37
		$3s\,3p^3\,(^5S^o_2)$	33326.05
		$4s\,(^3P^o_0)$	39683.16
		$4s\,(^3P^o_1)$	39760.28
		$4s\,(^3P^o_2)$	39955.05
		$4s\,(^1P^o_1)$	40991.88
		$3s\,3p^3\,(^3D^o_1)$	45276.19
		$3s\,3p^3\,(^3D^o_2)$	45293.63
		$3s\,3p^3\,(^3D^o_3)$	45321.85
		$4p\,(^1P_1)$	47284.06
		$3d\,(^1D^o_2)$	47351.55
		$4p\,(^3D_1)$	48020.07
		$4p\,(^3D_2)$	48102.32
		$4p\,(^3D_3)$	48264.29
		$4p\,(^3P_0)$	49028.29
		$4p\,(^3P_1)$	49060.60

Table 7.1 (continued)

Atomic number Z	Element (ground-state term); ionization limit [cm^{-1}]	Excited state	Energy level [cm^{-1}]
1	2	3	4
		$4p\,(^3P_2)$	49188.62
		$4p\,(^3S_1)$	49399.67
15	$P\,(3p^3 - {}^4S^o_{3/2})$ 84580.8	$3p^3\,(^2D^o_{3/2})$	11361.0
		$3p^3\,(^2D^o_{5/2})$	11376.6
		$3p^3\,(^2P^o_{1/2})$	18722.7
		$3p^3\,(^2P^o_{3/2})$	18748.0
		$4s\,(^4P_{1/2})$	55939.42
		$4s\,(^4P_{3/2})$	56090.63
		$4s\,(^4P_{5/2})$	56339.66
		$4s\,(^2P_{1/2})$	57876.57
		$4s\,(^2P_{3/2})$	58174.37
		$3s\,3p^4\,(^4P_{5/2})$	59534.55
		$3s\,3p^4\,(^4P_{3/2})$	59715.92
		$3s\,3p^4\,(^4P_{1/2})$	59820.37
		$4p\,(^2S^o_{1/2})$	64239.59
		$3p^2\,(^1D)\,4s\,(^2D_{3/2})$	65156.24
		$4s\,(^2D_{5/2})$	65157.13
		$4p\,(^4D^o_{1/2})$	65373.56
		$4p\,(^4D^o_{3/2})$	65450.13
		$4p\,(^4D^o_{5/2})$	65585.13
		$4p\,(^4D^o_{7/2})$	65788.46
		$4p\,(^4P^o_{1/2})$	66343.44
		$4p\,(^4P^o_{3/2})$	66360.28
		$4p\,(^4P^o_{5/2})$	66544.24
		$4p\,(^2D^o_{3/2})$	66813.27
		$4p\,(^4S^o_{3/2})$	66834.65
		$4p\,(^2D^o_{5/2})$	67113.87
16	$S\,(3p^4 - {}^3P_2)$ 83559	$3p^3\,(^4S^o)\,3p\,(^3P_1)$	396.05
		$3p\,(^3P_0)$	573.64
		$3p\,(^1D_2)$	9238.61
		$3p\,(^1S_0)$	22179.95
		$4s\,(^5S^o_2)$	52623.64
		$4s\,(^3S^o_1)$	55330.81
		$4p\,(^5P_1)$	63446.4
		$4p\,(^5P_2)$	63457.3
		$4p\,(^5P_3)$	63475.3
		$4p\,(^3P_1)$	64889.2
		$4p\,(^3P_0)$	64891.7
		$4p\,(^3P_2)$	64892.9
		$3p^3\,(^2D^o)\,4s'\,(^3D^o_1)$	67816.35
		$4s'\,(^3D^o_2)$	67825.19
		$4s'\,(^3D^o_3)$	67842.87
		$3d\,(^5D^o_4)$	67878.0
		$3d\,(^5D^o_0)$	67884.7
		$3d\,(^5D^o_1)$	67885.53

Table 7.1 (continued)

Atomic number Z	Element (ground-state term); ionization limit [cm^{-1}]	Excited state	Energy level [cm^{-1}]
1	2	3	4
		$3d\,(^5D^o_2)$	67887.80
		$3d\,(^5D^o_3)$	67890.02
17	$Cl\,(3p^5 - {}^2P^o_{3/2})$ 104591	$3p^4\,(^3P)\,3p\,(^2P^o_{1/2})$	882.35
		$4s\,(^4P_{5/2})$	71958.36
		$4s\,(^4P_{3/2})$	72488.57
		$4s\,(^4P_{1/2})$	72827.04
		$4s\,(^2P_{3/2})$	74225.85
		$4s\,(^2P_{1/2})$	74865.67
		$4p\,(^4P^o_{5/2})$	82918.89
		$4p\,(^4P^o_{3/2})$	83130.90
		$4p\,(^4P^o_{1/2})$	83364.93
		$4p\,(^4D^o_{7/2})$	83894.04
		$3p^4\,(^1D)\,4s'\,(^2D_{5/2})$	84120.26
		$4s'\,(^2D_{3/2})$	84121.87
		$4p\,(^4D^o_{5/2})$	84132.26
		$4p\,(^4D^o_{3/2})$	84485.31
		$4p\,(^2D^o_{5/2})$	84648.10
		$4p\,(^4D^o_{1/2})$	84688.64
		$4p\,(^2D^o_{3/2})$	84988.48
		$4p\,(^2S^o_{1/2})$	85244.33
		$4p\,(^2P^o_{3/2})$	85442.43
		$3s\,3p^6\,(^2S_{1/2})$	85678.9
		$4p\,(^4S^o_{3/2})$	85735.09
		$4p\,(^2S^o_{1/2})$	85917.94
18	$Ar\,(3p^6 - {}^1S_0)$ 127109.8	$3p^5\,(^2P^o_{3/2})\,4s\,[3/2]^o_2$	93143.76
		$4s\,[3/2]^o_1$	93750.60
		$3p^5\,(^2P^o_{1/2})\,4s'\,[1/2]^o_0$	94553.67
		$4s'\,[1/2]^o_1$	95399.83
		$4p\,[1/2]_1$	104102.10
		$4p\,[5/2]_3$	105462.76
		$4p\,[5/2]_2$	105617.27
		$4p\,[3/2]_1$	106087.26
		$4p\,[3/2]_2$	106237.55
		$4p\,[1/2]_0$	107054.27
		$4p'\,[3/2]_1$	107131.71
		$4p'\,[3/2]_2$	107289.70
		$4p'\,[1/2]_1$	107496.42
		$4p'\,[1/2]_0$	108722.62
19	$K\,(4s - {}^2S_{1/2})$ 35009.81	$4p\,(^2P^o_{1/2})$	12985.17
		$4p\,(^2P^o_{3/2})$	13042.88
		$5s\,(^2S_{1/2})$	21026.55
		$3d\,(^2D_{5/2})$	21534.68
		$3d\,(^2D_{3/2})$	21536.99
		$5p\,(^2P^o_{1/2})$	24701.38
		$5p\,(^2P^o_{3/2})$	24720.14

Table 7.1 (continued)

Atomic number Z	Element (ground-state term); ionization limit [cm^{-1}]	Excited state	Energy level [cm^{-1}]
1	2	3	4
		$4d\,(^2D_{5/2})$	27397.08
		$4d\,(^2D_{3/2})$	27398.15
		$6s\,(^2S_{1/2})$	27450.71
		$4f\,(^2F^o_{7/2,\,5/2})$	28127.8
		$6p\,(^2P^o_{1/2})$	28999.3
		$6p\,(^2P^o_{3/2})$	29007.7
20	Ca $(4s^2-{}^1S_0)$ 49306.0	$4p\,(^3P^o_0)$	15157.90
		$4p\,(^3P^o_1)$	15210.06
		$4p\,(^3P^o_2)$	15315.94
		$3d\,(^3D_1)$	20335.36
		$3d\,(^3D_2)$	20349.26
		$3d\,(^3D_3)$	20371.00
		$3d\,(^1D_2)$	21849.63
		$4p\,(^1P^o_1)$	23652.30
		$5s\,(^3S_1)$	31539.49
		$5s\,(^1S_0)$	33317.26
21	Sc $(3d\,4s^2-{}^2D_{3/2})$ 52922	$3d\,4s^2\,(^2D_{5/2})$	168.3
		$3d^2\,4s\,(^4F_{3/2})$	11520.0
		$3d^2\,4s\,(^4F_{5/2})$	11557.7
		$3d^2\,4s\,(^4F_{7/2})$	11610.3
		$3d^2\,4s\,(^4F_{9/2})$	11677.4
		$3d^2\,4s\,(^2F_{5/2})$	14926.1
		$3d^2\,4s\,(^2F_{7/2})$	15041.9
		$3d\,4s\,(^3D)\,4p\,(^4F^o_{3/2})$	15672.6
		$3d\,4s\,4p\,(^4F^o_{5/2})$	15756.6
		$3d\,4s\,4p\,(^4F^o_{7/2})$	15881.7
		$3d\,4s\,4p\,(^4D^o_{1/2})$	16009.8
		$3d\,4s\,4p\,(^4D^o_{3/2})$	16021.8
		$3d\,4s\,(^1D)\,4p\,(^2D^o_{5/2})$	16022.7
		$3d\,4s\,4p\,(^4F^o_{9/2})$	16026.6
		$3d\,4s\,4p\,(^2D^o_{3/2})$	16096.9
		$3d\,4s\,4p\,(^4D^o_{5/2})$	16141.1
		$3d\,4s\,4p\,(^4D^o_{7/2})$	16210.8
22	Ti $(3d^2\,4s^2-{}^3F_2)$ 55000	$3d^2\,4s^2\,(^3F_3)$	170.13
		$3d^2\,4s^2\,(^3F_4)$	386.87
		$3d^3\,(^4F)\,4s\,(^5F_1)$	6556.83
		$4s\,(^5F_2)$	6598.75
		$4s\,(^5F_3)$	6661.00
		$4s\,(^5F_4)$	6742.76
		$4s\,(^5F_5)$	6842.96
		$3d^2\,4s^2\,(^1D_2)$	7255.37
		$3d^2\,4s^2\,(^3P_0)$	8436.62
		$3d^2\,4s^2\,(^3P_1)$	8492.42
		$3d^2\,4s^2\,(^3P_2)$	8602.34
		$3d^3\,(^4F)\,4s\,(^3F_2)$	11531.76

Table 7.1 (continued)

Atomic number Z	Element (ground-state term); ionization limit [cm^{-1}]	Excited state	Energy level [cm^{-1}]
1	2	3	4
		$4s\,(^3F_3)$	11639.80
		$4s\,(^3F_4)$	11776.81
		$3d^2 4s^2\,(^1G_4)$	12118.39
23	V $(3d^3 4s^2 - {}^4F_{3/2})$ 54360	$3d^3 4s^2\,(^4F_{5/2})$	137.38
		$(^4F_{7/2})$	323.46
		$(^4F_{9/2})$	552.96
		$3d^4\,(^5D)\,4s\,(^6D_{1/2})$	2112.28
		$4s\,(^6D_{3/2})$	2153.21
		$4s\,(^6D_{5/2})$	2220.11
		$4s\,(^6D_{7/2})$	2311.35
		$4s\,(^6D_{9/2})$	2424.78
		$3d^4\,(^5D)\,4s\,(^4D_{1/2})$	8413.00
		$4s\,(^4D_{3/2})$	8476.23
		$4s\,(^4D_{5/2})$	8578.53
		$4s\,(^4D_{7/2})$	8715.76
		$3d^3 4s^2\,(^4P_{1/2})$	9544.63
		$(^4P_{3/2})$	9637.03
		$(^4P_{5/2})$	9824.61
24	Cr $(3d^5 4s - {}^7S_3)$ 54570	$3d^5\,(^6S)\,4s\,(^5S_2)$	7593.2
		$3d^4 4s^2\,(^5D_0)$	7750.8
		$(^5D_1)$	7810.8
		$(^5D_2)$	7927.5
		$(^5D_3)$	8095.2
		$(^5D_4)$	8307.6
25	Mn $(3d^5 4s^2 - {}^6S_{5/2})$ 59959.4	$3d^6\,(^5D)\,4s\,(^6D_{9/2})$	17052.3
		$4s\,(^6D_{7/2})$	17282.0
		$4s\,(^6D_{5/2})$	17451.5
		$4s\,(^6D_{3/2})$	17568.5
		$4s\,(^6D_{1/2})$	17637.1
		$3d^5 4s\,(^7S)\,4p\,(^8P^o_{5/2})$	18402.5
		$4p\,(^8P^o_{7/2})$	18531.6
		$4p\,(^8P^o_{9/2})$	18705.4
26	Fe $(3d^6 4s^2 - {}^5D_4)$ 63740	$3d^6 4s^2\,(^5D_3)$	415.93
		$(^5D_2)$	704.00
		$(^5D_1)$	888.13
		$(^5D_0)$	978.07
		$3d^7\,(^4F)\,4s\,(^5F_5)$	6928.27
		$4s\,(^5F_4)$	7376.76
		$4s\,(^5F_3)$	7728.06
		$4s\,(^5F_2)$	7985.78
		$4s\,(^5F_1)$	8154.71
		$3d^7\,(^4F)\,4s\,(^3F_4)$	11976.23
		$4s\,(^3F_3)$	12560.93
		$4s\,(^3F_2)$	12968.55
		$3d^7\,(^4P)\,4s\,(^5P_3)$	17550.17

Table 7.1 (continued)

Atomic number Z	Element (ground-state term); ionization limit [cm^{-1}]	Excited state	Energy level [cm^{-1}]
1	2	3	4
		$4s\,(^5P_2)$	17726.98
		$4s\,(^5P_1)$	17927.38
27	Co $(3d^7\,4s^2 - {}^4F_{9/2})$ 63400	$3d^7\,4s^2\,(^4F_{7/2})$	816.0
		$(^4F_{5/2})$	1406.8
		$(^4F_{3/2})$	1809.3
		$3d^8\,(^3F)\,4s\,(^4F_{9/2})$	3482.8
		$4s\,(^4F_{7/2})$	4142.7
		$4s\,(^4F_{5/2})$	4690.2
		$4s\,(^4F_{3/2})$	5075.8
		$3d^8\,(^3F)\,4s\,(^2F_{7/2})$	7442.4
		$4s\,(^2F_{5/2})$	8460.8
28	Ni $(3d^8\,4s^2 - {}^3F_4)$ 61600	$3d^9\,(^2D)\,4s\,(^3D_3)$	204.79
		$4s\,(^3D_2)$	879.81
		$3d^8\,4s^2\,(^3F_3)$	1332.15
		$4s\,(^3D_1)$	1713.08
		$3d^8\,4s^2\,(^3F_2)$	2216.52
		$3d^9\,(^2D)\,4s\,(^1D_2)$	3409.92
		$3d^8\,4s^2\,(^1D_2)$	13521.35
		$3d^{10}\,(^1S_0)$	14728.85
		$3d^8\,4s^2\,(^3P_2)$	15609.86
		$3d^8\,4s^2\,(^3P_1)$	15734.02
		$3d^8\,4s^2\,(^3P_0)$	16017.32
		$3d^8\,4s^2\,(^1G_4)$	22102.35
29	Cu $(3d^{10}\,4s - {}^2S_{1/2})$ 62317.4	$3d^9\,4s^2\,(^2D_{5/2})$	11202.56
		$3d^9\,4s^2\,(^2D_{3/2})$	13245.42
		$4p\,(^2P^o_{1/2})$	30535.30
		$4p\,(^2P^o_{3/2})$	30783.69
		$3d^9\,4s\,(^3D)\,4p'\,(^4P^o_{5/2})$	39018.65
		$4p'\,(^4P^o_{3/2})$	40114.0
		$4p'\,(^4F^o_{9/2})$	40909.14
		$4p'\,(^4P^o_{1/2})$	40943.7
		$4p'\,(^4F^o_{7/2})$	41153.43
		$4p'\,(^4F^o_{5/2})$	41562.89
		$4p'\,(^4F^o_{3/2})$	42302.5
		$5s\,(^2S_{1/2})$	43137.21
30	Zn $(3d^{10}\,4s^2 - {}^1S_0)$ 75769.3	$4p\,(^3P^o_0)$	32311.33
		$4p\,(^3P^o_1)$	32501.41
		$4p\,(^3P^o_2)$	32890.34
		$4p\,(^1P^o_1)$	46745.40
		$5s\,(^3S_1)$	53672.26
		$5s\,(^3S_0)$	55789.22
		$5p\,(^3P^o_0)$	61247.90
		$5p\,(^3P^o_1)$	61274.42
		$5p\,(^3P^o_2)$	61330.89
		$4d\,(^1D_2)$	62458.52

Table 7.1 (continued)

Atomic number Z	Element (ground-state term); ionization limit [cm^{-1}]	Excited state	Energy level [cm^{-1}]
1	2	3	4
		$4d\,(^3D_1)$	62768.76
		$4d\,(^3D_2)$	62772.02
		$4d\,(^3D_3)$	62776.99
		$5p\,(^1P^o_1)$	62910.43
31	Ga $(4s^2 4p - {}^2P^o_{1/2})$ 48387.6	$4p\,(^2P^o_{3/2})$	826.2
		$5s\,(^2S_{1/2})$	24788.6
		$5p\,(^2P^o_{1/2})$	33044.1
		$5p\,(^2P^o_{3/2})$	33155.0
		$4d\,(^2D_{3/2})$	34781.7
		$4d\,(^2D_{5/2})$	34787.9
32	Ge $(4p^2 - {}^3P_0)$ 63713.2	$4p^2\,(^3P_1)$	557.13
		$4p^2\,(^3P_2)$	1409.96
		$4p^2\,(^1D_2)$	7125.30
		$4p^2\,(^1S_0)$	16367.33
		$5s\,(^3P^o_0)$	37451.69
		$5s\,(^3P^o_1)$	37702.31
		$5s\,(^3P^o_2)$	39117.90
		$5s\,(^1P^o_1)$	40020.56
		$4s\,4p^3\,(^5S^o_2)$	41926.73
33	As $(4p^3 - {}^4S^o_{3/2})$ 78950	$4p^3\,(^2D^o_{3/2})$	10592
		$4p^3\,(^2D^o_{5/2})$	10915
		$4p^3\,(^2P^o_{1/2})$	18186
		$4p^3\,(^2P^o_{3/2})$	18647
		$5s\,(^4P_{1/2})$	50694
		$5s\,(^4P_{3/2})$	51610
		$5s\,(^4P_{5/2})$	52898
		$5s\,(^2P_{1/2})$	53136
		$5s\,(^2P_{3/2})$	54605
34	Se $(4p^4 - {}^3P_2)$ 78658.2	$4p^4\,(^3P_1)$	1989.50
		$4p^4\,(^3P_0)$	2534.36
		$4p^4\,(^1D_2)$	9576.15
		$4p^4\,(^1S_0)$	22446.20
		$5s\,(^5S^o_2)$	48182.2
		$5s\,(^3S^o_1)$	50996.9
		$5p\,(^5P_1)$	59242.8
		$5p\,(^5P_2)$	59287.9
		$5p\,(^5P_3)$	59391.4
		$5p\,(^3P_2)$	60677.5
		$5p\,(^3P_1)$	60622.4
		$5p\,(^3P_0)$	60696.1
35	Br $(4p^5 - {}^2P^o_{3/2})$ 95284.8	$4p^5\,(^2P^o_{1/2})$	3685
		$5s\,(^4P_{5/2})$	63429.8
		$5s\,(^4P_{3/2})$	64900.5
		$5s\,(^4P_{1/2})$	66877.2
		$5s\,(^2P_{3/2})$	67176.9

Table 7.1 (continued)

Atomic number Z	Element (ground-state term); ionization limit [cm^{-1}]	Excited state	Energy level [cm^{-1}]
1	2	3	4
		$5s\,(^2P_{1/2})$	68963.5
		$5p\,(^4P^o_{5/2})$	74665.7
		$5p\,(^4P^o_{3/2})$	75002.5
		$5p\,(^4D^o_{7/2})$	75514.8
		$5p\,(^4D^o_{5/2})$	75690.4
		$5p\,(^4P^o_{1/2})$	75807.3
		$4p^4\,(^1S)\,5s''\,(^2S_{1/2})$	75901.9
		$5p\,(^4D^o_{3/2})$	76736.4
		$4p^4\,(^1D)\,5s'\,(^2D_{3/2})$	77305.9
		$5s'\,(^2D_{5/2})$	77324.1
		$5p\,(^4D^o_{1/2})$	78069.3
36	Kr ($4p^6$ – $^1S_0$) 112914.5	$4p^5\,(^2P^o_{3/2})\,5s\,[3/2]^o_2$	79971.8
		$5s\,[3/2]^o_1$	80916.8
		$4p^5\,(^2P^o_{1/2})\,5s'\,[1/2]^o_0$	85191.7
		$5s'\,[1/2]^o_1$	85846.8
		$5p\,[1/2]_1$	91168.6
		$5p\,[5/2]_3$	92294.5
		$5p\,[5/2]_2$	92307.4
		$5p\,[3/2]_1$	92964.5
		$5p\,[3/2]_2$	93123.4
		$5p\,[1/2]_0$	94092.9
		$4d\,[1/2]^o_0$	96771.6
		$4d\,[1/2]_1$	97085.3
37	Rb ($5s$ – $^2S_{1/2}$) 33690.88	$5p\,(^2P^o_{1/2})$	12578.95
		$5p\,(^2P^o_{3/2})$	12816.55
		$4d\,(^2D_{5/2})$	19355.20
		$4d\,(^2D_{3/2})$	19355.65
		$6s\,(^2S_{1/2})$	20132.51
		$6p\,(^2P^o_{1/2})$	23715.08
		$6p\,(^2P^o_{3/2})$	23792.59
		$5d\,(^2D_{3/2})$	25700.54
		$5d\,(^2D_{5/2})$	25703.50
		$7s\,(^2S_{1/2})$	26311.44
		$4f\,(^2F^o_{7/2})$	26792.09
		$4f\,(^2F^o_{5/2})$	26792.12
38	Sr ($5s^2$ – $^1S_0$) 45932.1	$5p\,(^3P^o_0)$	14317.52
		$5p\,(^3P^o_1)$	14504.35
		$5p\,(^3P^o_2)$	14898.56
		$4d\,(^3D_1)$	18159.06
		$4d\,(^3D_2)$	18218.79
		$4d\,(^3D_3)$	18319.27
		$4d\,(^1D_2)$	20149.7
		$5p\,(^1P^o_1)$	21698.48
		$6s\,(^3S_1)$	29038.79
		$6s\,(^1S_0)$	30591.8

Table 7.1 (continued)

Atomic number Z	Element (ground-state term); ionization limit [cm^{-1}]	Excited state	Energy level [cm^{-1}]
1	2	3	4
39	Y ($4d5s^2$ – $^2D_{3/2}$) 50144	$4d$ ($^2D_{5/2}$)	530.4
		$5p$ ($^2P^o_{1/2}$)	10529.2
		$4d^2$ (^3F) $5s$ ($^4F_{3/2}$)	10937.3
		$5s$ ($^4F_{5/2}$)	11078.6
		$5s$ ($^4F_{7/2}$)	11278.0
		$5p$ ($^2P^o_{3/2}$)	11359.7
		$4d^2$ (^3F) $5s$ ($^4F_{9/2}$)	11532.1
40	Zr ($4d^2 5s^2$ – $^3F_2$) 55100	$4d^2 5s^2$ ($^3F_3$)	570.4
		$4d^2 5s^2$ ($^3F_4$)	1240.8
		$4d^2 5s^2$ ($^3P_2$)	4186.1
		$4d^2 5s^2$ ($^3P_0$)	4196.8
		$4d^2 5s^2$ ($^3P_1$)	4376.3
		$4d^3$ (^4F) $5s$ ($^5F_1$)	4870.5
		$5s$ ($^5F_2$)	5023.4
		$4d^2 5s^2$ ($^1D_2$)	5101.7
		$4d^3$ (^4F) $5s$ ($^5F_3$)	5249.1
		$5s$ ($^5F_4$)	5540.5
		$5s$ ($^5F_5$)	5888.9
		$4d^2 5s^2$ ($^1G_4$)	8057.3
41	Nb ($4d^4 5s$ – $^6D_{1/2}$) 55500	$5s$ ($^6D_{3/2}$)	154.2
		($^6D_{5/2}$)	392.0
		($^6D_{7/2}$)	695.2
		($^6D_{9/2}$)	1050.3
		$4d^3 5s^2$ ($^4F_{3/2}$)	1142.8
		($^4F_{5/2}$)	1586.9
		($^4F_{7/2}$)	2154.1
		($^4F_{9/2}$)	2805.4
		$4d^3 5s^2$ ($^4P_{1/2}$)	4998.2
		($^4P_{3/2}$)	5297.9
		($^4P_{5/2}$)	5965.4
42	Mo ($4d^5 5s$ – $^7S_3$) 57300	$4d^5$ (^6S) $5s$ ($^5S_2$)	10768.3
		$4d^4 5s^2$ ($^5D_0$)	10966.0
		($^5D_1$)	11142.8
		($^5D_2$)	11454.4
		($^5D_3$)	11858.5
		($^5D_4$)	12346.3
		$4d^5$ (^4G) $5s$ ($^5G_2$)	16641.1
		$5s$ ($^5G_3$)	16693.0
		$5s$ ($^5G_4$)	16747.7
		$5s$ ($^5G_5$)	16784.6
		$5s$ ($^5G_6$)	16784.0
		$4d^5$ (^4P) $5s$ ($^5P_3$)	18229.2
		$5s$ ($^5P_2$)	18356.5
		$5s$ ($^5P_1$)	18479.6

Table 7.1 (continued)

Atomic number Z	Element (ground-state term); ionization limit [cm^{-1}]	Excited state	Energy level [cm^{-1}]
1	2	3	4
43	Tc ($4d^5 5s^2$ – $^6S_{5/2}$) 58700	$4d^6$ (^5D) $5s$ ($^6D_{9/2}$)	2572.9
		($^6D_{7/2}$)	3250.9
		($^6D_{5/2}$)	3700.5
		($^6D_{3/2}$)	4002.6
		($^6D_{1/2}$)	4178.7
		$4d^6$ (^5D) $5s$ ($^4D_{7/2}$)	10516.5
		($^4D_{5/2}$)	11063.1
		($^4D_{3/2}$)	11578.6
		($^4D_{1/2}$)	11891.0
44	Ru ($4d^7 5s$ – $^5F_5$) 59410	$4d^7$ (^4F) $5s$ ($^5F_4$)	1190.6
		($^5F_3$)	2091.5
		($^5F_2$)	2713.2
		($^5F_1$)	3105.5
45	Rh ($4d^8 5s$ – $^4F_{9/2}$) 60200	$4d^8$ (^3F) $5s$ ($^4F_{7/2}$)	1530.0
		$5s$ ($^4F_{5/2}$)	2598.0
		$4d^9$ ($^2D_{5/2}$)	3309.9
		$5s$ ($^4F_{3/2}$)	3472.7
		$4d^9$ ($^2D_{3/2}$)	5658.0
		$4d^8$ (^3F) $5s$ ($^2F_{7/2}$)	5691.0
		$5s$ ($^2F_{5/2}$)	7791.2
46	Pd ($4d^{10}$ – $^1S_0$) 67236	$4d^9$ ($^2D_{5/2}$) $5s$ ($^3D_3$)	6564.1
		$5s$ ($^3D_2$)	7755.0
		$4d^9$ ($^2D_{3/2}$) $5s$ ($^3D_1$)	10093.9
		$4d^9$ ($^2D_{3/2}$) $5s$ ($^1D_2$)	11721.8
47	Ag ($4d^{10} 5s$ – $^2S_{1/2}$) 61106.6	$4d^{10}$ (^1S) $5p$ ($^2P^o_{1/2}$)	29552.0
		$4d^9 5s^2$ ($^2D_{5/2}$)	30242.3
		$5p$ ($^2P^o_{3/2}$)	30472.7
		$5s^2$ ($^2D_{3/2}$)	34714.2
		$6s$ ($^2S_{1/2}$)	42556.1
48	Cd ($4d^{10} 5s^2$ – $^1S_0$) 72540.1	$5p$ ($^3P^o_0$)	30113.99
		($^3P^o_1$)	30656.09
		($^3P^o_2$)	31826.95
		$5p$ ($^1P^o_1$)	43692.38
		$6s$ ($^3S_1$)	51483.98
		$6s$ ($^1S_0$)	53310.10
		$6p$ ($^3P^o_0$)	58391
		($^3P^o_1$)	58462
		($^3P^o_2$)	58636
		$5d$ ($^1D_2$)	59219.73
		$5d$ ($^3D_1$)	59485.77
		($^3D_2$)	59497.87
		($^3D_3$)	59515.98
49	In ($5p$ – $^2P^o_{1/2}$) 46670.11	$5p$ ($^2P^o_{3/2}$)	2212.6
		$6s$ ($^2S_{1/2}$)	24372.9

Table 7.1 (continued)

Atomic number Z	Element (ground-state term); ionization limit [cm^{-1}]	Excited state	Energy level [cm^{-1}]
1	2	3	4
		$6p\,(^2P^o_{1/2})$	31816.6
		$6p\,(^2P^o_{3/2})$	32114.8
		$5d\,(^2D_{3/2})$	32892.1
		$5d\,(^2D_{5/2})$	32915.4
50	Sn $(5p^2 - {}^3P_0)$ 59232.7	$5p^2\,(^3P_1)$	1692
		$5p^2\,(^3P_2)$	3428
		$5p^2\,(^1D_2)$	8613
		$5p^2\,(^1S_0)$	17163
		$6s\,(^3P^o_0)$	34640.8
		$6s\,(^3P^o_1)$	34914.3
		$6s\,(^3P^o_2)$	38628.9
		$6s\,(^1P^o_1)$	39257.1
		$5s\,5p^3\,(^5S^o_2)$	39625.5
51	Sb $(5p^3 - {}^4S^o_{3/2})$ 69430	$5p^3\,(^2D^o_{3/2})$	8512
		$5p^3\,(^2D^o_{5/2})$	9854
		$5p^3\,(^2P^o_{1/2})$	16396
		$5p^3\,(^2P^o_{3/2})$	18464
52	Te $(5p^4\,{}^3P_2 + 5p^4\,{}^1D_2)$ 72670	$5p^4\,(^3P_0) + 5p^4\,(^1S_0)$	4706.49
		$5p^4\,(^3P_1)$	4750.71
		$5p^4\,(^1D_2) + 5p^4\,(^3P_2)$	10557.88
		$5p^4\,(^1S_0) + 5p^4\,(^3P_0)$	23198.39
53	I $(5p^5 - {}^2P^o_{3/2})$ 84295.0	$5p^5\,(^2P^o_{1/2})$	7603.1
		$5p^4\,(^3P_2)\,6s\,[2]_{5/2}$	54633.5
		$6s\,[2]_{3/2}$	56092.9
		$6s\,[0]_{1/2}$	60896.2
		$6s\,[1]_{3/2}$	61819.8
		$6s\,[1]_{1/2}$	63186.7
		$6p\,[2]^o_{5/2}$	64906.3
		$6p\,[2]^o_{3/2}$	64990.0
		$6p\,[3]^o_{5/2}$	65644.5
		$6p\,[3]^o_{7/2}$	65670.0
		$6p\,[1]^o_{1/2}$	65857.0
		$5d\,[3]_{7/2}$	66015.0
		$5d\,[3]_{5/2}$	66020.5
		$5d\,[1]_{3/2}$	66355.1
		$6p\,[1]^o_{3/2}$	67062.1
		$5d\,[1]_{1/2}$	67298.3
		$5d\,[4]_{9/2}$	67726.4
54	Xe $(5p^6 - {}^1S_0)$ 97834.4	$5p^5\,(^2P^o_{3/2})\,6s\,[3/2]^o_2$	67068.05
		$6s\,[3/2]^o_1$	68045.66
		$5p^5\,(^2P^o_{1/2})\,6s'\,[1/2]^o_0$	76197.27
		$6s'\,[1/2]^o_1$	77185.54
		$6p\,[1/2]_1$	77269.64
		$6p\,[5/2]_2$	78120.29

Table 7.1 (continued)

Atomic number Z	Element (ground-state term); ionization limit [cm^{-1}]	Excited state	Energy level [cm^{-1}]
1	2	3	4
		$6p\,[5/2]_3$	78403.56
		$6p\,[3/2]_1$	78956.53
		$6p\,[3/2]_2$	79212.96
		$5d\,[1/2]^o_0$	79771.76
		$5d\,[1/2]^o_1$	79987.12
		$6p\,[1/2]_0$	80119.46
		$5d\,[7/2]^o_4$	80197.13
		$5d\,[3/2]^o_2$	80323.24
		$5d\,[7/2]^o_3$	80970.94
55	Cs ($6s$ – $^2S_{1/2}$) 31406.47	$6p\,(^2P^o_{1/2})$	11178.27
		$6p\,(^2P^o_{3/2})$	11732.31
		$5d\,(^2D_{3/2})$	14499.25
		$5d\,(^2D_{5/2})$	14596.84
		$7s\,(^2S_{1/2})$	18535.52
		$7p\,(^2P^o_{1/2})$	21765.36
		$7p\,(^2P^o_{3/2})$	21946.43
		$6d\,(^2D_{3/2})$	22588.82
		$6d\,(^2D_{5/2})$	22631.69
		$8s\,(^2S_{1/2})$	24317.15
		$4f\,(^2F_{7/2})$	24472.05
		$4f\,(^2F_{5/2})$	24472.23
56	Ba ($6s^2$ – $^1S_0$) 42034.8	$5d\,(^3D_1)$	9033.98
		$5d\,(^3D_2)$	9215.52
		$5d\,(^3D_3)$	9596.55
		$5d\,(^1D_2)$	11395.38
		$6p\,(^3P^o_0)$	12266.02
		$6p\,(^3P^o_1)$	12636.62
		$6p\,(^3P^o_2)$	13514.74
		$6p\,(^1P^o_1)$	18060.26
		$5d\,(^2D)\,6p'\,(^3F^o_2)$	22064.66
		$6p'\,(^3F^o_3)$	22947.44
		$5d^2\,(^1D_2)$	23062.1
		$6p'\,(^1D^o_2)$	23074.42
		$5d^2\,(^3P_0)$	23209.1
		$(^3P_1)$	23480.0
		$(^3P_2)$	23918.9
		$6p'\,(^3F^o_4)$	23757.08
57	La ($5d\,6s^2$ – $^2D_{3/2}$) 44980	$5d\,6s^2\,(^2D_{5/2})$	1053.16
		$5d^2\,(^3F)\,6s\,(^4F_{3/2})$	2668.19
		$6s\,(^4F_{5/2})$	3010.00
		$6s\,(^4F_{7/2})$	3494.53
		$6s\,(^4F_{9/2})$	4121.57
		$6s\,(^2F_{5/2})$	7011.91
		$5d^2\,(^3P)\,6s\,(^4P_{1/2})$	7231.41
		$6s\,(^4P_{3/2})$	7490.52

Table 7.1 (continued)

Atomic number Z	Element (ground-state term); ionization limit [cm^{-1}]	Excited state	Energy level [cm^{-1}]
1	2	3	4
		$6s\,(^4P_{5/2})$	7679.94
		$5d^2\,(^3F)\,6s\,(^2F_{7/2})$	8052.16
		$5d^2\,(^1D)\,6s\,(^2D_{3/2})$	8446.04
		$5d^2\,(^3P)\,6s\,(^2P_{1/2})$	9044.21
		$5d^2\,(^1D)\,6s\,(^2D_{5/2})$	9183.80
		$5d^2\,(^3P)\,6s\,(^2P_{3/2})$	9719.44
		$5d^2\,(^1G)\,6s\,(^2G_{9/2})$	9919.82
		$6s\,(^2G_{7/2})$	9960.90
		$5d^3\,(^4F_{3/2})$	12430.61
		$5d^3\,(^4F_{5/2})$	12787.40
		$5d^3\,(^4F_{7/2})$	13238.32
		$5d\,6s\,(^3D)\,6p\,(^4F^o_{3/2})$	13260.4
		$6p\,(^4F^o_{5/2})$	13631.0
		$5d^3\,(^4F_{9/2})$	13747.28
58	Ce $(4f5d\,6s^2 - {}^1G^o_4)$ 44670	$4f5d\,6s^2\,(^3F^o_2)$	228.85
		$6s^2\,(^3H^o_4)$	1279.42
		$6s^2\,(^3G^o_3)$	1388.94
		$6s^2\,(^3F^o_3)$	1663.12
		$6s^2\,(^3H^o_5)$	2208.66
		$4f\,(^2F^o)\,5d^2\,(^3F)\,6s\,(^4F)\,(^5H^o_3)$	2369.07
		$4f5d\,6s^2\,(^1D^o_2)$	2378.83
		$4f5d^2\,6s\,(^5H^o_4)$	2437.63
		$4f5d\,6s^2\,(^3F^o_4)$	3100.15
		$4f5d^2\,6s\,(^5I^o_4)$	3196.61
		$4f5d^2\,6s\,(^3G^o_5 + {}^5H^o_5)$	3210.58
		$4f5d\,6s^2\,(^3F^o_4 + {}^3G^o_4)$	3312.24
		$4f5d\,6s^2\,(^3D^o_1)$	3710.51
		$4f5d^2\,6s\,(^5I^o_5)$	3764.01
		$4f5d^2\,6s\,(^5D^o_0 + {}^3P^o_0)$	3974.50
		$4f5d\,6s^2\,(^3H^o_6)$	3976.10
		$4f5d^2\,6s\,(^3S^o_1 + {}^3P^o_1)$	4020.95
		$4f5d^2\,6s\,(^3G^o_3)$	4160.28
		$4f5d^2\,6s\,(^3G^o_4 + {}^5H^o_4)$	4173.49
		$4f5d\,6s^2\,(^3G^o_5)$	4199.37
		$4f5d^2\,6s\,(^3G^o_5) + 4f5d\,6s^2\,(^3G^o_5)$	4417.62
		$4f5d^2\,6s\,(^5I^o_6)$	4455.76
		$4f5d^2\,6s\,(^5H^o_6)$	4746.63
		$4f^2\,6s^2\,(^3H_4)$	4762.72
		$4f5d\,6s^2\,(^3D^o_2)$	4766.32
		$4f5d\,6s^2\,(^3D^o_3)$	5006.72
		$4f5d^2\,6s\,(^3S^o_1)$	5097.78
59	Pr $(4f^3\,6s^2 - {}^4I^o_{9/2})$ 44100	$4f^3\,6s^2\,(^4I^o_{11/2})$	1376.6
		$4f^3\,6s^2\,(^4I^o_{13/2})$	2846.7
		$4f^3\,6s^2\,(^4I^o_{15/2})$	4381.1
		$4f^2\,(^3H)\,5d\,6s^2\,(^4I_{9/2})$	4432.2

Table 7.1 (continued)

Atomic number Z	Element (ground-state term); ionization limit [cm^{-1}]	Excited state	Energy level [cm^{-1}]
1	2	3	4
		$4f^2 5d 6s^2 (^4K_{11/2})$	4866.5
		$4f^2 5d 6s^2 (^2H_{9/2})$	5822.9
		$4f^2 5d 6s (^4I_{11/2})$	6313.2
		$4f^2 5d 6s (^4H_{7/2})$	6535.5
		$4f^2 5d 6s^2 (^4K_{13/2})$	6603.6
		$4f^2 (^3H) 5d^2 (^3F) (^5L) 6s (^6L_{11/2})$	6714.2
		$4f^2 5d 6s^2 (^2H_{11/2})$	6892.9
		$4f^2 5d 6s^2 (^4G_{7/2})$	7617.4
		$4f^2 5d^2 (^5L) 6s (^6L_{13/2})$	7630.2
		$4f^2 5d 6s^2 (^4I_{13/2})$	7951.3
		$4f^2 5d^2 6s (^6I_{7/2})$	8013.1
		$4f^2 5d 6s^2 (^4H_{9/2})$	8029.2
		$4f^3 (^4I^o) 5d 6s (^3D) (^6L^o_{11/2})$	8080.5
		$4f^3 5d 6s (^6K^o_{9/2})$	8250.2
		$4f^2 (^3H) 5d^2 (^3F) (^5K) 6s (^6K_{9/2})$	8320.3
		$4f^2 5d 6s^2 (^4K_{15/2})$	8363.9
		$4f^2 5d^2 6s (^6I_{9/2})$	8643.8
		$4f^3 5d 6s (^6L^o_{13/2})$	8733.5
		$4f^2 5d^2 (^5K) 6s (^6K_{11/2})$	8829.1
		$4f^3 5d 6s (^6K^o_{11/2})$	8835.4
		$4f^2 5d^2 6s (^6I_{11/2})$	9268.7
60	Nd $(4f^4 6s^2 - {}^5I_4)$ 44560	$4f^4 6s^2 (^5I_5)$	1128.06
		$4f^4 6s^2 (^5I_6)$	2366.60
		$4f^4 6s^2 (^5I_7)$	3681.70
		$4f^4 6s^2 (^5I_8)$	5048.60
		$4f^3 (^4I^o) 5d 6s^2 (^5L^o_6)$	6764.21
		$4f^3 5d 6s^2 (^5K^o_5)$	6853.99
		$4f^3 5d 6s^2 (^5L^o_7)$	8402.49
		$4f^3 5d 6s^2 (^5K^o_6)$	8411.90
		$4f^4 (^5I) 5d 6s (^3D) (^7L_5)$	8475.36
		$4f^3 (^4I^o) 5d^2 (^3F) (^6M^o) 6s (^7M^o_6)$	8800.39
		$4f^3 5d 6s^2 (^5I^o_4)$	9083.81
		$4f^4 5d 6s (^7L_6)$	9115.09
		$4f^3 5d^2 6s (^7M^o_7)$	9692.28
		$4f^4 5d 6s (^7K_4)$	9814.68
		$4f^3 5d 6s^2 (^5H^o_3)$	9927.39
		$4f^4 5d 6s (^7L_7)$	9939.70
61	Pm $(4f^5 6s^2 - {}^6H^o_{5/2})$ 44800	$4f^5 6s^2 (^6H^o_{7/2})$	803.8
		$4f^5 6s^2 (^6H^o_{9/2})$	1748.8
		$4f^5 6s^2 (^6H^o_{11/2})$	2797.1
		$4f^5 6s^2 (^6H^o_{13/2})$	3919.0
		$4f^5 6s^2 (^6H^o_{15/2})$	5089.8
		$4f^5 6s^2 (^6F^o_{1/2})$	5249.5
		$4f^5 6s^2 (^6F^o_{3/2})$	5460.5
		$4f^5 6s^2 (^6F^o_{5/2})$	5872.8

Table 7.1 (continued)

Atomic number Z	Element (ground-state term); ionization limit [cm^{-1}]	Excited state	Energy level [cm^{-1}]
1	2	3	4
		$4f^5 6s^2 (^6F^o_{7/2})$	6562.9
		$4f^5 6s^2 (^6F^o_{9/2})$	7498.0
		$4f^5 6s^2 (^6F^o_{11/2})$	8609.2
		$4f^4 5d 6s^2 + 4f^5 6s 6p (J = 7/2)$	17104.7
		$4f^4 5d 6s^2 + 4f^5 6s 6p (J = 3/2)$	20006.0
62	Sm $(4f^6 6s^2 - {}^7F_0)$ 45520	$4f^6 6s^2 (^7F_1)$	292.6
		$4f^6 6s^2 (^7F_2)$	811.9
		$4f^6 6s^2 (^7F_3)$	1489.5
		$4f^6 6s^2 (^7F_4)$	2273.1
		$4f^6 6s^2 (^7F_5)$	3125.5
		$4f^6 6s^2 (^7F_6)$	4020.7
		$4f^6 (^7F) 5d (^8H) 6s (^9H_1)$	10801.1
		$4f^6 5d 6s (^9H_2)$	11044.9
		$4f^6 5d 6s (^9H_3)$	11406.5
		$4f^6 5d 6s (^9H_4)$	11877.5
		$4f^6 5d (^8D) 6s (^9D_2)$	12313.1
		$4f^6 5d (^8H) 6s (^9H_5)$	12445.3
		$4f^6 5d (^8D) 6s (^9D_3)$	12846.6
		$4f^6 5d (^8H) 6s (^7H_2)$	13050.0
		$4f^6 5d (^8H) 6s (^9H_6)$	13095.7
		$4f^6 5d (^8F) 6s (^9F_1 + {}^9G_1)$	13369.3
		$4f^6 5d (^8D) 6s (^9D_4)$	13458.5
		$4f^6 5d (^8H) 6s (^7H_3)$	13542.8
		$4f^6 5d (^8G) 6s (^9G_0)$	13551.2
		$4f^6 5d (^8G) 6s (^9G_2)$	13687.7
		$4f^6 5d (^8G) 6s (^9G_1)$	13732.5
		$4f^6 6s 6p (^3P^o) (^9G^o_0)$	13796.4
		$4f^6 5d (^8H) 6s (^9H_7)$	13814.9
		$4f^6 6s 6p (^3P^o) (^9G^o_1)$	13999.5
63	Eu $(4f^7 6s^2 - {}^8S^o_{7/2})$ 45735	$4f^7 (^8S^o) 5d (^9D^o) 6s (^{10}D^o_{5/2})$	12923.7
		$4f^7 5d 6s (^{10}D^o_{7/2})$	13048.9
		$4f^7 5d 6s (^{10}D^o_{9/2})$	13222.0
		$4f^7 5d 6s (^{10}D^o_{11/2})$	13457.2
		$4f^7 5d 6s (^{10}D^o_{13/2})$	13778.7
		$4f^7 (^8S^o) 6s 6p (^3P^o) (^{10}P_{7/2})$	14067.9
		$4f^7 6s 6p (^{10}P_{9/2})$	14563.6
		$4f^7 5d 6s (^8D^o_{3/2})$	15137.7
		$4f^7 5d 6s (^8D^o_{5/2})$	15248.8
		$4f^7 5d 6s (^8D^o_{7/2})$	15421.2
		$4f^7 6s 6p (^{10}P_{11/2})$	15581.6
		$4f^7 5d 6s (^8D^o_{9/2})$	15680.3
		$4f^7 6s 6p (^8P_{5/2})$	15890.5
		$4f^7 6s 6p (^8P_{7/2})$	15952.3
		$4f^7 5d 6s (^8D^o_{11/2})$	16079.8
		$4f^7 6s 6p (^8P_{9/2})$	16611.8

Table 7.1 (continued)

Atomic number Z	Element (ground-state term); ionization limit [cm^{-1}]	Excited state	Energy level [cm^{-1}]
1	2	3	4
		$4f^7 6s6p\,(^6P_{7/2})$	17340.6
		$4f^7 6s6p\,(^6P_{5/2})$	17707.4
		$4f^7 6s6p\,(^6P_{3/2})$	17945.5
64	Gd $(4f^7 5d6s^2 - {}^9D^o_2)$ 49603	$4f^7\,(^8S^o)\,5d6s^2\,(^9D^o_3)$	215.12
		$4f^7 5d6s^2\,(^9D^o_4)$	532.98
		$4f^7 5d6s^2\,(^9D^o_5)$	999.12
		$4f^7 5d6s^2\,(^9D^o_6)$	1719.09
		$4f^7 5d^2\,(^3F)\,(^{10}F^o)\,6s\,(^{11}F^o_2)$	6378.15
		$4f^7 5d^2 6s\,(^{11}F^o_3)$	6550.40
		$4f^7 5d^2 6s\,(^{11}F^o_4)$	6786.18
		$4f^7 5d6s^2\,(^7D^o_5)$	6976.51
		$4f^7 5d^2 6s\,(^{11}F^o_5)$	7103.42
		$4f^7 5d6s^2\,(^7D^o_4)$	7234.91
		$4f^7 5d6s^2\,(^7D^o_3)$	7426.71
		$4f^7 5d^2 6s\,(^{11}F^o_6)$	7480.35
		$4f^7 5d6s^2\,(^7D^o_2)$	7562.46
		$4f^7 5d6s^2\,(^7D^o_1)$	7653.93
		$4f^7 5d^2 6s\,(^{11}F^o_7)$	7947.29
		$4f^7 5d^2 6s\,(^{11}F^o_8)$	8498.43
		$4f^7 5d^2 6s\,(^9F^o_1)$	10222.23
		$4f^7 5d^2 6s\,(^9F^o_2)$	10359.91
		$4f^7 5d^2 6s\,(^9F^o_3)$	10576.41
		$4f^7 5d^2 6s\,(^9F^o_4)$	10883.51
		$4f^8 6s^2\,(^7F_6)$	10947.21
		$4f^7 5d^2 6s\,(^9F^o_5)$	11296.47
		$4f^7 5d^2\,(^3P)\,(^{10}P^o)\,6s\,(^{11}P^o_4)$	11685.59
		$4f^7 5d^2 6s\,(^9F^o_6)$	11830.39
		$4f^7 5d^2 6s\,(^{11}P^o_5)$	12057.16
		$4f^7 5d^2 6s\,(^{11}P^o_6)$	12345.97
		$4f^7 5d^2 6s\,(^9F^o_7)$	12486.55
		$4f^8 6s^2\,(^7F_5)$	12520.00
65	Tb $(4f^9 6s^2 - {}^6H^o_{15/2})$ 47300	$4f^8\,(^7F)\,5d6s^2\,(^8G_{13/2})$	285.50
		$4f^8 5d6s^2\,(^8G_{15/2})$	462.08
		$4f^8 5d6s^2\,(^8G_{11/2})$	509.85
		$4f^8 5d6s^2\,(J = 9/2)$	1371.05
		$4f^8 5d6s^2\,(^8D_{11/2})$	2310.09
		$4f^8 5d6s^2\,(^8G_{7/2})$	2419.48
		$4f^9 6s^2\,(^6H^o_{13/2})$	2771.68
		$4f^8 5d6s^2\,(^8G_{9/2})$	2840.17
		$4f^8 5d6s^2\,(^8G_{5/2})$	3174.58
		$4f^8 5d6s^2\,(^8G_{3/2})$	3705.82
		$4f^8 5d6s^2\,(^8F_{13/2})$	3719.71
		$4f^8 5d6s^2\,(^8D_{7/2})$	3819.85
		$4f^8 5d6s^2\,(^8G_{1/2})$	4018.21
		$4f^8 5d6s^2\,(^8H_{17/2})$	4646.83

Table 7.1 (continued)

Atomic number Z	Element (ground-state term); ionization limit [cm^{-1}]	Excited state	Energy level [cm^{-1}]
1	2	3	4
		$4f^9\,6s^2\,(^6H^o_{11/2})$	4670.46
		$4f^8\,5d\,6s^2\,(^8D_{5/2})$	4695.51
		$4f^8\,5d\,6s^2\,(^8F_{11/2})$	5353.37
		$4f^8\,5d\,6s^2\,(^8H_{15/2})$	5425.06
		$4f^8\,5d\,6s^2\,(^8D_{3/2})$	5483.98
		$4f^8\,5d\,6s^2\,(^8F_{9/2})$	5829.86
		$4f^9\,6s^2\,(^6H^o_{9/2})$	6174.93
		$4f^8\,5d\,6s^2\,(^8F_{1/2})$	6259.09
		$4f^8\,5d\,6s^2\,(^8H_{13/2})$	6351.75
		$4f^8\,5d\,6s^2\,(^8F_{7/2})$	6488.28
		$4f^8\,5d\,6s^2\,(^6F_{11/2})$	6674.16
		$4f^8\,5d\,6s^2\,(^8F_{5/2})$	6801.19
		$4f^8\,5d\,6s^2\,(^8F_{3/2})$	6849.72
		$4f^8\,5d\,6s^2\,(^8H_{11/2})$	6988.82
66	Dy $(4f^{10}\,6s^2 - {}^5I_8)$ 47900	$4f^{10}\,6s^2\,(^5I_7)$	4134.2
		$4f^{10}\,6s^2\,(^5I_6)$	7050.6
		$4f^9\,(^6H^o)\,5d\,6s^2\,(^7H^o_8)$	7565.6
		$4f^9\,5d\,6s^2\,(^7H^o_7)$	8519.2
		$4f^{10}\,6s^2\,(^5I_5)$	9211.6
		$4f^9\,5d\,6s^2\,(^7I^o_9)$	9990.9
		$4f^9\,5d\,6s^2\,(^7H^o_6)$	10088.8
		$4f^{10}\,6s^2\,(^5I_4)$	10925.2
		$4f^9\,5d\,6s^2\,(^7F^o_6)$	11673.5
		$4f^9\,5d\,6s^2\,(^7I^o_8)$	12007.1
		$4f^9\,5d\,6s^2\,(^7H^o_5)$	12298.6
		$4f^9\,5d\,6s^2\,(^7G^o_7)$	12655.1
		$4f^9\,5d\,6s^2\,(^7K^o_{10})$	12892.8
67	Ho $(4f^{11}\,6s^2 - {}^4I^o_{15/2})$ 48570	$4f^{11}\,6s^2\,(^4I^o_{13/2})$	5419.7
		$4f^{10}\,(^5I_8)\,5d\,6s^2\,(8,\,3/2)_{17/2}$	8378.9
		$4f^{10}\,5d\,6s^2\,(8,\,3/2)_{15/2}$	8427.1
		$4f^{11}\,6s^2\,(^4I^o_{11/2})$	8605.2
		$4f^{10}\,5d\,6s^2\,(8,\,3/2)_{13/2}$	9147.1
		$4f^{10}\,5d\,6s^2\,(8,\,3/2)_{19/2}$	9741.5
		$4f^{11}\,6s^2\,(^4I^o_{9/2})$	10695.7
		$4f^{10}\,(^5I_8)\,5d\,6s^2\,(8,\,5/2)_{21/2}$	11322.3
		$4f^{10}\,5d\,6s^2\,(8,\,5/2)_{17/2}$	11530.6
		$4f^{10}\,5d\,6s^2\,(8,\,5/2)_{19/2}$	11689.8
		$4f^{10}\,5d\,6s^2\,(8,\,5/2)_{15/2}$	12339.0
		$4f^{10}\,(^5I_7)\,5d\,6s^2\,(7,\,3/2)_{13/2}$	12344.5
		$4f^{10}\,5d\,6s^2\,(7,\,3/2)_{11/2}$	13082.9
68	Er $(4f^{12}\,6s^2 - {}^3H_6)$ 49260	$4f^{12}\,6s^2\,(^3F_4)$	5035.19
		$4f^{12}\,6s^2\,(^3H_5)$	6958.33
		$4f^{11}\,(^4I^o_{15/2})\,5d\,6s^2\,(15/2,\,3/2)^o_6$	7176.50
		$4f^{11}\,5d\,6s^2\,(15/2,\,3/2)^o_7$	7696.96
		$4f^{11}\,5d\,6s^2\,(15/2,\,3/2)^o_9$	8620.57

Table 7.1 (continued)

Atomic number Z	Element (ground-state term); ionization limit [cm^{-1}]	Excited state	Energy level [cm^{-1}]
1	2	3	4
		$4f^{11}5d6s^2(15/2, 3/2)^o_8$	9350.11
		$4f^{11}(^4I^o_{15/2})5d6s^2(15/2, 5/2)^o_{10}$	9655.85
		$4f^{11}5d6s^2(15/2, 5/2)^o_9$	10557.92
		$4f^{12}6s^2(^3H_4)$	10750.98
		$4f^{11}5d6s^2(15/2, 5/2)^o_5$	11401.20
		$4f^{11}5d6s^2(15/2, 5/2)^o_8$	11557.67
		$4f^{11}5d6s^2(15/2, 5/2)^o_6$	11799.78
		$4f^{11}5d6s^2(15/2, 5/2)^o_7$	11887.50
		$4f^{12}6s^2(^3F_3)$	12377.53
		$4f^{12}6s^2(^3F_2)$	13097.91
69	Tm $(4f^{13}6s^2 - {}^2F^o_{7/2})$ 49880	$4f^{13}6s^2(^2F^o_{5/2})$	8771.24
		$4f^{12}(^3H_6)5d6s^2(6, 3/2)_{9/2}$	13119.61
		$4f^{12}5d6s^2(6, 3/2)_{15/2}$	15271.00
		$4f^{12}5d6s^2(6, 3/2)_{11/2}$	15587.81
		$4f^{12}5d6s^2(6, 5/2)_{17/2}$	16456.91
		$4f^{13}(^2F^o_{7/2})6s6p(^3P^o_0)(7/2, 0)_{7/2}$	16742.24
		$4f^{12}5d6s^2(6, 5/2)_{7/2}$	16957.01
		$4f^{13}6s6p(^3P^o_1)(7/2, 1)_{7/2}$	17343.37
		$4f^{12}5d6s^2(6, 3/2)_{13/2}$	17454.82
		$4f^{13}6s6p(^3P^o_1)(7/2, 1)_{9/2}$	17613.66
		$4f^{13}6s6p(^3P^o_1)(7/2, 1)_{5/2}$	17752.63
70	Yb $(4f^{14}6s^2 - {}^1S_0)$ 50441	$4f^{14}6s6p(^3P^o_0)$	17288.44
		$4f^{14}6s6p(^3P^o_1)$	17992.01
		$4f^{14}6s6p(^3P^o_2)$	19710.39
		$4f^{13}(^2F^o_{7/2})5d6s^2(7/2, 3/2)^o_2$	23188.52
		$4f^{14}5d6s(^3D_1)$	24489.10
		$4f^{14}5d6s(^3D_2)$	24751.95
		$4f^{14}6s6p(^1P^o_1)$	25068.23
		$4f^{14}5d6s(^3D_3)$	25270.90
		$4f^{13}(^2F^o_{7/2})5d6s^2(7/2, 3/2)^o_5$	25859.68
71	Lu $(4f^{14}5d6s^2 - {}^2D_{3/2})$ 43762.4	$5d6s^2(^2D_{5/2})$	1993.9
		$6s^26p(^2P^o_{1/2})$	4136.0
		$6s^26p(^2P^o_{3/2})$	7476.3
		$5d6s(^3D)6p(^4F^o_{3/2})$	17427.3
		$5d6s(^3D)6p(^4F^o_{5/2})$	18504.6
		$5d^2(^3F)6s(^4F_{3/2})$	18851.3
		$5d^2(^3F)6s(^4F_{5/2})$	19403.3
72	Hf $(5d^26s^2 - {}^3F_2)$ 55600	$5d^26s^2(^3F_3)$	2356.7
		$5d^26s^2(^3F_4)$	4567.6
		$5d^26s^2(^3P_0)$	5521.8
		$5d^26s^2(^1D_2)$	5638.6
		$5d^26s^2(^3P_1)$	6572.6
		$5d^26s^2(^3P_2)$	8983.8
		$5d^26s^2(^1G_4)$	10532.5
		$5d^3(^4F)6s(^5F_1)$	14092.3

Table 7.1 (continued)

Atomic number Z	Element (ground-state term); ionization limit [cm^{-1}]	Excited state	Energy level [cm^{-1}]
1	2	3	4
		$6s\,(^5F_2)$	14740.7
		$6s\,(^5F_3)$	15673.3
		$6s\,(^5F_4)$	16766.6
		$6s\,(^5F_5)$	17901.3
73	Ta $(5d^3\,6s^2\ -\ ^4F_{3/2})$ 63600	$5d^3\,6s^2\,(^4F_{5/2})$	2010.1
		$(^4F_{7/2})$	3963.9
		$(^4F_{9/2})$	5621.0
		$5d^3\,6s^2\,(^4P_{1/2})$	6049.4
		$(^4P_{3/2})$	6068.9
		$(^4P_{5/2})$	9253.4
74	W $(5d^4\,6s^2\ -\ ^5D_0)$ 64000	$5d^4\,6s^2\,(^5D_1)$	1670.3
		$5d^5\,(^6S)\,6s\,(^7S_3)$	2951.3
		$5d^4\,6s^2\,(^5D_2)$	3325.5
		$(^5D_3)$	4830.0
		$(^5D_4)$	6219.3
		$5d^4\,6s^2\,(^3P_0)$	9528.1
75	Re $(5d^5\,6s^2\ -\ ^6S_{5/2})$ 64000	$5d^5\,6s^2\,(^4P_{5/2})$	11584.0
		$5d^6\,(^5D)\,6s\,(^6D_{9/2})$	11754.5
		$5d^5\,6s^2\,(^4P_{3/2})$	13826.1
		$5d^6\,(^5D)\,6s\,(^6D_{7/2})$	14216.9
		$5d^5\,6s^2\,(^4G_{5/2})$	14621.5
76	Os $(5d^6\,6s^2\ -\ ^5D_4)$ 70450	$5d^6\,6s^2\,(^5D_2)$	2740.5
		$(^5D_3)$	4159.3
		$5d^7\,(^4F)\,6s\,(^5F_5)$	5143.9
		$5d^6\,6s^2\,(^5D_1)$	5766.1
		$(^5D_0)$	6092.8
		$5d^7\,(^4F)\,6s\,(^5F_4)$	8742.8
77	Ir $(5d^7\,6s^2\ -\ ^4F_{9/2})$ 73000	$5d^8\,(^3F)\,6s\,(^4F_{9/2})$	2835.0
		$5d^7\,6s^2\,(^4F_{3/2})$	4078.9
		$(^4F_{5/2})$	5784.6
		$(^4F_{7/2})$	6323.9
		$5d^8\,(^3F)\,6s\,(^4F_{7/2})$	7106.6
		$(^4F_{5/2})$	9877.5
78	Pt $(5d^9\,6s\ -\ ^3D_3)$ 72300	$5d^9\,(^2D_{5/2})\,6s\,(^3D_2)$	776
		$5d^8\,6s^2\,(^3F_4)$	824
		$5d^{10}\,(^2S_0)$	6140
		$5d^8\,6s^2\,(^3P_2)$	6567
79	Au $(5d^{10}\,6s\ -\ ^2S_{1/2})$ 74409.0	$5d^9\,6s^2\,(^2D_{5/2})$	9161.18
		$(^2D_{3/2})$	21435
		$6p\,(^2P^o_{1/2})$	37359
		$(^2P^o_{3/2})$	41174
80	Hg $(5d^{10}\,6s^2\ -\ ^1S_0)$ 84184.1	$6p\,(^3P^o_0)$	37645.24
		$(^3P^o_1)$	39412.46

Table 7.1 (continued)

Atomic number Z	Element (ground-state term); ionization limit [cm^{-1}]	Excited state	Energy level [cm^{-1}]
1	2	3	4
		$(^3P^o_2)$	44043.14
		$6p\,(^1P^o_1)$	54068.90
		$7s\,(^3S_1)$	62350.54
		$7s\,(^1S_0)$	63928.34
		$5d^9 6s^2\,(^2D_{5/2})\,6p\,(^3P^o_2)$	68886.6
		$7p\,(^3P^o_0)$	69516.7
		$(^3P^o_1)$	69661.9
		$(^3P^o_2)$	71207.5
		$7p\,(^1P^o_1)$	71295.1
		$6d\,(^1D_2)$	71333.30
		$6d\,(^3D_1)$	71336.26
		$(^3D_2)$	71396.33
		$(^3D_3)$	71431.42
81	Tl $(6s^2 6p\ -\ ^2P^o_{1/2})$ 49266.7	$6p\,(^2P^o_{3/2})$	7793
		$7s\,(^2S_{1/2})$	26477
		$7p\,(^2P^o_{1/2})$	34160
		$7p\,(^2P^o_{3/2})$	35161
		$6d\,(^2D_{3/2})$	36118
		$6d\,(^2D_{5/2})$	36200
82	Pb $(6p^2 - (1/2,\ 1/2)_0)$ 59819.6	$6p^2\,(3/2,\ 1/2)_1$	7819.26
		$6p^2\,(3/2,\ 1/2)_2$	10650.33
		$6p^2\,(3/2,\ 3/2)_2$	21457.80
		$6p^2\,(3/2,\ 3/2)_0$	29466.83
		$6p\,7s\,(1/2,\ 1/2)^o_0$	34959.91
		$6p\,7s\,(1/2,\ 1/2)^o_1$	35287.22
		$6p\,7p\,(1/2,\ 1/2)_1$	42918.64
		$6p\,7p\,(1/2,\ 1/2)_0$	44400.89
		$6p\,7p\,(1/2,\ 3/2)_1$	44674.99
		$6p\,7p\,(1/2,\ 3/2)_2$	44809.36
		$6p\,6d\ -\ 1/2\,[5/2]^o_2$	45443.17
83	Bi $(6p^3\ -\ ^4S^o_{3/2})$ 58762	$6p^3\,(^2D^o_{3/2})$	11419.0
		$6p^3\,(^2D^o_{5/2})$	15437.7
		$6p^3\,(^2P^o_{1/2})$	21661.0
		$6p^2\,(^3P_0)\,7s\,(^4P_{1/2})$	32588.2
		$6p^3\,(^2P^o_{3/2})$	33164.70
		$7p\,(^2P^o_{1/2})$	41125.2
		$7p\,(^2P^o_{3/2})$	42940.6
		$6d\,(^2D_{3/2})$	43912.3
		$6d\,(^2D_{5/2})$	44817.1
		$6p^2\,(^3P_1)\,7s\,(^4P_{3/2})$	44865.1
		$7s\,(^2P_{1/2})$	45915.6
		$6p^2\,(^3P_0)\,8s\,(^4P_{1/2})$	47373.3
		$6p^2\,(^3P_2)\,7s\,(^4P_{5/2})$	48489.76
		$7s\,(^2P_{3/2})$	49460.80

Table 7.1 (continued)

Atomic number Z	Element (ground-state term); ionization limit [cm^{-1}]	Excited state	Energy level [cm^{-1}]
1	2	3	4
88	Ra $(6p^6 7s^2 - {}^1S_0)$	$7p\,({}^3P^o_0)$	13078.4
	42573.4	$6d\,({}^3D_1)$	13715.9
		$6d\,({}^3D_2)$	13994.0
		$7p\,({}^3P^o_1)$	13999.4
		$6d\,({}^3D_3)$	14707.4
		$7p\,({}^3P^o_2)$	16688.5
		$6d\,({}^1D_2)$	17081.5
		$7p\,({}^1P^o_1)$	20715.7
89	Ac $(6d 7s^2 - {}^2D_{3/2})$	$6d 7s^2\,({}^2D_{5/2})$	2231.4
	42000	$6d^2 7s\,({}^4F_{3/2})$	9217.3
		$6d^2 7s\,({}^4F_{5/2})$	9863.6
		$6d^2 7s\,({}^4F_{7/2})$	10906.0
		$6d^2 7s\,({}^4F_{9/2})$	12078.1
		$7s^2({}^1S)\,7p\,({}^2P^o_{1/2})$	–
		$7s^2 7p\,({}^2P^o_{3/2})$	–
		$6d 7s\,({}^3D)\,7p\,({}^4F^o_{3/2})$	13712.9
		$6d 7s 7p\,({}^4F^o_{5/2})$	14940.7
		$6d 7s 7p\,({}^4D^o_{1/2})$	17199.7
		$6d 7s 7p\,({}^4F^o_{7/2})$	17683.9
		$6d 7s 7p\,({}^2D^o_{3/2})$	17736.3
		$6d 7s 7p\,({}^2D^o_{5/2})$	17950.7
		$6d 7s 7p\,({}^4D^o_{3/2})$	19012.5
90	Th $(6d^2 7s^2 - {}^3F_2)$	$6d^2 7s^2\,({}^3P_0)$	2558.06
	49000	$6d^2 7s^2\,({}^3F_3)$	2869.26
		$6d^2 7s^2\,({}^3P_2)$	3687.99
		$6d^2 7s^2\,({}^3P_1)$	3865.48
		$6d^2 7s^2\,({}^3F_4)$	4961.66
		$6d^3 7s\,({}^5F_1)$	5563.14
		$6d^3 7s\,({}^5F_2)$	6362.40
		$6d^2 7s^2\,({}^1D_2)$	7280.13
		$6d^3 7s\,({}^5F_3)$	7502.29
		$5f 6d 7s^2\,({}^3H^o_4)$	7795.27
		$6d^2 7s^2\,({}^1G_4)$	8111.01
		$5f 6d 7s^2\,({}^3F^o_2)$	8243.60
		$6d^3 7s\,({}^5F_4)$	8800.25
		$6d^3 7s\,({}^5F_5)$	9804.81
		$5f 6d 7s^2\,({}^1G^o_4)$	10414.13
		$5f 6d 7s^2\,({}^3G^o_3)$	10526.54
		$6d 7s^2 7p\,({}^3F^o_2)$	10783.15
91	Pa $(5f^2 6d 7s^2 - {}^4K_{11/2})$	$5f^2\,({}^3H)\,6d 7s^2\,({}^4I_{9/2})$	825.42
	47000		1375.78
		$5f^2\,({}^3F)\,6d 7s^2\,({}^4G_{5/2})$	1618.33
			2840.25

Table 7.1 (continued)

Atomic number Z	Element (ground-state term); ionization limit [cm^{-1}]	Excited state	Energy level [cm^{-1}]
1	2	3	4
		$5f^2\,(^3F)\,6d\,7s^2\,(^4H_{7/2})$	2966.53
			3033.92
			3188.27
			3292.50
		$5f^2\,(^3H)\,6d\,7s^2\,(^2H_{9/2})$	3323.86
			3588.97
			3670.56
		$5f^2\,(^3H)\,6d\,7s^2\,(^4K_{13/2})$	3711.62
			3794.19
			3819.00
			3935.46
			3991.59
		$5f^2\,(^3H)\,6d\,7s^2\,(^4I_{11/2})$	4121.45
			4195.27
92	U $(5f^3\,6d\,7s^2\,-\,^5L^{\circ}_6)$ 49960	$5f^3\,6d\,7s^2\,(^5K^{\circ}_5)$	620.32
		$5f^3\,6d\,7s^2\,(^5L^{\circ}_7)$	3800.83
		$5f^3\,6d\,7s^2\,(^5H^{\circ}_3+{}^3G^{\circ}_3)$	3868.49
		$5f^3\,6d\,7s^2\,(^5K^{\circ}_6)$	4275.71
		$5f^3\,6d\,7s^2\,(^5I^{\circ}_4)$	4453.42
		$5f^3\,6d\,7s^2\,(^3I^{\circ}_5)$	5762.08
		$5f^3\,6d\,7s^2\,(^3P^{\circ}_0)$	5988.06
		$5f^3\,6d\,7s^2\,(^5H^{\circ}_4+{}^3H^{\circ}_4)$	5991.31
		$5f^3\,6d^2\,7s\,(^7M^{\circ}_6)$	6249.03
		$5f^3\,6d\,7s^2\,(J=6)^{\circ}_6$	7005.53
		$5f^4\,7s^2\,(^5I_4)$	7020.71
		$5f^3\,6d\,7s^2\,(^5G^{\circ}_3+{}^5H^{\circ}_3)$	7103.92
		$5f^3\,6d\,7s^2\,(^5G^{\circ}_2)$	7191.68
		$5f^3\,6d\,7s^2\,(^5K^{\circ}_7)$	7326.12
		$5f^3\,6d\,7s^2\,(^5L^{\circ}_8)$	7645.65
		$5f^3\,6d\,7s^2\,(^5I^{\circ}_5)$	7864.20
		$5f^3\,6d^2\,7s\,(^7M^{\circ}_7)$	8118.63
		$5f^3\,6d\,7s^2\,(J=4)^{\circ}_4$	8133.29
		$5f^3\,6d\,7s^2\,(^5G^{\circ}_2)$	8856.99
		$5f^3\,6d\,7s^2\,(^5H^{\circ}_3)$	8878.55
		$5f^4\,7s^2\,(^5I_5)$	10051.31
		$5f^3\,6d\,7s^2\,(^3L^{\circ}_7)$	10069.18
		$5f^3\,6d^2\,7s\,(^7L^{\circ}_5)$	10081.03
93	Np $(5f^4\,6d\,7s^2\,-\,^6L_{11/2})$ 50540	$5f^4\,(^5I)\,6d\,7s^2\,(^6K_{9/2})$	2033.94
		$5f^5\,7s^2\,(^6H^{\circ}_{5/2})$	2831.14
		$5f^4\,6d\,7s^2\,(^6I_{7/2})$	3450.97
		$5f^4\,6d\,7s^2\,(^6L_{13/2})$	3502.83
		$5f^4\,6d\,7s^2\,(^4G_{5/2})$	4615.71
		$5f^4\,6d\,7s^2\,(^6K_{11/2})$	5185.00
		$5f^5\,7s^2\,(^6H^{\circ}_{7/2})$	5456.13
		$5f^4\,6d\,7s^2\,(^6G_{3/2})$	6474.29

Table 7.1 (continued)

Atomic number Z	Element (ground-state term); ionization limit [cm^{-1}]	Excited state	Energy level [cm^{-1}]
1	2	3	4
		$5f^4 6d 7s^2 (^6I_{9/2})$	6643.50
		$5f^4 6d 7s^2 (^4H_{7/2}, {}^6I_{7/2})$	6892.08
		$5f^4 6d 7s^2 (^6L_{15/2})$	6903.45
		$5f^4 6d 7s^2 (^6H_{5/2})$	7015.18
		$5f^4 (^5I)\, 6d^2 (^3F)\, 7s\, (^8M_{11/2})$	7112.43
		$5f^4 6d 7s^2 (^6K_{13/2})$	7792.02
		$5f^5 7s^2 (^6H^\circ_{9/2})$	7805.80
		$5f^4 6d 7s^2 (^4H_{9/2})$	7871.86
		$5f^4 6d 7s^2 (^6H_{7/2})$	8278.30
		$5f^4 6d 7s^2 (^6I_{11/2})$	8706.56
		$5f^4 6d^2 7s (^8M_{13/2})$	8950.61
		$5f^4 6d 7s^2 (^6H_{7/2})$	9507.73
		$5f^4 6d 7s^2 (^6H_{5/2})$	9524.44
		$5f^4 6d 7s^2 (^6I_{11/2})$	9694.04
		$5f^5\, 7s^2 (^6H^\circ_{11/2})$	9854.96
		$5f^4 6d^2 7s (^8L_{9/2})$	9869.64
		$5f^4 6d 7s^2 (^4L_{13/2})$	9955.10
94	Pu $(5f^6 7s^2 - {}^7F_0)$ 49000	$5f^6 7s^2 (^7F_1)$	2203.6
		$5f^6 7s^2 (^7F_2)$	4299.6
		$5f^6 7s^2 (^7F_3)$	6144.4
		$5f^5 6d 7s^2 [5/2, 3/2]^\circ_4$	6313.6
		$5f^6 7s^2 (^7F_4)$	7774.5
		$5f^5 6d 7s^2 [5/2, 3/2]^\circ_2$	8767.9
		$5f^6 7s^2 (^7F_5)$	9179.1
		$5f^5 6d 7s^2 [7/2, 3/2]^\circ_5$	9386.6
		$5f^5 6d 7s^2 [5/2, 3/2]^\circ_3$	9724.1
		$5f^6 7s^2 (^5D_0)$	9772.4
		$5f^6 7s^2 (^7F_6)$	10238.3
		$5f^5 6d 7s^2 [5/2, 3/2]^\circ_1$	10486.7
		$5f^5 6d 7s^2 [5/2, 5/2]^\circ_0$	11746.9
		$5f^5 6d 7s^2 [7/2, 3/2]^\circ_3$	11840.5
		$5f^5 6d 7s^2 [7/2, 3/2]^\circ_4$	12159.2
		$5f^5 6d 7s^2 [5/2, 5/2]^\circ_1$	12177.7
		$5f^5 6d 7s^2 [7/2, 3/2]^\circ_2$	12322.4
		$5f^5 6d 7s^2 [9/2, 3/2]^\circ_6$	12351.3
		$5f^5 6d 7s^2 [5/2, 5/2]^\circ_2$	13517.4
		$5f^6 6d 7s (^9H_2)$	13528.3
		$5f^6 7s^2 (J = 1)$	13677.9
		$5f^5 6d 7s^2 [5/2, 5/2]^\circ_3$	13726.4
		$5f^5 6d 7s^2 [5/2, 5/2]^\circ_5$	14291.8
		$5f^6 6d 7s (^9H_2)$	14342.0
		$(J = 1)^\circ$	14416.1
		$5f^5 6d 7s^2 [5/2, 5/2]^\circ_4$	14853.1

Table 7.1 (continued)

Atomic number Z	Element (ground-state term); ionization limit [cm^{-1}]	Excited state	Energy level [cm^{-1}]
1	2	3	4
96	Cm $(5f^7 6d 7s^2 - {}^9D_2^o)$ 48600	$5f^7 6d 7s^2 ({}^9D_3^o)$	302.15
		$5f^7 6d 7s^2 ({}^9D_4^o)$	815.65
		$5f^8 7s^2 ({}^7F_6)$	1214.18
		$5f^7 6d 7s^2 ({}^9D_5^o)$	1764.26
		$5f^7 6d 7s^2 ({}^9D_6^o)$	3809.35
		$5f^8 7s^2 ({}^7F_4)$	4877.58
		$5f^8 7s^2 ({}^7F_5)$	5136.50
		$5f^8 7s^2 ({}^7F_3)$	7208.81
		$5f^8 7s^2 ({}^7F_2)$	7521.10
		$5f^8 7s^2 ({}^7F_1)$	8696.66
		$5f^8 7s^2 ({}^7F_0)$	8887.23
		$5f^7 6d 7s^2 ({}^7D_4^o)$	8958.45
		$5f^7 6d 7s^2 ({}^7D_5^o)$	9064.87
		$5f^7 7s^2 7p ({}^9P_3)$	9263.36
		$5f^7 6d 7s^2 ({}^7D_3^o)$	9458.05
		$5f^7 6d 7s^2 ({}^7D_2^o)$	9671.69
		$5f^7 7s^2 7p ({}^9P_4)$	9784.52
		$5f^7 6d 7s^2 ({}^7D_1^o)$	10133.86
		$5f^7 6d^2 7s ({}^{11}F_2^o)$	10144.93
		$5f^7 6d^2 7s ({}^{11}F_3^o)$	10484.87
		$5f^7 6d^2 7s ({}^{11}F_4^o)$	10971.17
		$5f^7 6d^2 7s ({}^{11}F_5^o)$	11641.68
		$5f^7 6d^2 7s ({}^{11}F_6^o)$	12534.98
		$5f^7 6d^2 7s ({}^{11}F_7^o)$	13720.28
97	Bk $(5f^9 7s^2 - {}^6H_{15/2}^o)$ 50200	$5f^9 7s^2 ({}^6F_{11/2}^o)$	5416.69
		$5f^9 7s^2 ({}^6F_{9/2}^o)$	5757.44
		$5f^9 7s^2 ({}^6H_{13/2}^o)$	6530.71
		$5f^8 6d 7s^2 ({}^8G_{13/2})$	9141.12
		$5f^8 6d 7s^2 ({}^8G_{11/2})$	9300.58
		$5f^9 7s^2 ({}^6H_{11/2}^o)$	9535.12
		$5f^8 6d 7s^2 ({}^8G_{15/2})$	10587.34
		$5f^9 7s^2 ({}^6H_{9/2}^o)$	10605.57
		$5f^8 6d 7s^2 ({}^8G_{9/2})$	10735.96
		$5f^8 6d 7s^2 ({}^8G_{7/2})$	13191.92
98	Cf $(5f^{10} 7s^2 - {}^5I_8)$ 50800	$5f^{10} 7s^2 ({}^5F_5)$	8516.4
		$5f^{10} 7s^2 ({}^5I_7)$	9078.14
		$5f^{10} 7s^2 ({}^3P_2)$	10589.2
		$5f^{10} 7s^2 ({}^5I_6)$	11074.4
		$5f^{10} 7s^2 ({}^5F_4)$	13965.7
		$5f^{10} 7s^2 ({}^5I_4)$	15375.5
		$5f^{10} 7s^2 ({}^5I_5)$	15846.14
99	Es $(5f^{11} 7s^2 - {}^4I_{15/2}^o)$ 51800	$5f^{11} 7s^2 ({}^2H_{11/2}^o)$	8759.25
		$5f^{11} 7s 7p ({}^6I_{15/2})$	17802.89
		$5f^{11} 7s 7p ({}^6I_{17/2})$	19209.15
		$5f^{10} 6d 7s^2 ({}^6I_{17/2})$	19367.93

7.2 Diagrams of Atomic Energy Levels and Grotrian Diagrams

Figures 7.1–43 reproduce the diagrams of energy levels and spectra (they are also called Grotrian diagrams) of neutral elements with atomic number Z = 1–30 (H – Zn) and some atoms from the I, II, and VIII groups of the periodic table, having valence electron shells of the type ns, ns^2 and np^6 (Kr, Rb, Sr, Cd, Xe, Cs, Ba, Hg and Tl). In the selection of diagrams to be presented here, the policy was to give visual examples of excited-state distributions for various electron configurations of atoms and to mark the most intense transitions in the observed emission or absorption spectra.

We have restricted our consideration to levels that are not too high, so that long Rydberg series and autoionizing states resulting from double electron excitation and located above the ionization limit have been excluded here. The position of atomic energy levels in diagrams (this usually implies the position of the multiplet centre of gravity, Sect. 7.1) is shown in units of cm^{-1} on the ordinate axis and the numbers above the horizontal lines of the levels mark the excitation energies in units of eV. Additional labels (primes) on the diagrams are used to denote valence electron configurations and terms of the parent atomic core. Fine-structure intervals [cm^{-1}] with the proper sign, appropriate to either normal (+) or inverted (–) multiplets, are placed near the level lines in rectangular frames.

Apart from the numerical value of the transition wavelength λ [Å], the values of corresponding absorption oscillator strengths f_{ik} are sometimes marked in parentheses. (Some of the f_{ik} values are given in abbreviated form; thus, for example, $2^{-3} \equiv 0.002$ and so on). The lower index occasionally added to the values of λ or f_{ik} is used to clarify the identification of the component of the transition multiplet. This index labels either the relevant value of the upper level total angular momentum J_k or both J_i and J_k for the lower and upper levels if they both have multiplet structure. An asterisk (*) occasionally used as a superscript to f_{ik} marks the multiplet weighted value (Sect. 7.3).

For the rare gas atoms (Ne – Xe) partial diagrams of energy levels belonging to the lower and upper components of multiplet sublevels (these components have been marked by bars under or above the relevant values of the total electron angular momentum quantum number) have been drawn for a given configuration and given quantum numbers nl of the valence electron. In addition, Figs. 7.11, 20, 34 and 39 show Grotrian diagrams for the lower $2p$ – $1s$ transitions (in Paschen notation), which also include transition probabilities A_{ki} (in $10^6\,s^{-1}$) and radiative lifetimes τ_k [ns].

The information presented in Figs. 7.1–43 was taken from available publications on atomic energy levels [7.1.1–10], from tables of spectral lines [7.2.1–4] and reference material on transition probabilities [7.3.1–10]. We also mention a series of measurements of A_{ki} and τ_k values in rare gas atoms [7.2.5]. The error in determining the values of T_k, λ, f_{ik}, A_{ki} and τ_k was accounted for in giving values with an uncertainty of ± 1 or 2 in the last digit.

Let us now consider certain selection rules dealing with the allowed radiative transitions between atomic states [7.2.6]:

1) The momentum and parity conservation rules in an atom lead to certain selection rules, which restrict possible changes in the states of the emitting atom. The most intense transitions, the electric dipole transitions (E1) between states $i(LSJM) \to k(L'S'J'M')$, are allowed provided that:

$$\Delta J = J - J' = 0, \pm 1 \quad \text{and} \quad 0 \nleftrightarrow 0\,;$$

$$\text{even term} \rightleftarrows \text{odd term.}$$

2) In the case of states which are constructed according to the LS-coupling scheme, there are additional selection rules:

$$\Delta S = S - S' = 0;$$
$$\Delta L = L - L' = 0, \pm 1 \quad \text{and} \quad 0 \nleftrightarrow 0\,.$$

It should be mentioned that by convention, all wavelengths in spectroscopy with $\lambda > 2000$ Å are referred to transitions in air and shorter wavelengths are referred to transitions in vacuum. If σ is a vacuum wavenumber equal to the transition energy $\Delta E/h$, the index of refraction n in air for such a wave can be derived by use of Edlén's dispersion formula [7.2.7]:

$$n = 1 + 10^{-4}\left(0.834213 + \frac{2.40603}{1.3 - 10^{-10}\sigma^2} + \frac{0.015997}{0.389 - 10^{-10}\sigma^2}\right),$$

where σ is in cm^{-1}. Knowing n, one may easily convert the transition wavelength in vacuum λ_{vac} to λ_{air}, the observed transition wavelength in air:

$$\lambda_{air} = \lambda_{vac}/n\,,$$
$$\lambda_{vac}[\text{Å}] = 10^8/\Delta E[\mathrm{cm}^{-1}] = 12398.5/\Delta E[\mathrm{eV}]\,.$$

Table 7.2 lists the numerical values of the correction $\Delta\lambda = \lambda_{vac} - \lambda_{air}$, which makes it possible to convert air to vacuum wavelengths. These values were derived using Edlén's dispersion formula for standard air.

The reader is reminded, finally, that electron terms of some complicated systems (for example, the iron-group atoms) are described for brevity by the prefixes "*a*", "*b*", "*c*",... for low-lying even-parity terms, "*z*", "*y*", "*x*",... for odd-parity terms, and "*e*", "*f*",... for high-lying even-parity terms.

References

7.2.1 A.R.Striganov, N.S.Sventitskii: *Tables of Spectral Lines of Neutral and Ionized Atoms* (IFI/Plenum, New York 1968)
7.2.2 A.R.Striganov, G.A.Odintsova: *Tables of Spectral Lines of Atoms and Ions* (Energoizdat, Moscow, 1982) (in Russian)
7.2.3 A.N.Zaidel, V.K.Prokof'ev, S.M.Raiskii, V.A.Slavnyi, E.Y.Shreider: *Tables of Spectral Lines*, 4th ed. (Nauka, Moscow 1977) (in Russian)
7.2.4 M.Outred: "Tables of atomic spectral lines for the 10^4 Å to $40 \cdot 10^3$ Å region", J. Phys. Chem. Ref. Data **7**, 1–262 (1978)
7.2.5 H.Horiguchi, R.S.F.Chang, D.W.Setser: "Radiative lifetimes of Xe ($6p$, $6p'$) states", J. Chem. Phys. **75**, 1207 (1981) (and references cited therein on previous experiments with Ne, Ar and Kr)
7.2.6 A.Hibbert: "Atomic Structure Theory", in *Progress in Atomic Spectroscopy*, Part A, ed. by W.Hanle, H.Kleinpoppen (Plenum, New York 1978) pp. 1–69
7.2.7 B.Edlén: Metrologia **2**, 71 (1966)

Table 7.2. Wavelength correction $\Delta\lambda$, calculated from the Edlén dispersion formulas: $\lambda_{vac} = \lambda_{air} + \Delta\lambda$, $\lambda_{air} = \lambda_{vac} - \Delta\lambda$

λ_{air} [Å]	$\Delta\lambda$ [Å]	λ_{air} [Å]	$\Delta\lambda$ [Å]	λ_{air} [Å]	$\Delta\lambda$ [Å]
2000	0.648	5000	1.39	11000	3.01
2100	0.667	5200	1.45	11200	3.07
2200	0.687	5400	1.50	11400	3.12
2300	0.708	5600	1.55	11600	3.18
2400	0.731	5800	1.61	11800	3.23
2500	0.754	6000	1.66	12000	3.28
2600	0.777	6200	1.72	12200	3.34
2700	0.801	6400	1.77	12400	3.39
2800	0.825	6600	1.82	12600	3.45
2900	0.850	6800	1.88	12800	3.50
3000	0.875	7000	1.93	13000	3.56
3100	0.900	7200	1.98	13200	3.61
3200	0.925	7400	2.04	13400	3.66
3300	0.950	7600	2.09	13600	3.72
3400	0.976	7800	2.15	13800	3.77
3500	1.00	8000	2.20	14000	3.83
3600	1.03	8200	2.25	14200	3.88
3700	1.05	8400	2.31	14400	3.94
3800	1.08	8600	2.36	14600	3.99
3900	1.10	8800	2.42	14800	4.05
4000	1.13	9000	2.47	15000	4.10
4100	1.16	9200	2.52	16000	4.37
4200	1.18	9400	2.58	17000	4.64
4300	1.21	9600	2.63	18000	4.92
4400	1.24	9800	2.69	19000	5.19
4500	1.26	10000	2.74	20000	5.46
4600	1.29	10200	2.80	30000	8.18
4700	1.32	10400	2.85	40000	10.9
4800	1.34	10600	2.90	50000	13.6
4900	1.37	10800	2.96	100000	27.3

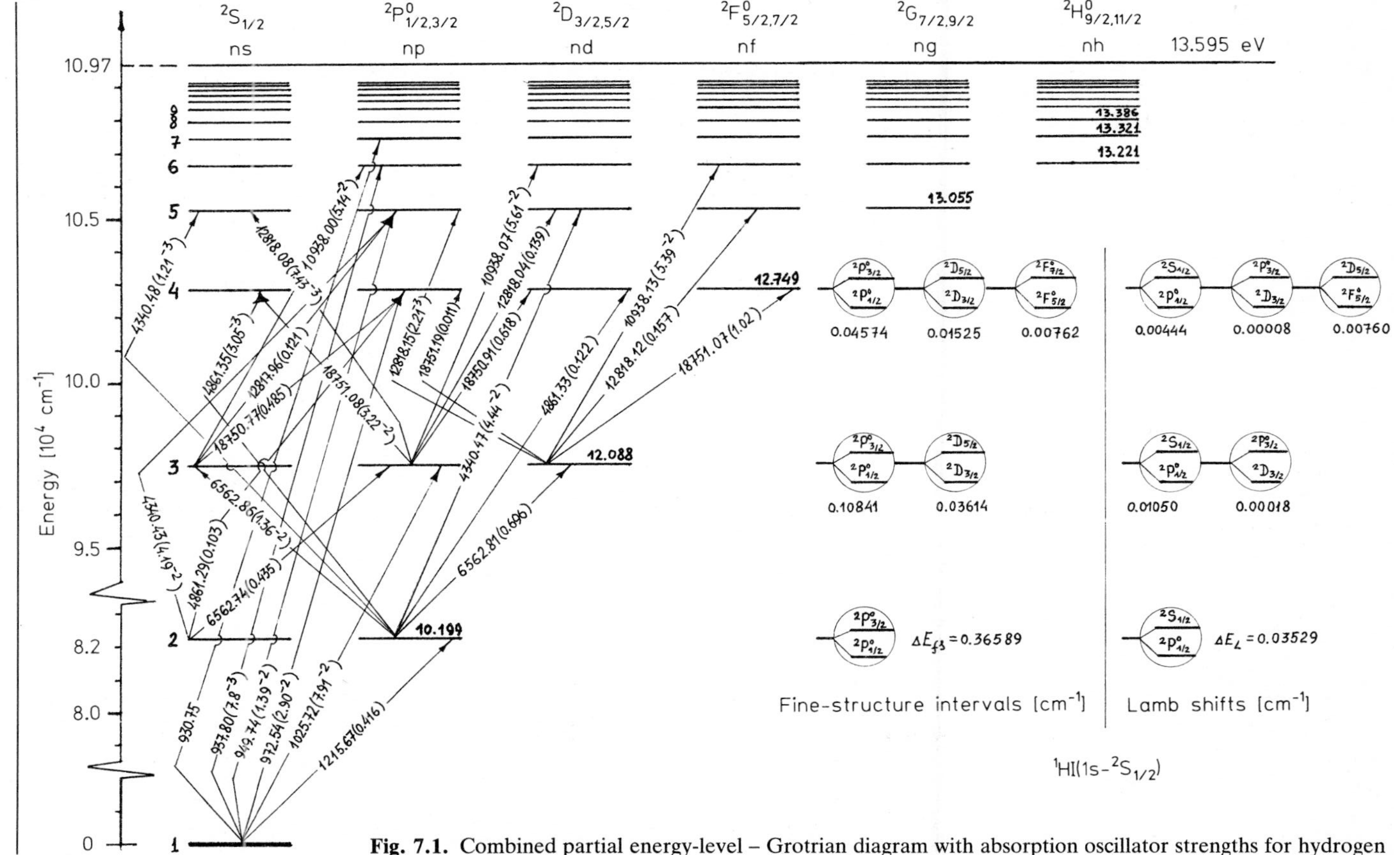

Fig. 7.1. Combined partial energy-level – Grotrian diagram with absorption oscillator strengths for hydrogen

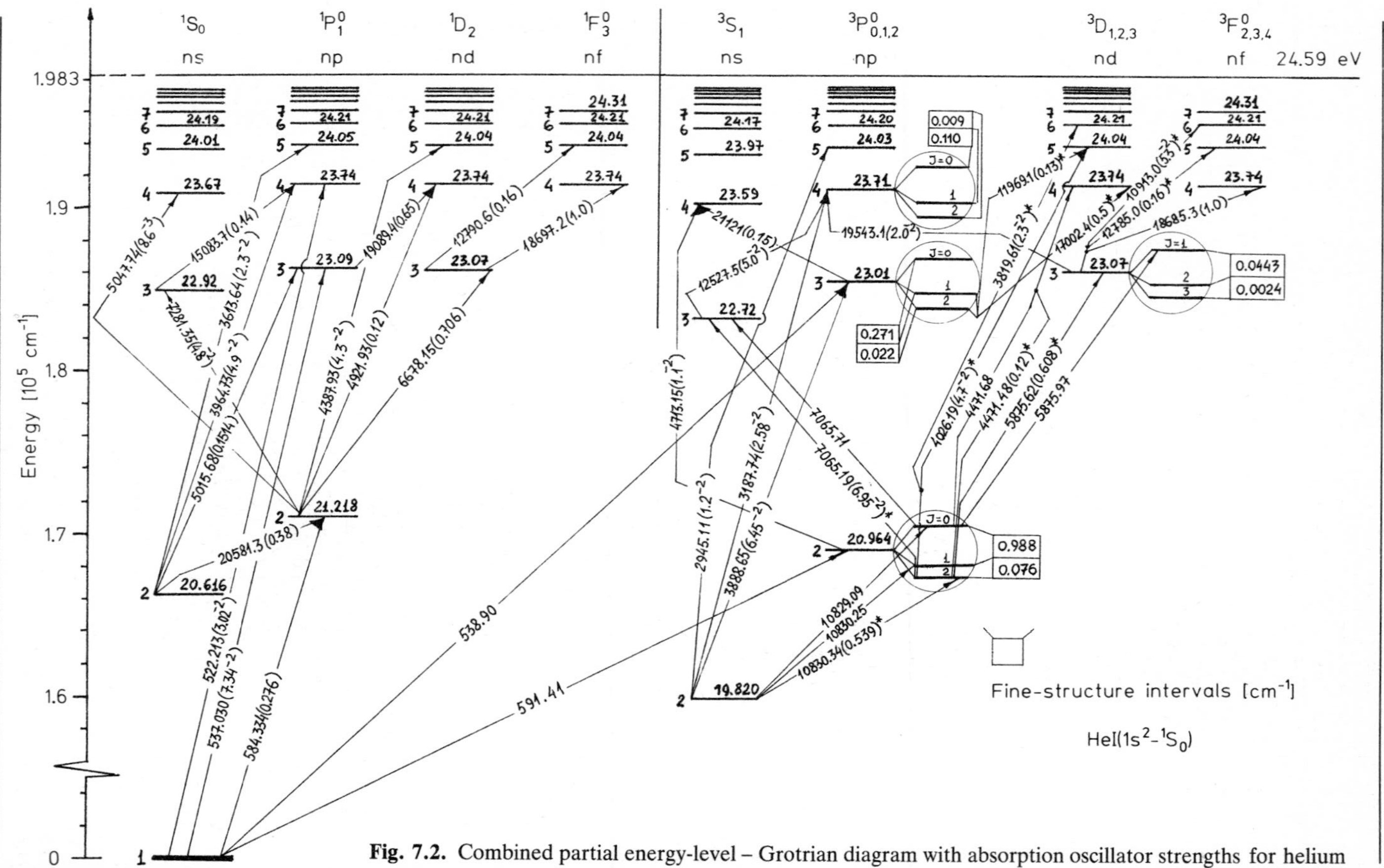

Fig. 7.2. Combined partial energy-level – Grotrian diagram with absorption oscillator strengths for helium

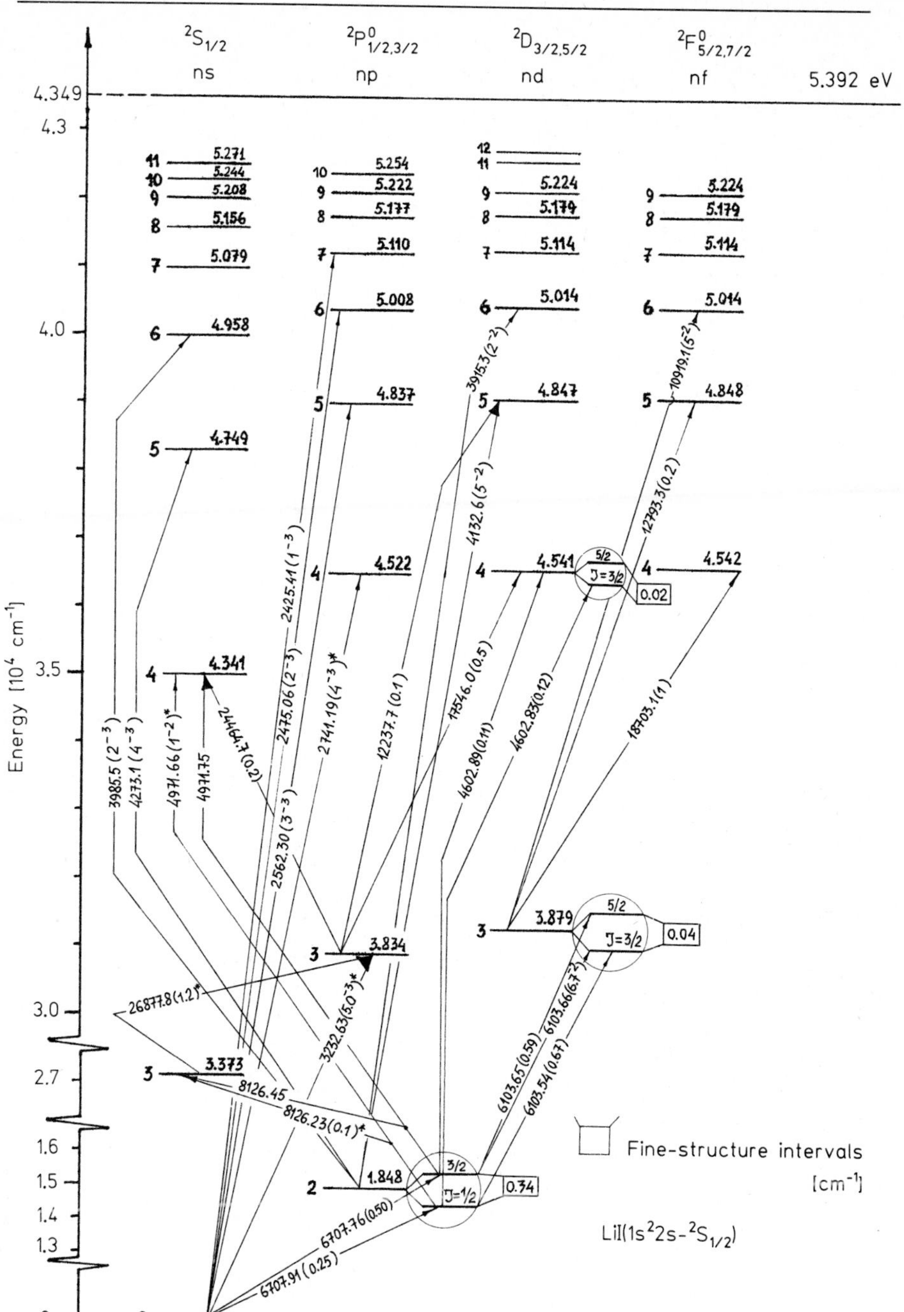

Fig. 7.3. Combined partial energy-level – Grotrian diagram with absorption oscillator strengths for lithium

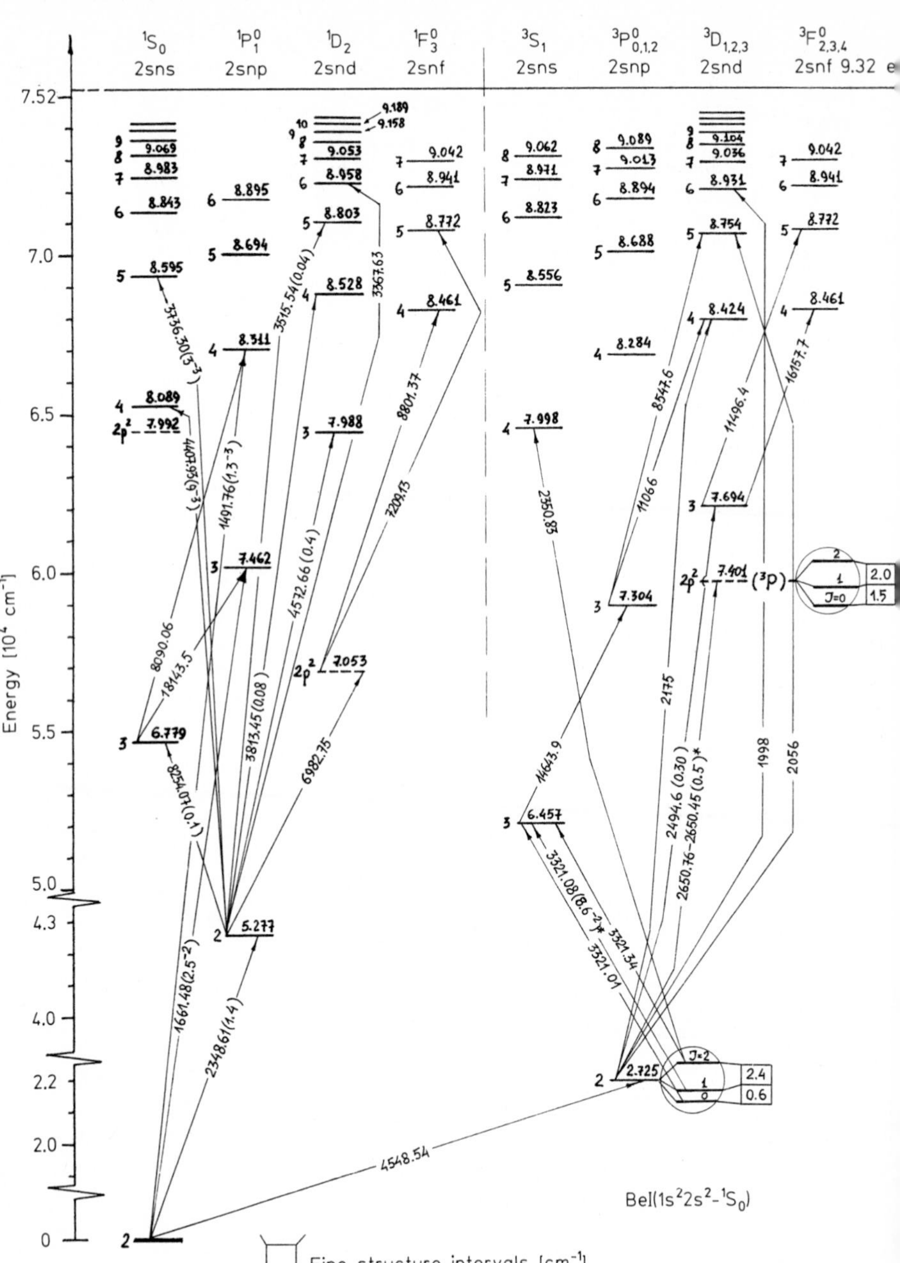

Fig. 7.4. Combined partial energy-level – Grotrian diagram with absorption oscillator strengths for beryllium

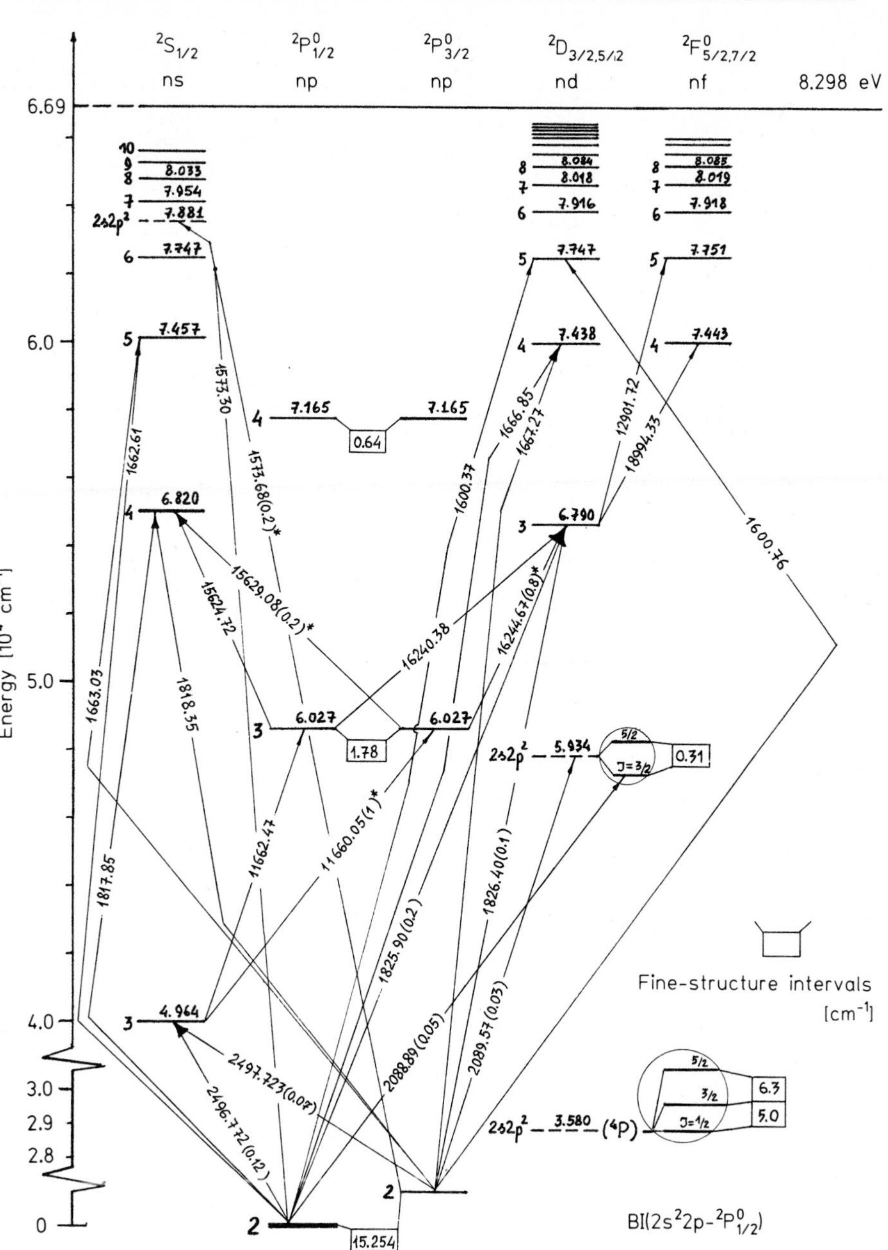

Fig. 7.5. Combined partial energy-level – Grotrian diagram with absorption oscillator strengths for boron

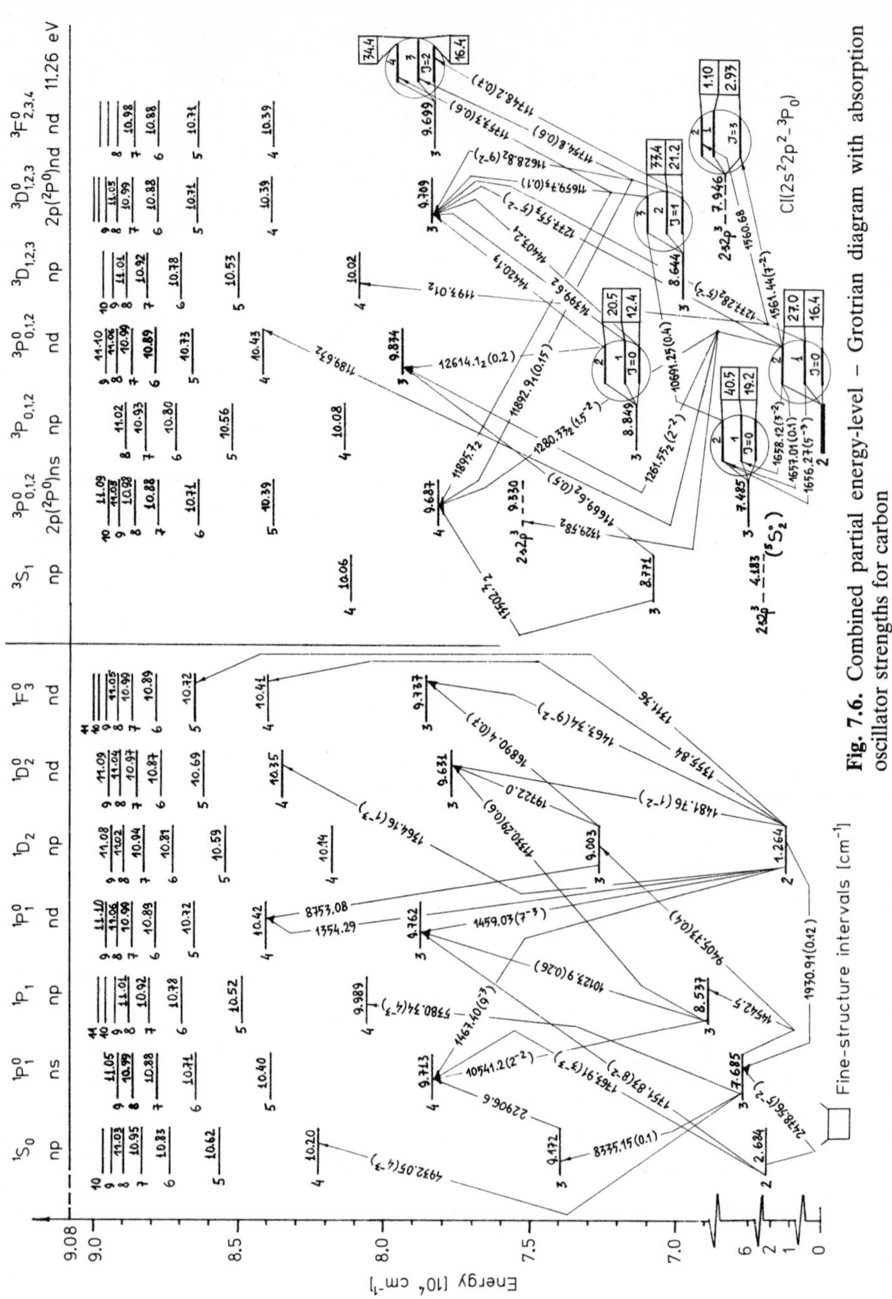

Fig. 7.6. Combined partial energy-level – Grotrian diagram with absorption oscillator strengths for carbon

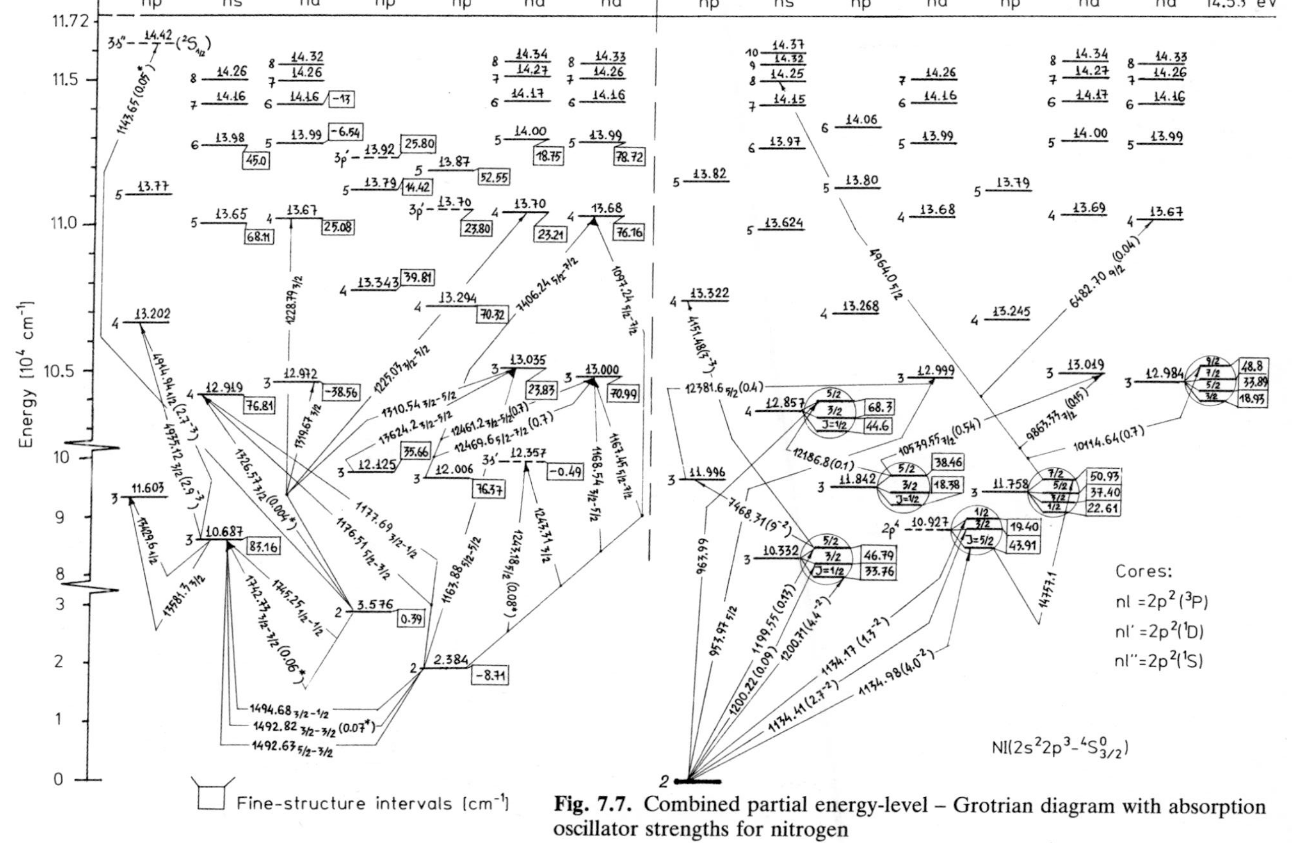

Fig. 7.7. Combined partial energy-level – Grotrian diagram with absorption oscillator strengths for nitrogen

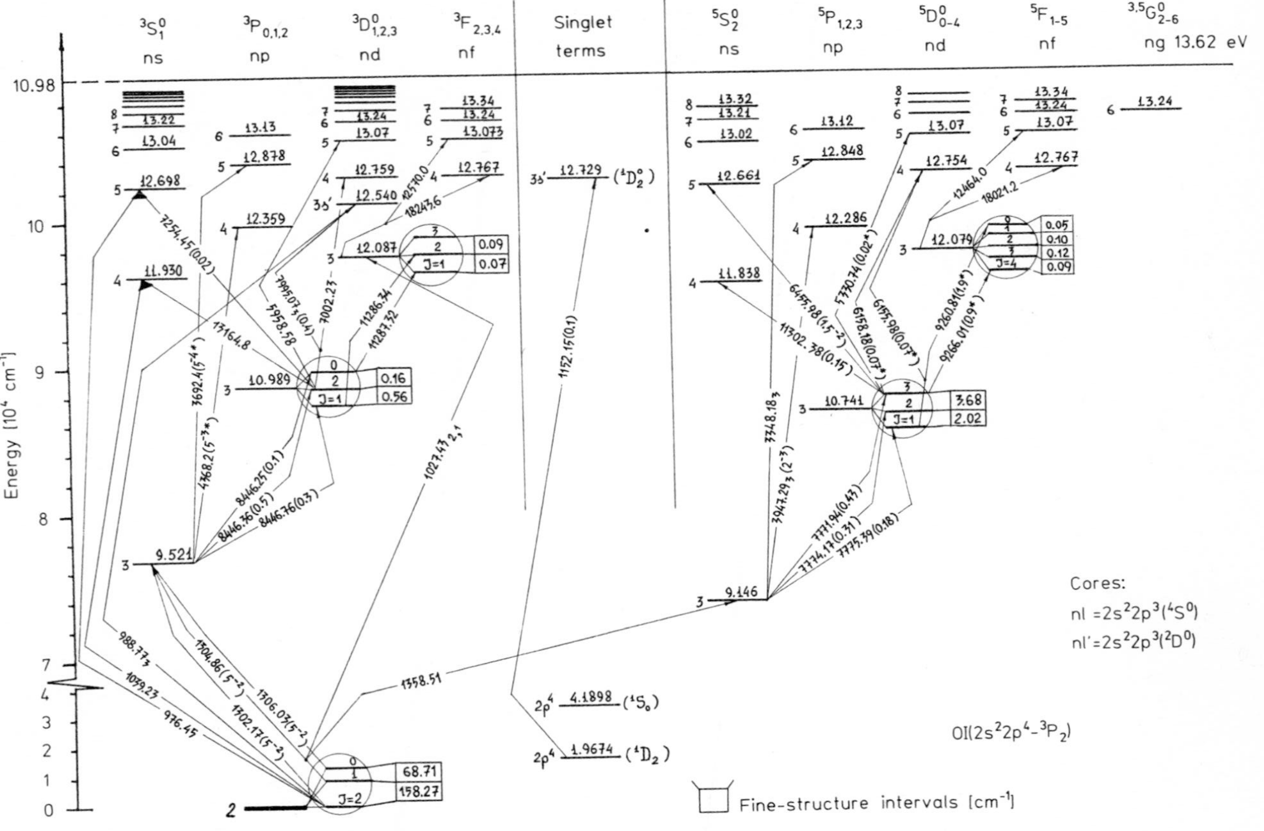

Fig. 7.8. Combined partial energy-level – Grotrian diagram with absorption oscillator strengths for oxygen

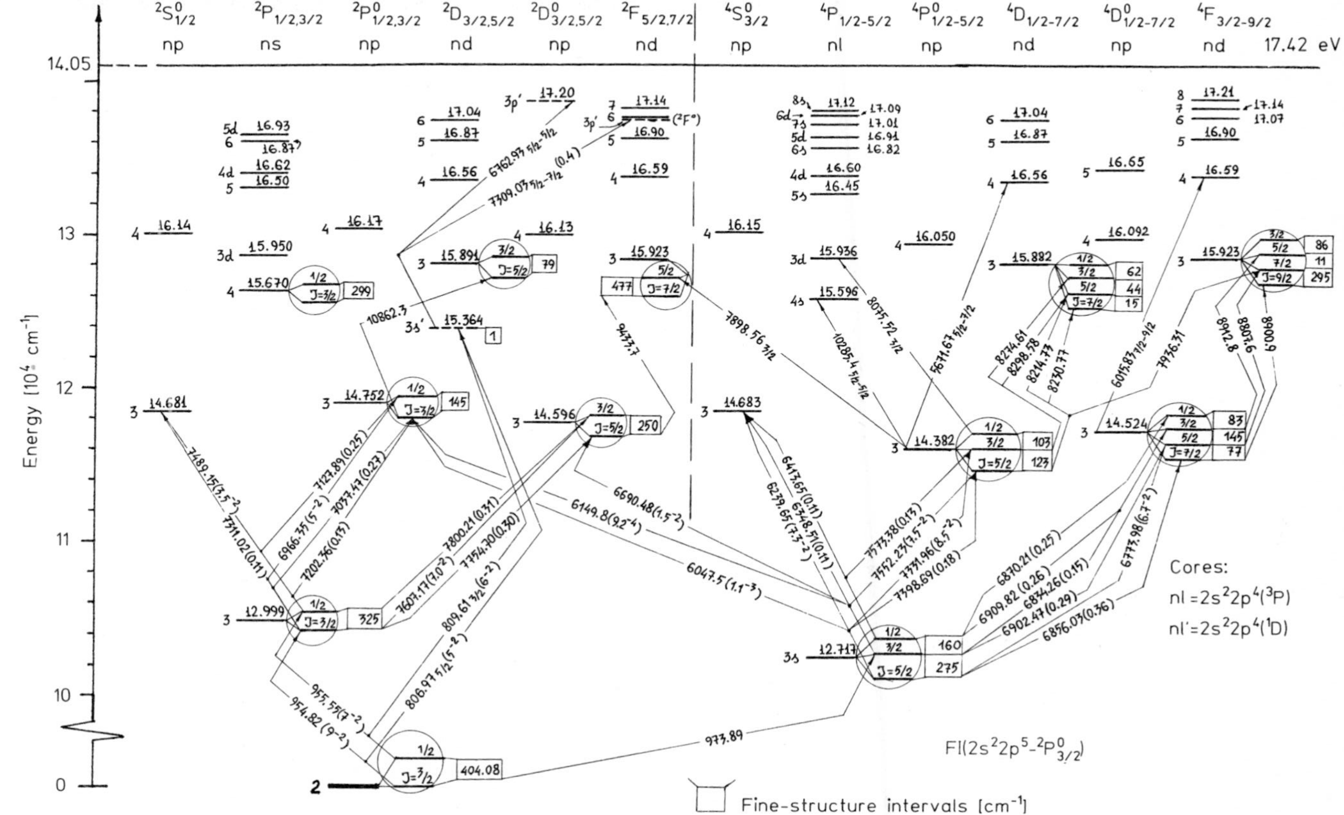

Fig. 7.9. Combined partial energy-level – Grotrian diagram with absorption oscillator strengths for fluorine

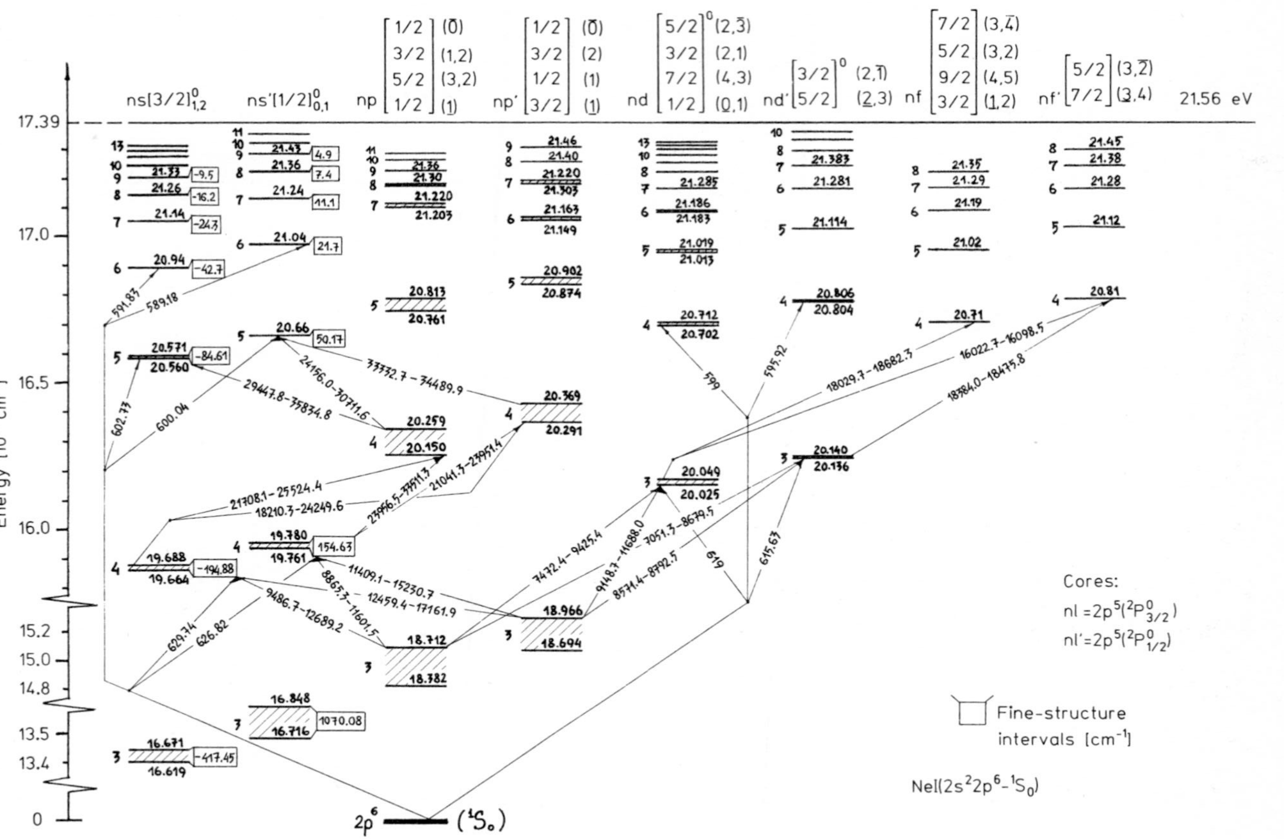

Fig. 7.10. Simplified energy-level diagram of neon with some prominent lines

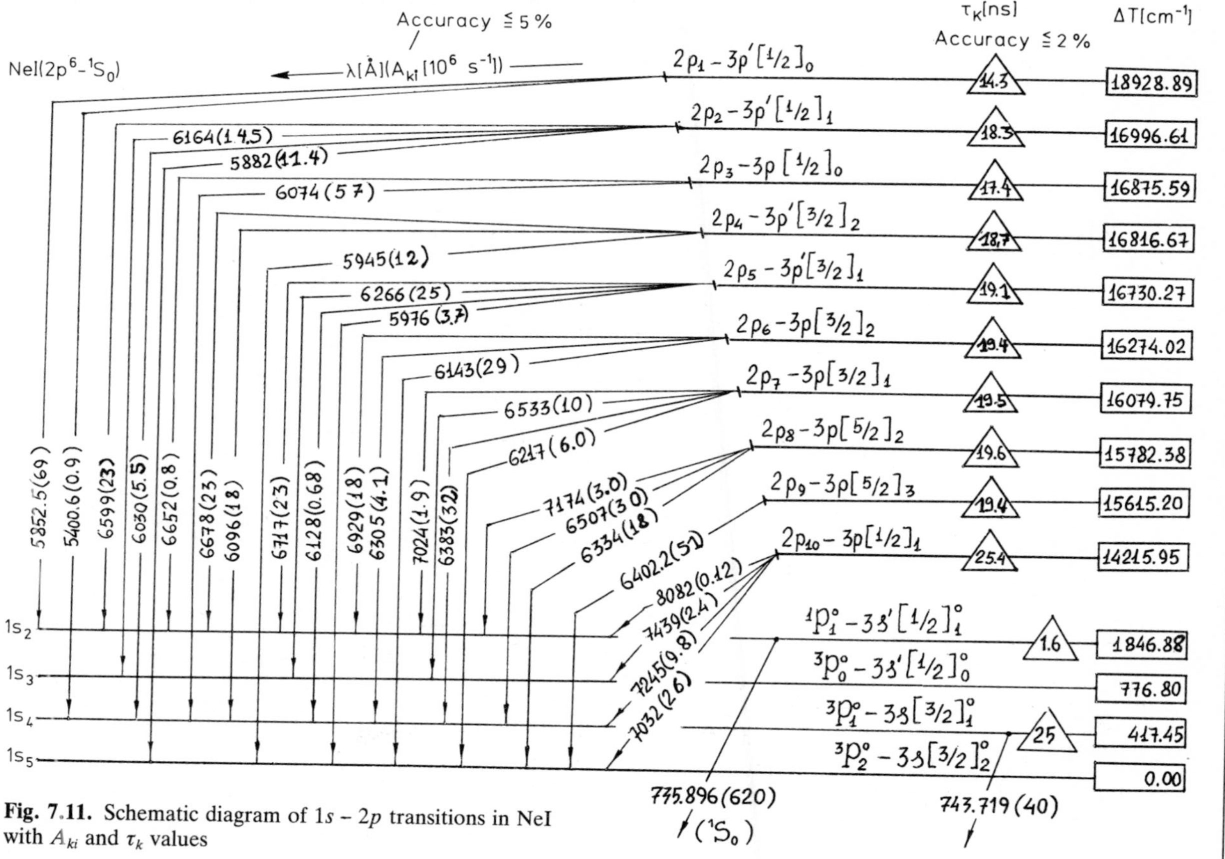

Fig. 7.11. Schematic diagram of 1s – 2p transitions in NeI with A_{ki} and τ_k values

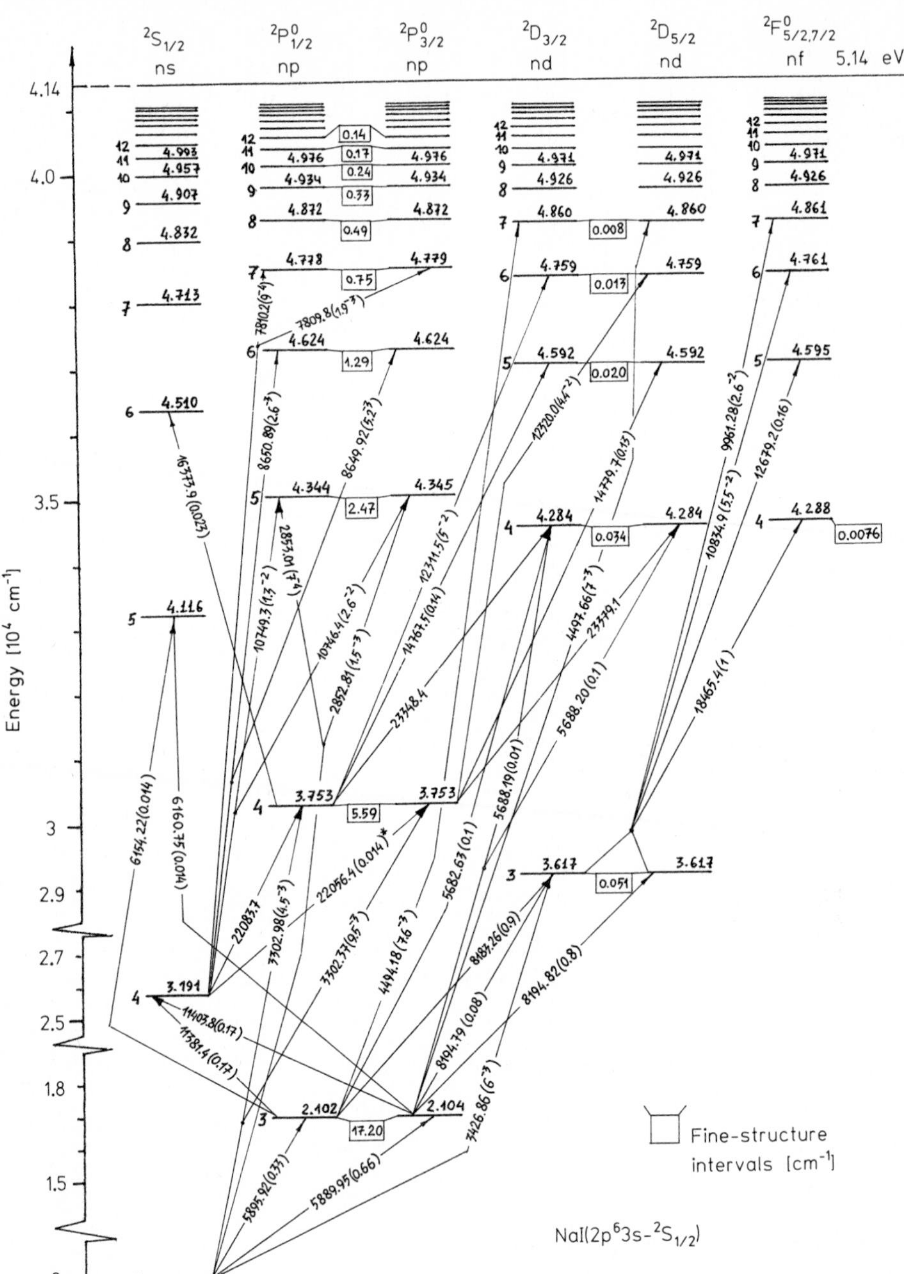

Fig. 7.12. Combined partial energy-level – Grotrian diagram with absorption oscillator strengths for sodium

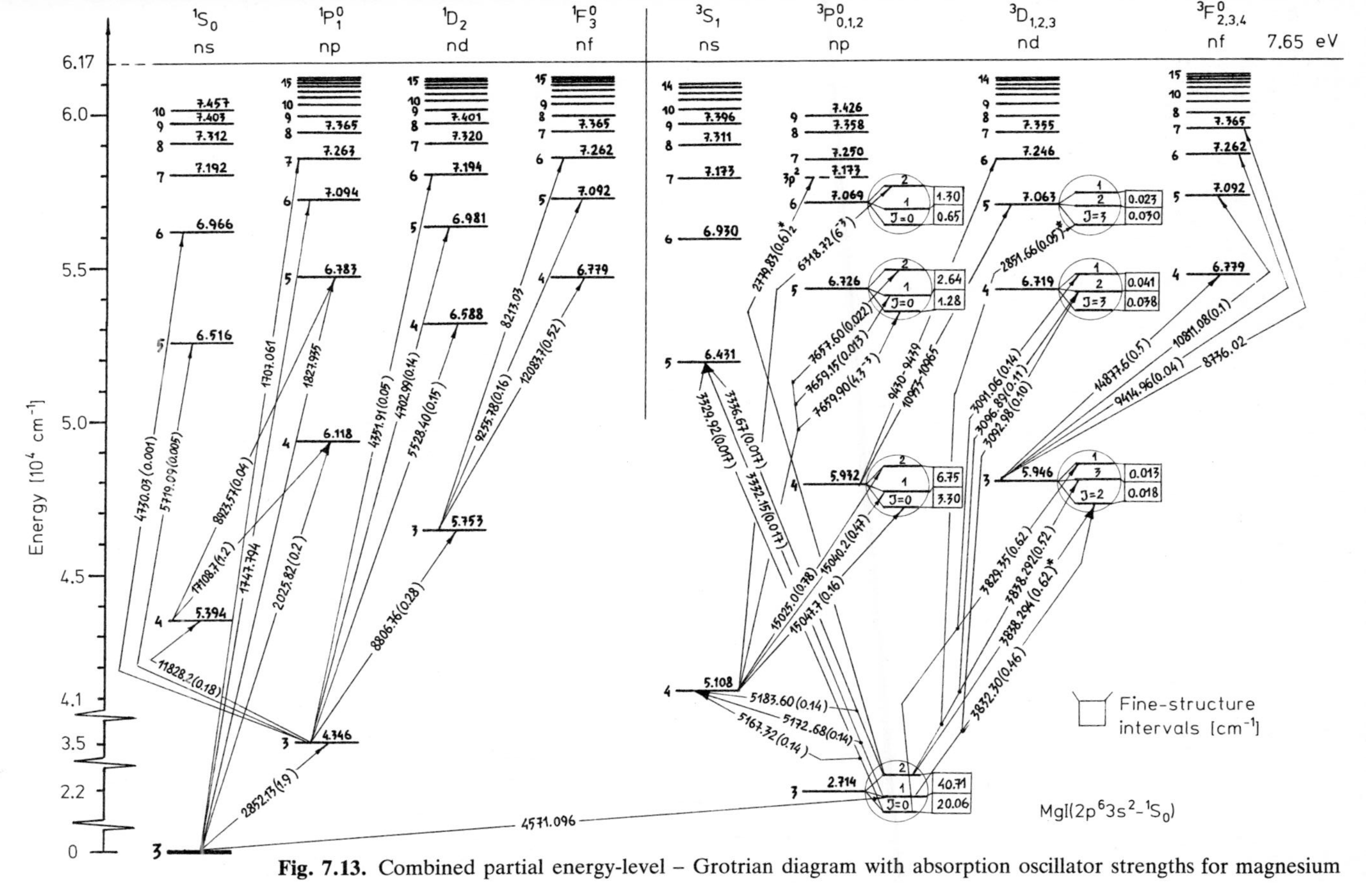

Fig. 7.13. Combined partial energy-level – Grotrian diagram with absorption oscillator strengths for magnesium

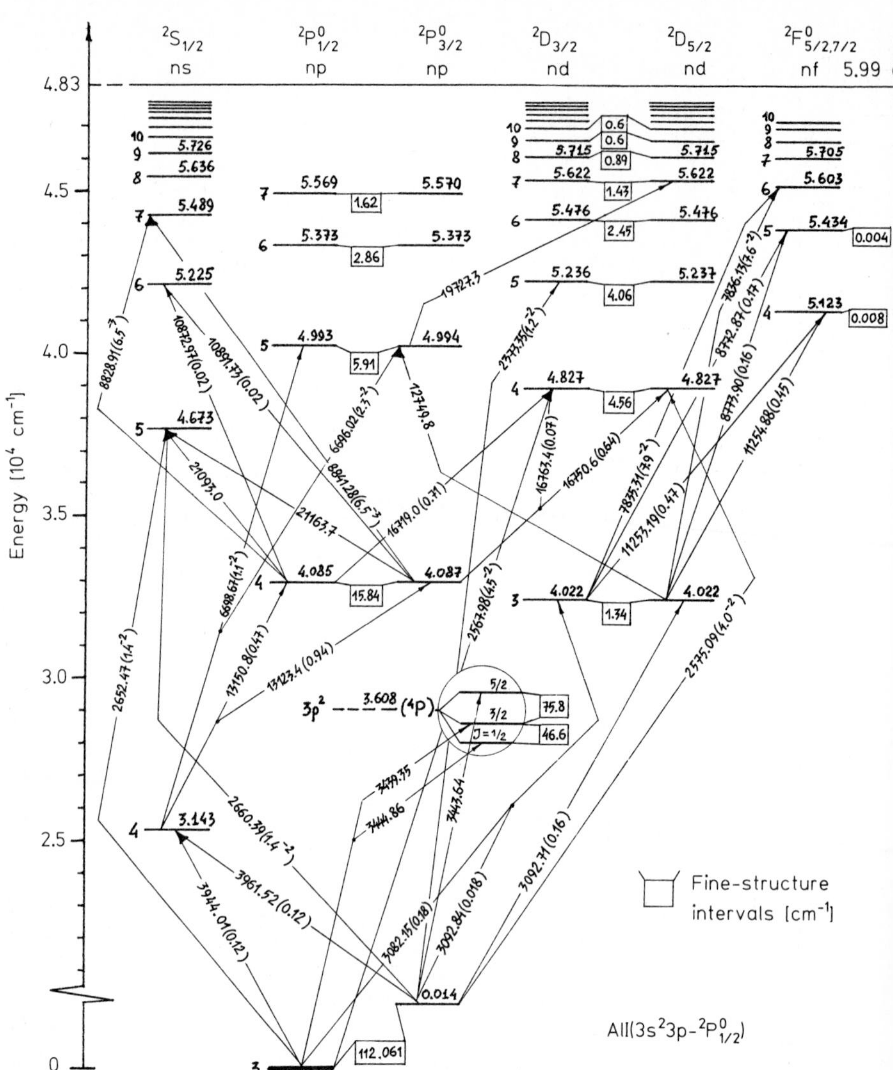

Fig. 7.14. Combined partial energy-level – Grotrian diagram with absorption oscillator strengths for aluminium

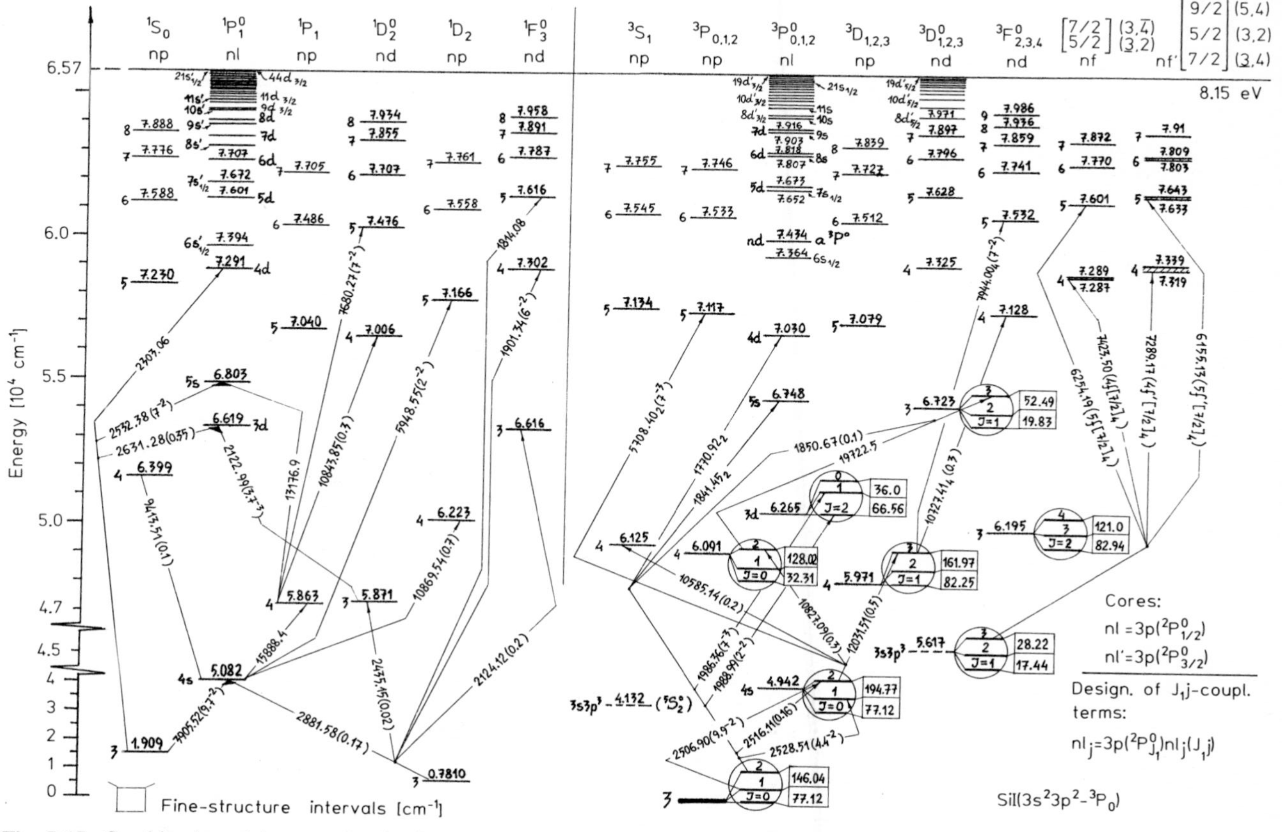

Fig. 7.15. Combined partial energy-level – Grotrian diagram with absorption oscillator strengths for silicon

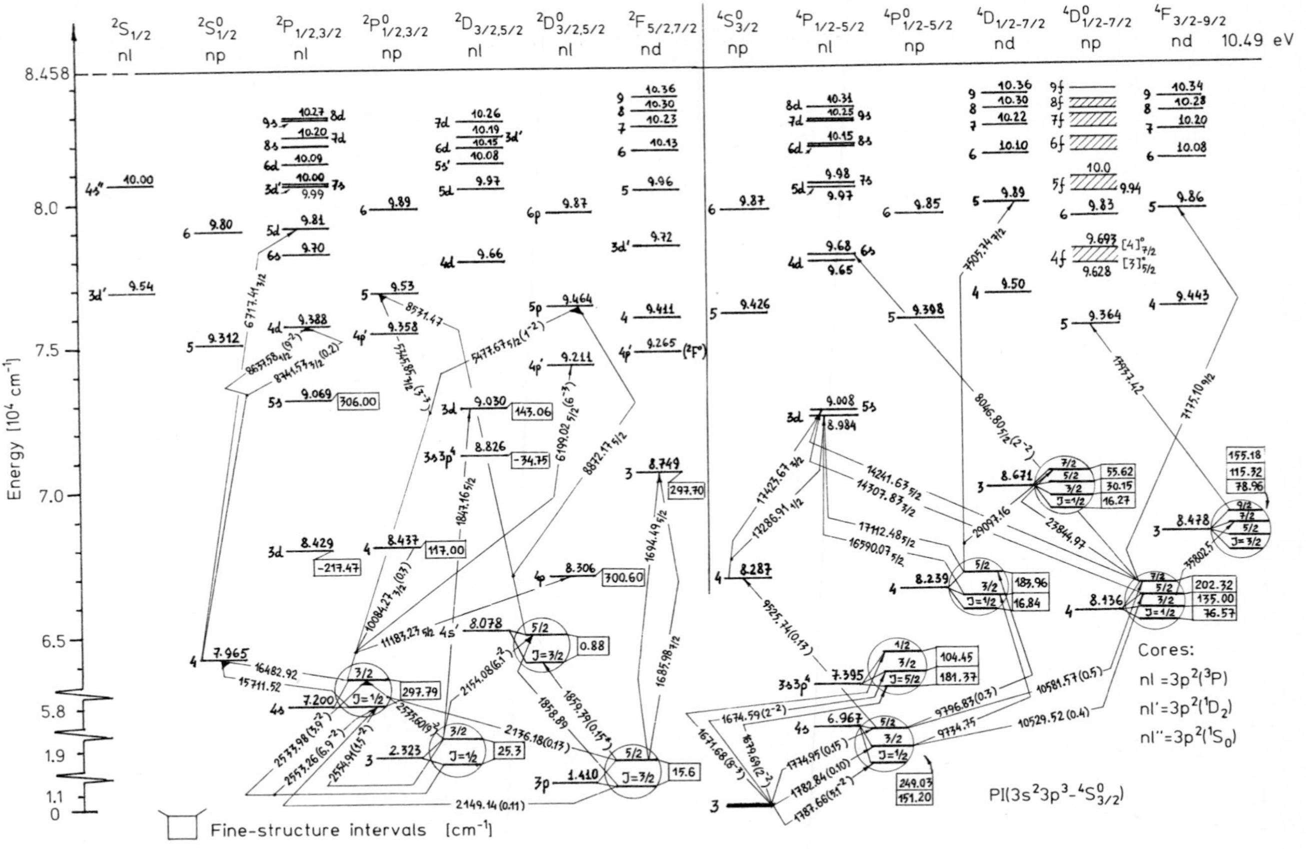

Fig. 7.16. Combined partial energy-level – Grotrian diagram with absorption oscillator strengths for phosphorus

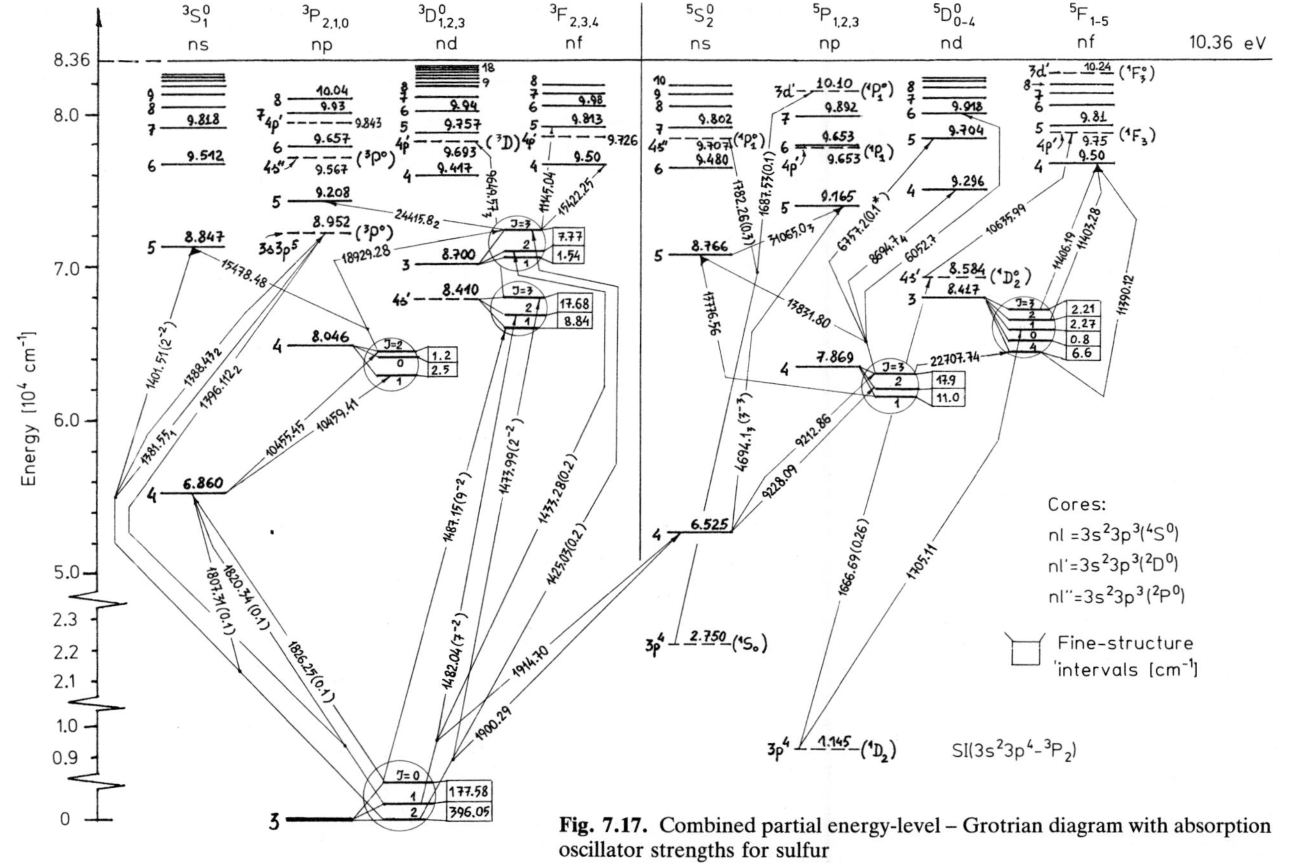

Fig. 7.17. Combined partial energy-level – Grotrian diagram with absorption oscillator strengths for sulfur

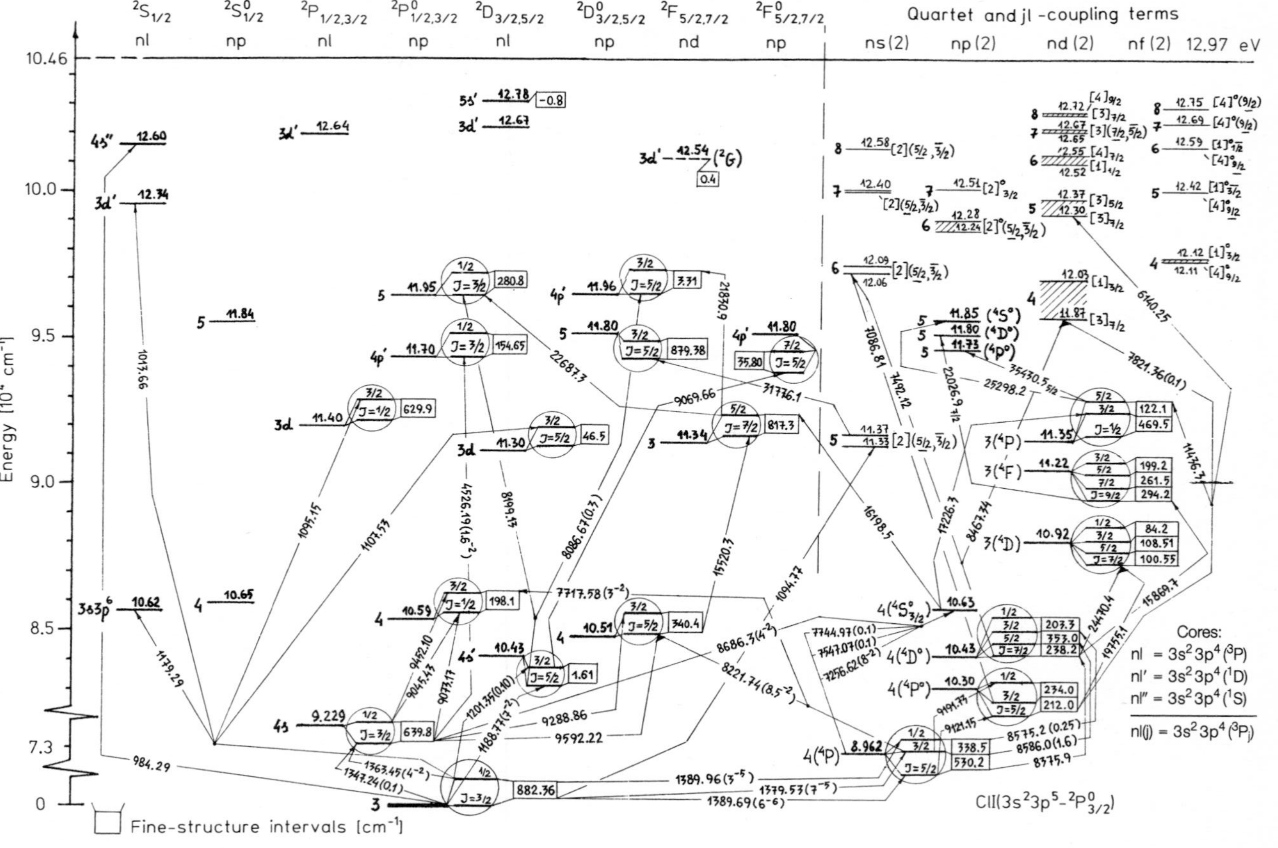

Fig. 7.18. Combined partial energy-level – Grotrian diagram with absorption oscillator strengths for chlorine

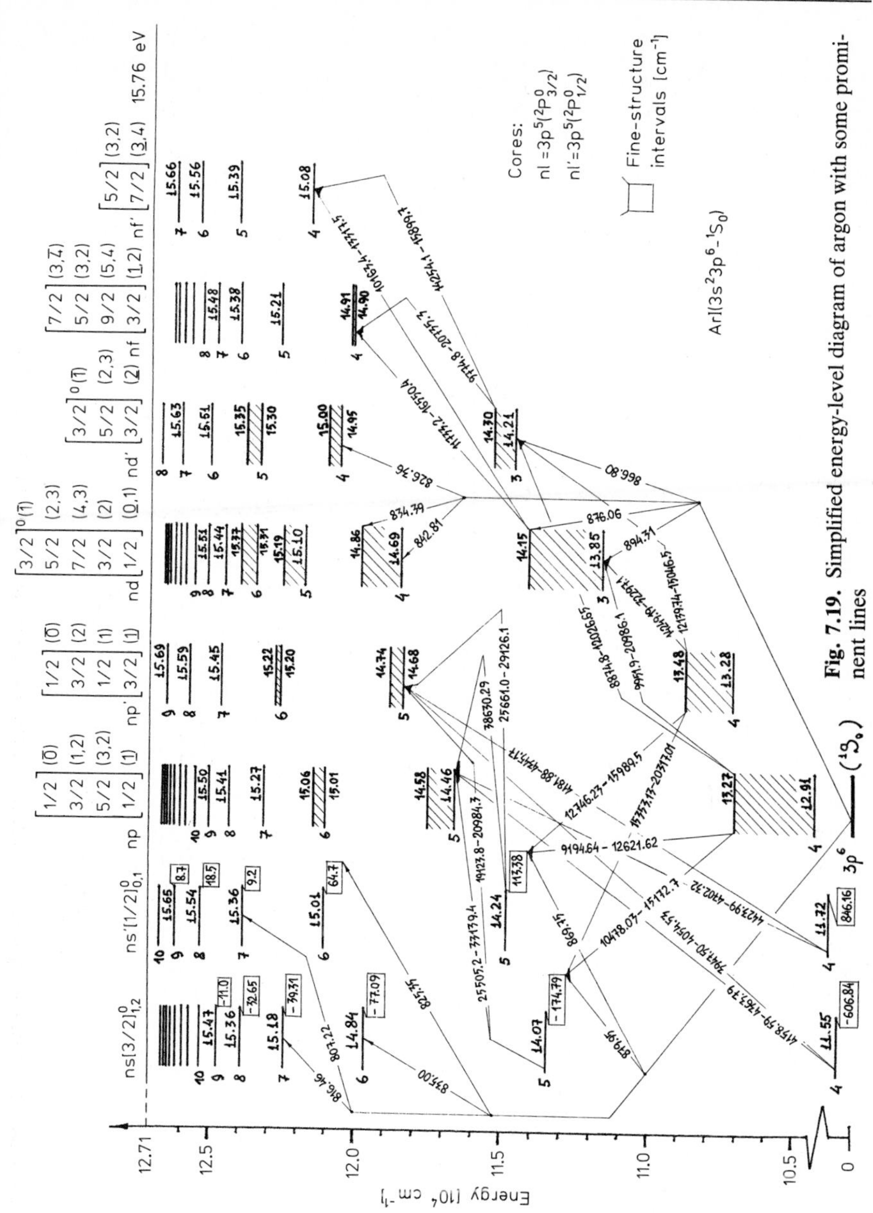

Fig. 7.19. Simplified energy-level diagram of argon with some prominent lines

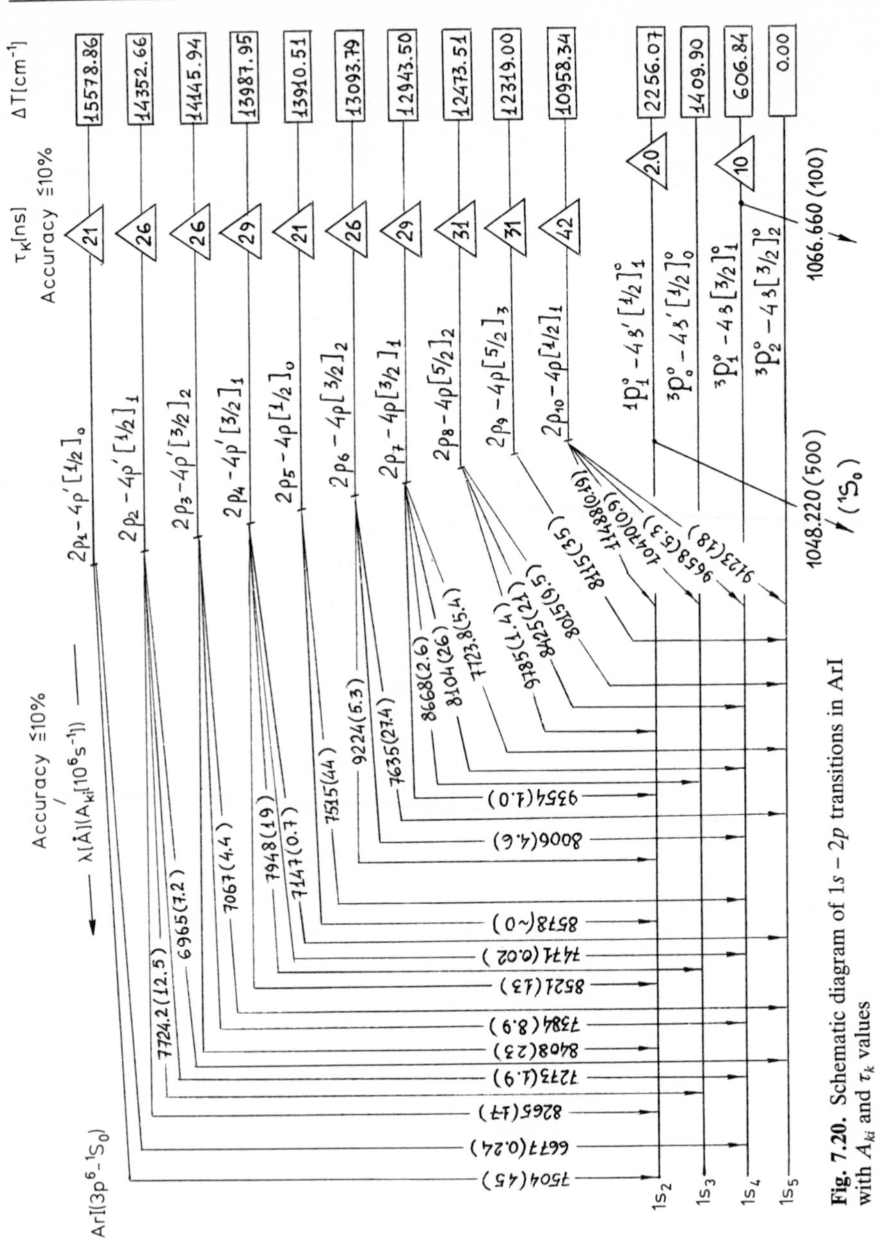

Fig. 7.20. Schematic diagram of $1s - 2p$ transitions in ArI with A_{ki} and τ_k values

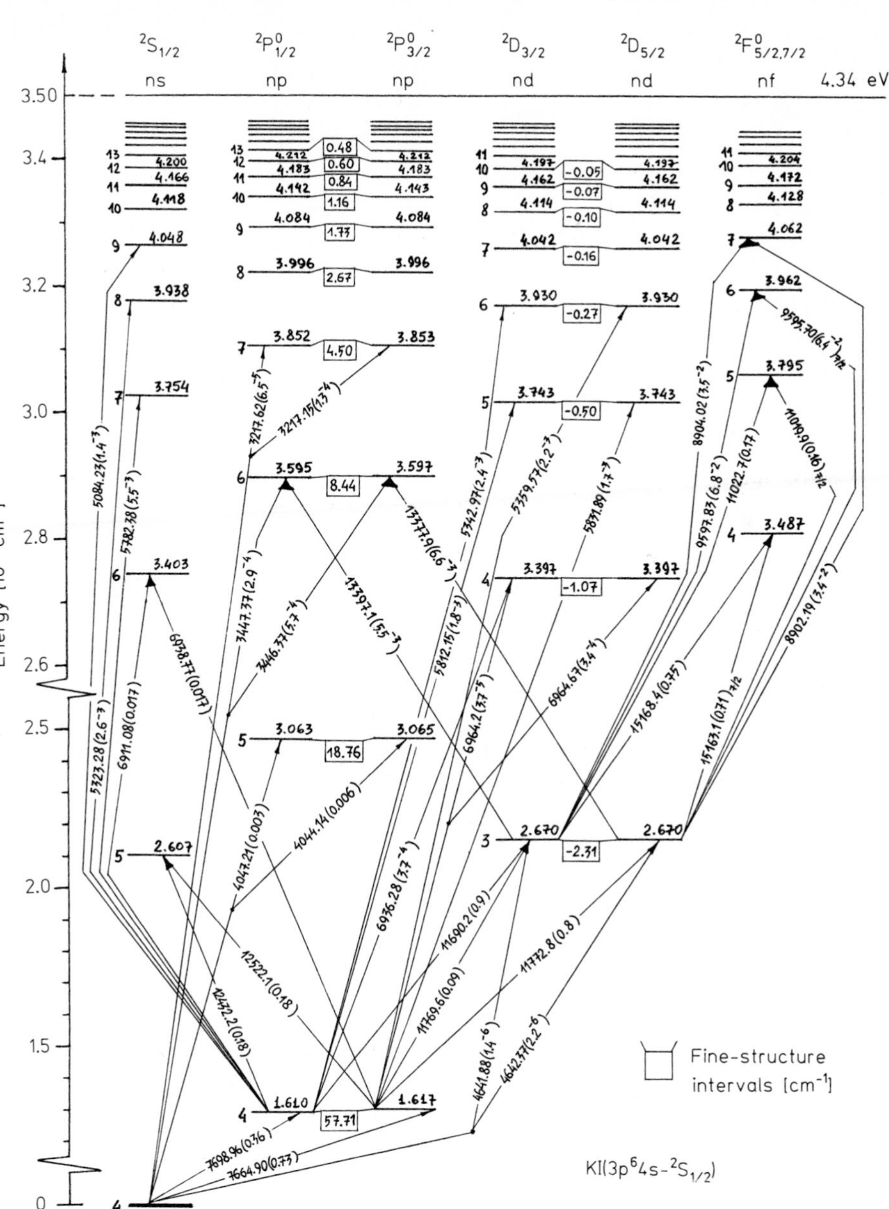

Fig. 7.21. Combined partial energy-level – Grotrian diagram with absorption oscillator strengths for potassium

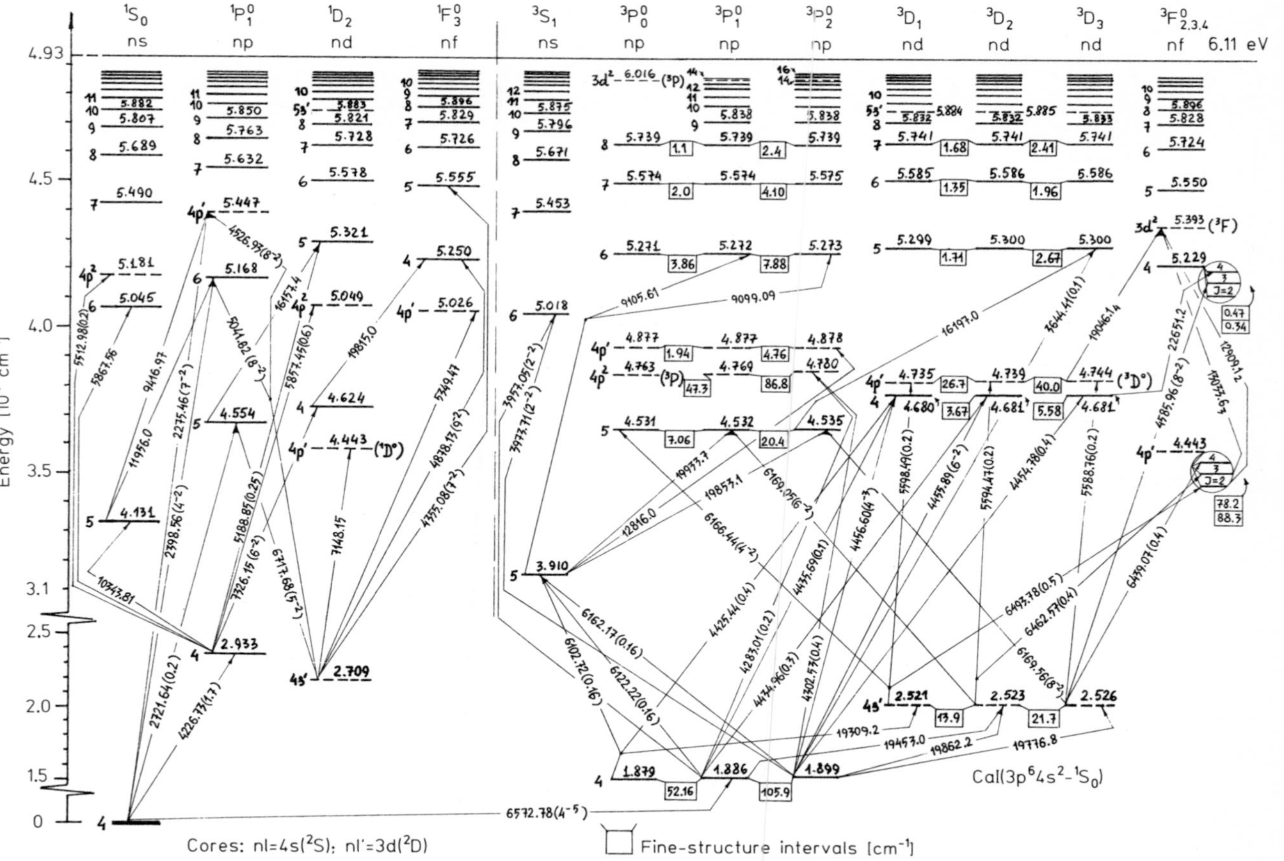

Fig. 7.22. Combined partial energy-level – Grotrian diagram with absorption oscillator strengths for calcium

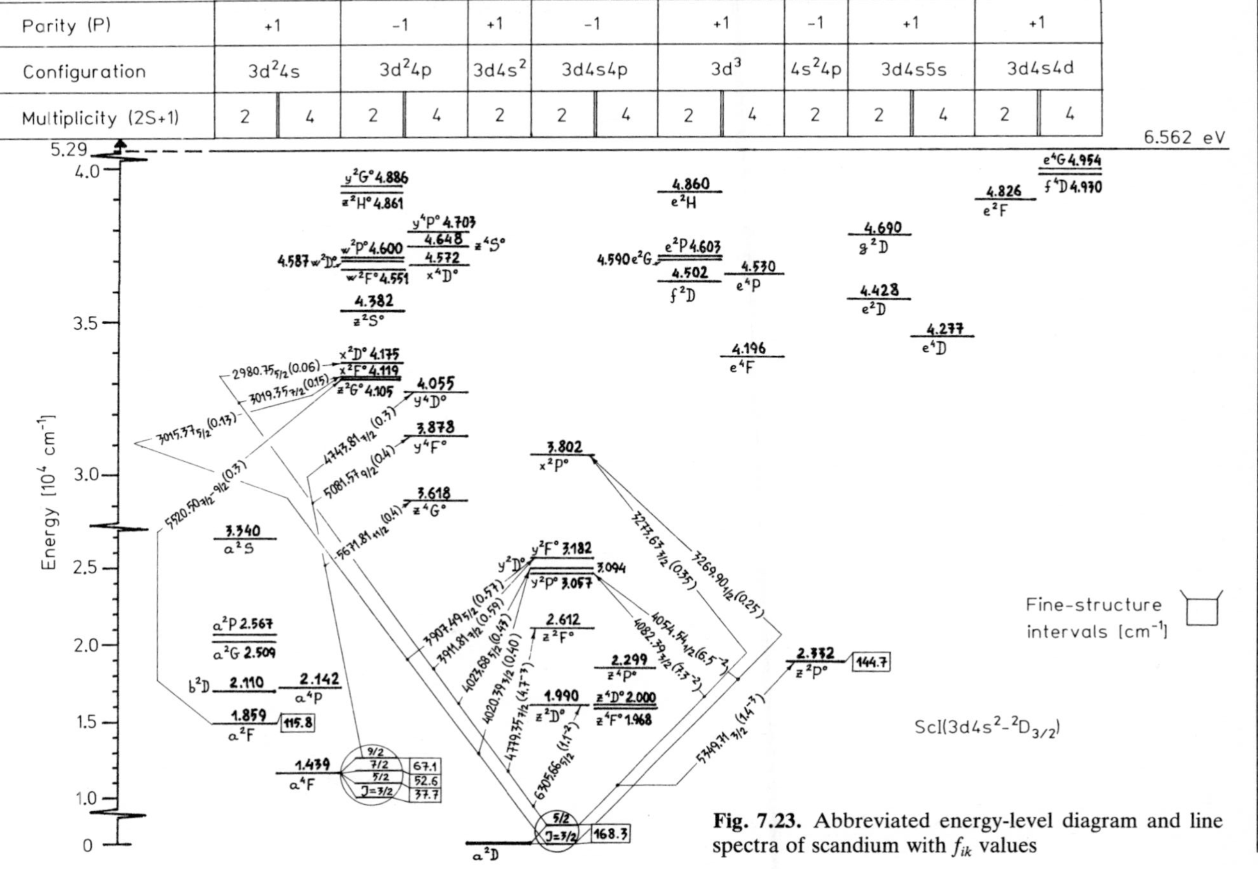

Fig. 7.23. Abbreviated energy-level diagram and line spectra of scandium with f_{ik} values

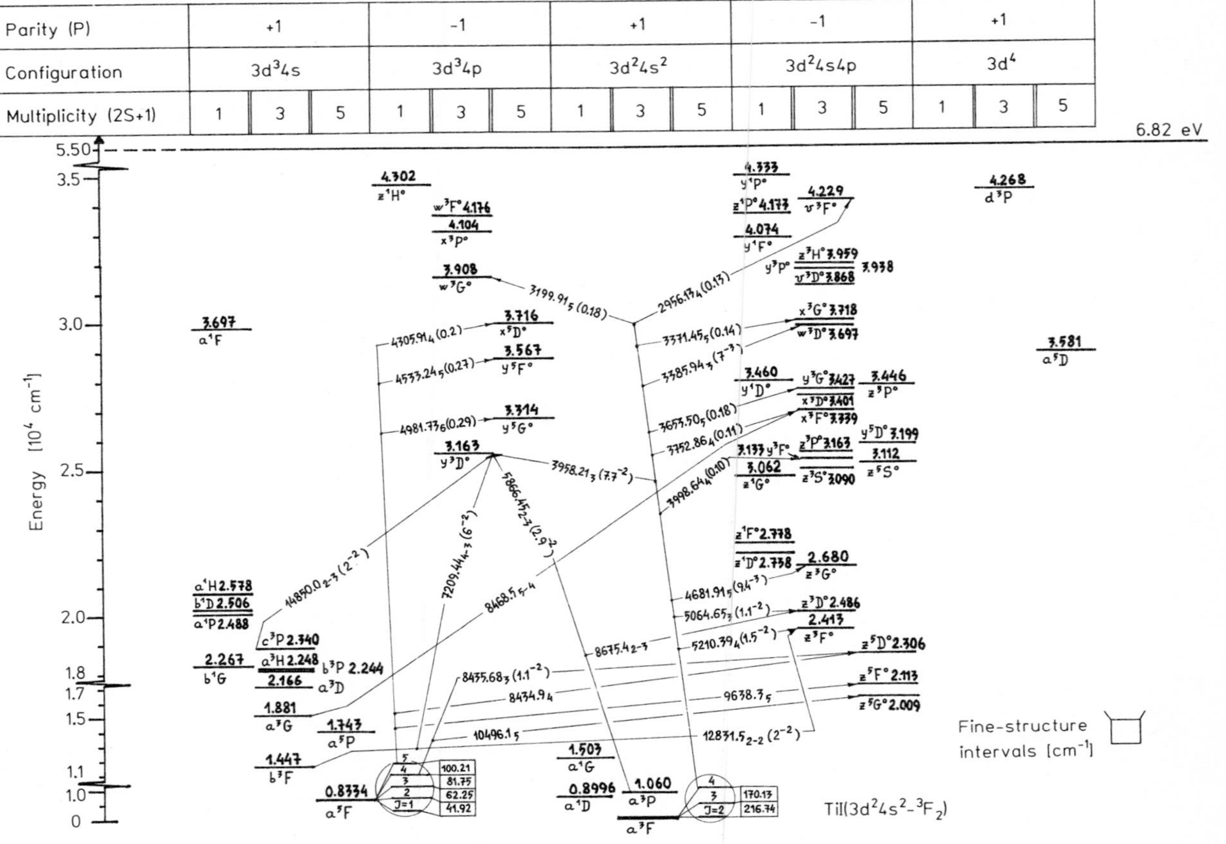

Fig. 7.24. Abbreviated energy-level diagram and line spectra of titanium with f_{ik} values

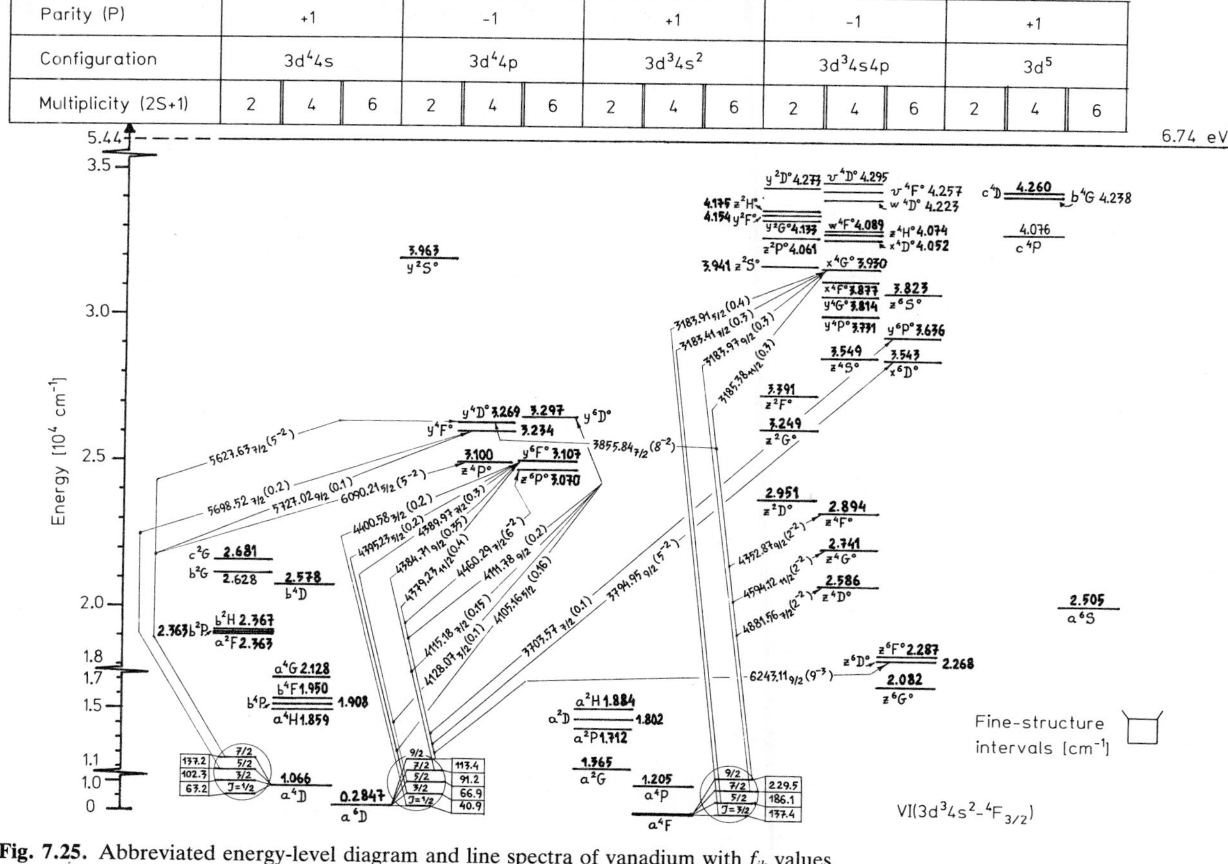

Fig. 7.25. Abbreviated energy-level diagram and line spectra of vanadium with f_{ik} values

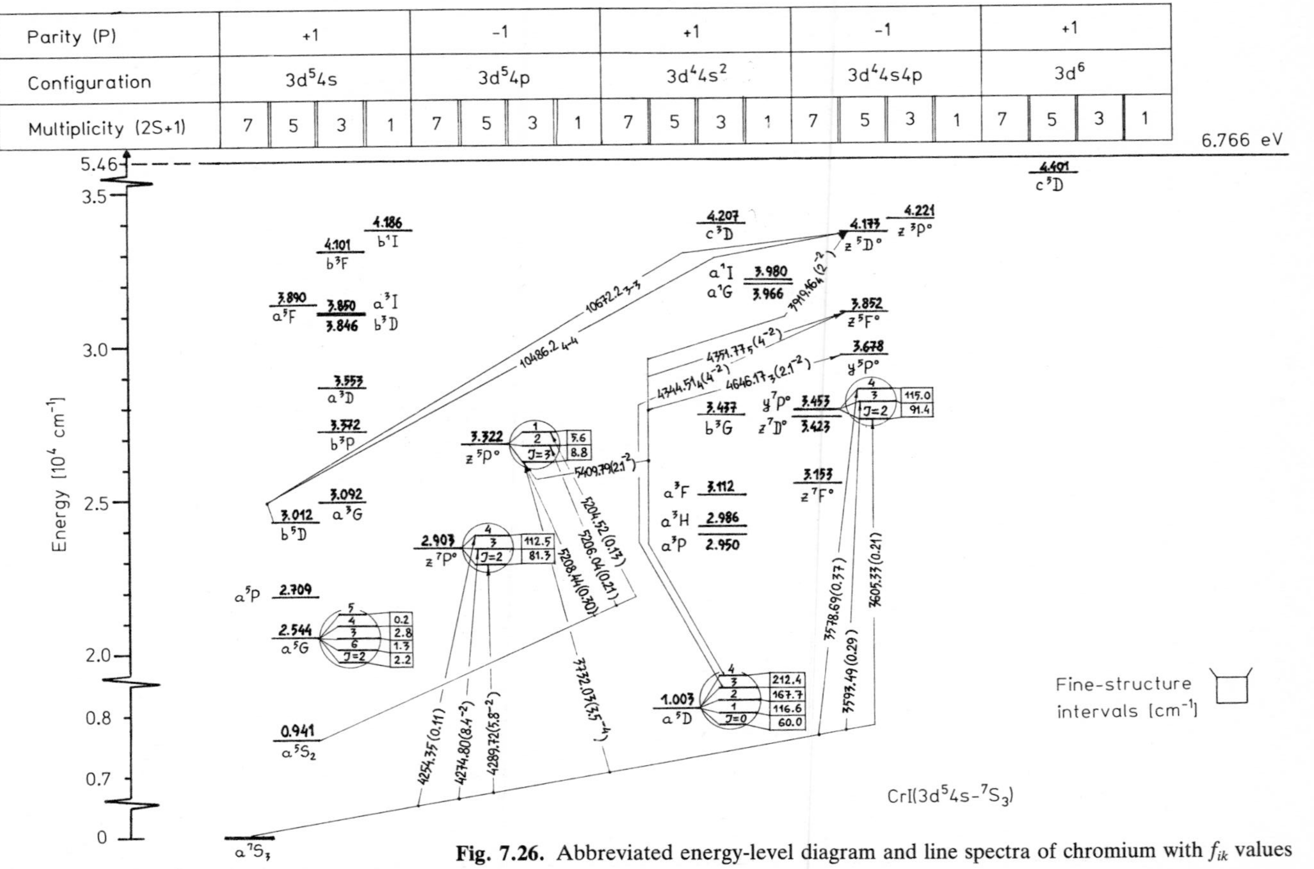

Fig. 7.26. Abbreviated energy-level diagram and line spectra of chromium with f_{ik} values

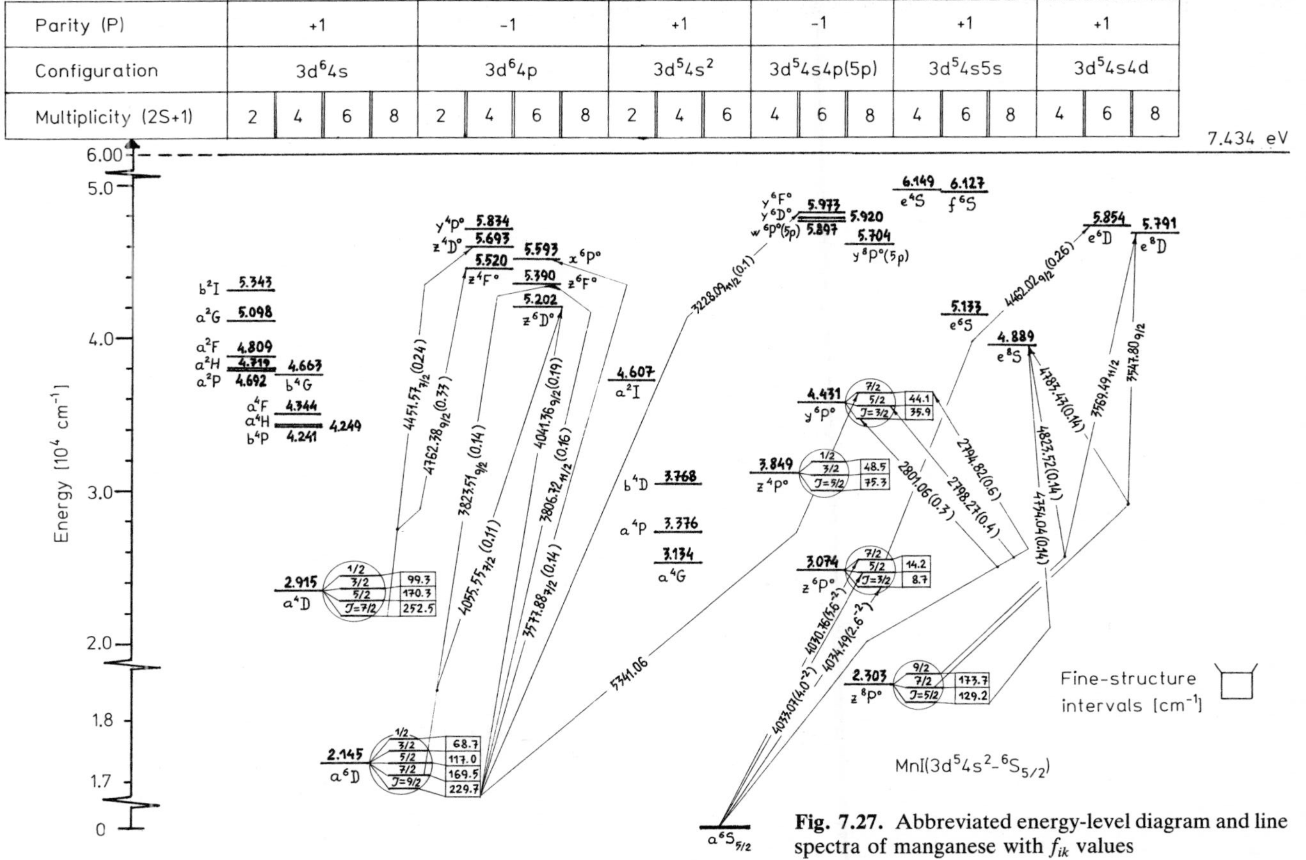

Fig. 7.27. Abbreviated energy-level diagram and line spectra of manganese with f_{ik} values

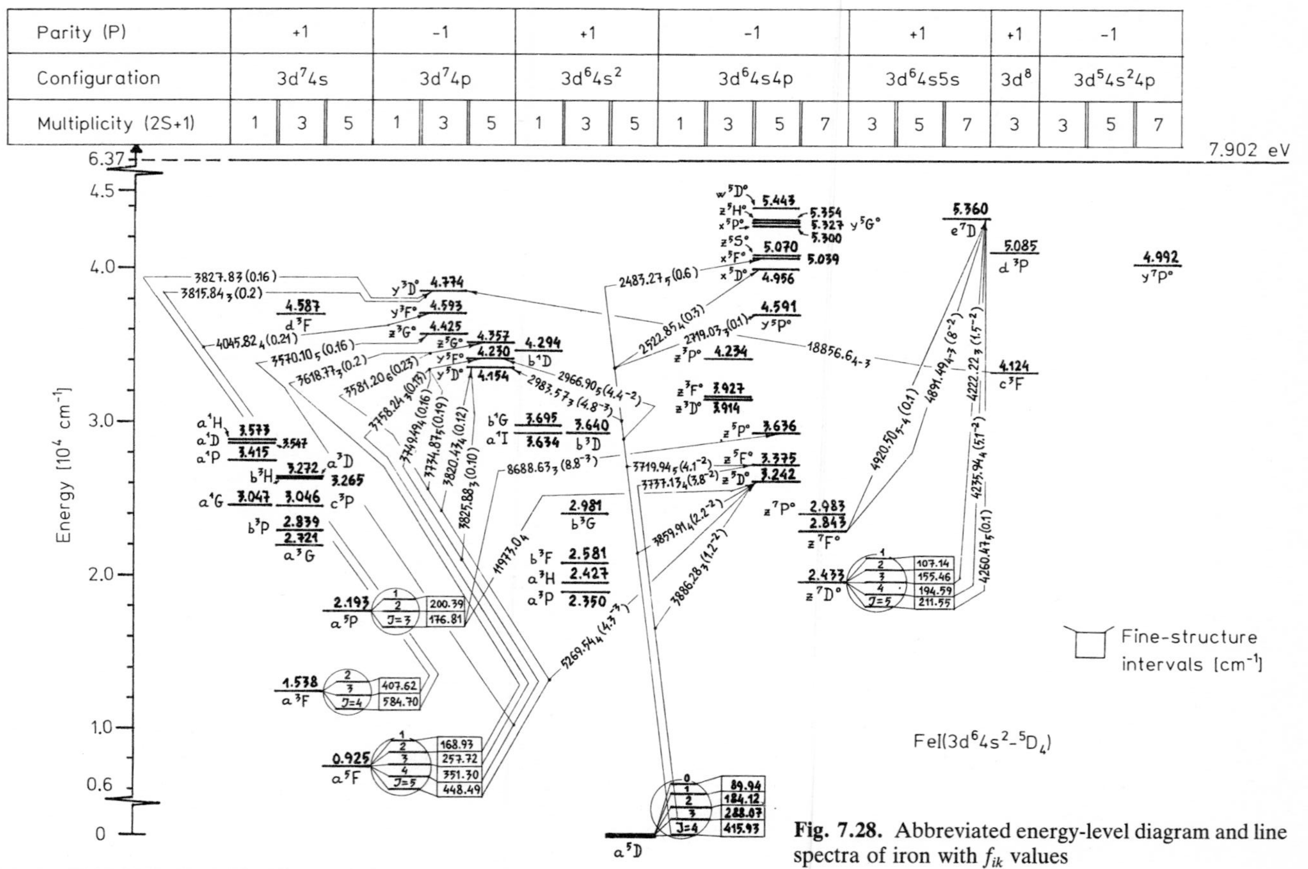

Fig. 7.28. Abbreviated energy-level diagram and line spectra of iron with f_{ik} values

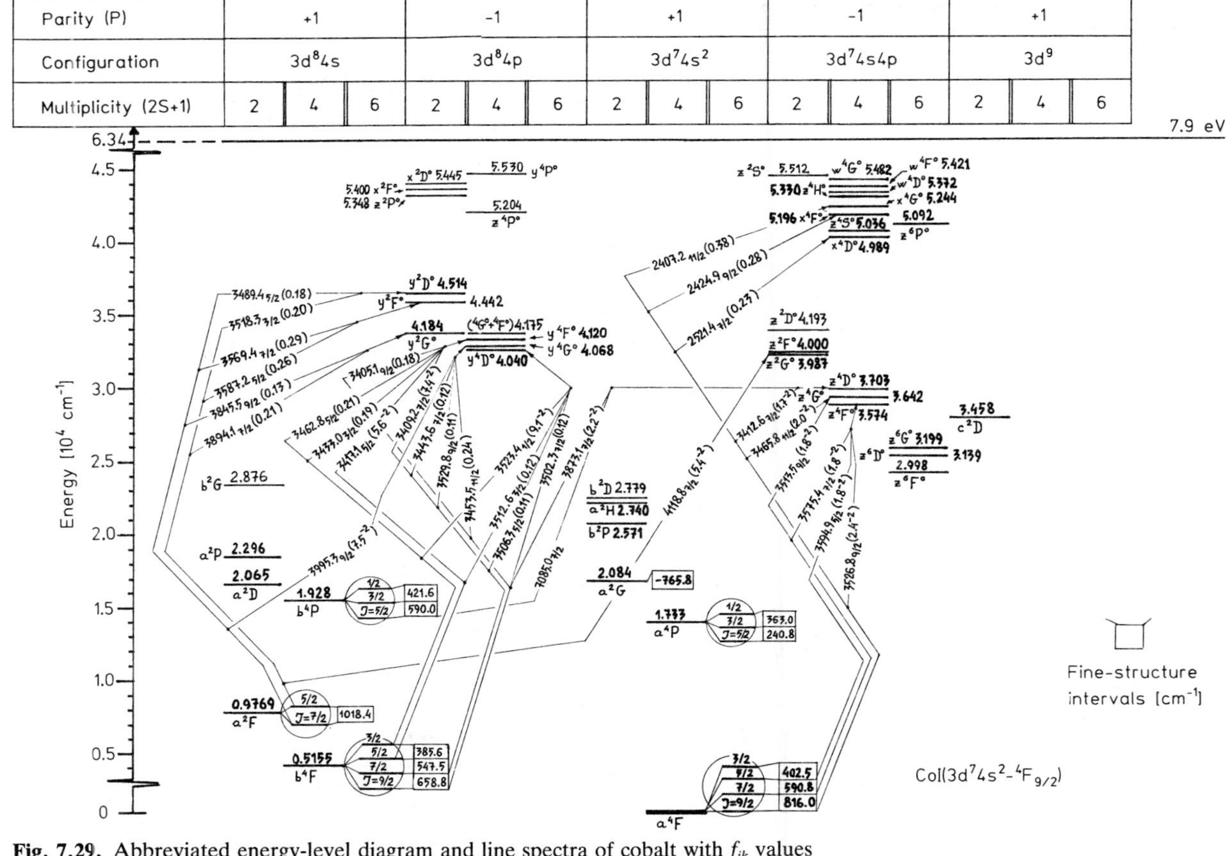

Fig. 7.29. Abbreviated energy-level diagram and line spectra of cobalt with f_{ik} values

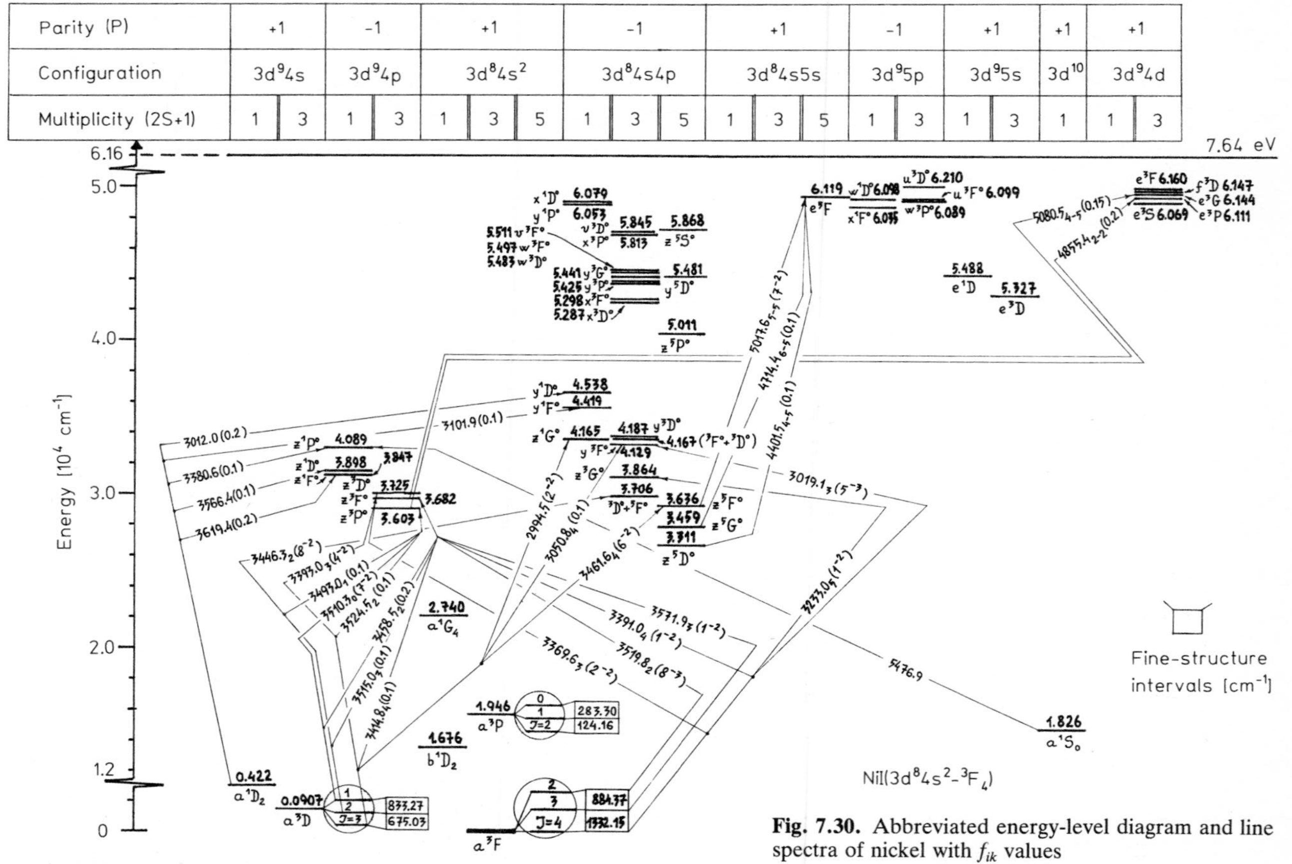

Fig. 7.30. Abbreviated energy-level diagram and line spectra of nickel with f_{ik} values

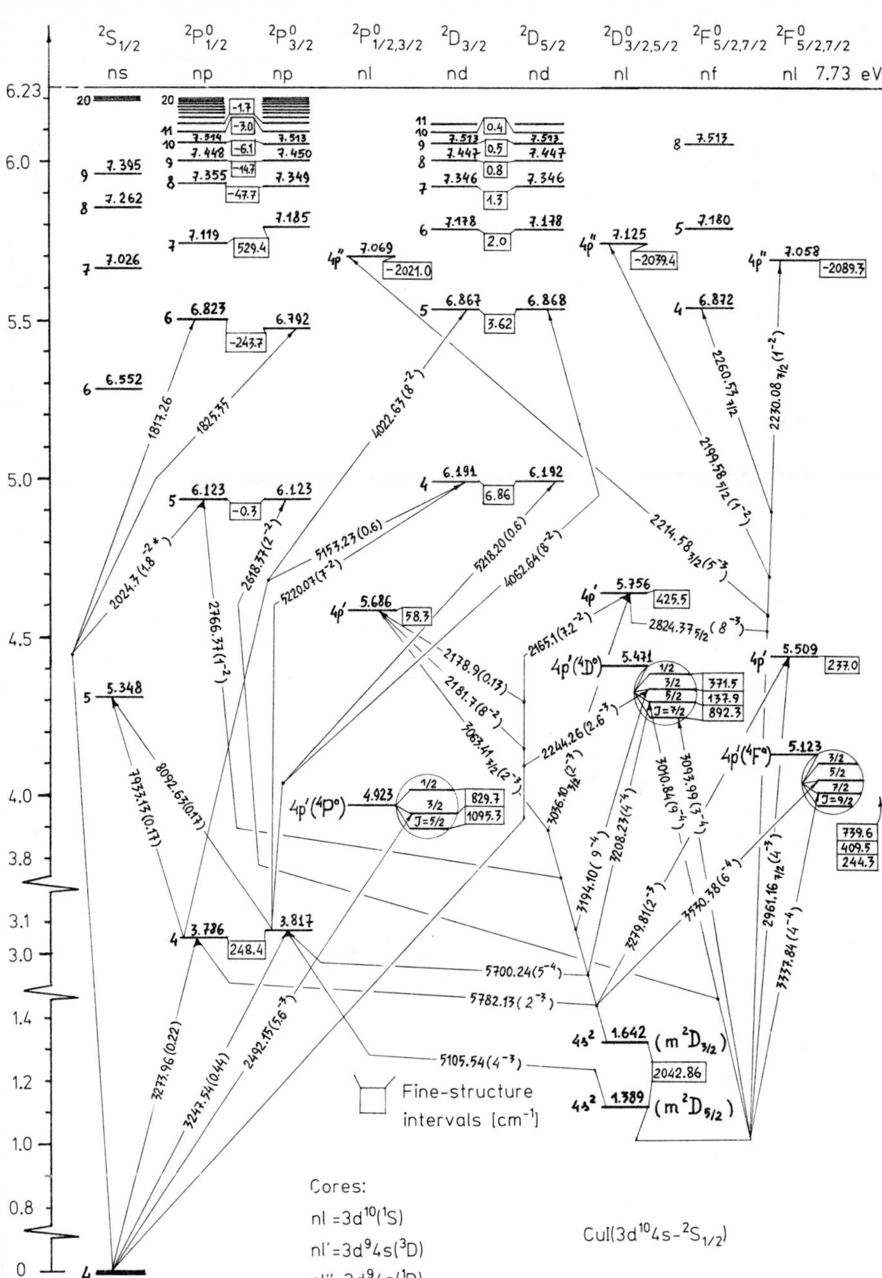

Fig. 7.31. Combined partial energy-level – Grotrian diagram with absorption oscillator strengths for copper

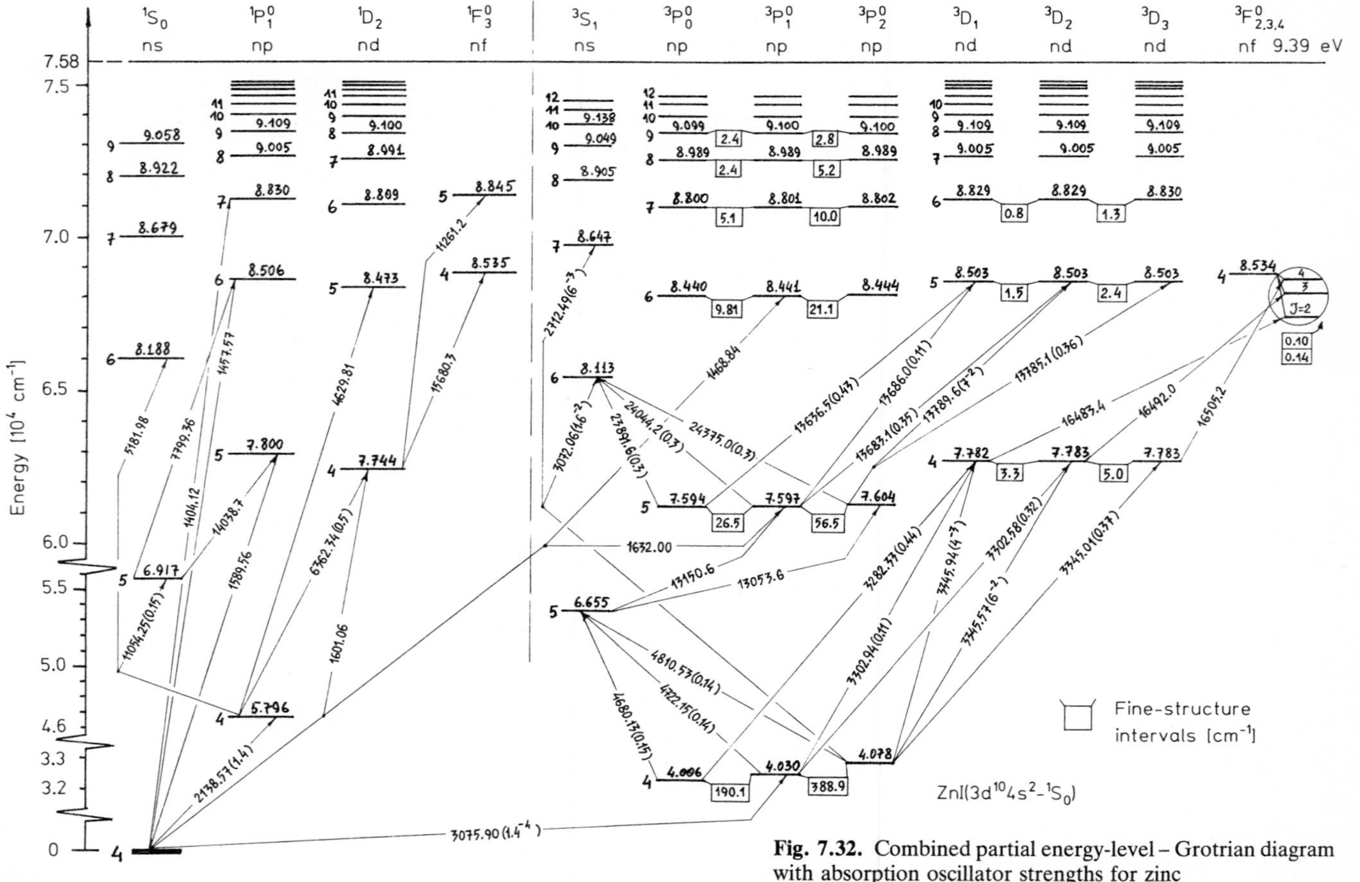

Fig. 7.32. Combined partial energy-level – Grotrian diagram with absorption oscillator strengths for zinc

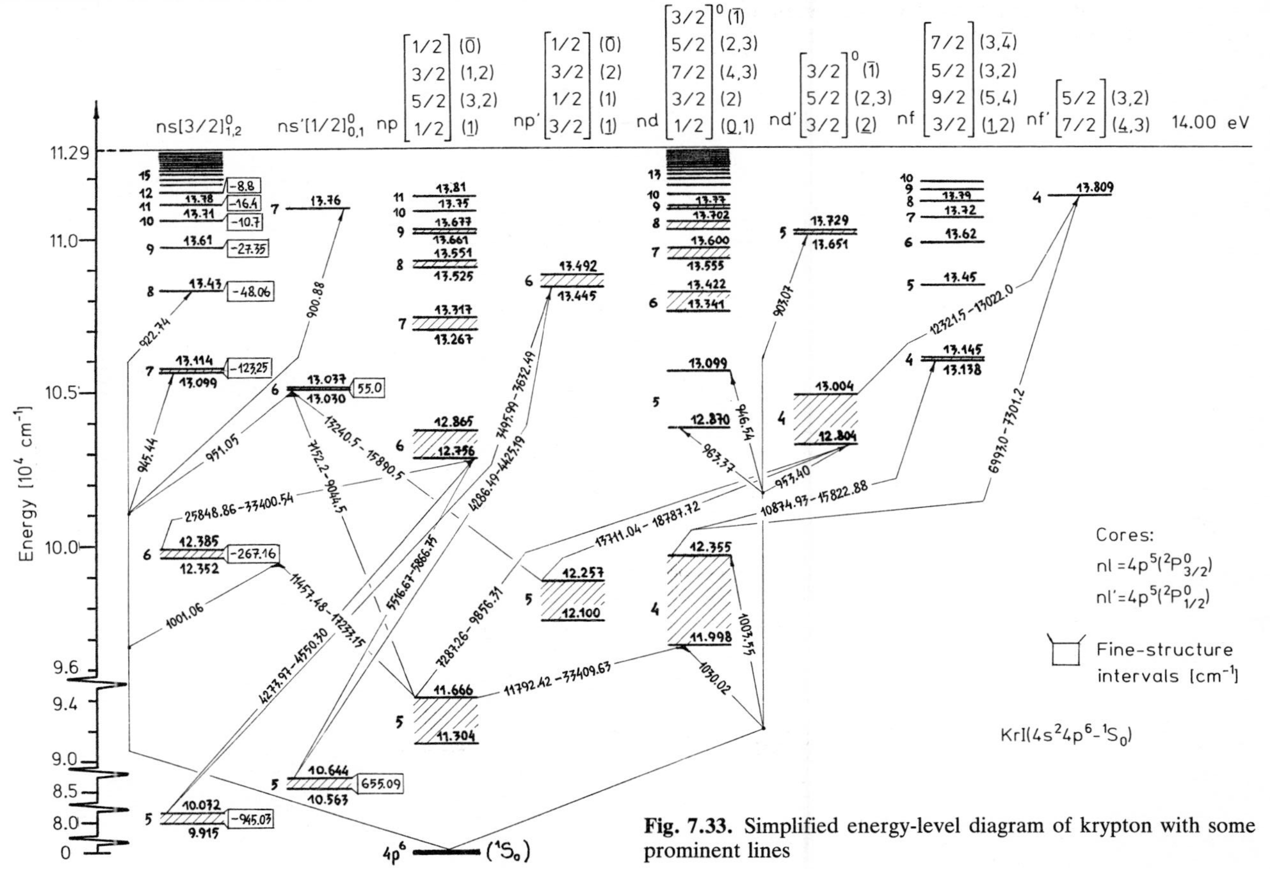

Fig. 7.33. Simplified energy-level diagram of krypton with some prominent lines

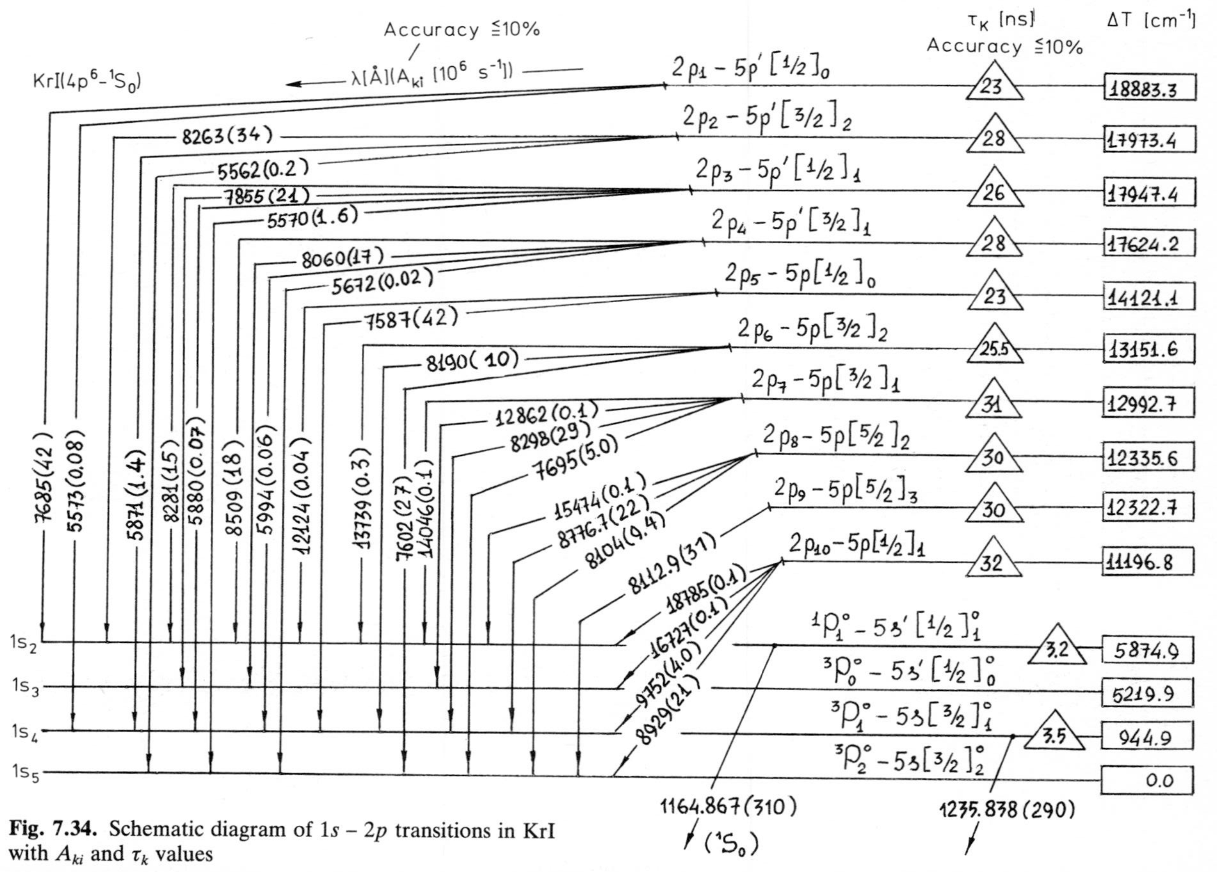

Fig. 7.34. Schematic diagram of $1s - 2p$ transitions in KrI with A_{ki} and τ_k values

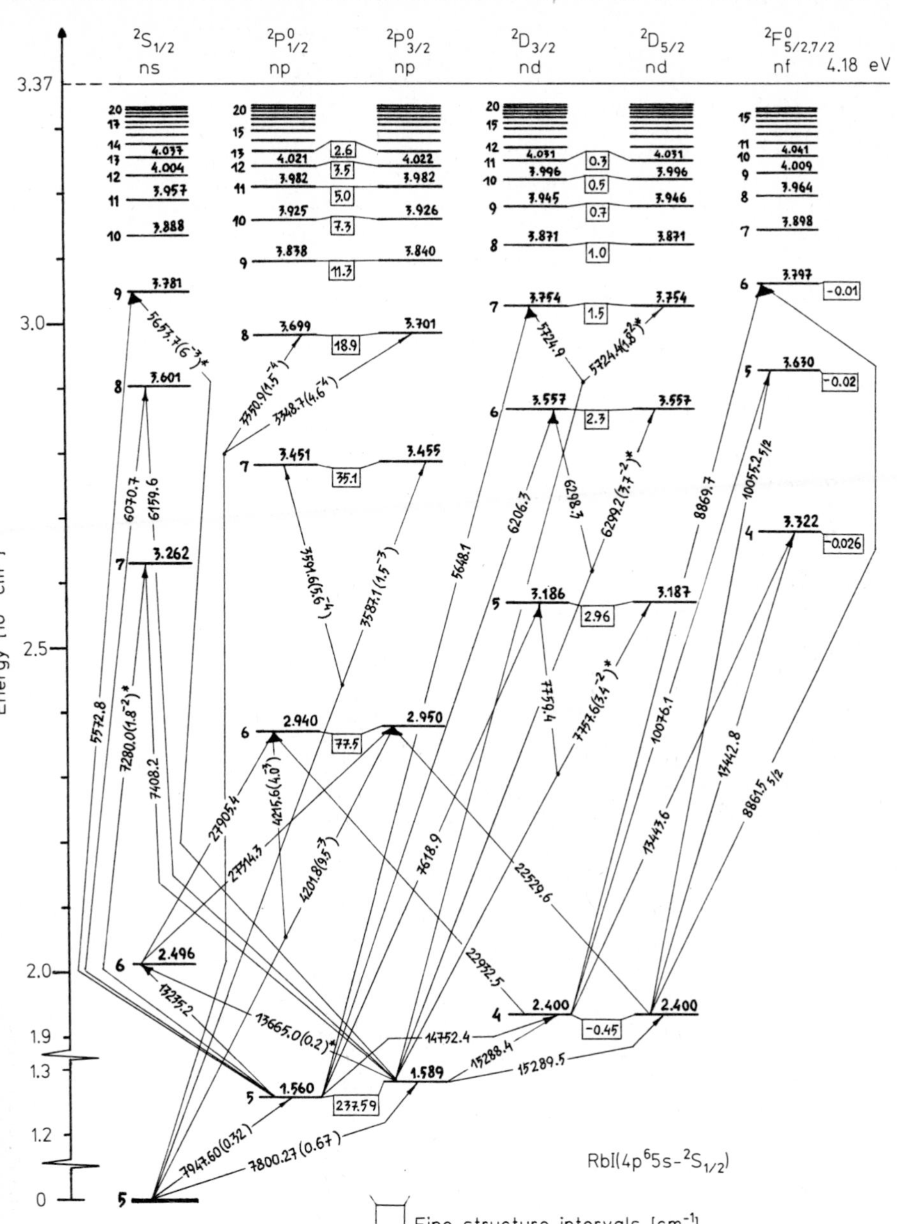

Fig. 7.35. Combined partial energy-level – Grotrian diagram with absorption oscillator strengths for rubidium

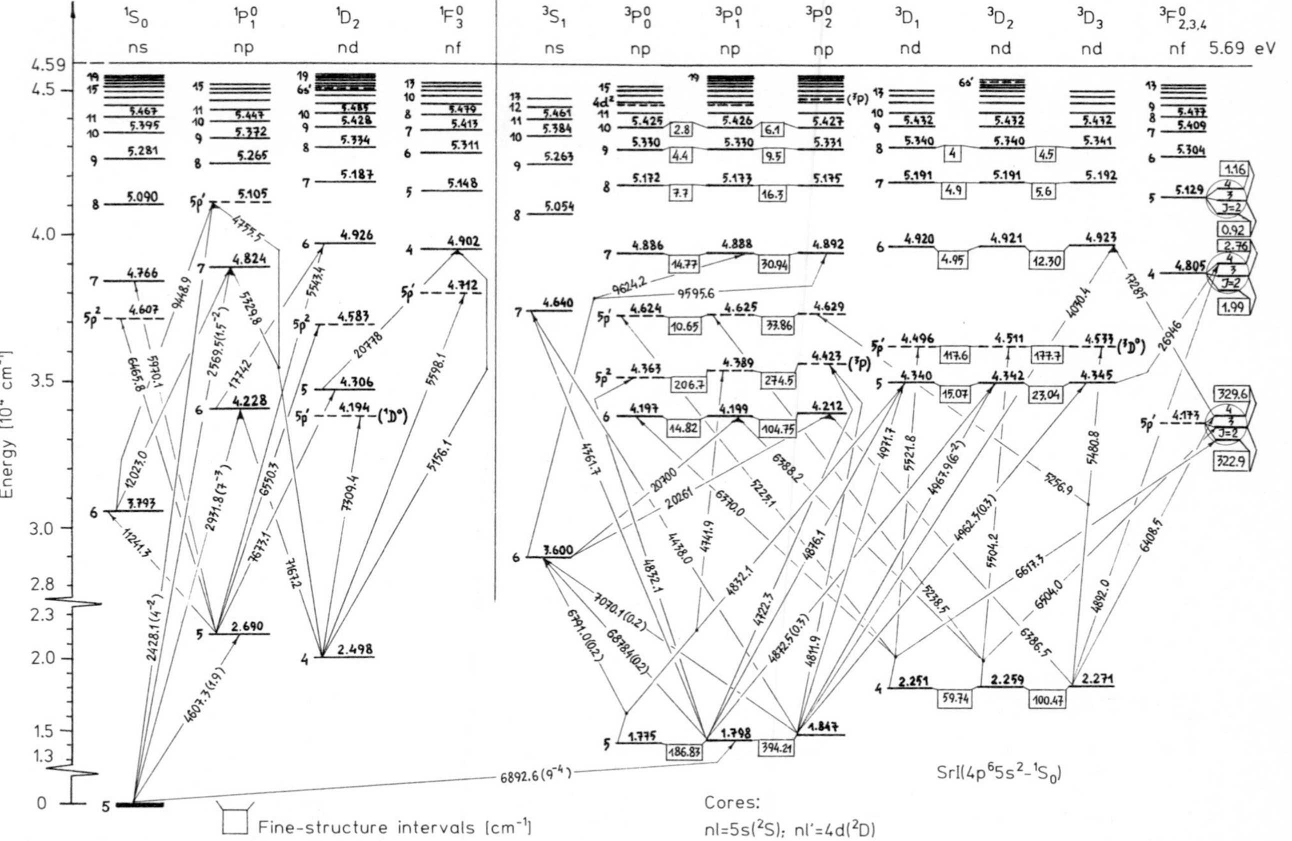

Fig. 7.36. Combined partial energy-level – Grotrian diagram with absorption oscillator strengths for strontium

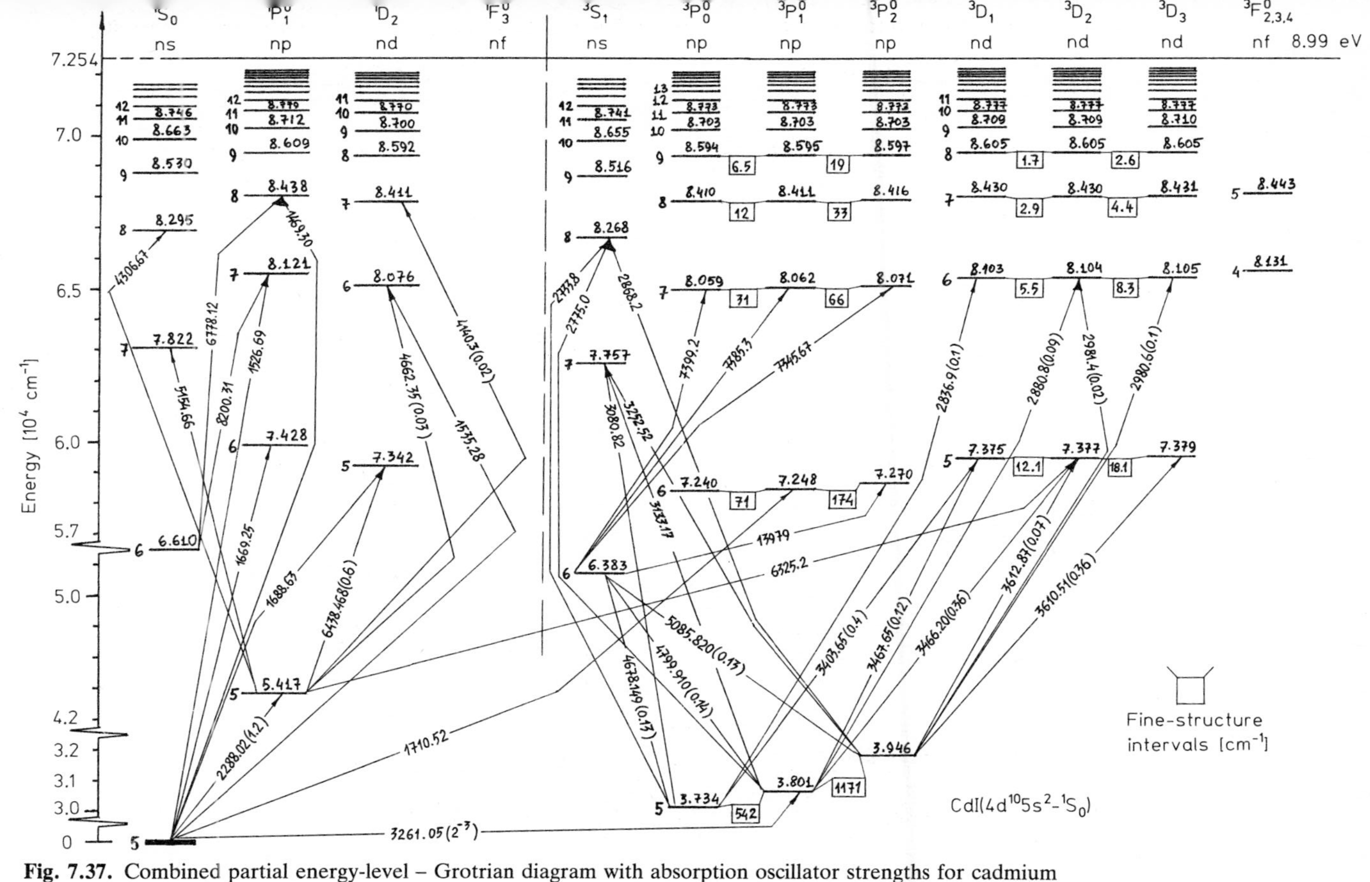

Fig. 7.37. Combined partial energy-level – Grotrian diagram with absorption oscillator strengths for cadmium

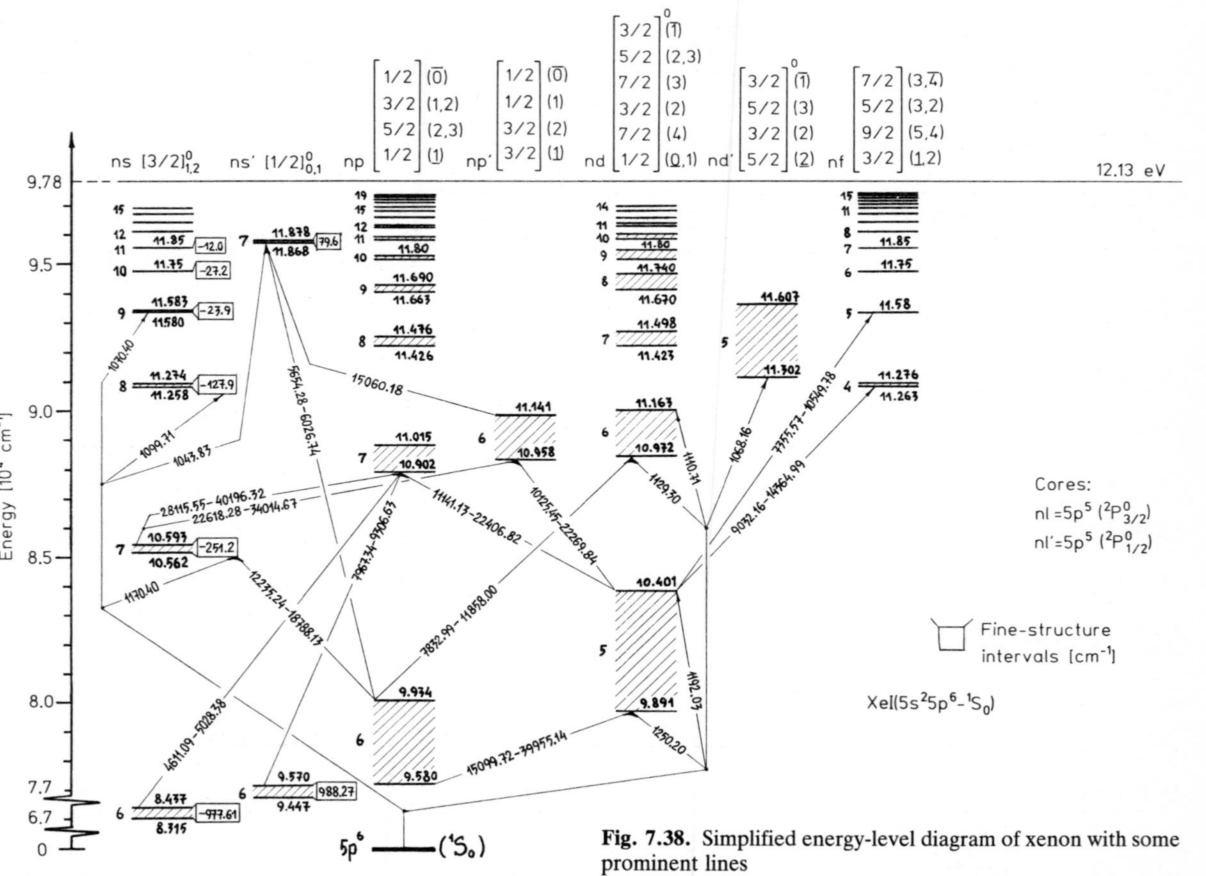

Fig. 7.38. Simplified energy-level diagram of xenon with some prominent lines

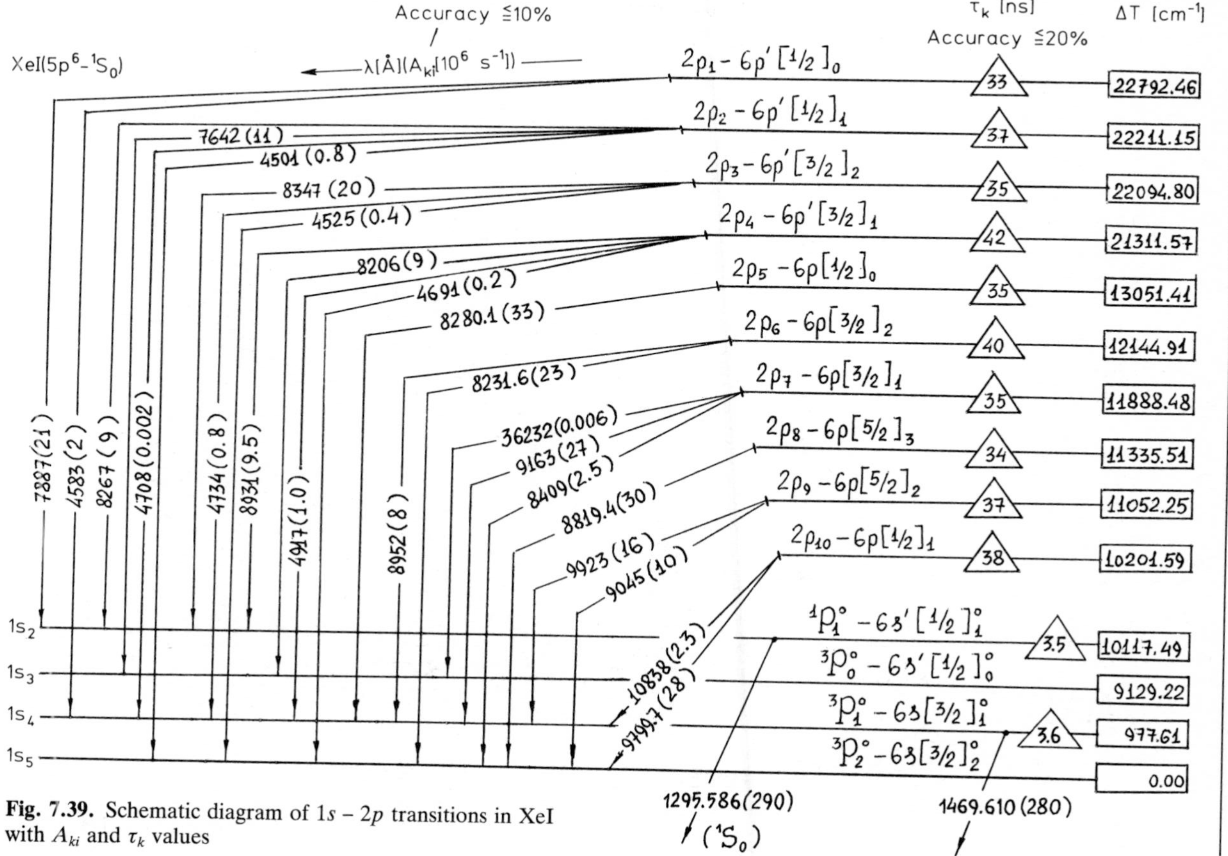

Fig. 7.39. Schematic diagram of 1s – 2p transitions in XeI with A_{ki} and τ_k values

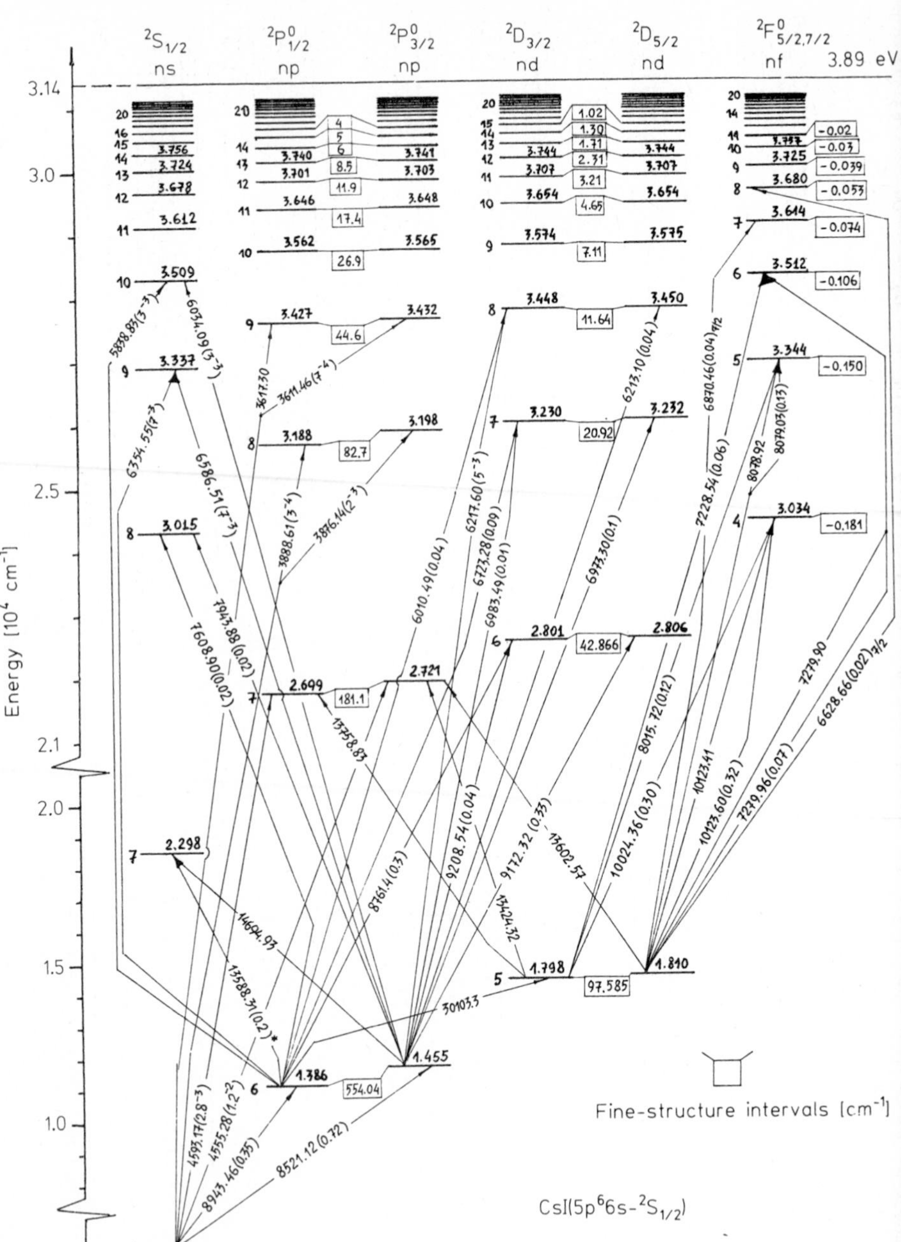

Fig. 7.40. Combined partial energy-level – Grotrian diagram with absorption oscillator strengths for caesium

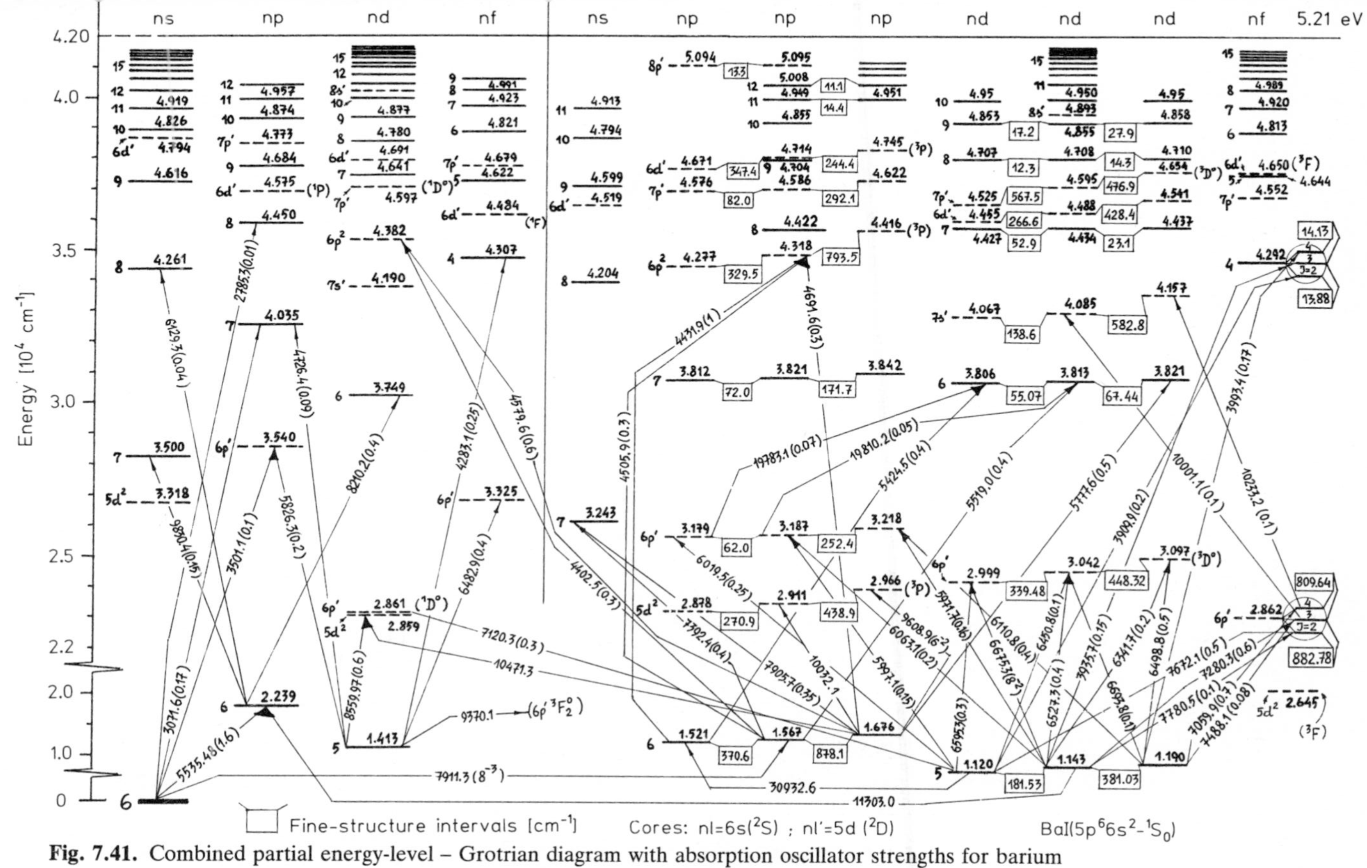

Fig. 7.41. Combined partial energy-level – Grotrian diagram with absorption oscillator strengths for barium

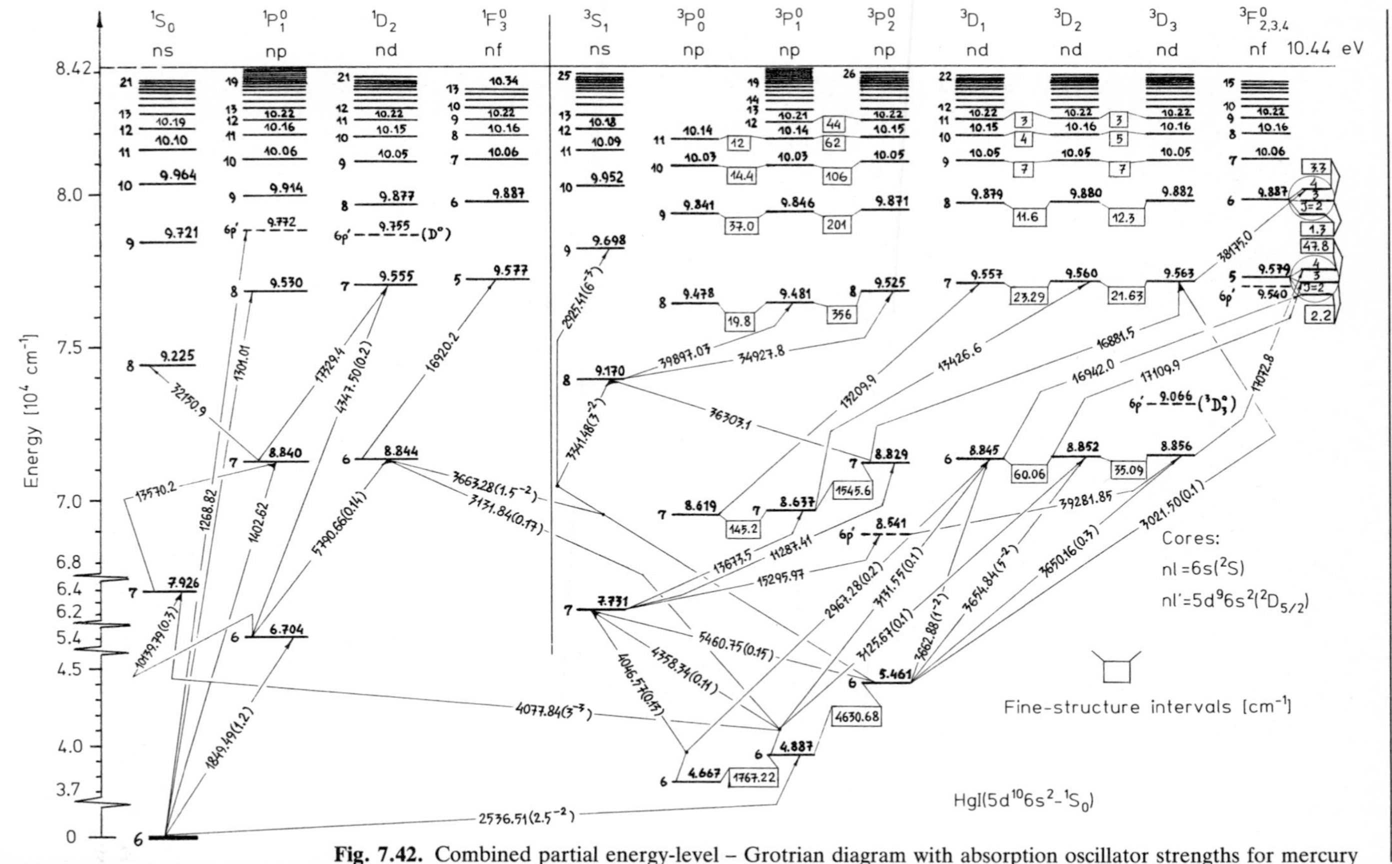

Fig. 7.42. Combined partial energy-level – Grotrian diagram with absorption oscillator strengths for mercury

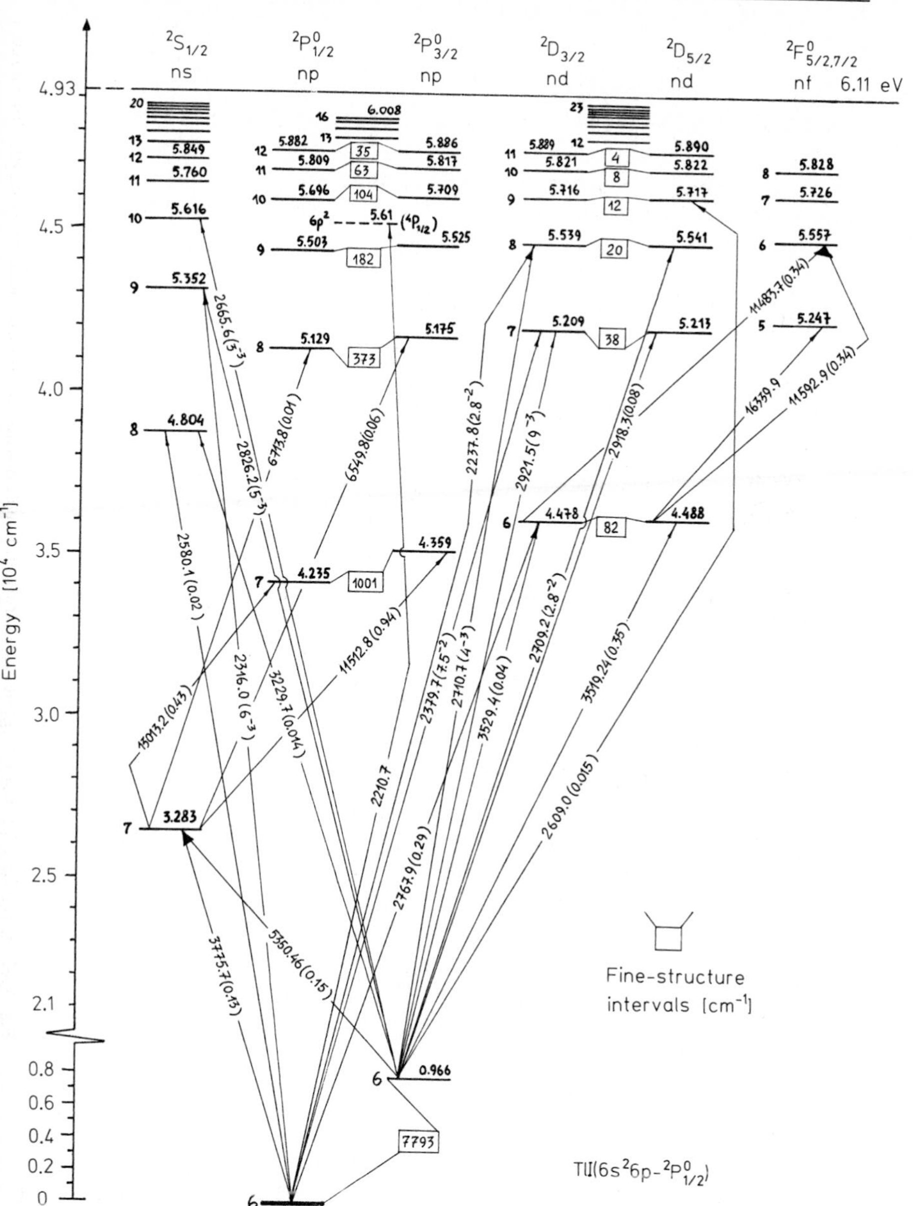

Fig. 7.43. Combined partial energy-level – Grotrian diagram with absorption oscillator strengths for thallium

7.3 Atomic Oscillator Strengths in Absorption

Table 7.3 gives the values of absorption oscillator strengths f_{ik} for some of the most intense optical transitions in atoms. These data partly supplement the values of f_{ik} cited in the Grotrian diagrams Figs. 7.1–43, and they also refer to elements which are not represented on these diagrams.

Let us draw attention to the definition of characteristics of allowed radiative transitions between atomic levels [7.3.1]:

1) The probability of spontaneous emission (E1) per unit time, A_{ki}, from the upper state k to the lower state i is given by

$$A_{ki} = \frac{4e^2\omega_{ik}^3}{3\hbar c^3(2J_k+1)} \cdot |\langle i|D|k\rangle|^2 ,$$

where $\omega_{ik} = (E_k - E_i)/\hbar$ is the frequency of the emitted photon, $\langle i|D|k\rangle$ is the matrix element of the atomic dipole moment, and J_k is the total electronic angular momentum of the atom in the state k.

2) The absorption oscillator strength $f_{ik}\,(i \to k)$ is defined as the dimensionless value

$$f_{ik} = \frac{2m_e}{3\hbar} \cdot \frac{\omega_{ik}}{(2J_i+1)} |\langle i|D|k\rangle|^2 = \frac{m_e c^3}{2e^2\omega_{ik}^2} \cdot \frac{(2J_k+1)}{(2J_i+1)} A_{ki} .$$

Here J_i is the total electronic angular momentum of an atom in the state i. If the transition wavelength $\lambda = 2\pi c/\omega_{ik}$ is introduced instead of ω_{ik} and measured in Å units, and if A_{ki} is measured in s^{-1}, we obtain the following table for conversion of the f_{ik} and A_{ki} values:

A_{ki} [s^{-1}]	f_{ik}
$A_{ki} = 1$	$6.6705 \cdot 10^{15} \dfrac{2J_i+1}{\lambda^2(2J_k+1)}$
$f_{ik} = 1.4992 \cdot 10^{-16}\lambda^2 \dfrac{2J_k+1}{2J_i+1}$	1

3) In a number of cases it is convenient to introduce the value of the averaged (over multiplets of the lower and upper terms) oscillator strength f_{ik}^{m}, which is connected with the oscillator strengths of the separate multiplet components by the relation

$$f_{ik}^{\mathrm{m}} = \frac{1}{\bar{\lambda}_{ik}\Sigma_{J_i}(2J_i+1)} \Sigma_{J_k,J_i}(2J_i+1) \cdot \lambda(J_i,J_k) \cdot f(J_i,J_k) .$$

Usually the wavelength differences for the lines within a multiplet are rather small, so that the wavelength factors may be neglected.

The main information about atomic transition probabilities is represented in the reference books [7.3.1–6], reviews [7.3.7, 8] and certain papers included in the bibliography [7.3.9, 10]. Taking into account the error in the determination of the f_{ik} values, the results included in Table 7.3 are grouped according to accuracy classes (see Introduction).

References

7.3.1 W.L.Wiese, M.W.Smith, B.M.Glennon: *Atomic Transition Probabilities – H through Ne, Vol. 1*, Nat. Stand. Ref. Data Ser. Nat. Bur. Stand. **4** (1966)

7.3.2 W.L.Wiese, M.W.Smith, B.M.Miles: *Atomic Transition Probabilities – Na through Ca, Vol. 2*, Nat. Stand. Ref. Data Ser. Nat. Bur. Stand. **22** (1969)

7.3.3 W.L.Wiese, B.M.Glennon: "Atomic Transition Probabilities", in *American Institute of Physics Handbook*, ed. by D.E.Gray, 3rd ed. (McGraw-Hill, New York 1972) Chap. 7, pp. 200–263

7.3.4 W.L.Wiese, G.A.Martin: "Transition Probabilities", in *Wavelengths and Transition Probabilities for Atoms and Atomic Ions*, Nat. Stand. Ref. Data Ser. Nat. Bur. Stand. **68**, Part II, 359–406 (1980)

7.3.5 J.R.Fuhr, G.A.Martin, W.L.Wiese, S.M.Younger: "Atomic transition probabilities for Fe, Co and Ni", J. Phys. Chem. Ref. Data **10**, 305 (1981) (and references cited therein on previous data compilations of NBS Atomic Transition Probabilities Data Center)

7.3.6 G.A.Kasabov, V.V.Eliseev: *Spectroscopy Tables for Low-Temperature Plasma* (Atomizdat, Moscow 1973) (in Russian)

7.3.7 W.L.Wiese: "Atomic Transition Probabilities and Lifetimes", in *Progress in Atomic Spectroscopy*, Part B, ed. by W.Hanle, H.Kleinpoppen (Plenum, New York 1979) pp. 1101–1155

7.3.8 M.C.E.Huber, R.J.Sandeman: "Transition probabilities and their accuracy", Phys. Scr. **22**, 373 (1980)

7.3.9 J.R.Fuhr, B.J.Miller, G.A.Martin: *Bibliography on Atomic Transition Probabilities: 1914 through October 1977*, Nat. Bur. Stand. (U.S.) Spec. Publ. **505** (1978)

7.3.10 B.J.Miller, J.R.Fuhr, G.A.Martin: *Bibliography on Atomic Transition Probabilities: November 1977 through March 1980*, Nat. Bur. Stand. (U.S.) Rep. NBS-SP-505/1 (US Gov't Printing Office, Washington, D.C. 1980)

Table 7.3. Transition probabilities A_{ki} and absorption oscillator strengths f_{ik} for low-lying atomic states (i labels lower state, k upper state)

Atom (ground state)	Transition ($i-k$)	Wavelength λ [Å]	Transition probability A_{ki} [$10^8\,s^{-1}$]	Oscillator strength f_{ik} (and accuracy)	g_k/g_i
1	2	3	4	5	6
$H(1\,^2S)$	$1S-2P$	1215.67	6.262	0.416 (A)	6/2
	$3P$	1025.72	1.672	0.0791 (A)	6/2
	$4P$	972.54	0.681	0.0290 (A)	6/2
	$2S-3P$	6562.74	0.225	0.435 (A)	6/2
	$4P$	4861.29	0.0967	0.103 (A)	6/2
	$2P-3S$	6562.86	0.0631	0.0136 (A)	2/6
	$3D$	6562.81	0.647	0.696 (A)	10/6
	$4S$	4861.35	0.0258	0.00305 (A)	2/6
	$4S$	4861.33	0.206	0.122 (A)	10/6
$He(1\,^1S_0)$	$1\,^1S-2\,^1P$	584.33	17.99	0.276 (A)	3/1
	$3\,^1P$	537.03	5.66	0.073 (B)	3/1
	$4\,^1P$	522.21	2.46	0.030 (B)	3/1
	$2\,^3S-2\,^3P$	10830	0.1022	0.539 (A)	9/3
	$3\,^3P$	3888.65	0.0948	0.0645 (A)	9/3
	$4\,^3P$	3187.75	0.0564	0.0258 (A)	9/3
	$2\,^1S-2\,^1P$	20581.3	0.020	0.376 (B)	3/1
	$3\,^1P$	5015.68	0.1338	0.151 (A)	3/1
	$4\,^1P$	3964.73	0.069	0.049 (B)	3/1
	$2\,^3P-3\,^3S$	7065.2	0.2786	0.0695 (A)	3/9
	$3\,^3D$	5875.7	0.7053	0.608 (A)	15/9
	$4\,^3S$	4713.2	0.095	0.011 (B)	3/9
	$4\,^3D$	4471.5	0.25	0.12 (B)	15/9
	$2\,^1P-3\,^1S$	7281.35	0.1829	0.0485 (A)	1/3
	$3\,^1D$	6678.15	0.6339	0.706 (A)	5/3
	$4\,^1S$	5047.74	0.0675	0.0086 (B)	1/3
	$4\,^1D$	4921.93	0.20	0.12 (B)	5/3
$Li(2\,^2S_{1/2})$	$2\,^2S-2\,^2P^o_{1/2}$	6707.91	0.366	0.247 (A)	2/2
	$2\,^2P^o_{3/2}$	6707.76	0.366	0.494 (A)	4/2
	$3\,^2P^o$	3232.6	0.01	0.005 (C)	6/2
	$2\,^2P^o_{1/2}-3\,^2S$	8126.2	0.012	0.11 (C)	2/2
	$3\,^2D$	6103.5	0.60	0.67 (C)	4/2
	$2\,^2P^o_{3/2}-3\,^2S$	8126.5	0.023	0.11 (C)	2/4
	$3\,^2D_{3/2}$	6103.7	0.12	0.067 (C)	4/4
	$3\,^2D_{5/2}$	6103.7	0.71	0.59 (C)	6/4
$Be(2\,^1S_0)$	$2\,^1S-2\,^1P^o$	2348.6	5.40	1.34 (B)	3/1
	$2\,^1P^o-3\,^1S$	8254.1	0.38	0.13 (D)	1/3
	$3\,^1D$	4572.7	0.79	0.41 (D)	5/3
	$2\,^3P^o-3\,^3S$	3321.2	1.56	0.086 (B)	3/9
$B(2\,^2P^o_{1/2})$	$2\,^2P^o_{1/2}-3\,^2S$	2496.77	1.3	0.12 (C)	2/2
	$2s\,2p^2\,(^2D_{3/2})$	2088.9	0.35	0.046 (D)	4/2
	$3\,^2D_{3/2}$	1825.9	2.0	0.2 (D)	4/2
	$2\,^2P^o_{3/2}-3\,^2S$	2497.72	1.5	0.07 (C)	2/4
	$2s\,2p^2\,(^2D)$	2089.6	0.15	0.025 (D)	10/4
	$3\,^2D$	1826.4	0.8	0.1 (D)	10/4

Table 7.3 (continued)

Atom (ground state)	Transition $(i - k)$	Wavelength λ [Å]	Transition probability A_{ki} [10^8 s^{-1}]	Oscillator strength f_{ik} (and accuracy)	g_k/g_i
1	2	3	4	5	6
C $(2p^2 - {}^3P_0)$	$2\,{}^3P_0 - 2p^3\,{}^3D_1^o$	1560.3	0.8	0.1 (E)	3/1
	$2\,{}^3P_1 - 3s\,{}^3P_2^o$	1656.3	0.8	0.05 (E)	5/3
	$3s\,{}^3P_0^o$	1657.9	3.2	0.04 (E)	1/3
	$2\,{}^3P_2 - 3s\,{}^3P_2^o$	1657.01	2.4	0.1 (E)	5/5
	$3s\,{}^3P_1^o$	1658.1	1.3	0.03 (E)	3/5
	$2p^3\,{}^3D_3^o$	1561.4	1.2	0.06 (E)	7/5
	$2\,{}^1D_2 - 3s\,{}^1P_1^o$	1930.9	3.0	0.10 (D)	3/5
	$2\,{}^1S_0 - 3d\,{}^1P_1^o$	1751.8	0.9	0.13 (E)	3/1
	$3s\,{}^1P_1^o$	2478.6	0.2	0.06 (D)	3/1
N $(2p^3 - {}^4S_{3/2}^o)$	$2\,{}^4S_{3/2}^o - 3s\,{}^4P$	1200	4.2	0.27 (C)	12/4
	$2p^4\,{}^4P$	1134	1.36	0.080 (B)	12/4
	$2p^3\,{}^2D^o - 3s\,{}^2P$	1493	3.4	0.07 (C)	6/10
	$3s'\,{}^2D$	1243	3.6	0.08 (C)	10/10
	$2p^3\,{}^2P^o - 3s\,{}^2P$	1743	1.2	0.06 (C)	6/6
	$4s\,{}^2P$	1328	0.15	0.004 (C)	6/6
	$3s'\,{}^2D$	1412	0.40	0.02 (C)	10/6
	$3s''\,{}^2S$	1144	8	0.05 (D)	2/6
O $(2p^4 - {}^3P_2)$	$2p^4\,{}^3P - 3s\,{}^3S^o$	1304	6.0	0.05 (C)	3/9
	$3d\,{}^3D^o$	1027	0.4	0.01 (D)	15/9
	$3s\,{}^5S_2^o - 3p\,{}^5P$	7774	1.0	2.7 (C)	15/5
	$3s\,{}^3S_1^o - 3p\,{}^3P$	8446.5	0.28	0.9 (C)	9/3
F $(2p^5 - {}^2P_{3/2}^o)$	$2p^5\,{}^2P_{1/2}^o - 3s\,{}^2P_{1/2}$	955.5	5.1	0.07 (D)	2/2
	$3s\,{}^2P_{3/2}$	958.5	1.3	0.035 (D)	4/2
	$2p^5\,{}^2P_{3/2}^o - 3s\,{}^2P_{3/2}$	954.8	6.4	0.09 (D)	4/4
	$3s\,{}^2P_{1/2}$	951.9	2.6	0.02 (D)	2/4
	$3s\,{}^4P_{5/2} - 3p\,{}^4D_{7/2}^o$	6856.0	0.4	0.4 (E)	8/6
	$3p\,{}^4S_{3/2}^o$	6239.7	0.25	0.1 (E)	4/6
	$3s\,{}^2P_{3/2} - 3p\,{}^2P_{3/2}^o$	7037.5	0.3	0.2 (E)	4/4
	$3p\,{}^2D_{5/2}^o$	7754.7	0.3	0.4 (E)	6/4
	$3s\,{}^2P_{1/2} - 3p\,{}^2P_{1/2}^o$	7127.9	0.4	0.3 (E)	2/2
Ne $(2p^6 - {}^1S_0)$	$2\,{}^1S_0 - 3s\,[3/2]_1^o$	743.72	0.40	0.010 (C)	3/1
	$3s'\,[1/2]_1^o$	735.90	6.2	0.15 (C)	3/1
	$4s\,[3/2]_1^o$	629.74	0.48	0.008 (D)	3/1
	$4s'\,[1/2]_1^o$	626.82	0.74	0.013 (D)	3/1
	$3d\,[3/2]_1^o$	619.10	0.33	0.013 (D)	3/1
	$3d\,[1/2]_1^o$	618.67	0.93	0.016 (D)	3/1
	$3d'\,[3/2]_1^o$	615.63	0.38	0.0065 (D)	3/1
Na $(3s - {}^2S_{1/2})$	$3\,{}^2S_{1/2} - 3\,{}^2P_{1/2}^o$	5895.92	0.610	0.318 (A)	2/2
	$3\,{}^2P_{3/2}^o$	5889.95	0.612	0.637 (A)	4/2
	$4\,{}^2P_{1/2}^o$	3303.0	0.028	0.0046 (D)	2/2
	$4\,{}^2P_{3/2}^o$	3302.4	0.028	0.092 (D)	4/2
	$3\,{}^2P_{1/2}^o - 4\,{}^2S_{1/2}$	11381.5	0.089	0.17 (D)	2/2
	$3\,{}^2D_{3/2}$	8183.26	0.45	0.91 (D)	4/2

Table 7.3 (continued)

Atom (ground state)	Transition ($i-k$)	Wavelength λ [Å]	Transition probability A_{ki} [$10^8\,s^{-1}$]	Oscillator strength f_{ik} (and accuracy)	g_k/g_i
1	2	3	4	5	6
	$3\,^2P^o_{3/2} - 4\,^2S_{1/2}$	11403.8	0.176	0.17 (*D*)	2/4
	$3\,^2D_{5/2}$	8194.82	0.52	0.78 (*D*)	6/4
	$3\,^2D_{3/2}$	8194.79	0.086	0.087 (*D*)	4/4
Mg ($3s^2 - {}^1S_0$)	$3\,^1S_0 - 3\,^1P^o_1$	2852.1	5.3	1.9 (*D*)	3/1
	$3\,^3P^o_1$	4571.1	$2.8 \cdot 10^{-6}$	$2.6 \cdot 10^{-6}$ (*D*)	3/1
	$3\,^3P^o_0 - 4\,^3S_1$	5167.3	0.116	0.14 (*C*)	3/1
	$3\,^3P^o_1 - 4\,^3S_1$	5172.7	0.35	0.14 (*C*)	3/3
	$3\,^3P^o_2 - 4\,^3S_1$	5183.6	0.58	0.14 (*C*)	3/5
Al ($3p - {}^2P^o_{1/2}$)	$3\,^2P^o_{1/2} - 4\,^2S_{1/2}$	3944.0	0.50	0.12 (*D*)	2/2
	$3\,^2D_{3/2}$	3082.2	0.63	0.18 (*D*)	4/2
	$5\,^2S_{1/2}$	2652.5	0.13	0.014 (*D*)	2/2
	$4\,^2D_{3/2}$	2568.0	0.23	0.045 (*D*)	4/2
	$3\,^2P^o_{3/2} - 4\,^2S_{1/2}$	3961.5	1.0	0.12 (*D*)	2/4
	$3\,^2D_{3/2}$	3092.8	0.12	0.017 (*D*)	4/4
	$3\,^2D_{5/2}$	3092.7	0.74	0.16 (*D*)	6/4
	$5\,^2S_{1/2}$	2660.4	0.26	0.014 (*D*)	2/4
	$4\,^2D_{3/2}$	2575.4	0.044	0.0044 (*D*)	4/4
	$4\,^2D_{5/2}$	2575.1	0.28	0.042 (*D*)	6/4
Si ($3p^2 - {}^3P_0$)	$3\,^3P_0 - 4s\,^3P^o_1$	2514.3	0.61	0.17 (*D*)	3/1
	$3\,^3P_1 - 4s\,^3P^o_0$	2524.1	1.81	0.058 (*D*)	1/3
	$4s\,^3P^o_1$	2519.2	0.46	0.044 (*D*)	3/3
	$4s\,^3P^o_2$	2506.9	0.63	0.099 (*C*)	5/3
	$3\,^3P_2 - 4s\,^3P^o_1$	2528.5	0.77	0.044 (*C*)	3/5
	$4s\,^3P^o_2$	2516.1	1.64	0.16 (*C*)	5/5
P ($3p^3 - {}^4S^o_{3/2}$)	$3\,^4S^o_{3/2} - 4s\,^4P_{1/2}$	1787.7	2.1	0.050 (*D*)	2/4
	$4s\,^4P_{3/2}$	1782.8	2.1	0.10 (*D*)	4/4
	$4s\,^4P_{5/2}$	1775.0	2.2	0.15 (*D*)	6/4
	$3p^4\,^4P_{5/2}$	1679.7	0.39	0.025 (*D*)	6/4
	$3p^4\,^4P_{3/2}$	1674.6	0.40	0.016 (*D*)	4/4
	$3p^4\,^4P_{1/2}$	1671.7	0.39	0.008 (*D*)	2/4
S ($3p^4 - {}^3P_2$)	$3\,^3P_2 - 4s\,^3S^o_1$	1807.3	3.8	0.11 (*D*)	3/5
	$3\,^3P_1 - 4s\,^3S^o_1$	1820.3	2.2	0.11 (*D*)	3/3
	$3\,^3P_0 - 4s\,^3S^o_1$	1826.2	0.72	0.11 (*D*)	3/1
Cl ($3p^5 - {}^2P^o_{3/2}$)	$3\,^2P^o_{3/2} - 4s\,^2P_{3/2}$	1347.2	4.2	0.11 (*D*)	4/4
	$4s\,^2P_{1/2}$	1335.7	1.7	0.023 (*D*)	2/4
	$3\,^2P^o_{1/2} - 4s\,^2P_{3/2}$	1363.4	0.75	0.042 (*D*)	4/2
	$4s\,^2P_{1/2}$	1351.7	3.2	0.088 (*D*)	2/2
Ar ($3p^6 - {}^1S_0$)	$3\,^1S_0 - 4s\,[3/2]^o_1$	1066.6	1.0	0.051 (*D*)	3/1
	$4s'\,[1/2]^o_1$	1048.2	5.0	0.25 (*D*)	3/1
K ($4s - {}^2S_{1/2}$)	$4\,^2S_{1/2} - 4\,^2P^o_{1/2}$	7699.0	0.39	0.35 (*C*)	2/2
	$4\,^2P^o_{3/2}$	7664.9	0.40	0.70 (*C*)	4/2
	$5\,^2P^o_{1/2}$	4047.2	0.012	0.003 (*D*)	2/2
	$5\,^2P^o_{3/2}$	4044.1	0.012	0.006 (*D*)	4/2

Table 7.3 (continued)

Atom (ground state)	Transition $(i - k)$	Wavelength λ [Å]	Transition probability A_{ki} [$10^8\,s^{-1}$]	Oscillator strength f_{ik} (and accuracy)	g_k/g_i
1	2	3	4	5	6
	$4\,^2P^o_{1/2} - 5\,^2S_{1/2}$	12432.2	0.078	0.18 (D)	2/2
	$4\,^2D_{3/2}$	11690.2	0.22	0.90 (D)	4/2
	$4\,^2P^o_{3/2} - 5\,^2S_{1/2}$	12522.1	0.15	0.18 (D)	2/4
	$4\,^2D_{3/2}$	11769.6	0.043	0.09 (D)	4/4
	$4\,^2D_{5/2}$	11772.8	0.26	0.81 (D)	6/4
Ca ($4s^2 - {}^1S_0$)	$4\,^1S_0 - 4\,^1P^o_1$	4226.7	2.2	1.7 (C)	3/1
	$4\,^3P^o_1$	6572.8	$2.2 \cdot 10^{-5}$	$4.2 \cdot 10^{-5}$ (D)	3/1
	$4\,^3P^o_0 - 4\,^3D_1$	4425.4	0.50	0.40 (D)	3/1
	$5\,^3S_1$	6102.7	0.096	0.16 (D)	3/1
	$4\,^3P^o_1 - 4\,^3D_1$	4435.7	0.34	0.10 (D)	3/3
	$4\,^3D_2$	4435.0	0.67	0.33 (D)	5/3
	$5\,^3S_1$	6122.2	0.29	0.16 (D)	3/3
	$4\,^3P^o_2 - 4\,^3D_1$	4456.6	0.025	0.004 (D)	3/5
	$4\,^3D_2$	4455.9	0.20	0.059 (D)	5/5
	$4\,^3D_3$	4454.8	0.87	0.36 (D)	7/5
	$5\,^3S_1$	6162.2	0.48	0.16 (D)	3/5
Cu ($4s - {}^2S_{1/2}$)	$4\,^2S_{1/2} - 4\,^2P^o_{1/2}$	3274.0	1.37	0.22 (C)	2/2
	$4\,^2P^o_{3/2}$	3247.5	1.39	0.44 (C)	4/2
Zn ($4s^2 - {}^1S_0$)	$4\,^1S_0 - 4\,^1P^o_1$	2138.6	7.4	1.5 (C)	3/1
	$4\,^3P^o_1$	3075.9	$3.3 \cdot 10^{-4}$	$1.4 \cdot 10^{-4}$ (C)	3/1
	$4\,^3P^o_0 - 5\,^3S_1$	4680.1	0.14	0.14 (D)	3/1
	$4\,^3D_1$	3282.3	0.90	0.44 (C)	3/1
	$4\,^3P^o_1 - 5\,^3S_1$	4722.2	0.43	0.14 (D)	3/3
	$4\,^3D_1$	3302.9	0.67	0.11 (C)	3/3
	$4\,^3D_2$	3302.6	1.2	0.33 (C)	5/3
	$4\,^3P^o_2 - 5\,^3S_1$	4810.5	0.67	0.14 (D)	3/5
	$4\,^3D_1$	3345.9	0.045	0.0045 (C)	3/5
	$4\,^3D_2$	3345.6	0.40	0.067 (C)	5/5
	$4\,^3D_3$	3345.0	1.7	0.40 (C)	7/5
Ga ($4p - {}^2P^o_{1/2}$)	$4\,^2P^o_{1/2} - 5\,^2S_{1/2}$	4033.0	0.49	0.12 (D)	2/2
	$4\,^2D_{3/2}$	2874.2	1.2	0.30 (D)	4/2
	$6\,^2S_{1/2}$	2659.9	0.12	0.013 (D)	2/2
	$5\,^2D_{3/2}$	2450.1	0.28	0.050 (D)	4/2
	$4\,^2P^o_{3/2} - 5\,^2S_{1/2}$	4172.0	0.92	0.12 (D)	2/4
	$4\,^2D_{3/2}$	2944.2	0.27	0.035 (D)	4/4
	$4\,^2D_{5/2}$	2943.6	1.4	0.27 (D)	6/4
	$6\,^2S_{1/2}$	2719.7	0.23	0.013 (D)	2/4
	$5\,^2D_{5/2}$	2500.2	0.34	0.048 (D)	6/4
Kr ($4p^6 - {}^1S_0$)	$4\,^1S_0 - 5s\,[3/2]^o_1$	1235.8	2.2	0.15 (C)	3/1
	$5s'\,[1/2]^o_1$	1164.6	2.2	0.14 (C)	3/1
Rb ($5s - {}^2S_{1/2}$)	$5\,^2S_{1/2} - 5\,^2P^o_{1/2}$	7947.6	0.34	0.32 (C)	2/2
	$5\,^2P^o_{3/2}$	7800.3	0.37	0.67 (C)	4/2
	$6\,^2P^o_{1/2}$	4215.5	0.015	0.004 (D)	2/2
	$6\,^2P^o_{3/2}$	4201.8	0.018	0.0095 (D)	4/2

Table 7.3 (continued)

Atom (ground state)	Transition $(i-k)$	Wavelength λ [Å]	Transition probability A_{ki} [$10^8\,s^{-1}$]	Oscillator strength f_{ik} (and accuracy)	g_k/g_i
1	2	3	4	5	6
	$5\,^2P^o_{1/2} - 6\,^2S_{1/2}$	13237.3	0.071	0.19 (*D*)	2/2
	$5\,^2P^o_{3/2} - 6\,^2S_{1/2}$	13667.0	0.26	0.364 (*D*)	2/4
Sr $(5s^2 - {}^1S_0)$	$5\,^1S_0 - 5\,^1P^o_1$	4607.3	2.1	2.0 (*C*)	3/1
	$5\,^3P^o_1$	6892.6	$4.0 \cdot 10^{-4}$	$8.6 \cdot 10^{-4}$ (*D*)	3/1
	$5\,^3P^o_0 - 6\,^3S_1$	6791.0	0.087	0.18 (*D*)	3/1
	$5\,^3P^o_1 - 6\,^3S_1$	6878.4	0.24	0.17 (*D*)	3/3
	$5\,^3D_2$	4872.5	0.52	0.31 (*D*)	5/3
	$5\,^3P^o_2 - 6\,^3S_1$	7070.1	0.40	0.18 (*D*)	3/5
	$5\,^3D_2$	4967.9	0.16	0.06 (*D*)	5/5
	$5\,^3D_3$	4962.2	0.64	0.33 (*D*)	7/5
Ag $(5s - {}^2S_{1/2})$	$5\,^2S_{1/2} - 5\,^2P^o_{1/2}$	3382.9	1.3	0.22 (*C*)	2/2
	$5\,^2P^o_{3/2}$	3280.7	1.4	0.45 (*C*)	4/2
Cd $(5s^2 - {}^1S_0)$	$5\,^1S_0 - 5\,^1P^o_1$	2288.0	6.0	1.4 (*C*)	3/1
	$5\,^3P^o_1$	3261.1	0.0041	0.0019 (*D*)	3/1
	$5\,^3P^o_0 - 6\,^3S_1$	4678.1	0.13	0.13 (*D*)	3/1
	$5\,^3D_1$	3403.7	0.77	0.40 (*D*)	3/1
	$5\,^3P^o_1 - 6\,^3S_1$	4799.9	0.41	0.14 (*D*)	3/3
	$5\,^3D_1$	3467.7	0.67	0.12 (*D*)	3/3
	$5\,^3D_2$	3466.2	1.2	0.36 (*D*)	5/3
	$5\,^3P^o_2 - 6\,^3S_1$	5085.8	0.56	0.13 (*D*)	3/5
	$5\,^3D_2$	3612.9	0.35	0.068 (*D*)	5/5
	$5\,^3D_3$	3610.5	1.3	0.36 (*D*)	7/5
In $(5p - {}^2P^o_{1/2})$	$5\,^2P^o_{1/2} - 6\,^2S_{1/2}$	4101.8	0.56	0.14 (*D*)	2/2
	$5\,^2D_{3/2}$	3039.4	1.3	0.36 (*D*)	4/2
	$7\,^2S_{1/2}$	2753.9	0.15	0.017 (*D*)	2/2
	$6\,^2D_{3/2}$	2560.2	0.23	0.045 (*D*)	4/2
	$5\,^2P^o_{3/2} - 6\,^2S_{1/2}$	4511.3	1.02	0.157 (*C*)	2/4
	$5\,^2D_{3/2}$	3258.6	0.38	0.06 (*D*)	4/4
	$5\,^2D_{5/2}$	3256.1	1.3	0.31 (*D*)	6/4
	$7\,^2S_{1/2}$	2932.6	1.1	0.007 (*D*)	2/4
	$6\,^2D_{3/2}$	2713.9	0.15	0.017 (*D*)	4/4
	$6\,^2D_{5/2}$	2710.3	0.36	0.059 (*D*)	6/4
Xe $(5p^6 - {}^1S_0)$	$5\,^1S_0 - 6s\,[3/2]^o_1$	1469.6	2.8	0.27 (*C*)	3/1
	$6s'\,[1/2]^o_1$	1295.6	2.9	0.22 (*C*)	3/1
	$5d\,[1/2]^o_1$	1250.2	0.14	0.010 (*E*)	3/1
	$5d\,[3/2]^o_1$	1192.0	6.2	0.40 (*D*)	3/1
	$7s\,[3/2]^o_1$	1170.4	1.6	0.10 (*D*)	3/1
	$6d\,[1/2]^o_1$	1129.3	0.044	0.0025 (*D*)	3/1
	$6d\,[3/2]^o_1$	1110.7	1.5	0.083 (*D*)	3/1
	$5d'\,[3/2]^o_1$	1068.2	4.0	0.20 (*D*)	3/1
Cs $(6s - {}^2S_{1/2})$	$6\,^2S_{1/2} - 6\,^2P^o_{1/2}$	8943.5	0.33	0.39 (*C*)	2/2
	$6\,^2P^o_{3/2}$	8521.1	0.37	0.81 (*C*)	4/2
	$7\,^2P^o_{1/2}$	4593.2	0.008	$2.5 \cdot 10^{-3}$ (*D*)	2/2
	$7\,^2P^o_{3/2}$	4555.3	0.019	0.012 (*D*)	4/2

Table 7.3 (continued)

Atom (ground state)	Transition ($i - k$)	Wavelength λ [Å]	Transition probability A_{ki} [10^8 s^{-1}]	Oscillator strength f_{ik} (and accuracy)	g_k/g_i
1	2	3	4	5	6
	$6\,^2P^o_{1/2} - 5\,^2D_{3/2}$	30103	0.0092	0.25 (D)	4/2
	$7\,^2S_{1/2}$	13588	0.076	0.21 (D)	2/2
	$6\,^2D_{3/2}$	8761.4	0.13	0.30 (D)	4/2
	$6\,^2P^o_{3/2} - 5\,^2D_{3/2}$	36130	0.0011	0.021 (D)	4/4
	$5\,^2D_{5/2}$	34900	0.0074	0.20 (D)	6/4
	$7\,^2S_{1/2}$	14695	0.11	0.17 (D)	2/4
	$6\,^2D_{3/2}$	9208.5	0.031	0.040 (D)	4/4
	$6\,^2D_{5/2}$	9172.3	0.17	0.33 (D)	6/4
Ba ($6s^2 - {}^1S_0$)	$6\,^1S_0 - 6\,^1P^o_1$	5535.5	1.2	1.6 (C)	3/1
	$6\,^3P^o_1$	7911.3	0.0030	0.084 (D)	3/1
	$6\,^3P^o_0 - 7\,^3S_1$	7195.2	0.23	0.53 (D)	3/1
	$6\,^3D_1$	5424.5	0.53	0.4 (D)	3/1
	$6\,^3P^o_1 - 7\,^3S_1$	7392.4	0.48	0.39 (D)	3/3
	$6\,^3D_2$	5519.0	0.50	0.38 (D)	5/3
	$6\,^3P^o_2 - 7\,^3S_1$	7905.7	0.60	0.34 (D)	3/5
	$6\,^3D_2$	5800.2	0.10	0.05 (D)	5/5
	$6\,^3D_3$	5777.6	0.64	0.45 (D)	7/5
Au ($6s - {}^2S_{1/2}$)	$6\,^2S_{1/2} - 6\,^2P^o_{1/2}$	2676.0	1.1	0.12 (D)	2/2
	$6\,^2P^o_{3/2}$	2428.0	1.5	0.26 (D)	4/2
Hg ($6s^2 - {}^1S_0$)	$6\,^1S_0 - 6\,^1P^o_1$	1849.5	7.6	1.17 (B)	3/1
	$6\,^3P^o_1$	2536.5	0.085	0.0245 (B)	3/1
	$6\,^3P^o_0 - 7\,^3S_1$	4046.6	0.18	0.13 (D)	3/1
	$6\,^3D_1$	2967.3	0.45	0.18 (D)	3/1
	$6\,^3P^o_1 - 7\,^3S_1$	4358.3	0.40	0.11 (D)	3/3
	$6\,^3D_1$	3131.6	0.7	0.1 (D)	3/3
	$6\,^3D_2$	3125.7	0.5	0.12 (D)	5/3
	$6\,^3P^o_2 - 7\,^3S_1$	5460.7	0.56	0.15 (D)	3/5
	$6\,^3D_1$	3662.9	0.083	0.01 (D)	3/5
	$6\,^3D_2$	3654.8	0.25	0.05 (D)	5/5
	$6\,^3D_3$	3650.2	1.1	0.30 (D)	7/5
	$6\,^1P^o_1 - 7\,^1S_0$	10139.8	0.53	0.27 (D)	1/3
Tl ($6p - {}^2P^o_{1/2}$)	$6\,^2P^o_{1/2} - 7\,^2S_{1/2}$	3775.7	0.625	0.133 (C)	2/2
	$6\,^2D_{3/2}$	2767.9	1.3	0.29 (D)	4/2
	$6\,^2P^o_{3/2} - 7\,^2S_{1/2}$	5350.5	0.705	0.151 (C)	2/4
	$6\,^2D_{3/2}$	3529.4	0.22	0.041 (D)	4/4
	$6\,^2D_{5/2}$	3519.2	1.2	0.34 (D)	6/4
	$7\,^2S_{1/2} - 7\,^2P^o_{1/2}$	13013.2	0.171	0.434 (A)	2/2
	$7\,^2P^o_{3/2}$	11512.8	0.237	0.942 (A)	4/2
Pb ($6p^2 - {}^3P_0$)	$6\,^3P_0 - 7s\,^3P^o_1$	2833.1	0.6	0.21 (D)	3/1
	$6\,^3P_1 - 7s\,^3P^o_0$	3683.5	1.5	0.10 (D)	1/3
	$6\,^3P_2 - 7s\,^3P^o_1$	4057.8	0.9	0.13 (D)	3/5
	$6d\,^3F^o_3$	2802.0	1.6	0.26 (D)	7/5
Bi ($6p^3 - {}^4S^o_{3/2}$)	$6\,^4S^o_{3/2} - 7s\,^4P_{1/2}$	3067.7	2.1	0.15 (D)	2/4
	$6\,^2D^o_{3/2} - 7s\,^4P_{1/2}$	4722.5	0.12	0.020 (D)	2/4

7.4 Lifetimes of Resonant Excited States in Atoms

The radiative lifetime of an excited state k is inversely proportional to the total probability of radiative transitions to the low-lying levels, that is

$$\tau_k \equiv (\Sigma_i A_{ki})^{-1} ,$$

where A_{ki} is the Einstein coefficient, which determines the probability of spontaneous emission per unit time from the upper level k into a definite lower level i. The τ_k value characterizes the mean time in which the population of the upper state, $N_k(t)$, drops to 1/e of its value at the beginning of the radiative decay process at time $t = 0$:

$$N_k(t) = N_k(0) \exp(-t/\tau_k) ,$$

in the absence of collision processes, and absorption and induced emission, which de- and repopulate the levels.

In Table 7.4, values of radiative lifetimes of low-lying resonantly excited atomic states are summarized. In the fourth column of the table, the excitation energy of a level with respect to the ground state is given, or the position of the multiplet centre of gravity $\overline{T}$ is specified (Sect. 7.1). To compile Table 7.4 we used the recommended values of τ_k for some elements [7.4.1–12] and the data of particular publications listed in the bibliographies [7.4.13–16]. The numerical values of atomic radiative lifetimes are grouped with respect to the usual accuracy classes.

References

7.4.1 R.T.Thompson, R.G.Fowler: "Lifetimes for the upper states in atomic helium", J. Quant. Spectrosc. Radiat. Transfer **15**, 1017 (1975)

7.4.2 M.E.M.Head, C.E.Head, T.N.Lawrence: "Review of Experimental Lifetimes: Third Period Elements", in *Beam-Foil Spectroscopy, Vol. 1, Atomic Structure and Lifetimes*, ed. by I.A.Sellin, D.J.Pegg (Plenum, New York 1976) pp. 147–154

7.4.3 C.E.Head, T.N.Lawrence, M.E.M.Head: "Review of Experimental Lifetimes: Fourth Period Elements", in *Beam-Foil Spectroscopy, Vol. 1, Atomic Structure and Lifetimes*, ed. by I.A.Sellin, D.J.Pegg (Plenum, New York 1976) pp. 155–163

7.4.4 Ya.F.Verolaynen, A.Ya.Nickolaich: "Radiative lifetimes of excited first-group atoms", Usp. Fiz. Nauk **137**, 305 (1982)

7.4.5 N.P.Penkin: "Experimental determination of electronic transition probabilities and lifetimes of the excited atomic and ionic states", in *Atomic Physics 6*, Proc. 6th Int. Conf. At. Phys., August 17–22, 1978, Riga, USSR, ed. by R.Damburg (Zinātne, Riga and Plenum, New York 1979) pp. 33–64

7.4.6 C.E.Theodosiou: "Lifetimes of singly excited states in He *I*", Phys. Rev. **A30**, 2910 (1984)

7.4.7 A.Gaupp, P.Kuske, H.J.Andrä: "Accurate lifetime measurements of the lowest $^2P_{1/2}$ states in neutral lithium and sodium", Phys. Rev. **A26**, 3351 (1982)

7.4.8 C.E.Theodosiou: "Lifetimes of alkali-metal-atom Rydberg states", Phys. Rev. **A30**, 2881 (1984)

7.4.9 S.A.Kandela, H.Schmoranzer: "Precision lifetime measurement of fine structure states in the Ne *I* $2p^5 3p$ configuration", Phys. Lett. **86A**, 101 (1981)

7.4.10 G.Inoue, J.K.Ku, D.W.Setser: "Laser-induced fluorescence study of Xe($6p$, $6p'$, $7p$ and $6d$) states in Ne and Ar: radiative lifetimes and collisional deactivation rate constants", J. Chem. Phys. **81**, 5760 (1984)

7.4.11 P.Hannaford, R.M.Lowe: "Determination of atomic lifetimes using laser-induced fluorescence from sputtered metal vapor", Opt. Engineering **22**, 532 (1983)

7.4.12 S.Salih, J.E.Lawler: "Ru *I*: radiative livetimes and stellar abundances", J. Opt. Soc. Am. **B2**, 422 (1985)

7.4.13 J.R.Fuhr, B.J.Miller, G.A.Martin: *Bibliography on Atomic Transition Probabilities: 1914 through October 1977*, Nat. Bur. Stand. (U.S.) Spec. Publ. **505** (1978)

7.4.14 B.J.Miller, J.R.Fuhr, G.A.Martin: *Bibliography on Atomic Transition Probabilities: November 1977 through March 1980*, Nat. Bur. Stand. (U.S.) Rep. NBS-SP-505/1 (US Gov't Printing Office, Washington, D.C. 1980)

7.4.15 A.L.Osherovitch, Ya.F.Verolaynen: "Delayed Coincidence Method in Atomic and Molecular Spectroscopy", in *Problems of Atmospheric Optics*, ed. by A.L.Osherovitch (Leningrad State University Printing Office, Leningrad 1979) pp. 80–154

7.4.16 W.L.Wiese: "Atomic Transition Probabilities and Lifetimes", in *Progress in Atomic Spectroscopy*, Part B, ed. by W.Hanle, H.Kleinpoppen (Plenum, New York 1979) pp. 1101–1155

Table 7.4. Radiative lifetimes of low-lying atomic states

Atomic number Z	Element (ground-state term)	Excited state	Excitation energy [eV] for the centre of gravity of the multiplet	Lifetime τ [ns]
1	2	3	4	5
1	$H(1s - {}^2S_{1/2})$	$2p\,({}^2P^o)$	10.20	1.60 (A)
		$3p\,({}^2P^o)$	12.09	5.30 (A)
		$3s\,({}^2S)$	12.09	159 (A)
		$3d\,({}^2D)$	12.09	15.5 (A)
		$4p\,({}^2P^o)$	12.75	12.7 (A)
		$4s\,({}^2S)$	12.75	227 (A)
		$4d\,({}^2D)$	12.75	36.0 (A)
		$4f\,({}^2F^o)$	12.75	72.5 (A)
		$5p\,({}^2P^o)$	13.06	23.8 (A)
		$5s\,({}^2S)$	13.06	352 (A)
		$5d\,({}^2D)$	13.06	69.5 (A)
		$5f\,({}^2F^o)$	13.06	140 (A)
		$5g\,({}^2G)$	13.06	235 (A)
		$6p\,({}^2P^o)$	13.22	40.7 (A)
		$6s\,({}^2S)$	13.22	534 (A)
		$6d\,({}^2D)$	13.22	119 (A)
		$6f\,({}^2F^o)$	13.22	240 (A)
		$6g\,({}^2G)$	13.22	403 (A)
		$6h\,({}^2H^o)$	13.22	608 (A)
2	$He(1s^2 - {}^1S_0)$	$2p\,({}^3P^o)$	20.96	97.85 (A)
		$2p\,({}^1P^o)$	21.22	0.555 (A)
		$3s\,({}^3S)$	22.72	36.0 (B)
		$3s\,({}^1S)$	22.92	54.2 (B)
		$3p\,({}^3P^o)$	23.01	94.7 (A)
		$3d\,({}^3D)$	23.07	14.2 (B)
		$3d\,({}^1D)$	23.07	15.8 (B)
		$3p\,({}^1P^o)$	23.09	1.724 (A)
		$4s\,({}^3S)$	23.59	64 (C)
		$4s\,({}^1S)$	23.67	87 (C)
		$4p\,({}^3P^o)$	23.71	150 (C)
		$4d\,({}^3D)$	23.74	31.5 (B)
		$4d\,({}^1D)$	23.74	37.4 (B)
		$4p\,({}^1P^o)$	23.74	3.97 (B)
		$4f\,({}^3F^o)$	23.74	72 (C)
		$4f\,({}^1F^o)$	23.74	72 (C)
3	$Li(2s - {}^2S_{1/2})$	$2p\,({}^2P^o)$	1.85	27.3 (A)
		$3s\,({}^2S)$	3.37	27.9 (A)
		$3p\,({}^2P^o)$	3.83	200 (C)
		$3d\,({}^2D)$	3.88	14.7 (B)
		$4s\,({}^2S)$	4.34	56 (C)
		$4p\,({}^2P^o)$	4.52	450 (C)
		$4d\,({}^2D)$	4.54	33 (C)

Table 7.4 (continued)

Atomic number Z	Element (ground-state term)	Excited state	Excitation energy [eV] for the centre of gravity of the multiplet	Lifetime τ [ns]
1	2	3	4	5
		$4f\,(^2F^o)$	4.54	72 (*D*)
		$5s\,(^2S)$	4.75	103 (*C*)
		$5p\,(^2P^o)$	4.84	670 (*D*)
		$5d\,(^2D)$	4.85	64 (*C*)
		$5f\,(^2F^o)$	4.85	140 (*D*)
		$6s\,(^2S)$	4.96	175 (*C*)
		$6p\,(^2P^o)$	5.01	1500 (*C*)
		$6d\,(^2D)$	5.01	108 (*C*)
		$6f\,(^2F^o)$	5.01	240 (*D*)
4	Be $(2s^2 - {}^1S_0)$	$2p\,(^1P^o)$	5.28	1.9 (*C*)
		$3s\,(^3S)$	6.46	6.4 (*C*)
		$3d\,(^3D)$	7.69	5.2 (*C*)
5	B $(2p - {}^2P^o_{1/2})$	$3s\,(^2S)$	4.96	3.6 (*C*)
		$2s\,2p^2\,(^2D)$	5.93	20 (*D*)
		$3p\,(^2P^o)$	6.03	50 (*D*)
		$3d\,(^2D)$	6.79	4.3 (*D*)
6	C $(2p^2 - {}^3P_0)$	$3s\,(^3P^o_0)$	7.48	3.1 (*C*)
		$3s\,(^1P^o_1)$	7.68	2.9 (*D*)
		$2s\,2p^3\,(^3D^o)$	7.95	8.0 (*C*)
7	N $(2p^3 - {}^4S^o_{3/2})$	$3s\,(^4P)$	10.33	2.5 (*D*)
		$3s\,(^2P)$	10.68	2.2 (*D*)
		$2s\,2p^4\,(^4P)$	10.92	7.2 (*C*)
8	O $(2p^4 - {}^3P_2)$	$3s\,(^3S^o_1)$	9.52	1.8 (*D*)
		$4p\,(^5P)$	12.29	193 (*C*)
		$4p\,(^3P)$	12.36	160 (*C*)
		$3s'\,(^3D^o)$	12.54	5.0 (*C*)
		$3s'\,(^1D^o_2)$	12.73	2.0 (*C*)
		$4d\,(^5D^o)$	12.75	96 (*C*)
9	F $(2p^5 - {}^2P^o_{3/2})$	$3s\,(^4P)$	12.72	7 (*E*)
		$3s\,(^2P)$	13.0	3.5 (*D*)
		$3p\,(^4P^o_{5/2})$	14.37	33 (*C*)
		$3p\,(^4D^o)$	14.52	29 (*C*)
		$3p\,(^2D^o)$	14.60	31.5 (*C*)
		$3p\,(^4S^o_{3/2})$	14.68	22 (*C*)
		$3p\,(^2P^o)$	14.75	25 (*C*)
10	Ne $(2p^6 - {}^1S_0)$	$3s\,[3/2]^o_1\,(1s_4)$	16.67	25 (*D*)
		$3s'\,[1/2]^o_1\,(1s_2)$	16.85	1.6 (*D*)
		$3p\,[1/2]_1\,(2p_{10})$	18.38	25.4 (*B*)
		$3p\,[5/2]_3\,(2p_9)$	18.56	19.4 (*B*)
		$3p\,[5/2]_2\,(2p_8)$	18.58	19.6 (*B*)
		$3p\,[3/2]_1\,(2p_7)$	18.61	19.5 (*B*)

Table 7.4 (continued)

Atomic number Z	Element (ground-state term)	Excited state	Excitation energy [eV] for the centre of gravity of the multiplet	Lifetime τ [ns]
1	2	3	4	5
		$3p\,[3/2]_2\ (2p_6)$	18.64	19.4 (*B*)
		$3p'\,[3/2]_1\ (2p_5)$	18.69	19.1 (*B*)
		$3p'\,[3/2]_2\ (2p_4)$	18.70	18.7 (*B*)
		$3p\,[1/2]_0\ (2p_3)$	18.71	17.4 (*B*)
		$3p'\,[1/2]_1\ (2p_2)$	18.73	18.3 (*B*)
		$3p'\,[1/2]_0\ (2p_1)$	18.97	14.3 (*B*)
11	$\mathrm{Na}\,(3s - {}^2S_{1/2})$	$3p\,({}^2P^o_{1/2})$	2.103	16.4 (*A*)
		$3p\,({}^2P^o_{3/2})$	2.104	16.3 (*A*)
		$4s\,({}^2S_{1/2})$	3.19	38 (*D*)
		$3d\,({}^2D)$	3.62	20 (*C*)
		$4p\,({}^2P^o)$	3.75	110 (*C*)
		$5s\,({}^2S_{1/2})$	4.12	75 (*C*)
		$4d\,({}^2D)$	4.28	54 (*C*)
		$4f\,({}^2F^o)$	4.29	70 (*D*)
		$5p\,({}^2P^o)$	4.34	360 (*C*)
		$6s\,({}^2S_{1/2})$	4.51	150 (*C*)
		$5d\,({}^2D)$	4.59	110 (*C*)
		$5f\,({}^2F^o)$	4.59	135 (*D*)
		$6p\,({}^2P^o)$	4.62	860 (*C*)
		$7s\,({}^2S_{1/2})$	4.71	270 (*C*)
		$6d\,({}^2D)$	4.76	190 (*D*)
		$6f\,({}^2F^o)$	4.76	235 (*D*)
		$7p\,({}^2P^o)$	4.78	1500 (*C*)
		$7d\,({}^2D)$	4.86	320 (*D*)
12	$\mathrm{Mg}\,(3s^2 - {}^1S_0)$	$3p\,({}^1P^o_1)$	4.35	2.1 (*C*)
		$4s\,({}^3S_1)$	5.11	10 (*C*)
		$4s\,({}^1S_0)$	5.39	47 (*C*)
		$3d\,({}^1D_2)$	5.75	80 (*C*)
		$3d\,({}^3D)$	5.95	6.0 (*C*)
		$4p\,({}^1P^o_1)$	6.12	9.7 (*B*)
		$5s\,({}^1S_0)$	6.52	100 (*C*)
		$4d\,({}^1D_2)$	6.59	55 (*C*)
		$4d\,({}^3D)$	6.72	16 (*C*)
		$6s\,({}^3S_1)$	6.97	52 (*C*)
		$5d\,({}^1D_2)$	6.98	50 (*C*)
		$5d\,({}^3D)$	7.06	34 (*C*)
		$3p^2\,({}^3P^o)$	7.17	2.2 (*C*)
		$6d\,({}^3D)$	7.25	56 (*C*)
		$7d\,({}^3D)$	7.35	92 (*C*)
13	$\mathrm{Al}\,(3s^2 3p - {}^2P^o_{1/2})$	$4s\,({}^2S_{1/2})$	3.14	6.8 (*B*)
		$3d\,({}^2D)$	4.02	13.4 (*B*)

Table 7.4 (continued)

Atomic number Z	Element (ground-state term)	Excited state	Excitation energy [eV] for the centre of gravity of the multiplet	Lifetime τ [ns]
1	2	3	4	5
14	Si $(3p^2-{}^3P_0)$	$4s\,({}^3P^o)$	4.942	5.9 (C)
		$4s\,({}^1P_1^o)$	5.082	4.1 (C)
		$3s\,3p^3\,({}^3D^o)$	5.617	21 (C)
		$3d\,({}^1D_2^o)$	5.871	21 (C)
		$3d\,({}^1P_1^o)$	6.619	8.1 (C)
		$3d\,({}^1F_3^o)$	6.616	3.0 (D)
15	P $(3p^3-{}^4S_{3/2}^o)$	$4s\,({}^4P)$	6.97	4.0 (D)
		$3d\,({}^2F)$	8.75	4 (D)
16	S $(3p^4-{}^3P_2)$	$4s'\,({}^3D^o)$	8.41	11 (D)
		$3s\,3p^5\,({}^3P^o)$	8.95	2.8 (D)
17	Cl $(3p^5-{}^2P_{3/2}^o)$	$4s\,({}^2P)$	9.23	1.0 (D)
		$4s'\,({}^2D)$	10.43	4.1 (D)
18	Ar $(3p^6-{}^1S_0)$	$4s\,[3/2]_1^o\,(1s_4)$	11.62	10 (D)
		$4s'\,[1/2]_1^o\,(1s_2)$	11.83	2.0 (D)
		$4p\,[1/2]_1\,(2p_{10})$	12.91	42 (C)
		$4p\,[5/2]_3\,(2p_9)$	13.08	31 (C)
		$4p\,[5/2]_2\,(2p_8)$	13.09	31 (C)
		$4p\,[3/2]_1\,(2p_7)$	13.15	29 (C)
		$4p\,[3/2]_2\,(2p_6)$	13.17	26 (C)
		$4p\,[1/2]_0\,(2p_5)$	13.27	21 (C)
		$4p'\,[3/2]_1\,(2p_4)$	13.28	29 (C)
		$4p'\,[3/2]_2\,(2p_3)$	13.30	26 (C)
		$4p'\,[1/2]_1\,(2p_2)$	13.33	26 (C)
		$4p'\,[1/2]_0\,(2p_1)$	13.48	21 (C)
		$5p\,[1/2]_1\,(3p_{10})$	14.46	200 (D)
		$5p\,[5/2]_3\,(3p_9)$	14.50	160 (D)
		$5p\,[5/2]_2\,(3p_8)$	14.51	150 (D)
		$5p\,[3/2]_1\,(3p_7)$	14.52	195 (C)
		$5p\,[3/2]_2\,(3p_6)$	14.53	220 (C)
		$5p\,[1/2]_0\,(3p_5)$	14.58	100 (D)
		$5p'\,[3/2]_1\,(3p_4)$	14.68	–
		$5p'\,[1/2]_1\,(3p_3)$	14.69	200 (D)
		$5p'\,[3/2]_2\,(3p_2)$	14.69	180 (C)
		$5p'\,[1/2]_0\,(3p_1)$	14.74	107 (C)
19	K $(4s-{}^2S_{1/2})$	$4p\,({}^2P^o)$	1.61	27 (C)
		$5s\,({}^2S_{1/2})$	2.61	46 (D)
		$3d\,({}^2D)$	2.67	42 (C)
		$5p\,({}^2P^o)$	3.06	130 (C)
		$4d\,({}^2D)$	3.40	280 (D)
		$6s\,({}^2S_{1/2})$	3.40	75 (D)
		$4f\,({}^2F^o)$	3.49	65 (D)

Table 7.4 (continued)

Atomic number Z	Element (ground-state term)	Excited state	Excitation energy [eV] for the centre of gravity of the multiplet	Lifetime τ [ns]
1	2	3	4	5
		$6p\,(^2P^o)$	3.60	310 (C)
		$5d\,(^2D)$	3.74	720 (D)
		$7s\,(^2S_{1/2})$	3.75	160 (D)
		$5f\,(^2F^o)$	3.80	120 (D)
		$7p\,(^2P^o)$	3.85	570 (C)
		$6d\,(^2D)$	3.93	1100 (D)
		$6f\,(^2F^o)$	3.96	190 (D)
20	Ca $(4s^2-{}^1S_0)$	$4p\,(^3P_1^o)$	1.89	$4\cdot10^5$ (D)
		$4p\,(^1P_1^o)$	2.93	4.6 (C)
		$5s\,(^3S_1)$	3.91	11 (C)
		$3d\,4p\,(^1D_2^o)$	4.44	17 (C)
		$5p\,(^1P_1^o)$	4.55	19 (C)
		$4d\,(^1D_2)$	4.62	80 (D)
		$4d\,(^3D_1)$	4.68	12 (C)
		$3d\,4p\,(^1F_3^o)$	5.03	60 (C)
		$6s\,(^1S_0)$	5.05	13 (C)
		$6p\,(^1P_1^o)$	5.17	15 (C)
		$4p^2\,(^1S_0)$	5.18	89 (B)
		$4f\,(^3F_3^o)$	5.23	42 (C)
		$5d\,(^1D_2)$	5.32	25 (B)
21	Sc $(3d\,4s^2-{}^2D_{3/2})$	$4s\,4p\,-\,z\,^4F^o_{3/2}$	1.943	$2\cdot10^3$ (D)
		$4s\,4p\,-\,z\,^2D^o_{5/2}$	1.987	420 (C)
		$z\,^2D^o_{3/2}$	1.996	510 (C)
		$4s\,4p\,-\,z\,^4D^o_{5/2}$	2.001	1000 (C)
		$4s^2\,4p\,-\,z\,^2P^o$	2.332	1500 (C)
		$4s\,4p\,-\,z\,^2F^o_{5/2}$	2.608	700 (C)
		$z\,^2F^o_{3/2}$	2.614	1000 (C)
		$4s\,4p\,-\,y\,^2P^o$	3.057	22 (C)
		$4s\,4p\,-\,y\,^2D^o$	3.094	5.7 (C)
		$4s\,4p\,-\,y\,^2F^o$	3.182	5.5 (C)
22	Ti $(3d^2\,4s^2-{}^3F_2)$	$y\,^3F_2^o$	3.113	23 (D)
		$y\,^3F_3^o$	3.129	20 (C)
		$y\,^3F_4^o$	3.148	21 (C)
		$y\,^3D_2^o$	3.154	18 (C)
		$y\,^3D_3^o$	3.179	18 (C)
		$y\,^5G_5^o$	3.319	15 (D)
		$y\,^3G_3^o$	3.409	10 (C)
		$y\,^3G_5^o$	3.441	10.5 (C)
23	V $(3d^3\,4s^2-{}^4F_{3/2})$	$z\,^4G^o_{7/2}$	2.723	12 (D)
		$z\,^4G^o_{11/2}$	2.767	12 (D)
		$z\,^4F^o_{9/2}$	2.916	7 (D)

Table 7.4 (continued)

Atomic number Z	Element (ground-state term)	Excited state	Excitation energy [eV] for the centre of gravity of the multiplet	Lifetime τ [ns]
1	2	3	4	5
		$y\,^6F^o_{7/2}$	3.099	11 (*D*)
		$y\,^6D^o_{9/2}$	3.315	9.0 (*C*)
		$x\,^4G^o$	3.930	6 (*D*)
		$z\,^4I^o_{11/2}$	4.627	15 (*D*)
		$z\,^4I^o_{13/2}$	4.638	14 (*D*)
		$z\,^4I^o_{15/2}$	4.652	12 (*C*)
		$y\,^4H^o_{9/2}$	4.652	12 (*D*)
		$x\,^4H^o_{13/2}$	4.771	9 (*D*)
		$y\,^2I^o_{11/2}$	4.836	7.4 (*D*)
		$w\,^4H^o_{13/2}$	5.026	9 (*D*)
24	Cr $(3d^5 4s - {}^7S_3)$	$4p - z\,^7P^o_2$	2.889	32 (*B*)
		$z\,^7P^o_3$	2.900	32 (*B*)
		$z\,^7P^o_4$	2.914	31.6 (*A*)
		$4p - z\,^5P^o_3$	3.321	17.5 (*C*)
		$z\,^5P^o_2$	3.322	16.7 (*C*)
		$z\,^5P^o_1$	3.323	17.3 (*C*)
		$4s\,4p - y\,^7P^o_2$	3.438	6.3 (*C*)
		$y\,^7P^o_3$	3.449	7.1 (*B*)
		$y\,^7P^o_4$	3.464	6.9 (*B*)
		$4s\,4p - y\,^5P^o_1$	3.648	73 (*C*)
		$y\,^5P^o_2$	3.668	71 (*C*)
		$y\,^5P^o_3$	3.698	63 (*C*)
		$4d - e\,^7D_{4,5}$	5.240	16 (*C*)
		$6s - f\,^7S_3$	5.659	43 (*C*)
		$4p - y\,^5H^o$	5.663	9.5 (*D*)
		$4p - y\,^3F^o$	5.705	11 (*C*)
		$4s\,5s - f\,^7D$	5.796	9.5 (*C*)
25	Mn $(3d^5 4s^2 - {}^6S_{5/2})$	$4p - z\,^6P^o_{7/2}$	3.075	55 (*C*)
		$5s - e\,^8S_{7/2}$	4.889	8.3 (*C*)
		$5s - e\,^6S_{5/2}$	5.133	18 (*C*)
		$3d^6 4p - z\,^6F^o$	5.390	17.5 (*C*)
		$3d^6 4p - z\,^4F^o$	5.520	15.5 (*C*)
		$3d^6 4p - z\,^4D^o$	5.693	12.5 (*C*)
		$4d - e\,^6D$	5.854	16.5 (*C*)
26	Fe $(3d^6 4s^2 - {}^5D_4)$	$z\,^5D^o_4$	3.211	81 (*C*)
		$z\,^5D^o$	3.242	83 (*C*)
		$z\,^5F^o_5$	3.332	62 (*C*)
		$z\,^5F^o$	3.375	64 (*C*)
		$z\,^5P^o_3$	3.603	44 (*B*)
		$y\,^5D^o_4$	4.103	6.2 (*C*)
		$z\,^3P^o_2$	4.209	37 (*C*)
		$y\,^5F^o$	4.230	8.2 (*C*)

Table 7.4 (continued)

Atomic number Z	Element (ground-state term)	Excited state	Excitation energy [eV] for the centre of gravity of the multiplet	Lifetime τ [ns]
1	2	3	4	5
		$z\,^3P^o_1$	4.260	94 (C)
		$z\,^3P^o_0$	4.284	110 (C)
		$z\,^5G^o$	4.357	11 (C)
		$y\,^5P^o_2$	4.607	7.7 (C)
		$e\,^5G_6$	6.264	13 (C)
27	Co ($3d^7 4s^2 - {}^4F_{9/2}$)	$4s4p - z\,^4F^o_{9/2}$	3.514	63 (C)
		$z\,^4F^o_{7/2}$	3.568	77 (C)
		$z\,^4F^o_{5/2}$	3.622	73 (C)
		$4s4p - z\,^4D^o_{7/2}$	3.632	27 (C)
		$z\,^4D^o_{5/2}$	3.713	31 (C)
		$3d^8 4p - y\,^4G^o$	4.068	8.4 (C)
		$3d^8 4p - y\,^4F^o$	4.120	7.2 (C)
		$3d^8 4p - y\,^2G^o$	4.184	9.0 (C)
		$3d^8 4p - y\,^2F^o$	4.442	6.5 (C)
28	Ni ($3d^8 4s^2 - {}^3F_4$)	$z\,^3P^o$	3.603	10 (C)
		$z\,^5F^o_4$	3.606	30 (C)
		$z\,^3F^o_3$	3.635	12 (C)
		$z\,^3F^o_4$	3.655	15 (C)
		$z\,^3D^o_2$	3.706	17 (C)
		$z\,^3F^o_2$	3.796	14 (C)
		$z\,^1F^o_3$	3.847	11 (C)
		$z\,^1D^o_2$	3.898	16 (C)
		$y\,^3F^o_4$	4.088	16 (C)
		$y\,^3F^o_3$	4.105	9.1 (C)
		$y\,^3F^o_2$	4.167	8.6 (C)
		$y\,^1D^o_2$	4.538	7.4 (C)
29	Cu ($3d^{10} 4s - {}^2S_{1/2}$)	$4p\,(^2P^o_{1/2})$	3.786	7 (D)
		$4p\,(^2P^o_{3/2})$	3.817	7.2 (C)
		$4s4p'\,(^4P^o_{3/2})$	4.92	320 (B)
		$5s\,(^2S_{1/2})$	5.35	22 (D)
		$4s4p'\,(^4D^o_{3/2})$	5.47	370 (C)
		$4s4p'\,(^2F^o_{7/2})$	5.51	420 (C)
		$4s4p'\,(^2P^o_{3/2})$	5.69	12 (C)
		$4s4p'\,(^2D^o_{3/2})$	5.76	20 (C)
		$4s4p'\,(^2D^o_{5/2})$	5.76	170 (C)
		$5p\,(^2P^o_{3/2})$	6.12	35 (D)
		$4d\,(^2D)$	6.19	13 (C)
		$6s\,(^2S_{1/2})$	6.55	50 (C)
		$5d\,(^2D)$	6.87	28 (C)
		$7s\,(^2S_{1/2})$	7.03	93 (C)
		$6d\,(^2D)$	7.18	53 (C)

Table 7.4 (continued)

Atomic number Z	Element (ground-state term)	Excited state	Excitation energy [eV] for the centre of gravity of the multiplet	Lifetime τ [ns]
1	2	3	4	5
30	Zn $(3d^{10}4s^2 - {}^1S_0)$	$4s\,4p\,({}^1P_1^o)$	5.796	1.40 (B)
		$4s\,5s\,({}^3S_1)$	6.655	8.1 (C)
		$4s\,5s\,({}^1S_0)$	6.917	41 (A)
		$4s\,4d\,({}^1D_2)$	7.744	22 (B)
		$4s\,4d\,({}^3D)$	7.783	6.3 (C)
		$4s\,6s\,({}^3S_1)$	8.113	22 (C)
		$4s\,6s\,({}^1S_0)$	8.188	103 (B)
		$4s\,5d\,({}^1D_2)$	8.473	86 (C)
		$4s\,5d\,({}^3D)$	8.503	14 (C)
31	Ga $(4s^2 4p - {}^2P_{1/2}^o)$	$5s\,({}^2S_{1/2})$	3.073	6.2 (C)
		$5p\,({}^2P^o)$	4.106	50 (C)
		$4d\,({}^2D_{3/2})$	4.312	4.7 (C)
		$4d\,({}^2D_{5/2})$	4.313	5.8 (C)
		$6s\,({}^2S_{1/2})$	4.660	28 (C)
32	Ge $(4p^2 - {}^3P_0)$	$5s\,({}^3P_0^o)$	4.643	3.6 (D)
		$5s\,({}^3P_1^o)$	4.675	3.7 (D)
		$5s\,({}^3P_2^o)$	4.850	4.0 (D)
		$5s\,({}^1P_1^o)$	4.962	4.0 (D)
33	As $(4p^3 - {}^4S_{3/2}^o)$	$4p^2\,({}^3P)\,5s\,({}^4P)$	6.460	3.7 (C)
		$5s\,({}^2P)$	6.710	3.4 (C)
		$4p^2\,({}^1D)\,5s'\,({}^2D_{5/2})$	7.540	2.9 (C)
34	Se $(4p^4 - {}^3P_2)$	$4p^3\,({}^4S^o)\,5s\,({}^3S_1^o)$	6.323	1.7 (D)
		$4p^3\,({}^2D^o)\,5s'\,({}^1D_2^o)$	7.870	1.9 (D)
		$4p^3\,({}^2P^o)\,5s''\,({}^3P_2^o)$	8.885	2.6 (D)
35	Br $(4p^5 - {}^2P_{3/2}^o)$	$5p\,({}^4S_{3/2}^o)$	9.754	43 (C)
36	Kr $(4p^6 - {}^1S_0)$	$5s\,[3/2]_1^o\,(1s_4)$	9.24	4.4 (C)
		$5s'\,[1/2]_1^o\,(1s_2)$	9.83	4.5 (C)
		$5p\,[1/2]_1\,(2p_{10})$	10.13	32 (C)
		$5p\,[5/2]_3\,(2p_9)$	10.20	30 (C)
		$5p\,[5/2]_2\,(2p_8)$	10.34	30 (C)
		$5p\,[3/2]_1\,(2p_7)$	10.48	31 (C)
		$5p\,[3/2]_2\,(2p_6)$	10.51	25.5 (C)
		$5p\,[1/2]_0\,(2p_5)$	10.54	23 (C)
		$5p'\,[3/2]_1\,(2p_4)$	10.56	28 (C)
		$5p'\,[1/2]_1\,(2p_3)$	10.63	26 (C)
		$5p'\,[3/2]_2\,(2p_2)$	10.69	28 (C)
		$5p'\,[1/2]_0\,(2p_1)$	10.72	23 (C)
		$6p\,[5/2]_2\,(3p_8)$	12.79	160 (D)
		$6p\,[3/2]_1\,(3p_7)$	12.809	145 (D)
		$6p\,[3/2]_2\,(3p_6)$	12.815	115 (C)
		$6p'\,[3/2]_2\,(3p_2)$	13.46	125 (C)

Table 7.4 (continued)

Atomic number Z	Element (ground-state term)	Excited state	Excitation energy [eV] for the centre of gravity of the multiplet	Lifetime τ [ns]
1	2	3	4	5
37	Rb ($5s - {}^2S_{1/2}$)	$5p\,({}^2P^o_{1/2})$	1.560	28.5 (*C*)
		$5p\,({}^2P^o_{3/2})$	1.589	26.5 (*C*)
		$4d\,({}^2D)$	2.40	85 (*C*)
		$6s\,({}^2S_{1/2})$	2.496	46 (*C*)
		$6p\,({}^2P^o_{1/2})$	2.940	125 (*C*)
		$6p\,({}^2P^o_{3/2})$	2.950	112 (*B*)
		$5d\,({}^2D)$	3.19	230 (*C*)
		$7s\,({}^2S_{1/2})$	3.262	90 (*C*)
		$4f\,({}^2F^o)$	3.32	59 (*D*)
		$7p\,({}^2P^o_{1/2})$	3.451	270 (*C*)
		$7p\,({}^2P^o_{3/2})$	3.455	250 (*C*)
		$6d\,({}^2D)$	3.56	270 (*C*)
		$8s\,({}^2S_{1/2})$	3.601	155 (*C*)
		$5f\,({}^2F^o)$	3.63	100 (*D*)
		$7d\,({}^2D)$	3.75	360 (*C*)
		$6f\,({}^2F^o)$	3.80	165 (*D*)
38	Sr ($5s^2 - {}^1S_0$)	$5p\,({}^3P^o_1)$	1.80	$2.1 \cdot 10^4$ (*C*)
		$5p\,({}^1P^o_1)$	2.690	6.2 (*D*)
		$5d\,({}^3D)$	4.343	16.3 (*A*)
		$5p^2\,({}^3P_2)$	4.423	7.9 (*A*)
		$5p^2\,({}^1D_2)$	4.583	9.5 (*C*)
		$4f\,({}^1F^o_3)$	4.902	34 (*B*)
39	Y ($4d\,5s^2 - {}^2D_{3/2}$)	$5s\,5p - z\,{}^2D^o_{5/2}$	1.992	130 (*C*)
		$z\,{}^2D^o_{3/2}$	2.002	170 (*C*)
		$5s\,5p - z\,{}^2F^o_{5/2}$	2.669	39 (*B*)
		$z\,{}^2F^o_{7/2}$	2.717	62 (*B*)
		$5s\,5p - y\,{}^2D^o_{3/2}$	2.992	5.5 (*C*)
		$y\,{}^2P^o_{3/2}$	3.035	15.5 (*B*)
		$y\,{}^2F^o_{5/2}$	3.040	6.4 (*C*)
		$y\,{}^2P^o_{1/2}$	3.062	14 (*C*)
		$y\,{}^2D^o_{5/2}$	3.068	6.1 (*C*)
		$y\,{}^2F^o_{7/2}$	3.087	6.7 (*B*)
40	Zr ($4d^2\,5s^2 - {}^3P_2$)	$5s\,5p - z\,{}^3G^o_3$	2.709	120 (*B*)
		$z\,{}^3G^o_4$	2.746	110 (*B*)
		$z\,{}^3G^o_5$	2.798	107 (*B*)
		$4d^3\,5p - y\,{}^5G^o$	3.293	13 (*C*)
		$4d^3\,5p - y\,{}^3G^o_3$	3.190	16 (*B*)
		$y\,{}^3G^o_4$	3.225	21.5 (*B*)
		$y\,{}^3G^o_5$	3.277	27.6 (*B*)
41	Nb ($4d^4\,5s - {}^6D_{1/2}$)	$5s\,5p - z\,{}^6F^o_{7/2}$	2.469	550 (*C*)
		$z\,{}^6F^o_{9/2}$	2.533	1050 (*C*)
		$5s\,5p - z\,{}^6D^o$	2.513	160 (*C*)

Table 7.4 (continued)

Atomic number Z	Element (ground-state term)	Excited state	Excitation energy [eV] for the centre of gravity of the multiplet	Lifetime τ [ns]
1	2	3	4	5
		$5s\,5p - z\,^4D^o$	2.597	95 (C)
		$4d^4\,5p - y\,^6F^o$	3.098	8.0 (C)
		$4d^4\,5p - x\,^6D^o$	3.393	7.5 (C)
		$5s\,5p - y\,^6D^o_{9/2}$	3.400	15 (C)
42	Mo $(4d^5\,5s - {}^7S_3)$	$5p - z\,^7P^o$	3.224	16 (C)
		$5p - z\,^5P^o$	3.577	21 (C)
		$5s\,5p - z\,^7D^o_2$	3.863	36 (C)
		$5s\,5p - y\,^7P^o$	3.923	7.0 (D)
		$5s\,5p - z\,^7D^o_3$	3.925	21 (C)
		$z\,^7D^o_4$	3.983	120 (C)
44	Ru $(4d^7\,5s - {}^5F_5)$	$4d^7\,(^4F)\,5p - z\,^5D^o$	3.415	16 (C)
		$5p - z\,^5F^o$	3.491	12 (C)
		$5p - z\,^3G^o_5$	3.533	12 (C)
		$z\,^3G^o_4$	3.706	16 (C)
		$5p - z\,^5G^o$	3.734	10 (C)
		$5p - z\,^3F^o_4$	3.763	12 (C)
		$z\,^3F^o_3$	4.016	10 (C)
45	Rh $(4d^8\,5s - {}^4F_{9/2})$	$4d^8\,(^3F)\,5p - z\,^4D^o$	3.556	10.3 (C)
		$5p - z\,^4G^o_{9/2}$	3.539	10.1 (C)
		$z\,^4G^o_{7/2}$	3.856	8.6 (C)
		$5p - z\,^4F^o$	3.769	9.1 (C)
		$5p - z\,^2G^o_{9/2}$	3.920	8.2 (C)
		$z\,^2G^o_{7/2}$	4.097	9.2 (C)
		$5p - z\,^2F^o_{7/2}$	3.968	8.7 (C)
		$z\,^2F^o_{5/2}$	4.209	7.3 (C)
		$5p - z\,^2D^o_{5/2}$	3.973	8.1 (C)
		$z\,^2D^o_{3/2}$	4.199	7.0 (C)
		$4d^8\,(^3P)\,5p - z\,^4P^o$	4.391	12 (C)
46	Pd $(4d^{10} - {}^1S_0)$	$4d^9\,(^2D_{5/2})\,5p\,(^3P^o_2)$	4.224	7 (D)
		$(^3P^o_1)$	4.486	7.5 (C)
		$5p\,(^3F^o_4)$	4.455	7.1 (C)
		$5p\,(^3D^o_3)$	4.636	7.0 (C)
		$4d^9\,(^2D_{3/2})\,5p'\,(^3D^o_1)$	5.005	4.9 (C)
		$5p'\,(^1P^o_1)$	5.063	5.0 (C)
47	Ag $(4d^{10}\,5s - {}^2S_{1/2})$	$5p\,(^2P^o_{1/2})$	3.664	7.9 (C)
		$5p\,(^2P^o_{3/2})$	3.778	6.7 (B)
		$5d\,(^2D)$	6.05	13 (C)
		$7s\,(^2S_{1/2})$	6.43	41 (C)
		$6d\,(^2D)$	6.72	30 (D)
		$8s\,(^2S_{1/2})$	6.89	140 (C)
		$7d\,(^2D)$	7.03	61 (C)
		$8d\,(^2D)$	7.20	110 (C)

Table 7.4 (continued)

Atomic number Z	Element (ground-state term)	Excited state	Excitation energy [eV] for the centre of gravity of the multiplet	Lifetime τ [ns]
1	2	3	4	5
48	Cd $(5s^2 - {}^1S_0)$	$5p\,({}^1P_1^o)$	5.42	1.7 (*D*)
		$6s\,({}^3S_1)$	6.38	8.2 (*C*)
		$5d\,({}^1D_2)$	7.34	20 (*D*)
		$5d\,({}^3D)$	7.38	6.4 (*C*)
		$7s\,({}^3S_1)$	7.76	22 (*C*)
		$6d\,({}^3D)$	8.10	16 (*C*)
		$8s\,({}^3S_1)$	8.27	48 (*C*)
		$7d\,({}^3D)$	8.43	31 (*C*)
		$8d\,({}^3D)$	8.60	56 (*C*)
49	In $(5p - {}^2P_{1/2}^o)$	$6s\,({}^2S_{1/2})$	3.022	7.4 (*C*)
		$6p\,({}^2P^o)$	3.969	55 (*C*)
		$5d\,({}^2D)$	4.080	7.0 (*C*)
		$7s\,({}^2S_{1/2})$	4.501	20 (*C*)
50	Sn $(5p^2 - {}^3P_0)$	$5p\,({}^2P_{1/2}^o)\,6s\,({}^3P_0^o)$	4.295	6.0 (*D*)
		$6s\,({}^3P_1^o)$	4.329	4.8 (*C*)
		$5p\,({}^2P_{3/2}^o)\,6s'\,({}^3P_2^o)$	4.789	4.4 (*C*)
		$6s'\,({}^1P_1^o)$	4.867	4.5 (*C*)
		$5p\,({}^2P_{1/2}^o)\,5d\,({}^3D_2^o)$	5.473	5.3 (*C*)
		$5d\,({}^3D_1^o)$	5.518	4.1 (*C*)
		$5d\,({}^3D_3^o)$	5.527	5.8 (*C*)
51	Sb $(5p^3 - {}^4S_{3/2}^o)$	$5p^2\,({}^3P)\,6s\,({}^4P_{1/2})$	5.36	5.0 (*C*)
		$6s\,({}^4P_{3/2})$	5.70	5.0 (*C*)
		$6s\,({}^4P_{5/2})$	5.99	4.8 (*C*)
		$6s\,({}^2P_{3/2})$	6.12	4.0 (*C*)
		$5p^2\,({}^1D)\,6s'\,({}^2D_{3/2})$	6.85	3.7 (*D*)
		$6s'\,({}^2D_{5/2})$	6.91	3.8 (*C*)
		$5p^2\,({}^3P)\,7s\,({}^4P_{1/2})$	7.141	12 (*C*)
		$7s\,({}^4P_{3/2})$	7.559	10 (*C*)
		$7s\,({}^4P_{5/2})$	7.875	10 (*C*)
		$7s\,({}^2P_{1/2})$	7.611	7.9 (*C*)
		$7s\,({}^2P_{3/2})$	7.663	8.4 (*C*)
		$5p^2\,({}^1S)\,6s''\,({}^2S_{1/2})$	8.140	24 (*C*)
		$5p^2\,({}^3P)\,8s\,({}^4P_{5/2})$	8.512	20 (*C*)
52	Te $(5p^4 - {}^3P_2)$	$6s\,({}^5S_2^o)$	5.487	72 (*C*)
		$6s\,({}^3S_1^o)$	5.784	2.4 (*D*)
53	I $(5p^5 - {}^2P_{3/2}^o)$	$6s\,({}^4P_{5/2})$	6.774	58 (*C*)
		$6s\,({}^4P_{3/2})$	7.665	58 (*C*)
		$6s\,({}^2P_{3/2})$	6.955	3 (*D*)
		$6p\,({}^4P_{3/2}^o)$	8.315	39 (*C*)
		$6p\,({}^4D_{5/2}^o)$	8.993	42 (*C*)
		$6p\,({}^2P_{3/2}^o)$	9.058	50 (*C*)

Table 7.4 (continued)

Atomic number Z	Element (ground-state term)	Excited state	Excitation energy [eV] for the centre of gravity of the multiplet	Lifetime τ [ns]
1	2	3	4	5
54	Xe $(5p^6 - {}^1S_0)$	$6s\,[3/2]^o_1\,(1s_4)$	8.44	3.6 (*C*)
		$6s'\,[1/2]^o_1\,(1s_2)$	9.57	3.5 (*C*)
		$6p\,[1/2]_1\,(2p_{10})$	10.03	38 (*C*)
		$6p\,[5/2]_2\,(2p_9)$	10.13	37 (*C*)
		$6p\,[5/2]_3\,(2p_8)$	10.26	34 (*C*)
		$6p\,[3/2]_1\,(2p_7)$	10.34	35 (*C*)
		$6p\,[3/2]_2\,(2p_6)$	10.37	40 (*C*)
		$6p\,[1/2]_0\,(2p_5)$	10.41	35 (*C*)
		$6p'\,[3/2]_1\,(2p_4)$	10.44	42 (*C*)
		$6p'\,[3/2]_2\,(2p_3)$	10.46	35 (*C*)
		$6p'\,[1/2]_1\,(2p_2)$	10.47	37 (*C*)
		$6p'\,[1/2]_0\,(2p_1)$	10.49	33 (*C*)
		$7p\,[1/2]_1\,(3p_{10})$	10.90	150 (*C*)
		$7p\,[5/2]_2\,(3p_9)$	10.95	100 (*C*)
		$7p\,[5/2]_3\,(3p_8)$	10.97	130 (*C*)
		$7p\,[3/2]_2\,(3p_7)$	10.996	120 (*C*)
		$7p\,[3/2]_1\,(3p_6)$	11.003	130 (*C*)
		$7p\,[1/2]_0\,(3p_5)$	11.015	80 (*D*)
55	Cs $(6s - {}^2S_{1/2})$	$6p\,({}^2P^o)$	1.432	31 (*C*)
		$7s\,({}^2S_{1/2})$	2.298	48 (*D*)
		$7p\,({}^2P^o_{1/2})$	2.699	157 (*B*)
		$7p\,({}^2P^o_{3/2})$	2.721	135 (*B*)
		$6d\,({}^2D)$	2.804	60 (*C*)
		$8s\,({}^2S_{1/2})$	3.015	95 (*C*)
		$4f\,({}^2F^o)$	3.034	50 (*D*)
		$8p\,({}^2P^o_{1/2})$	3.188	330 (*C*)
		$8p\,({}^2P^o_{3/2})$	3.198	305 (*C*)
		$7d\,({}^2D)$	3.231	90 (*C*)
		$9s\,({}^2S_{1/2})$	3.337	170 (*D*)
		$5f\,({}^2F^o)$	3.344	90 (*D*)
		$8d\,({}^2D)$	3.449	150 (*C*)
56	Ba $(6s^2 - {}^1S_0)$	$6p\,({}^1P^o_1)$	2.239	8.5 (*C*)
		$5d\,6p\,({}^3D^o_1)$	3.00	17 (*B*)
		$5d\,6p\,({}^3D^o_3)$	3.10	10 (*C*)
		$5d\,6p\,({}^3P^o_1)$	3.19	12 (*C*)
		$5d\,6p\,({}^1F^o_3)$	3.32	44 (*C*)
		$5d\,6p\,({}^1P^o_1)$	3.54	12.2 (*B*)
		$7p\,({}^1P^o_1)$	4.04	14 (*C*)
		$8p\,({}^1P^o_1)$	4.45	9.5 (*C*)
		$7d\,({}^1D_2)$	4.64	22 (*C*)

Table 7.4 (continued)

Atomic number Z	Element (ground-state term)	Excited state	Excitation energy [eV] for the centre of gravity of the multiplet	Lifetime τ [ns]
1	2	3	4	5
60	Nd $(4f^4 6s^2 - {}^5I_4)$	$5d^2 6s\,({}^7K^o_3)$	2.105	615 (*C*)
		$5d\,6s^2\,({}^5H^o)$	2.347	80 (*C*)
		$5d^2 6s\,({}^5H^o_4)$	2.286	320 (*C*)
		$6s\,6p\,({}^5K^o)$	2.778	14 (*C*)
64	Gd $(4f^7 5d\,6s^2 - {}^9D^o_2)$	$5d\,({}^9D^o)\,6s\,7p\,({}^9D_{2\text{–}4})$	2.212	130 (*C*)
		$({}^9D_{5,6})$	2.241	105 (*C*)
		$6s\,6p\,({}^9F_2)$	2.155	600 (*D*)
		$({}^9F_4)$	2.229	440 (*C*)
		$({}^9F_5)$	2.295	540 (*D*)
		$5d\,({}^7D^o)\,6s\,6p\,({}^9D_3)$	2.798	15.6 (*C*)
		$({}^9D_4)$	2.829	20.5 (*C*)
		$({}^9D_5)$	2.876	61 (*C*)
		$({}^9D_6)$	2.880	54 (*C*)
66	Dy $(4f^{10} 6s^2 - {}^5I_8)$	$6s\,6p\,({}^3P^o_1)\,({}^7I^o_9)$	1.98	1200 (*C*)
		$4f^9\,({}^6H^o)\,5d\,6s^2\,({}^7K^o_9)$	2.19	2000 (*C*)
		$6s\,6p\,({}^3P^o_2)\,({}^5K^o_9)$	2.20	2700 (*D*)
		$4f^9\,({}^6F^o)\,5d\,6s^2\,({}^7H^o_7)$	2.34	1000 (*C*)
		$({}^5H^o_7)$	2.69	115 (*C*)
		$4f^9\,({}^6H^o)\,5d\,6s^2\,({}^5K^o_7)$	2.70	73 (*C*)
		$4f^{10}\,({}^5I_7)\,6s\,6p\,({}^3P^o_2)\,({}^7K^o_9)$	2.71	500 (*D*)
		$4f^{10}\,({}^5I_7)\,6s\,6p\,({}^3P^o_2)\,({}^7I^o_8)$	2.72	1200 (*C*)
		$6s\,6p\,({}^1P^o_1)\,({}^5K^o_9)$	2.94	4.8 (*C*)
		$4f^9\,({}^6H^o)\,5d^2\,({}^3F)\,({}^8K^o)\,6s\,({}^9K^o_8)$	2.95	12 (*C*)
		$6s\,6p\,({}^1P^o_1)\,({}^5I^o_8)$	2.96	8 (*C*)
67	Ho $(4f^{11} 6s^2 - {}^4I^o_{15/2})$	$4f^{10}\,5d\,6s^2\,({}^4I_{13/2})$	2.51	400 (*C*)
		$({}^4K_{13/2})$	2.85	109 (*C*)
		$({}^4L_{15/2})$	2.91	250 (*C*)
		$4f^{11}\,6s\,6p\,({}^6K_{17/2})$	2.91	102 (*C*)
		$6s\,6p\,(13/2,\,2)_{13/2}$	2.97	38 (*C*)
		$(15/2,\,1)_{13/2}$	2.98	11 (*C*)
		$4f^{10}\,5d^2\,6s\,({}^8I_{17/2})$	3.01	70 (*C*)
		$4f^{11}\,6s\,6p\,({}^4K_{17/2})$	3.02	6.3 (*C*)
		$4f^{11}\,6s\,6p\,({}^4I_{15/2})$	3.06	6.0 (*C*)
		$4f^{10}\,5d\,6s^2\,(J = 13/2)$	3.07	25 (*C*)
		$4f^{10}\,5d\,6s^2\,({}^6H_{15/2})$	3.13	155 (*C*)
70	Yb $(4f^{14} 6s^2 - {}^1S_0)$	$6s\,6p\,({}^3P^o_1)$	2.231	850 (*D*)
72	Hf $(5d^2 6s^2 - {}^3F_2)$	$6s\,6p - z\,{}^5G^o_2$	2.233	1100 (*C*)
		$z\,{}^5G^o_4$	2.599	2200 (*D*)
		$6s\,6p - z\,{}^5F^o_1$	2.695	290 (*C*)
		$z\,{}^5F^o_2$	2.784	370 (*C*)
		$z\,{}^5F^o_4$	3.073	410 (*C*)

Table 7.4 (continued)

Atomic number Z	Element (ground-state term)	Excited state	Excitation energy [eV] for the centre of gravity of the multiplet	Lifetime τ [ns]
1	2	3	4	5
		$6s\,6p - z\,^1F_3^o$	2.932	120 (C)
		$6s\,6p - z\,^5D_2^o$	3.178	103 (C)
		$z\,^5D_3^o$	3.262	80 (C)
		$z\,^5D_4^o$	3.412	170 (C)
		$6s\,6p - z\,^3S_1^o$	3.478	41 (C)
73	Ta ($5d^3\,6s^2 - {}^4F_{3/2}$)	$6s^2\,(a\,^3F)\,6p - z\,^4D_{1/2}^o$	2.294	570 (C)
		$z\,^4D_{5/2}^o$	2.624	910 (C)
		$6s^2\,(a\,^3P)\,6p - z\,^2D_{3/2}^o$	2.575	680 (C)
		$5d^3\,6s\,(a\,^5F)\,6p - z\,^4F_{3/2}^o$	2.710	450 (C)
		$z\,^4F_{5/2}^o$	2.897	390 (C)
		$6p - z\,^6F_{7/2}^o$	3.296	260 (C)
		$z\,^6F_{9/2}^o$	3.439	265 (C)
		$z\,^6F_{11/2}^o$	3.764	300 (C)
		$6p - z\,^6D_{3/2}^o$	3.067	170 (C)
		$z\,^6D_{5/2}^o$	3.322	110 (C)
		$z\,^6D_{7/2}^o$	3.444	120 (C)
		$z\,^6D_{9/2}^o$	3.567	140 (C)
74	W ($5d^4\,6s^2 - {}^5D_0$)	$6s\,6p - z\,^7D_1^o$	2.660	275 (C)
		$z\,^7D_2^o$	2.971	250 (C)
		$z\,^7D_3^o$	3.247	160 (C)
		$z\,^7D_4^o$	3.570	190 (C)
		$z\,^7D_5^o$	3.691	700 (C)
		$z\,^5F_1^o$	3.222	820 (C)
		$z\,^5F_2^o$	3.430	180 (C)
		$z\,^5F_3^o$	3.613	260 (C)
		$z\,^5F_4^o$	3.897	480 (C)
		$z\,^7P_2^o$	3.252	74 (C)
		$z\,^7P_3^o$	3.408	86 (C)
		$z\,^7P_4^o$	3.458	63 (C)
		$z\,^5P_1^o$	3.496	130 (C)
		$z\,^5P_2^o$	3.644	80 (C)
		$z\,^5P_3^o$	3.792	60 (C)
75	Re ($5d^5\,6s^2 - {}^6S_{5/2}$)	$6s\,(a\,^7S)\,6p - z\,^8P_{7/2}^o$	2.535	860 (C)
		$6p - z\,^6P_{5/2}^o$	3.577	40 (C)
		$z\,^6P_{7/2}^o$	3.582	30 (C)
		$z\,^6P_{3/2}^o$	3.591	50 (C)
77	Ir ($5d^7\,6s^2 - {}^4F_{9/2}$)	$6s\,({}^5F)\,6p - z\,^6D_{9/2}^o$	3.262	580 (C)
		$z\,^6D_{7/2}^o$	3.785	820 (C)
		$z\,^6D_{5/2}^o$	4.100	170 (C)
		$z\,^6D_{3/2}^o$	4.025	100 (C)

Table 7.4 (continued)

Atomic number Z	Element (ground-state term)	Excited state	Excitation energy [eV] for the centre of gravity of the multiplet	Lifetime τ [ns]
1	2	3	4	5
		$6s\,(^5F)\,6p - z\,^6F^o_{11/2}$	3.528	340 (*C*)
		$z\,^6F^o_{9/2}$	4.031	340 (*C*)
		$z\,^6F^o_{7/2}$	4.200	28 (*C*)
		$z\,^6F^o_{5/2}$	4.330	95 (*C*)
		$z\,^6F^o_{3/2}$	4.512	90 (*C*)
		$6s\,(^5F)\,6p - z\,^6G^o_{11/2}$	4.238	70 (*C*)
		$z\,^6G^o_{9/2}$	4.349	30 (*C*)
		$z\,^6G^o_{7/2}$	4.390	40 (*C*)
78	Pt $(5d^9\,6s - {}^3D_3)$	$5d^8\,6s\,6p - z\,^5D^o_4$	3.74	680 (*C*)
		$5d^9\,6p\,(1^o_2)$	4.04	14 (*D*)
		$5d^8\,6s\,6p\,(2^o_5)$	4.18	130 (*C*)
		$5d^9\,6p\,(3^o_3)$	4.23	17 (*D*)
		$5d^8\,6s\,6p\,(4^o_3)$	4.38	27 (*C*)
79	Au $(5d^{10}\,6s - {}^2S_{1/2})$	$6p\,(^2P^o_{1/2})$	4.63	6.0 (*C*)
		$6p\,(^2P^o_{3/2})$	5.105	4.6 (*C*)
		$5d^9\,6s\,6p\,(^4F^o_{7/2})$	5.646	83 (*C*)
		$6d\,(^2D)$	7.687	11.5 (*C*)
		$8s\,(^2S_{1/2})$	8.027	53 (*C*)
		$7d\,(^2D)$	8.368	28 (*C*)
		$8d\,(^2D_{5/2})$	8.680	49 (*C*)
80	Hg $(5d^{10}\,6s^2 - {}^1S_0)$	$6p\,(^3P^o_1)$	4.887	118 (*B*)
		$6p\,(^1P^o_1)$	6.704	–
		$7s\,(^3S_1)$	7.731	9.5 (*D*)
		$7s\,(^1S_0)$	7.926	30 (*B*)
		$7p\,(^3P^o_2)$	8.829	170 (*D*)
		$6d\,(^3D_1)$	8.845	7.2 (*C*)
		$6d\,(^3D_2)$	8.852	9.1 (*C*)
		$6d\,(^3D_3)$	8.856	8.3 (*C*)
		$6d\,(^1D_2)$	8.844	11 (*B*)
		$8s\,(^3S_1)$	9.170	21 (*C*)
		$8s\,(^1S_0)$	9.225	84 (*C*)
		$7d\,(^1D_2)$	9.555	40 (*B*)
81	Tl $(6s^2\,6p - {}^2P^o_{1/2})$	$7s\,(^2S_{1/2})$	3.283	7.6 (*C*)
		$7p\,(^2P^o_{1/2})$	4.235	61.9 (*B*)
		$7p\,(^2P^o_{3/2})$	4.359	48.4 (*B*)
		$6d\,(^2D)$	4.484	7.0 (*C*)
		$8s\,(^2S_{1/2})$	4.804	20 (*C*)
		$7d\,(^2D)$	5.212	18 (*C*)
		$9s\,(^2S_{1/2})$	5.352	43 (*C*)
		$9p\,(^2P^o)$	5.518	16.4 (*C*)
		$8d\,(^2D)$	5.540	47 (*C*)

Table 7.4 (continued)

Atomic number Z	Element (ground-state term)	Excited state	Excitation energy [eV] for the centre of gravity of the multiplet	Lifetime τ [ns]
1	2	3	4	5
82	Pb $(6p^2 - {}^3P_0)$	$7s\,({}^3P_1^o)$	4.375	5.8 (C)
		$6d\,({}^3F_2^o)$	5.634	26 (C)
		$6d\,({}^3D_2^o)$	5.711	4.2 (C)
		$6d\,({}^3D_1^o)$	5.712	3.8 (C)
		$6d\,({}^3F_3^o)$	5.744	6.1 (C)
		$7s\,({}^3P_2^o)$	5.975	5.9 (C)
		$8s\,({}^3P_1^o)$	6.036	12.7 (C)
		$7s\,({}^1P_1^o)$	6.130	5.0 (C)
83	Bi $(6p^3 - {}^4S_{3/2}^o)$	$7s\,({}^4P_{1/2})$	4.04	4.6 (C)
		$6d\,({}^2D_{3/2})$	5.44	28 (C)
		$6d\,({}^2D_{5/2})$	5.557	4 (D)
		$7s\,({}^4P_{3/2})$	5.563	6 (D)
		$7s\,({}^2P_{1/2})$	5.693	5.8 (C)
		$8s\,({}^4P_{1/2})$	5.874	8 (D)
		$7s\,({}^4P_{5/2})$	6.012	5.2 (C)
		$7s\,({}^2P_{3/2})$	6.132	5.1 (C)
		$7d\,({}^2D_{5/2})$	6.343	7.8 (C)
92	U $(5f^3 6d 7s^2 - {}^5L_6^o)$	$6d\,7s\,7p\,({}^7M_7)$	2.095	220 (C)
		$6d^2\,7p\,({}^7N_7)$	3.458	11 (C)

7.5 Energy Levels and Lifetimes for Metastable States in Atoms

Excited states of atoms, which do not decay into lower states according to the selection rules for electric dipole radiative transitions (Sect. 7.2), are referred to as metastable states.

Rigorous selection rules restricting the possible changes of state for the emitting particle arise from the momentum and parity conservation rules for the closed system "emitter + radiative field" [7.5.1]. One rule requires the initial-state angular momentum of the particle, J_i, the final-state angular momentum, J_f, and the photon angular momentum $j\,(j = 1, 2, 3\ldots)$ $(\boldsymbol{J}_i - \boldsymbol{J}_f = \boldsymbol{j})$ to be added together in such a way that they fulfil the relations

$$|J_i - J_f| \leqslant j \leqslant J_i + J_f \,,$$

$$M_i - M_f = M_j \,,$$

where M_i, M_f and M_j are the projections of the momenta $\boldsymbol{J}_i$, $\boldsymbol{J}_f$ and $\boldsymbol{j}$ on the quantization axis. In addition, the parities P_i and P_f of the initial and final states of the emitter, as well as the parity of the radiated photon $P_{\rm ph}$ must obey the condition

$$P_{\rm f}P_{\rm ph} = P_{\rm i} \quad \text{or} \quad P_{\rm i}P_{\rm f} = P_{\rm ph} \, .$$

The parity of an electric-type photon is $P_{\rm ph}\,({\rm E}j) = (-1)^j$ and the parity of a magnetic-type photon is $P_{\rm ph}\,({\rm M}j) = (-1)^{j+1}$; thus the parity selection rule for electric multipole radiation is

$$P_{\rm i}P_{\rm f} \overset{\rm E}{=} (-1)^j \, .$$

Similarly, for magnetic multipole radiation we obtain

$$P_{\rm i}P_{\rm f} \overset{\rm M}{=} (-1)^{j+1} \, .$$

It should be mentioned that the parity of the electron subsystem in the atom is $P = (-1)^{\Sigma l_i}$, where l_i is the orbital angular momentum of the ith electron.

In addition to these rigorous selection rules, there are other approximate rules which reflect the properties of a given emitter. Thus, in the common case of an atom whose terms are formed according to the LS-coupling scheme, there are additional selection rules for the most intense dipole ($j = 1$) and quadrupole ($j = 2$) radiative transitions [7.5.2]:

$$\begin{aligned}
&\text{M1: } \Delta S = \Delta L = 0;\ \Delta J = 0{,}1 \quad \text{and} \quad 0 \nleftrightarrow 0;\\
&\text{E2: } \Delta S = 0;\ \Delta L = 0, 1, 2 \quad \text{and} \quad 0 \nleftrightarrow 0, 1;\ \Delta J = 0, 1, 2\\
&\qquad\quad \text{and} \quad 0 \nleftrightarrow 0, 1, \quad \text{and} \quad 1/2 \nleftrightarrow 1/2;\\
&\text{M2: } \Delta S = 1;\ \Delta L = 0, 1 \quad \text{and} \quad 0 \nleftrightarrow 0;\ \Delta J = 0, 1, 2 \quad \text{and} \quad 0 \nleftrightarrow 0, 1,\\
&\qquad\quad \text{and} \quad 1/2 \nleftrightarrow 1/2; \quad \text{or} \quad \Delta S = 0;\ \Delta L = 0, 1, 2 \quad \text{and} \quad 0 \nleftrightarrow 0, 1 \, .
\end{aligned}$$

Restrictions on photon emission from the metastable atomic states are removed when relativistic and other types of interaction between the electrons and between the electrons and the nucleus are taken into account. As a result, one-photon transitions of the type M1, E2, M2, two-photon decay 2E1 accompanied by the emission of two correlated photons, and so on, are induced.

Typical information about the metastable levels of the atomic particles C, N, Si, P and the chalcogens is found in Figs. 7.44, 45. The numerical parameters quoted in the figures include the excitation energies, transition wavelengths, transition probabilities and radiative decay lifetimes (see the central inserts for details).

Table 7.5 presents a more complete set of values of the excitation energy T_k for low-lying metastable atomic states, the transition wavelengths λ (the decay channel with the largest Einstein coefficient is indicated first, then the next in order, in parentheses) and the spontaneous emission lifetimes $\tau_{\rm m}$ for these states, obtained mainly by means of calculation.

We have made use here of reviews [7.5.3–5], data compilations [7.5.6–8] and some informative journal publications [7.5.9–12], and finally, the extensive bibliographies [7.5.13–15]. The data concerning the radiative lifetimes τ_m are grouped into accuracy classes (see the Introduction) in accordance with the error estimation of the method used to obtain them. The numerical values of excitation energy and transition wavelength are rounded off so that possible corrections would probably involve only the last significant figure within the range ± 1.

References

7.5.1 L.D.Landau, E.M.Lifshitz: *Relativistic Quantum Theory, Part I* (Pergamon, London 1971) Chap. V

7.5.2 A.Hibbert: "Atomic Structure Theory", in *Progress in Atomic Spectroscopy*, Part A, ed. by W.Hanle, H.Kleinpoppen (Plenum, New York 1978) pp. 1–69

7.5.3 R.H.Garstang: "Forbidden Transitions", in *Atomic and Molecular Processes*, ed. by D.R.Bates (Academic, New York 1962) Chap. 1

7.5.4 R.Marrus, P.J.Mohr: "Forbidden transitions in one- and two-electron atoms", Adv. At. Mol. Phys. **14**, 181 (1978)

7.5.5 K.Schofield: "Critically evaluated rate constants for gaseous reactions of several electronically excited species", J. Phys. Chem. Ref. Data **8**, 723 (1979)

7.5.6 W.L.Wiese, M.W.Smith, B.M.Glennon: *Atomic Transition Probabilities – H through Ne, Vol. I*, Nat. Stand. Ref. Data Ser. Nat. Bur. Stand. **4** (1966)

7.5.7 W.L.Wiese, M.W.Smith, B.M.Miles: *Atomic Transition Probabilities – Na through Ca, Vol. II*, Nat. Stand. Ref. Data Ser. Nat. Bur. Stand. **22** (1969)

7.5.8 M.W.Smith, W.L.Wiese: J. Phys. Chem. Ref. Data **2**, 85 (1973)

7.5.9 R.H.Garstang: J. Res. Nat. Bur. Stand. **68**, 61 (1964)

7.5.10 D.Layzer, R.H.Garstang: Annu. Rev. Astron. Astrophys. **6**, 449 (1968)

7.5.11 B.Warner: Z. Astrophys. **69**, 399 (1968)

7.5.12 S.J.Czyzak, T.K.Krueger: Mon. Not. Roy. Astron. Soc. **126**, 177 (1963)

7.5.13 J.R.Fuhr, B.J.Miller, G.A.Martin: *Bibliography on Atomic Transition Probabilities: 1914 through October 1977*, Nat. Bur. Stand. (U.S.) Spec. Publ. **505** (1978)

7.5.14 B.J.Miller, J.R.Fuhr, G.A.Martin: *Bibliography on Atomic Transition Probabilities: November 1977 through March 1980*, Nat. Bur. Stand. (U.S.) Rep. NBS-SP-505/1 (US Gov't Print. Office, Washington, D.C. 1980)

7.5.15 M.Kafatos, J.P.Lynch: Astrophys. J. Suppl. Ser. **42**, 611 (1980) (and references therein)

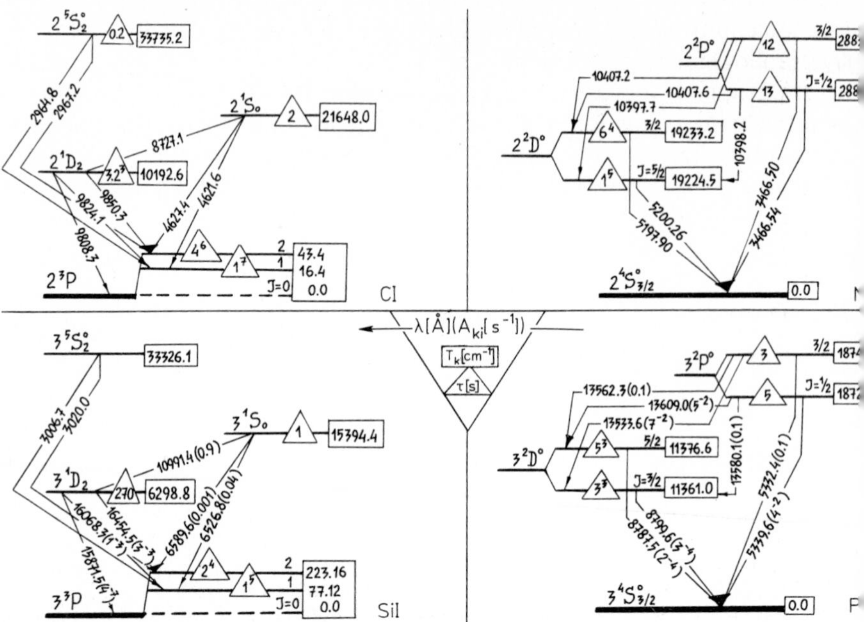

Fig. 7.44. Schematic diagram of metastable energy levels with λ and τ_k values for elements C, N, Si and P

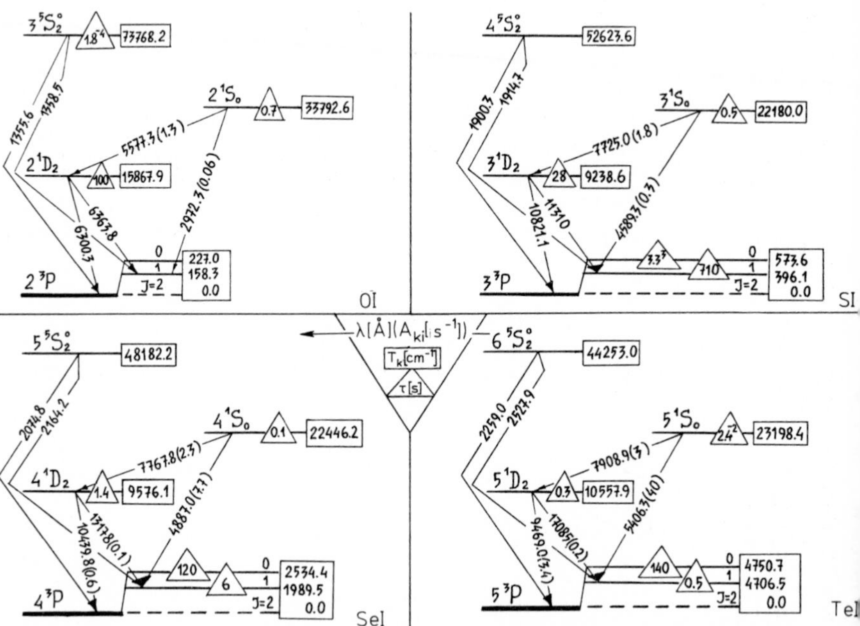

Fig. 7.45. Schematic diagram of metastable energy levels with λ and τ_k values for chalcogens

Table 7.5. Excitation energies and spontaneous emission lifetimes of metastable atoms

Atomic number Z	Element (ground-state term)	Metastable level	Excitation energy $[cm^{-1}]$	Excitation energy [eV]	Lower state for radiative decay of metastable level	Wavelength of radiative transition to lower state λ [Å]	Radiative lifetime τ_m [s] (and accuracy)
1	$H(1\,^2S_{1/2})$	$2\,^2S_{1/2}$	82258.96	10.20	$1\,^2S_{1/2}$	$(2\,E1)$	$0.1215\,(A)$
2	$He(1\,^1S_0)$	$2\,^3S_1$	159856.1	19.82	$1\,^1S_0$	625.6	$7900\,(B)$
		$2\,^1S_0$	166277.6	20.62	$1\,^1S_0$	601.4	$0.020\,(C)$
		$2\,^3P_2^o$	169086.9	20.96	$1\,^1S_0$	591.4	$A_{M2} = 0.22\,s^{-1}$
4	$Be(2\,^1S_0)$	$2\,^3P_2^o$	21981.3	2.726	$2\,^1S_0$	4548	$5.6 \cdot 10^3\,(D)$
5	$B(2\,^2P_{1/2}^o)$	$2\,^2P_{3/2}^o$	15.25	0.0019	$2\,^2P_{1/2}^o$	$6.6 \cdot 10^6$	$3.2 \cdot 10^7\,(D)$
		$2s\,2p^2 - {}^4P_{1/2-5/2}$	28877	3.580	$2\,^2P^o$	3460	–
6	$C(2\,^3P_0)$	$2\,^3P_1$	16.4	0.0020	$2\,^3P_0$	$6.1 \cdot 10^6$	$1.3 \cdot 10^7\,(D)$
		$2\,^3P_2$	43.4	0.0054	$2\,^3P_1$	$3.7 \cdot 10^6$	$3.7 \cdot 10^6\,(D)$
		$2\,^1D_2$	10192.6	1.264	$2\,^3P$	9850–9808	$3200\,(D)$
		$2\,^1S_0$	21648.0	2.684	$2\,^1D_2\,(2\,^3P)$	8727 (4620)	$2\,(D)$
		$2s\,2p^3 - {}^5S_2^o$	33735.2	4.183	$2\,^3P$	2966	$0.18\,(C)$
7	$N(2\,^4S_{3/2}^o)$	$2\,^2D_{5/2}^o$	19224.5	2.384	$2\,^4S_{3/2}^o$	5200	$1.6 \cdot 10^5\,(D)$
		$2\,^2D_{3/2}^o$	19233.2	2.385	$2\,^4S_{3/2}^o$	5198	$4.4 \cdot 10^4\,(D)$
		$2\,^2P_{1/2}^o$	28838.9	3.576	$2\,^2D^o\,(2\,^4S_{3/2}^o)$	10400 (3467)	$12.6\,(D)$
		$2\,^2P_{3/2}^o$	28839.3	3.576	$2\,^2D^o\,(2\,^4S_{3/2}^o)$	10400 (3466)	$11.4\,(D)$
8	$O(2\,^3P_2)$	$2\,^3P_1$	158.3	0.0196	$2\,^3P_2$	$6.3 \cdot 10^5$	–
		$2\,^3P_0$	227.0	0.028	$2\,^3P_1\,(2\,^3P_2)$	$1.5 \cdot 10^6$	–
		$2\,^1D_2$	15867.9	1.967	$2\,^3P$	6300	$150\,(C)$
		$2\,^1S_0$	33792.6	4.190	$2\,^1D_2\,(2\,^3P_1)$	5577 (2972)	$0.7\,(C)$
		$3\,^5S_2^o$	73768.2	9.146	$2\,^3P$	1357	$1.8 \cdot 10^{-4}\,(C)$
9	$F(2\,^2P_{3/2}^o)$	$2\,^2P_{1/2}^o$	404.08	0.0501	$2\,^2P_{3/2}^o$	$2.5 \cdot 10^5$	$660\,(D)$
10	$Ne(2\,^1S_0)$	$3s\,[3/2]_2^o\,(^3P_2^o)$	134041.8	16.62	$2\,^1S_0$	746	$20\,(E)$
		$3s'\,[1/2]_0^o\,(^3P_0^o)$	134818.6	16.72	$2\,^3P_1^o$	$2.8 \cdot 10^5$	$400\,(E)$

Table 7.5 (continued)

Atomic number Z	Element (ground-state term)	Metastable level	Excitation energy $[cm^{-1}]$	[eV]	Lower state for radiative decay of metastable level	Wavelength of radiative transition to lower state λ [Å]	Radiative lifetime τ_m [s] (and accuracy)
12	Mg $(3\,^1S_0)$	$3\,^3P_1^o$	21870.5	2.712	$3\,^1S_0$	4571	0.0022 (C)
		$3\,^3P_2^o$	21911.2	2.717	$3\,^1S_0$	4563	$5 \cdot 10^3$ (E)
13	Al $(3\,^2P_{1/2}^o)$	$3\,^2P_{3/2}^o$	112.06	0.0139	$3\,^2P_{1/2}^o$	$8.9 \cdot 10^5$	$8 \cdot 10^4$ (D)
		$3s\,3p^2 - {}^4P_{1/2-5/2}$	29097	3.608	$3\,^2P^o$	3440	–
14	Si $(3\,^3P_0)$	$3\,^3P_1$	77.12	0.0096	$3\,^3P_0$	$1.3 \cdot 10^6$	$1 \cdot 10^5$ (E)
		$3\,^3P_2$	223.16	0.0277	$3\,^3P_1$	$6.8 \cdot 10^5$	$2 \cdot 10^4$ (E)
		$3\,^1D_2$	6298.8	0.781	$3\,^3P_2\,(3\,^3P_1)$	16455	270 (E)
		$3\,^1S_0$	15394.4	1.909	$3\,^1D_2\,(3\,^3P_1)$	10991 (6527)	1.1 (E)
		$3\,^5S_2^o$	33326.1	4.132	$3\,^3P$	3010	–
15	P $(3\,^4S_{3/2}^o)$	$3\,^2D_{3/2}^o$	11361.0	1.409	$3\,^4S_{3/2}^o$	8800	$3.4 \cdot 10^3$ (D)
		$3\,^2D_{5/2}^o$	11376.6	1.411	$3\,^4S_{3/2}^o\,(3\,^2D_{3/2}^o)$	8787	$5.2 \cdot 10^3$ (D)
		$3\,^2P_{1/2}^o$	18722.7	2.321	$3\,^2D^o\,(3\,^4S_{3/2}^o)$	13600 (5340)	5.1 (D)
		$3\,^2P_{3/2}^o$	18748.0	2.324	$3\,^2D^o\,(3\,^4S_{3/2}^o)$	13560 (5330)	3.4 (D)
16	S $(3\,^3P_2)$	$3\,^3P_1$	396.1	0.0491	$3\,^3P_2$	$2.5 \cdot 10^5$	710 (D)
		$3\,^3P_0$	573.6	0.0711	$3\,^3P_1$	$5.6 \cdot 10^5$	3300 (D)
		$3\,^1D_2$	9238.6	1.145	$3\,^3P_2\,(3\,^3P_1)$	10821 (11310)	28 (D)
		$3\,^1S_0$	22180.0	2.750	$3\,^1D_2\,(3\,^3P_1)$	7725 (4589)	0.47 (D)
		$4\,^5S_2^o$	52623.6	6.525	$3\,^3P$	1900	–
17	Cl $(3\,^2P_{3/2}^o)$	$3\,^2P_{1/2}^o$	882.4	0.109	$3\,^2P_{3/2}^o$	$1.1 \cdot 10^5$	82 (D)
		$4\,^4P_{5/2-1/2}$	72280	8.962	$3\,^2P^o$	1380	–
18	Ar $(3\,^1S_0)$	$4s\,[3/2]_2^o\,(^3P_2^o)$	93143.8	11.55	$3\,^1S_0$	1073.6	60 (E)
		$4s'\,[1/2]_0^o\,(^3P_0^o)$	94553.7	11.72	$4\,^3P_1^o$	$1.2 \cdot 10^5$	50 (E)
20	Ca $(4\,^1S_0)$	$4\,^3P_0^o$	15157.9	1.879		6595	–
		$4\,^3P_1^o$	15210.1	1.886	$4\,^1S_0$	6573	$0.5 \cdot 10^{-3}$ (D)
		$4\,^3P_2^o$	15315.9	1.899		6527	–
		$3\,^1D_2$	21849.6	2.709	$4\,^3P^o\,(4\,^1S_0)$	15057 (4575)	$2.3 \cdot 10^{-3}$ (D)

		$4\,^3F_4$	386.9	0.0480	$4\,^3F_3$	$4.6 \cdot 10^5$	–
		$4\,^5F_{1-5}$	6721.4	0.833	$4\,^3F$	15000	–
23	V $(4\,^4F_{3/2})$	$4\,^4F_{5/2}$	137.4	0.0170	$4\,^4F_{3/2}$	$7.3 \cdot 10^5$	–
		$4\,^4F_{7/2}$	323.4	0.0400	$4\,^4F_{5/2}$	$5.4 \cdot 10^5$	–
		$4\,^4F_{9/2}$	553.0	0.0686	$4\,^4F_{7/2}$	$4.4 \cdot 10^5$	–
24	Cr $(4\,^7S_3)$	$4\,^5S_2$	7593.2	0.941	$4\,^7S_3$	13166	–
		$4\,^5D_{0-4}$	8090.2	1.003	$4\,^5S_2\,(4\,^7S_3)$	$2 \cdot 10^5$ (12000)	–
25	Mn $(4\,^6S_{5/2})$	$4\,^6D_{9/2}$	17052.3	2.114		5863	3.4 (*E*)
		$4\,^6D_{7/2}$	17282.0	2.143		5785	3.2 (*E*)
		$4\,^6D_{5/2}$	17451.5	2.164	$4\,^6S_{5/2}$	5729	3.0 (*E*)
		$4\,^6D_{3/2}$	17568.5	2.178		5690	2.9 (*E*)
		$4\,^6D_{1/2}$	17637.1	2.187		5668	2.9 (*E*)
26	Fe $(4\,^5D_4)$	$4\,^5D_3 - 4\,^5D_0$	415.9–978.1	0.0516–0.121	$4\,^5D$	$2.4 \cdot 10^5 - 5.4 \cdot 10^5$	400–1700 (*E*)
		$4\,^5F_5$	6928.3	0.859	$4\,^5D_4$	14430	50 (*E*)
		$4\,^5P_3$	17550.2	2.176	$4\,^5D_4\,(4\,^5D_3)$	5696	4 (*E*)
		$4\,^3P_2$	18378	2.279	$4\,^5D_3\,(4\,^5D_1)$	5566	2.5 (*E*)
27	Co $(4\,^4F_{9/2})$	$4\,^4F_{7/2}$	816.0	0.101	$4\,^4F_{9/2}$	$1.2 \cdot 10^5$	–
		$4\,^4F_{5/2}$	1406.8	0.174	$4\,^4F_{7/2}$	$1.7 \cdot 10^5$	–
		$4\,^4F_{3/2}$	1809.3	0.224	$4\,^4F_{5/2}$	$2.5 \cdot 10^5$	–
28	Ni $(4\,^3F_4)$	$4\,^3F_3$	1332.2	0.165	$4\,^3F_4$	$7.5 \cdot 10^4$	16 (*E*)
		$4\,^3F_2$	2216.5	0.275	$4\,^3F_3$	$1.1 \cdot 10^5$	40 (*E*)
		$3d^8 4s^2 - b\,^1D_2$	13521.3	1.676	$4\,^3F_3\,(4\,^3F_2)$	8202 (8843)	1.7 (*E*)
29	Cu $(4\,^2S_{1/2})$	$4\,^2D_{5/2}$	11202.6	1.389	$4\,^2S_{1/2}$	8924	–
		$4\,^2D_{3/2}$	13245.4	1.642	$4\,^2S_{1/2}$	7548	–
30	Zn $(4\,^1S_0)$	$4\,^3P^o_0$	32311.3	4.006		3094	–
		$4\,^3P^o_1$	32501.4	4.030	$4\,^1S_0$	3076	$25 \cdot 10^{-6}$ (*D*)
		$4\,^3P^o_2$	32890.3	4.078		3040	$1 \cdot 10^3$ (*E*)
31	Ga $(4\,^2P^o_{1/2})$	$4\,^2P^o_{3/2}$	826.2	0.102	$4\,^2P^o_{1/2}$	$1.2 \cdot 10^5$	200 (*D*)
32	Ge $(4\,^3P_0)$	$4\,^3P_1$	557.1	0.069	$4\,^3P_0$	$1.8 \cdot 10^5$	320 (*D*)
		$4\,^3P_2$	1410.0	0.175	$4\,^3P_1$	$1.2 \cdot 10^5$	120 (*D*)
		$4\,^1D_2$	7125.3	0.883	$4\,^3P_2\,(4\,^3P_1)$	17490 (15220)	6.7 (*D*)
		$4\,^1S_0$	16367.3	2.029	$4\,^1D_2\,(4\,^3P_1)$	10817 (6323)	0.46 (*D*)

Table 7.5 (continued)

Atomic number Z	Element (ground-state term)	Metastable level	Excitation energy [cm^{-1}]	[eV]	Lower state for radiative decay of metastable level	Wavelength of radiative transition to lower state λ [Å]	Radiative lifetime τ_m [s] (and accuracy)
33	As $(4\,^4S^o_{3/2})$	$4\,^2D^o_{3/2}$	10592	1.313	$4\,^4S^o_{3/2}$	9439	13 (D)
		$4\,^2D^o_{5/2}$	10915	1.353	$4\,^4S^o_{3/2}\,(4\,^2D^o_{3/2})$	9160 $(3 \cdot 10^5)$	180 (D)
		$4\,^2P^o_{1/2}$	18186	2.255	$4\,^4S^o_{3/2}\,(4\,^2D^o_{3/2})$	5497 (13165)	0.85 (D)
		$4\,^2P^o_{3/2}$	18647	2.312	$4\,^4S^o_{3/2}\,(4\,^2D^o_{3/2})$	5361 (12411)	0.37 (D)
34	Se $(4\,^3P_2)$	$4\,^3P_1$	1989.5	0.247	$4\,^3P_2$	$5.0 \cdot 10^4$	5.9 (D)
		$4\,^3P_0$	2534.4	0.314	$4\,^3P_1\,(4\,^3P_2)$	$1.8 \cdot 10^5\,(3.9 \cdot 10^4)$	115 (D)
		$4\,^1D_2$	9576.1	1.187	$4\,^3P_2\,(4\,^3P_1)$	10440 (13177)	1.4 (D)
		$4\,^1S_0$	22446.2	2.783	$4\,^3P_1\,(4\,^1D_2)$	4887 (7768)	0.10 (D)
35	Br $(4\,^2P^o_{3/2})$	$4\,^2P^o_{1/2}$	3685	0.457	$4\,^2P^o_{3/2}$	27129	1.0 (D)
36	Kr $(4\,^1S_0)$	$5s\,[3/2]^o_2\,(^3P^o_2)$	79971.8	9.915	$4\,^1S_0$	1250.4	85 (E)
		$5s'\,[1/2]^o_0\,(^3P^o_0)$	85191.7	10.56	$5\,^3P^o_1$	$2.3 \cdot 10^4$	0.5 (E)
38	Sr $(5\,^1S_0)$	$5\,^3P_1$	14504.4	1.798	$5\,^1S_0$	6893	$21 \cdot 10^{-6}\,(C)$
		$4\,^1D_2$	20149.7	2.498	$5\,^1S_0$	4961	$1.7 \cdot 10^{-2}\,(D)$
39	Y $(4\,^2D_{3/2})$	$4\,^2D_{5/2}$	530.4	0.0658	$4\,^2D_{3/2}$	$1.9 \cdot 10^5$	–
40	Zr $(5\,^3F_2)$	$5\,^3F_3$	570.4	0.0707	$5\,^3F_2$	$1.8 \cdot 10^5$	–
		$5\,^3F_4$	1240.8	0.154	$5\,^3F_3$	$1.5 \cdot 10^5$	–
42	Mo $(5\,^7S_3)$	$5\,^5S_2$	10768.3	1.335	$5\,^7S_3$	9284	–
		$5\,^5D_{0-4}$	11831.7	1.467	$5\,^5S_2\,(5\,^7S_3)$	$9 \cdot 10^4$ (8449)	–
48	Cd $(5\,^1S_0)$	$5\,^3P^o_0$	30114.0	3.734	$5\,^1S_0$	3320	–
		$5\,^3P^o_1$	30656.1	3.800	$5\,^1S_0$	3261	$2.40 \cdot 10^{-6}\,(B)$
		$5\,^3P^o_2$	31827.0	3.946	$5\,^1S_0$	3141	130 (D)
49	In $(5\,^2P^o_{1/2})$	$5\,^2P^o_{3/2}$	2212.6	0.274	$5\,^2P^o_{1/2}$	$4.5 \cdot 10^4$	10 (D)
50	Sn $(5\,^3P_0)$	$5\,^3P_1$	1692	0.210	$5\,^3P_0$	$5.9 \cdot 10^4$	12 (D)
		$5\,^3P_2$	3428	0.425	$5\,^3P_1$	$5.8 \cdot 10^4$	16 (D)
		$5\,^1D_2$	8613	1.068	$5\,^3P_2\,(5\,^3P_1)$	$1.9 \cdot 10^4$ (14444)	1.0 (D)

51	Sb ($5\,^4S^o_{3/2}$)	$5\,^2D^o_{3/2}$	8512	1.055	$5\,^4S^o_{3/2}$	11745	0.90 (D)
		$5\,^2D^o_{5/2}$	9854	1.222	$5\,^4S^o_{3/2}$ ($5\,^2D^o_{3/2}$)	10145 ($7.5 \cdot 10^4$)	10 (D)
		$5\,^2P^o_{1/2}$	16396	2.033	$5\,^4S^o_{3/2}$ ($5\,^2D^o_{3/2}$)	6097.5 (12681)	0.21 (D)
		$5\,^2P^o_{3/2}$	18464	2.289	$5\,^4S^o_{3/2}$ ($5\,^2D^o_{3/2}$)	5414 (10045)	0.089 (D)
52	Te ($5\,^3P_2$)	$5\,^3P_0$	4706.5	0.584	$5\,^3P_2$	21250	140 (D)
		$5\,^3P_1$	4750.7	0.589	$5\,^3P_2$	21040	0.45 (D)
		$5\,^1D_2$	10557.9	1.309	$5\,^3P_2$ ($5\,^3P_1$)	9469 (17220)	0.28 (D)
		$5\,^1S_0$	23198.4	2.876	$5\,^3P_1$ ($5\,^1D_2$)	5419 (7909)	0.024 (D)
53	I ($5\,^2P^o_{3/2}$)	$5\,^2P^o_{1/2}$	7603.1	0.943	$5\,^2P^o_{3/2}$	13149	0.12 (D)
54	Xe ($5\,^1S_0$)	$6s\,[3/2]^o_2\,(^3P^o_2)$	67068.0	8.315	$5\,^1S_0$	1491.0	150 (E)
		$6s'\,[1/2]^o_0\,(^3P^o_0)$	76197.3	9.447	$6\,^3P^o_1$	12264	0.08 (E)
56	Ba ($6\,^1S_0$)	$5\,^3D_{1-3}$	9034.0–9596.6	1.120–1.190	$6\,^1S_0$	11066–10418	–
		$5\,^1D_2$	11395.4	1.413	$6\,^1S_0$	8773	0.2 (D)
74	W ($6\,^5D_0$)	$6\,^5D_1$	1670.3	0.207	$6\,^5D$	$6.0 \cdot 10^4$	–
		$6\,^7S_3$	2951.3	0.366		$3.4 \cdot 10^4$	–
		$6\,^5D_2$	3325.5	0.412		$3.0 \cdot 10^4$	–
		$6\,^5D_3$	4830.0	0.599		$2.1 \cdot 10^4$	–
		$6\,^5D_4$	6219.3	0.771		$1.6 \cdot 10^4$	–
79	Au ($6\,^2S_{1/2}$)	$6\,^2D_{5/2}$	9161.2	1.136	$6\,^2S_{1/2}$	10913	–
		$6\,^2D_{3/2}$	21435	2.658		4664	–
80	^{199}Hg ($6\,^1S_0$)	$6\,^3P^o_0$	37645.1	4.667	$6\,^1S_0$	2656	1.45 (B)
		$6\,^3P^o_1$	39412.3	4.887		2537	–
		$6\,^3P^o_2$	44043.0	5.461		2270	6.5 (D)
81	Tl ($6\,^2P^o_{1/2}$)	$6\,^2P^o_{3/2}$	7793	0.966	$6\,^2P^o_{1/2}$	12829	0.23 (D)
82	Pb ($6\,^3P_0$)	$6\,^3P_1$	7819.3	0.969	$6\,^3P_0$	12785	0.14 (D)
		$6\,^3P_2$	10650.3	1.320	$6\,^3P_0$ ($6\,^3P_1$)	9387 (3532)	2.6 (D)
		$6\,^1D_2$	21457.8	2.660	$6\,^3P_1$ ($6\,^3P_2$)	7330 (9250)	0.037 (D)
		$6\,^1S_0$	29466.8	3.653	$6\,^3P_1$ ($6\,^3P_2$)	4618 (5313)	0.011 (D)
83	Bi ($6\,^4S^o_{3/2}$)	$6\,^2D^o_{3/2}$	11419.0	1.416	$6\,^4S^o_{3/2}$	8755	0.032 (D)
		$6\,^2D^o_{5/2}$	15437.7	1.914	$6\,^4S^o_{3/2}$ ($6\,^2D^o_{3/2}$)	6476 (24900)	0.12 (D)
		$6\,^2P^o_{1/2}$	21661.0	2.686	$6\,^4S^o_{3/2}$ ($6\,^2D^o_{3/2}$)	4615 (9761)	0.016 (D)
		$6\,^2P^o_{3/2}$	33164.8	4.112	$6\,^2D^o_{3/2}$ ($6\,^2D^o_{5/2}$)	4597 (5640)	$5.8 \cdot 10^{-3}$ (D)

7.6 Lifetimes of Atomic Rydberg States

In Tables 7.6, 7 the values of radiative lifetimes for highly excited (Rydberg) states of hydrogen and alkali atoms are presented. Regularities in the behaviour of the τ_k value for corresponding spectral series, i.e. sets of excited states with the same orbital and spin angular momenta, can be regarded as a specific feature of radiative transitions between such states.

When values of the electron principal quantum number n are large, the radiative lifetime of a highly excited hydrogen atom can be approximated as

$$\tau_k = \tau_0 n^3 + \tau_1 n ,$$

where τ_0, τ_1 are parameters depending on the orbital quantum number l of the electron. Included in Table 7.6 are values of the parameters τ_0, τ_1 obtained in processing the calculated data for the radiative lifetime of atomic hydrogen levels with n = 8–12 [7.6.1].

In the case of multielectron atoms, the τ_k value for highly excited states may be introduced in the following way:

$$\tau_k = \tau_0 n^{*3} = \tau_0 (n - \delta_l)^3 ,$$

where n^* is the effective principal quantum number of a valence electron, δ_l is the quantum defect of a level (Sect. 5.2), and τ_0 is constant for a given atomic spectral series. Values of τ_0 for alkali atoms cited in Table 7.7 were obtained after processing the data for measured radiative lifetimes of atomic Rydberg states [7.6.2, 3].

The numerical values of these parameters are grouped into accuracy classes (see the Introduction) according to their estimated error.

References

7.6.1 A.Lindgrad, S.E.Nielsen: At. Data Nucl. Data Tables **19**, 533 (1977)

7.6.2 Ya.F.Verolaynen, A.Ya.Nickolaich: "Radiative lifetimes of excited first-group atoms", Usp. Fiz. Nauk **137**, 305 (1982)

7.6.3 C.E.Theodosiou: "Lifetimes of alkali-metal-atom Rydberg states", Phys. Rev. **A30**, 2881 (1984)

Table 7.6. Parameters τ_0, τ_1 characterizing the radiative lifetimes of hydrogen Rydberg states

State	*s*	*p*	*d*	*f*	*g*
$l =$	0	1	2	3	4
τ_0 [ns]	1.75 (*A*)	0.185 (*A*)	0.528 (*A*)	1.05 (*A*)	1.76 (*A*)
τ_1 [ns]	26 (*C*)	0.16 (*E*)	0.9 (*E*)	2.7 (*E*)	4.8 (*D*)

State	*h*	*i*	*k*	*l*	*m*
$l =$	5	6	7	8	9
τ_0 [ns]	2.67 (*A*)	3.76 (*A*)	5.08 (*B*)	6.59 (*A*)	8.2 (*B*)
τ_1 [ns]	8 (*D*)	11 (*D*)	13 (*E*)	9 (*E*)	23 (*E*)

Table 7.7. Parameter τ_0 [ns] characterizing the radiative lifetimes of Rydberg states in alkali atoms

Atom	Valence electron state						
	s	*p*	*d*	*f*	*g*	*h*	*i*
Li	0.84 (*D*)	3.4 (*C*)	0.47 (*D*)	1.1 (*C*)	–	–	–
Na	1.36 (*C*)	2.7 (*C*)	0.93 (*C*)	1.0 (*B*)	1.76 (*B*)	2.7 (*B*)	3.8 (*B*)
K	1.21 (*C*)	3.9 (*D*)	2.6 (*D*)	0.76 (*C*)	1.77 (*B*)	–	–
Rb	1.18 (*C*)	2.9 (*D*)	1.4 (*D*)	0.66 (*C*)	–	–	–
Cs	1.3 (*D*)	3.4 (*D*)	0.7 (*D*)	0.67 (*C*)	1.7 (*C*)	–	–

8. Spectroscopic Characteristics of Atomic Positive Ions

Numerical data are presented for low-lying terms of singly charged atomic positive ions, radiative lifetimes of low-lying resonant excited states and typical parameters of ionic metastable states. The optical transition ($n \leqslant 2$) wavelengths, transition probabilities and radiative lifetimes of atomic ions isoelectronic to hydrogen and helium are also compiled over a wide range of nuclear charge variation.

8.1 Low-Lying Terms of Singly Ionized Atoms

Table 8.1 gives values of the excitation energy T for low-lying levels of singly ionized atoms. These states characterize, as a rule, some of the first terms of major electron configurations for a given atomic ion. The energy levels have been derived from the optical spectra of ions. They are listed in order of increasing excitation energy from the level of the ground state. The term and configuration of the excited valence electrons are also provided. The symbol of the ground-state term and the value of the optical limit (ionization limit) for a series of transitions converging to the ground-state level of the doubly ionized atom are given for convenience in a separate column of Table 8.1.

The data included in Table 8.1 were selected from [7.1.1–10].

Table 8.1. Low-lying energy levels of singly ionized atoms

Atomic number Z	Positive ion (ground-state term); ionization limit [cm^{-1}]	Excited state	Energy level [cm^{-1}]
1	2	3	4
2	$He^+ (1s - {}^2S_{1/2})$ 438908.89	$2p\,({}^2P^o_{1/2})$	329179.299
		$2s\,({}^2S_{1/2})$	329179.768
		$2p\,({}^2P^o_{3/2})$	329185.157
		$3p\,({}^2P^o_{1/2})$	390140.832
		$3s\,({}^2S_{1/2})$	390140.971
		$3d\,({}^2D_{3/2})$	390142.564
		$3p\,({}^2P^o_{3/2})$	390142.567
		$3d\,({}^2D_{5/2})$	390143.143
3	$Li^+ (1s^2 - {}^1S_0)$ 610078	$2s\,({}^3S_1)$	476034.6
		$2s\,({}^1S_0)$	491374.6
		$2p\,({}^3P^o_1)$	494261.2
		$2p\,({}^3P^o_2)$	494263.4
		$2p\,({}^3P^o_0)$	494266.6
		$2p\,({}^1P^o_1)$	501808.6
4	$Be^+ (2s - {}^2S_{1/2})$ 146883	$2p\,({}^2P^o_{1/2})$	31928.8
		$2p\,({}^2P^o_{3/2})$	31935.3
		$3s\,({}^2S_{1/2})$	88231.9
		$3p\,({}^2P^o_{1/2})$	96495.4
		$3p\,({}^2P^o_{3/2})$	96497.3
		$3d\,({}^2D_{3/2})$	98054.6
		$3d\,({}^2D_{5/2})$	98055.1
5	$B^+ (2s^2 - {}^1S_0)$ 202887	$2p\,({}^3P^o_0)$	37337
		$2p\,({}^3P^o_1)$	37342
		$2p\,({}^3P^o_2)$	37358
		$2p\,({}^1P^o_1)$	73396.6
		$2p^2\,({}^3P_0)$	98913
		$2p^2\,({}^3P_1)$	98922
		$2p^2\,({}^3P_2)$	98934
6	$C^+ (2p - {}^2P^o_{1/2})$ 196665	$2p\,({}^2P^o_{3/2})$	63.4
		$2s\,2p^2\,({}^4P_{1/2})$	43003.7
		$2s\,2p^2\,({}^4P_{3/2})$	43025.7
		$2s\,2p^2\,({}^4P_{5/2})$	43054.0
		$2s\,2p^2\,({}^2D_{5/2})$	74930.6
		$2s\,2p^2\,({}^2D_{3/2})$	74933.1
		$2s\,2p^2\,({}^2S_{1/2})$	96494.2
7	$N^+ (2p^2 - {}^3P_0)$ 238750	$2p\,({}^3P_1)$	48.7
		$2p\,({}^3P_2)$	130.8
		$2p\,({}^1D_2)$	15316.2
		$2p\,({}^1S_0)$	32688.8
		$2s\,2p^3\,({}^5S^o_2)$	46784.6
		$2s\,2p^3\,({}^3D^o_3)$	92237
		$({}^3D^o_2)$	92250
		$({}^3D^o_1)$	92252
		$2s\,2p^3\,({}^3P^o_1)$	109216.6

Table 8.1 (continued)

Atomic number Z	Positive ion (ground-state term); ionization limit [cm^{-1}]	Excited state	Energy level [cm^{-1}]
1	2	3	4
		$(^3P^o_2)$	109217.6
		$(^3P^o_0)$	109223.5
		$2s\,2p^3\,(^1D^o_2)$	144187.9
		$2p\,3s\,(^3P^o_0)$	148908.6
		$(^3P^o_1)$	148940.2
		$(^3P^o_2)$	149076.5
		$2p\,3s\,(^1P^o_1)$	149187.8
8	$O^+\,(2p^3-{}^4S^o_{3/2})$ 283240	$2p\,(^2D^o_{5/2})$	26808
		$2p\,(^2D^o_{3/2})$	26829
		$2p\,(^2P^o_{3/2})$	40467
		$2p\,(^2P^o_{1/2})$	40468
		$2s\,2p^4\,(^4P_{5/2})$	119838
		$(^4P_{3/2})$	120001
		$(^4P_{1/2})$	120083
		$2s\,2p^4\,(^2D_{5/2})$	165988
		$(^2D_{3/2})$	165996
		$3s\,(^4P_{1/2})$	185235.4
		$3s\,(^4P_{3/2})$	185340.7
		$3s\,(^4P_{5/2})$	185499.2
9	$F^+\,(2p^4-{}^3P_2)$ 282059	$2p\,(^3P_1)$	341.0
		$2p\,(^3P_0)$	489.9
		$2s^2\,2p^4\,(^1D_2)$	20873.4
		$2s^2\,2p^4\,(^1S_0)$	44918.1
		$2s\,2p^5\,(^3P^o_2)$	164797.9
		$(^3P^o_1)$	165106.7
		$(^3P^o_0)$	165279.2
		$3s\,(^5S^o_2)$	176493.9
		$3s\,(^3S^o_1)$	182864.4
		$3p\,(^5P_1)$	202449.3
		$3p\,(^5P_2)$	202460.7
		$3p\,(^5P_3)$	202480.3
		$3p\,(^3P)$	207702.0
10	$Ne^+\,(2p^5-{}^2P^o_{3/2})$ 330389	$2p^4\,(^3P)\,2p\,(^2P^o_{1/2})$	780.3
		$2s\,2p^6\,(^2S_{1/2})$	217047.6
		$3s\,(^4P_{5/2})$	219130.8
		$3s\,(^4P_{1/2})$	219647.5
		$3s\,(^4P_{3/2})$	219948.4
		$3s\,(^2P_{3/2})$	224087.0
		$3s\,(^2P_{1/2})$	224699.3
		$3p\,(^4P^o_{5/2})$	246192.5
		$2p^4\,(^1D)\,3s'\,(^2D_{5/2})$	246394.1
		$3s'\,(^2D_{3/2})$	246397.5
		$3p\,(^4P^o_{3/2})$	246415.0
		$3p\,(^4P^o_{1/2})$	246597.7
		$3p\,(^4D^o_{7/2})$	249108.6

Table 8.1 (continued)

Atomic number Z	Positive ion (ground-state term); ionization limit [cm^{-1}]	Excited state	Energy level [cm^{-1}]
1	2	3	4
		$({}^4D^o_{5/2})$	249446.0
		$({}^4D^o_{3/2})$	249695.5
		$({}^4D^o_{1/2})$	249839.6
		$3p\,({}^2D^o_{5/2})$	251011.2
		$({}^2D^o_{3/2})$	251522.1
		$3p\,({}^2S^o_{1/2})$	252798.5
		$3p\,({}^4S^o_{3/2})$	252953.5
		$3p\,({}^2P^o_{3/2})$	254165.0
		$({}^2P^o_{1/2})$	254292.2
11	$Na^+\,(2p^6-{}^1S_0)$ 381390	$2p^5\,({}^2P^o_{3/2})\,3s\,[3/2]^o_2$	264924.3
		$3s\,[3/2]^o_1$	265689.6
		$2p^5\,({}^2P^o_{1/2})\,3s'\,[1/2]^o_0$	266281.6
		$3s'\,[1/2]^o_1$	268763.0
		$3p\,[1/2]_1$	293220.3
		$3p\,[5/2]_3$	297248.8
		$3p\,[5/2]_2$	297635.6
		$3p\,[3/2]_1$	298165.4
		$3p\,[3/2]_2$	299190.0
		$3p'\,[3/2]_1$	299885.4
		$3p'\,[3/2]_2$	300103.9
		$3p\,[1/2]_0$	300387.8
		$3p'\,[1/2]_1$	300507.1
		$3p'\,[1/2]_0$	308860.8
12	$Mg^+\,(3s-{}^2S_{1/2})$ 121267.6	$3p\,({}^2P^o_{1/2})$	35669.3
		$3p\,({}^2P^o_{3/2})$	35760.9
		$4s\,({}^2S_{1/2})$	69805.0
		$3d\,({}^2D_{5/2})$	71490.2
		$3d\,({}^2D_{3/2})$	71491.1
		$4p\,({}^2P^o_{1/2})$	80619.5
		$4p\,({}^2P^o_{3/2})$	80650.0
13	$Al^+\,(3s^2-{}^1S_0)$ 151863	$3p\,({}^3P^o_0)$	37393.0
		$3p\,({}^3P^o_1)$	37453.9
		$3p\,({}^3P^o_2)$	37577.8
		$3p\,({}^1P^o_1)$	59852.0
		$3p^2\,({}^1D_2)$	85481.3
		$4s\,({}^3S_1)$	91274.5
		$3p^2\,({}^3P_0)$	94085.0
		$3p^2\,({}^3P_1)$	94147.5
		$3p^2\,({}^3P_2)$	94268.7
		$4s\,({}^1S_0)$	95350.6
		$3d\,({}^3D_3)$	95549.4
		$3d\,({}^3D_2)$	95550.5
		$3d\,({}^3D_1)$	95551.4

Table 8.1 (continued)

Atomic number Z	Positive ion (ground-state term); ionization limit [cm^{-1}]	Excited state	Energy level [cm^{-1}]
1	2	3	4
14	Si^+ ($3p - {}^2P^o_{1/2}$) 131838	$3p$ (${}^2P^o_{3/2}$)	287.2
		$3s\,3p^2$ (${}^4P_{1/2}$)	42824.3
		(${}^4P_{3/2}$)	42932.6
		(${}^4P_{5/2}$)	43107.9
		$3s\,3p^2$ (${}^2D_{3/2}$)	55309.4
		(${}^2D_{5/2}$)	55325.2
		$4s$ (${}^2S_{1/2}$)	65500.5
		$3s\,3p^2$ (${}^2S_{1/2}$)	76665.4
		$3d$ (${}^2D_{3/2}$)	79338.5
		$3d$ (${}^2D_{5/2}$)	79355.0
		$4p$ (${}^2P^o_{1/2}$)	81191.3
		$4p$ (${}^2P^o_{3/2}$)	81251.3
		$3s\,3p^2$ (${}^2P_{1/2}$)	83802.0
		(${}^2P_{3/2}$)	84004.3
15	P^+ ($3p^2 - {}^3P_0$) 159451	$3p^2$ (${}^3P_1$)	164.9
		$3p^2$ (${}^3P_2$)	469.1
		$3p^2$ (${}^1D_2$)	8882.3
		$3p^2$ (${}^1S_0$)	21575.6
		$3s\,3p^3$ (${}^5S^o_2$)	45697.4
		$3s\,3p^3$ (${}^3D^o_1$)	65251.4
		(${}^3D^o_2$)	65272.3
		(${}^3D^o_3$)	65307.2
		$3s\,3p^3$ (${}^3P^o_2$)	76764.1
		(${}^3P^o_1$)	76812.3
		(${}^3P^o_0$)	76823.1
		$3s\,3p^3$ (${}^1D^o_2$)	77710.2
		$3p\,4s$ (${}^3P^o_0$)	86597.5
		$3p\,4s$ (${}^3P^o_1$)	86744.0
		$3p\,4s$ (${}^3P^o_2$)	87124.6
		$3p\,3d$ (${}^3F^o_2$)	87804.1
		$3p\,3d$ (${}^3F^o_3$)	87966.8
		$3p\,3d$ (${}^3F^o_4$)	88192.1
		$3p\,4s$ (${}^1P^o_1$)	88893.2
		$3p\,4p$ (${}^1P_1$)	101635.7
		$3s\,3p^3$ (${}^1P^o_1$)	102798.3
16	S^+ ($3p^3 - {}^4S^o_{3/2}$) 188233	$3p$ (${}^2D^o_{3/2}$)	14853
		$3p$ (${}^2D^o_{5/2}$)	14885
		$3p$ (${}^2P^o_{1/2}$)	24525
		$3p$ (${}^2P^o_{3/2}$)	24572
		$3s\,3p^4$ (${}^4P_{5/2}$)	79395.4
		(${}^4P_{3/2}$)	79756.8
		(${}^4P_{1/2}$)	79962.6
		$3s\,3p^4$ (${}^2D_{3/2}$)	97890.7
		(${}^2D_{5/2}$)	97918.9

Table 8.1 (continued)

Atomic number Z	Positive ion (ground-state term); ionization limit [cm^{-1}]	Excited state	Energy level [cm^{-1}]
1	2	3	4
		$3d\,(^2P_{3/2})$	105599.1
		$3d\,(^2P_{1/2})$	106044.2
		$4s\,(^4P_{1/2})$	109560.7
		$4s\,(^4P_{3/2})$	109831.6
		$3d\,(^4F_{3/2})$	110177.0
		$4s\,(^4P_{5/2})$	110268.6
		$3d\,(^4F_{5/2})$	110313.4
		$3d\,(^4F_{7/2})$	110508.7
		$3d\,(^4F_{9/2})$	110766.6
17	$Cl^+\,(3p^4 - {}^3P_2)$ 192070	$3p\,(^3P_1)$	696.0
		$3p\,(^3P_0)$	996.5
		$3p\,(^1D_2)$	11653.6
		$3p\,(^1S_0)$	27878.0
		$3s\,3p^5\,(^3P_2^o)$	93367.6
		$(^3P_1^o)$	93999.9
		$(^3P_0^o)$	94333.8
		$4s\,(^5S_2^o)$	107879.7
		$3d\,(^5D_4^o)$	110296.8
		$3d\,(^5D_3^o)$	110297.7
		$3d\,(^5D_2^o)$	110300.6
		$3d\,(^5D_1^o)$	110303.1
		$3d\,(^5D_0^o)$	110304.5
		$4s\,(^3S_1^o)$	112609.4
		$3p^3\,(^2D)\,3d'\,(^1P_1^o)$	115657.8
18	$Ar^+\,(3p^5 - {}^2P_{3/2}^o)$ 222848.2	$3p\,(^2P_{1/2}^o)$	1432.0
		$3s\,3p^6\,(^2S_{1/2})$	108722.5
		$3d\,(^4D_{7/2})$	132327.36
		$3d\,(^4D_{5/2})$	132481.21
		$3d\,(^4D_{3/2})$	132630.73
		$3d\,(^4D_{1/2})$	132737.70
		$4s\,(^4P_{5/2})$	134241.74
		$4s\,(^4P_{3/2})$	135086.00
		$4s\,(^4P_{1/2})$	135601.73
		$4s\,(^2P_{3/2})$	138243.64
		$4s\,(^2P_{1/2})$	139258.34
19	$K^+\,(3p^6 - {}^1S_0)$ 255100	$3p^5\,(^2P_{3/2}^o)\,4s\,[3/2]_2^o$	162507.0
		$4s\,[3/2]_1^o$	163237.0
		$3d\,[1/2]_0^o$	163436.3
		$3d\,[1/2]_1^o$	164496.1
		$3d\,[3/2]_2^o$	164932.3
		$3p^5\,(^2P_{1/2}^o)\,4s'\,[1/2]_0^o$	165149.5
		$4s'\,[1/2]_1^o$	166461.5
		$3d\,[7/2]_3^o$	170835.4
		$3d\,[5/2]_2^o$	171526.8

Table 8.1 (continued)

Atomic number Z	Positive ion (ground-state term); ionization limit [cm^{-1}]	Excited state	Energy level [cm^{-1}]
1	2	3	4
20	Ca^+ ($4s-{}^2S_{1/2}$) 95751.9	$3d\,({}^2D_{3/2})$	13650.2
		$3d\,({}^2D_{5/2})$	13710.9
		$4p\,({}^2P^o_{1/2})$	25191.5
		$4p\,({}^2P^o_{3/2})$	25414.4
		$5s\,({}^2S_{1/2})$	52166.9
		$4d\,({}^2D_{3/2})$	56839.2
		$4d\,({}^2D_{5/2})$	56858.5
		$5p\,({}^2P^o_{1/2})$	60533.0
		$5p\,({}^2P^o_{3/2})$	60611.3
21	Sc^+ ($3d\,4s-{}^3D_1$) 103237	$3d\,4s\,({}^3D_2)$	67.7
		$3d\,4s\,({}^3D_3)$	177.8
		$3d\,4s\,({}^1D_2)$	2540.9
		$3d^2\,({}^3F_2)$	4802.9
		$3d^2\,({}^3F_3)$	4883.6
		$3d^2\,({}^3F_4)$	4987.8
		$3d^2\,({}^1D_2)$	10944.6
		$4s^2\,({}^1S_0)$	11736.4
		$3d^2\,({}^3P_0)$	12074.1
		$3d^2\,({}^3P_1)$	12101.5
		$3d^2\,({}^3P_2)$	12154.4
		$3d^2\,({}^1G_4)$	14261.3
22	Ti^+ ($3d^2\,4s-{}^4F_{3/2}$) 109490	$3d^2\,({}^3F)\,4s\,({}^4F_{5/2})$	94.1
		$4s\,({}^4F_{7/2})$	225.7
		$4s\,({}^4F_{9/2})$	393.4
		$3d^3\,({}^4F_{3/2})$	908.0
		$3d^3\,({}^4F_{5/2})$	983.9
		$3d^3\,({}^4F_{7/2})$	1087.3
		$3d^3\,({}^4F_{9/2})$	1215.8
		$3d^2\,4s\,({}^2F_{5/2})$	4628.6
		$3d^2\,4s\,({}^2F_{7/2})$	4897.6
		$3d^2\,({}^1D)\,4s\,({}^2D_{3/2})$	8710.4
		$4s\,({}^2D_{5/2})$	8744.2
		$3d^3\,({}^2G_{7/2})$	8997.7
		$3d^3\,({}^2G_{9/2})$	9118.3
		$3d^3\,({}^4P_{1/2})$	9363.6
		$3d^3\,({}^4P_{3/2})$	9395.7
		$3d^3\,({}^4P_{5/2})$	9518.1
23	V^+ ($3d^4-{}^5D_0$) 118200	$3d^4\,({}^5D_1)$	36.0
		$({}^5D_2)$	106.6
		$({}^5D_3)$	208.9
		$({}^5D_4)$	339.2
		$3d^3\,({}^4F)\,4s\,({}^5F_1)$	2604.8
		$4s\,({}^5F_2)$	2687.0
		$4s\,({}^5F_3)$	2808.7
		$4s\,({}^5F_4)$	2968.2

Table 8.1 (continued)

Atomic number Z	Positive ion (ground-state term); ionization limit [cm^{-1}]	Excited state	Energy level [cm^{-1}]
1	2	3	4
		$4s\,(^5F_5)$	3162.8
		$3d^3\,(^4F)\,4s\,(^3F_2)$	8640.2
		$4s\,(^3F_3)$	8842.0
		$4s\,(^3F_4)$	9097.8
		$3d^4\,(^3P_0)$	11295.6
		$(^3P_1)$	11514.8
		$(^3P_2)$	11908.3
		$3d^4\,(^3H_4)$	12545.1
		$(^3H_5)$	12621.5
		$(^3H_6)$	12706.2
24	$Cr^+\,(3d^5 - {}^6S_{5/2})$ 133000	$3d^4\,(^5D)\,4s\,(^6D_{1/2})$	11962.0
		$4s\,(^6D_{3/2})$	12032.7
		$4s\,(^6D_{5/2})$	12148.0
		$4s\,(^6D_{7/2})$	12304.0
		$4s\,(^6D_{9/2})$	12496.8
		$3d^4\,(^5D)\,4s\,(^4D_{1/2})$	19528.4
		$4s\,(^4D_{3/2})$	19631.3
		$4s\,(^4D_{5/2})$	19798.0
		$4s\,(^4D_{7/2})$	20024.2
		$3d^5\,(^4G_{5/2})$	20512.6
		$(^4G_{11/2})$	20512.7
		$(^4G_{7/2})$	20518.3
		$(^4G_{9/2})$	20519.8
		$3d^5\,(^4P_{5/2})$	21822.9
		$(^4P_{1/2})$	21824.2
		$(^4P_{3/2})$	21824.8
25	$Mn^+\,(3d^5\,4s - {}^7S_3)$ 126145	$3d^5\,(^6S)\,4s\,(^5S_2)$	9473.0
		$3d^6\,(^5D_4)$	14325.9
		$(^5D_3)$	14593.8
		$(^5D_2)$	14781.2
		$(^5D_1)$	14901.2
		$(^5D_0)$	14959.8
		$3d^5\,(^4G)\,4s\,(^5G_6)$	27547.2
		$4s\,(^5G_5)$	27571.2
		$4s\,(^5G_4)$	27583.6
		$4s\,(^5G_3)$	27588.5
		$4s\,(^5G_2)$	27589.3
		$3d^6\,(^3P_2)$	29869.5
		$3d^5\,(^4P)\,4s\,(^5P_3)$	29889.5
		$4s\,(^5P_2)$	29919.4
		$4s\,(^5P_1)$	29951.4
		$3d^6\,(^3H_6)$	30523.7
		$3d^6\,(^3H_5)$	30679.5
		$3d^6\,(^3P_1)$	30685.1
		$3d^6\,(^3H_4)$	30796.1
		$3d^6\,(^3P_0)$	31022.0

Table 8.1 (continued)

Atomic number Z	Positive ion (ground-state term); ionization limit [cm^{-1}]	Excited state	Energy level [cm^{-1}]
1	2	3	4
26	Fe^+ $(3d^6 4s - {}^6D_{9/2})$	$3d^6\,({}^5D)\,4s\,({}^6D_{7/2})$	384.79
	130560	$4s\,({}^6D_{5/2})$	667.68
		$4s\,({}^6D_{3/2})$	862.61
		$4s\,({}^6D_{1/2})$	977.05
		$3d^7\,({}^4F_{9/2})$	1872.57
		$({}^4F_{7/2})$	2430.10
		$({}^4F_{5/2})$	2837.95
		$({}^4F_{3/2})$	3117.46
		$3d^6\,({}^5D)\,4s\,({}^4D_{7/2})$	7955.30
		$4s\,({}^4D_{5/2})$	8391.94
		$4s\,({}^4D_{3/2})$	8680.45
		$4s\,({}^4D_{1/2})$	8846.77
		$3d^7\,({}^4P_{5/2})$	13474.41
		$({}^4P_{3/2})$	13673.18
		$({}^4P_{1/2})$	13904.82
27	Co^+ $(3d^8 - {}^3F_4)$	$3d^8\,({}^3F_3)$	950.5
	137790	$3d^8\,({}^3F_2)$	1597.3
		$3d^7\,({}^4F)\,4s\,({}^5F_5)$	3350.6
		$4s\,({}^5F_4)$	4029.0
		$4s\,({}^5F_3)$	4560.8
		$4s\,({}^5F_2)$	4950.2
		$4s\,({}^5F_1)$	5204.8
		$4d^7\,({}^4F)\,4s\,({}^3F_4)$	9813.0
		$4s\,({}^3F_3)$	10708.5
		$4s\,({}^3F_2)$	11322.0
		$3d^8\,({}^1D_2)$	11651.5
		$3d^8\,({}^3P_2)$	13260.8
		$3d^8\,({}^3P_1)$	13404.5
		$3d^8\,({}^3P_0)$	13593.3
		$3d^7\,({}^4P)\,4s\,({}^5P_3)$	17771.7
		$4s\,({}^5P_2)$	18031.7
		$4s\,({}^5P_1)$	18338.8
		$3d^8\,({}^1G_4)$	19190.1
28	Ni^+ $(3d^9 - {}^2D_{5/2})$	$3d^9\,({}^2D_{3/2})$	1506.94
	146541.6	$3d^8\,({}^3F)\,4s\,({}^4F_{9/2})$	8393.90
		$4s\,({}^4F_{7/2})$	9330.04
		$4s\,({}^4F_{5/2})$	10115.66
		$4s\,({}^4F_{3/2})$	10663.89
		$3d^8\,({}^3F)\,4s\,({}^2F_{7/2})$	13550.39
		$4s\,({}^2F_{5/2})$	14995.57
		$3d^8\,({}^3P)\,4s\,({}^4P_{5/2})$	23108.3
		$3d^8\,({}^1D)\,4s\,({}^2D_{3/2})$	23796.2
		$4s\,({}^4P_{3/2})$	24788.2
		$4s\,({}^4P_{1/2})$	24835.9
		$4s\,({}^2D_{5/2})$	25036.4

Table 8.1 (continued)

Atomic number Z	Positive ion (ground-state term); ionization limit [cm^{-1}]	Excited state	Energy level [cm^{-1}]
1	2	3	4
29	Cu^+ ($3d^{10}$ – $^1S_0$) 163669	$3d^9$ ($^2D_{5/2}$) $4s$ ($^3D_3$)	21928.6
		$4s$ ($^3D_2$)	22847.0
		$3d^9$ ($^2D_{3/2}$) $4s$ ($^3D_1$)	23998.3
		$3d^9$ ($^2D_{3/2}$) $4s$ ($^1D_2$)	26264.5
30	Zn^+ ($3d^{10}4s$ – $^2S_{1/2}$) 144892	$4p$ ($^2P^o_{1/2}$)	48481
		$4p$ ($^2P^o_{3/2}$)	49354
		$4s^2$ ($^2D_{5/2}$)	62722
		($^2D_{3/2}$)	65441
		$5s$ ($^2S_{1/2}$)	88437
		$4d$ ($^2D_{3/2}$)	96909
		$4d$ ($^2D_{5/2}$)	96960
31	Ga^+ ($4s^2$ – $^1S_0$) 165460	$4p$ ($^3P^o_0$)	47370
		$4p$ ($^3P^o_1$)	47820
		$4p$ ($^3P^o_2$)	48750
		$4p$ ($^1P^o_1$)	70700
		$5s$ ($^3S_1$)	102940
		$5s$ ($^1S_0$)	106660
		$4p^2$ ($^1D_2$)	107720
32	Ge^+ ($4p$ – $^2P^o_{1/2}$) 128521.3	$4p$ ($^2P^o_{3/2}$)	1767.4
		$4s\,4p^2$ ($^4P_{1/2}$)	51575.9
		($^4P_{3/2}$)	52290.9
		($^4P_{5/2}$)	53366.7
		$5s$ ($^2S_{1/2}$)	62403.1
		$4p^2$ ($^2D_{3/2}$)	65015.7
		($^2D_{5/2}$)	65184.8
		$5p$ ($^2P^o_{1/2}$)	79006.9
		$5p$ ($^2P^o_{3/2}$)	79366.6
		$4d$ ($^2D_{3/2}$)	80836.9
		$4d$ ($^2D_{5/2}$)	81012.7
		$6s$ ($^2S_{1/2}$)	94784.5
33	As^+ ($4p^2$ – $^3P_0$) 149900	$4p^2$ ($^3P_1$)	1063.5
		($^3P_2$)	2541.3
		$4p^2$ ($^1D_2$)	10095.8
		$4p^2$ ($^1S_0$)	22598.6
		$4s\,4p^3$ ($^5S^o_2$)	54817.1
34	Se^+ ($4p^3$ – $^4S^o_{3/2}$) 170700	$4p^3$ ($^2D^o_{3/2}$)	13168
		($^2D^o_{5/2}$)	13784
		$4p^3$ ($^2P^o_{1/2}$)	23038
		($^2P^o_{3/2}$)	23895
		$4s\,4p^4$ ($^4P_{5/2}$)	83877
		($^4P_{3/2}$)	85579
		($^4P_{1/2}$)	86438
35	Br^+ ($4p^4$ – $^3P_2$) 175900	$4p^4$ ($^3P_1$)	3140
		$4p^4$ ($^3P_0$)	3840

Table 8.1 (continued)

Atomic number Z	Positive ion (ground-state term); ionization limit [cm^{-1}]	Excited state	Energy level [cm^{-1}]
1	2	3	4
		$4p^4\,(^1D_2)$	11410
		$5s\,(^5S^o_2)$	93927.5
36	$Kr^+\,(4p^5 - {}^2P^o_{3/2})$ 196475	$4p^5\,(^2P^o_{1/2})$	5370.1
		$4s\,4p^6\,(^2S_{1/2})$	109000.4
		$5s\,(^4P_{5/2})$	112828.3
		$5s\,(^4P_{3/2})$	115092.0
		$5s\,(^4P_{1/2})$	117603.0
		$5s\,(^2P_{3/2})$	118474.3
		$4d\,(^4D_{7/2})$	120209.9
		$4d\,(^4D_{5/2})$	120426.9
		$4d\,(^4D_{3/2})$	121000.4
		$5s\,(^2P_{1/2})$	121002.1
		$4d\,(^4D_{1/2})$	121779.5
37	$Rb^+\,(4p^6 - {}^1S_0)$ 220105	$5s\,[3/2]^o_2$	133341.4
		$5s\,[3/2]^o_1$	134869.5
		$5s\,[1/2]^o_0$	138794.2
		$5s\,[1/2]^o_1$	140609.8
		$4d\,(^3P^o_0)$	143022.1
		$4d\,(^3P^o_1)$	143461.8
		$4d\,(^3P^o_2)$	143955.7
		$4d\,(^3F^o_3)$	146834.4
		$4d\,(^3F^o_2)$	148688.5
		$4d\,(^3D^o_3)$	151878.7
38	$Sr^+\,(5s - {}^2S_{1/2})$ 88964	$4d\,(^2D_{3/2})$	14555.9
		$4d\,(^2D_{5/2})$	14836.2
		$5p\,(^2P^o_{1/2})$	23715.2
		$5p\,(^2P^o_{3/2})$	24516.6
		$6s\,(^2S_{1/2})$	47736.5
39	$Y^+\,(5s^2 - {}^1S_0)$ 98700	$4d\,(^2D)\,5s\,(^3D_1)$	840.2
		$5s\,(^3D_2)$	1045.1
		$5s\,(^3D_3)$	1449.7
		$4d\,(^2D)\,5s\,(^1D_2)$	3296.2
		$4d^2\,(^3F_2)$	8003.1
		$(^3F_3)$	8328.0
		$(^3F_4)$	8743.3
40	$Zr^+\,(4d^2\,5s - {}^4F_{3/2})$ 105900	$4d^2\,(^3F)\,5s\,(^4F_{5/2})$	314.7
		$5s\,(^4F_{7/2})$	763.4
		$5s\,(^4F_{9/2})$	1322.9
		$4d^3\,(^4F_{3/2})$	2572.2
		$(^4F_{5/2})$	2895.0
		$(^4F_{7/2})$	3299.6
		$(^4F_{9/2})$	3757.7
		$4d^2\,(^1D)\,5s\,(^2D_{3/2})$	4248.3
		$5s\,(^2D_{5/2})$	4505.5

Table 8.1 (continued)

Atomic number Z	Positive ion (ground-state term); ionization limit [cm^{-1}]	Excited state	Energy level [cm^{-1}]
1	2	3	4
41	Nb^+ ($4d^4 - {}^5D_0$) 115500	$4d^4$ (${}^5D_1$)	159.0
		(${}^5D_2$)	438.4
		(${}^5D_3$)	801.4
		(${}^5D_4$)	1224.9
		$4d^3$ (${}^4F$) $5s$ (${}^5F_1$)	2356.8
		(${}^5F_2$)	2629.1
		(${}^5F_3$)	3029.6
		(${}^5F_4$)	3542.5
		(${}^5F_5$)	4146.0
		$4d^4$ (${}^3P_0$)	5562.3
		(${}^3P_1$)	6192.3
		(${}^3P_2$)	7261.3
42	Mo^+ ($4d^5 - {}^6S_{5/2}$) 130300	$4d^4$ (${}^5D$) $5s$ (${}^6D_{1/2}$)	11783.4
		(${}^6D_{3/2}$)	12034.1
		(${}^6D_{5/2}$)	12417.3
		(${}^6D_{7/2}$)	12900.3
		(${}^6D_{9/2}$)	13460.7
		$4d^5$ (${}^4G_{5/2}$)	15199.2
		(${}^4G_{7/2}$)	15330.6
		(${}^4G_{9/2}$)	15427.7
		(${}^4G_{11/2}$)	15447.0
		$4d^5$ (${}^4P_{5/2}$)	15691.2
		(${}^4P_{3/2}$)	15699.2
		(${}^4P_{1/2}$)	15890.1
		$4d^5$ (${}^4D_{1/2}$)	16796.1
		(${}^4D_{7/2}$)	16946.8
		(${}^4D_{3/2}$)	17174.1
		(${}^4D_{5/2}$)	17344.1
43	Tc^+ ($4d^5 5s - {}^7S_3$) 123000	$4d^6$ (${}^5D_4$)	3461.3
		(${}^5D_3$)	4217.2
		(${}^5D_2$)	4669.2
		(${}^5D_1$)	4961.1
		(${}^5D_0$)	5101.0
		$4d^5$ (${}^6S$) $5s$ (${}^5S_2$)	12617.2
44	Ru^+ ($4d^7 - {}^4F_{9/2}$) 135200	$4d^7$ (${}^4F_{7/2}$)	1523
		(${}^4F_{5/2}$)	2494
		(${}^4F_{3/2}$)	3105
		$4d^7$ (${}^4P_{5/2}$)	8257
		(${}^4P_{3/2}$)	8477
		$4d^6$ (${}^5D$) $5s$ (${}^6D_{9/2}$)	9151
		$4d^7$ (${}^4P_{1/2}$)	9373
45	Rh^+ ($4d^8 - {}^3F_4$) 145800	$4d^8$ (${}^3F_3$)	2401
		(${}^3F_2$)	3581
		$4d^8$ (${}^1D_2$)	8164

Table 8.1 (continued)

Atomic number Z	Positive ion (ground-state term); ionization limit [cm^{-1}]	Excited state	Energy level [cm^{-1}]
1	2	3	4
		$4d^8\,(^3P_1)$	10515
		$(^3P_0)$	10761
		$(^3P_2)$	11644
46	$Pd^+\,(4d^9 - {}^2D_{5/2})$ 156700	$4d^9\,(^2D_{3/2})$	3540
		$4d^8\,(^3F)\,5s\,(^4F_{9/2})$	25081
		$(^4F_{7/2})$	27094
		$(^4F_{5/2})$	28927
		$(^4F_{3/2})$	29946
47	$Ag^+\,(4d^{10} - {}^1S_0)$ 173300	$4d^9\,(^2D_{5/2})\,5s\,(^3D_3)$	39164
		$5s\,(^3D_2)$	40741
		$4d^9\,(^2D_{3/2})\,5s\,(^3D_1)$	43739
		$4d^9\,(^2D_{3/2})\,5s\,(^1D_2)$	46046
48	$Cd^+\,(5s - {}^2S_{1/2})$ 136374.7	$5p\,(^2P^o_{1/2})$	44136.1
		$5p\,(^2P^o_{3/2})$	46618.5
		$4d^9\,(^2D_{5/2})\,5s^2\,(^2D_{5/2})$	69258.9
		$4d^9\,(^2D_{3/2})\,5s^2\,(^2D_{3/2})$	74893.7
		$6s\,(^2S_{1/2})$	82990.7
49	$In^+\,(5s^2 - {}^1S_0)$ 152195	$5p\,(^3P^o_0)$	42270
		$5p\,(^3P^o_1)$	43350
		$5p\,(^3P^o_2)$	45830
		$5p\,(^1P^o_1)$	63033.8
		$6s\,(^3S_1)$	93919.0
50	$Sn^+\,(5p - {}^2P^o_{1/2})$ 118017	$5p\,(^2P^o_{3/2})$	4251.5
		$5s\,5p^2\,(^4P_{1/2})$	46464
		$(^4P_{3/2})$	48368
		$(^4P_{5/2})$	50730
		$6s\,(^2S_{1/2})$	56886
51	$Sb^+\,(5p^2 - {}^3P_0)$ 133327.5	$5p^2\,(^3P_1)$	3055
		$(^3P_2)$	5659
		$5p^2\,(^1D_2)$	12790
		$5p^2\,(^1S_0)$	23905
52	$Te^+\,(5p^3 - {}^4S^o_{3/2})$ 150000	$5p^3\,(^2D^o_{3/2})$	10222.38
		$(^2D^o_{5/2})$	12421.85
		$5p^3\,(^2P^o_{1/2})$	20546.59
		$(^2P^o_{3/2})$	24032.10
53	$I^+\,(5p^4\,{}^3P_2 + 5p^4\,{}^1D_2)$ 154304	$5p^4\,(^3P_0) + 5p^4\,(^1S_0)$	6448
		$5p^4\,(^3P_1)$	7086.9
		$5p^4\,(^1D_2) + 5p^4\,(^3P_2)$	13727
		$5p^4\,(^1S_0) + 5p^4\,(^3P_0)$	29501
54	$Xe^+\,(5p^5 - {}^2P^o_{3/2})$ 169200	$5p^5\,(^2P^o_{1/2})$	10537
		$5s\,5p^6\,(^2S_{1/2})$	90873.8
		$6s\,(^4P_{5/2})$	93068.4
		$6s\,(^4P_{3/2})$	95064.3

Table 8.1 (continued)

Atomic number Z	Positive ion (ground-state term); ionization limit [cm^{-1}]	Excited state	Energy level [cm^{-1}]
1	2	3	4
		$5d\,(^4D_{5/2})$	95396.7
		$5d\,(^4D_{7/2})$	95437.6
		$5d\,(^4D_{3/2})$	96033.4
		$5d\,(^4D_{1/2})$	96858.1
55	$Cs^+\,(5p^6 - {}^1S_0)$ 187000	$5p^5\,(^2P^o_{3/2})\,5d\,[7/2]^o_3$	105949.7
		$5d\,[1/2]^o_1$	106222.8
		$6s\,[3/2]^o_2$	107392.3
		$5d\,[1/2]^o_0$	107563.1
		$6s\,[3/2]^o_1$	107905.0
		$5d\,[3/2]^o_2$	108304.2
		$5d\,[3/2]^o_1$	110945.2
		$5d\,[7/2]^o_4$	112236.5
		$5d\,[5/2]^o_2$	112795.1
		$5d\,[5/2]^o_3$	113716.6
56	$Ba^+\,(6s - {}^2S_{1/2})$ 80686.9	$5d\,(^2D_{3/2})$	4873.85
		$5d\,(^2D_{5/2})$	5674.82
		$6p\,(^2P^o_{1/2})$	20261.56
		$6p\,(^2P^o_{3/2})$	21952.42
		$7s\,(^2S_{1/2})$	42355.18
57	$La^+\,(5d^2 - {}^3F_2)$ 89200	$5d^2\,(^3F_3)$	1016.1
		$5d^2\,(^1D_2) + 5d\,6s\,(^1D_2)$	1394.5
		$5d\,6s\,(^3D_1)$	1895.1
		$5d^2\,(^3F_4)$	1970.7
		$5d\,6s\,(^3D_2)$	2591.6
		$5d\,6s\,(^3D_3)$	3250.3
		$5d^2\,(^3P_0)$	5249.7
		$5d^2\,(^3P_1)$	5718.1
		$5d^2\,(^3P_2)$	6227.4
		$6s^2\,(^1S_0)$	7394.6
		$5d^2\,(^1G_4)$	7473.3
58	$Ce^+\,(4f5d^2 - {}^4H^o_{7/2})$ 87000	$4f\,(^2F^o)\,5d^2\,(^3F)\,(^4I^o_{9/2} + {}^2G^o_{9/2})$	987.61
		$4f5d^2\,(^3F)\,(^4I^o_{9/2})$	1410.30
		$4f5d^2\,(J = 7/2)^o$	1873.93
		$4f5d^2\,(J = 1/2)^o$	2140.49
		$4f5d\,(^1G^o)\,6s\,(J = 9/2)^o$	2382.25
		$4f5d^2\,(^4I^o_{11/2})$	2563.23
		$4f5d^2\,(^4H^o_{9/2})$	2581.26
		$4f5d\,(^3F^o)\,6s\,(^4F^o_{3/2})$	2595.64
		$4f5d\,(^3F^o)\,6s\,(J = 5/2)^o$	2634.67
		$4f5d\,(^1G^o)\,6s\,(J = 7/2)^o$	2641.56
		$4f5d^2\,(^4H^o_{11/2})$	2879.70
		$4f5d\,(^3F^o)\,6s\,(^4F^o_{5/2})$	3363.43
		$4f5d^2\,(^2S^o_{1/2})$	3508.47
		$4f5d\,(^1G^o)\,6s\,(J = 9/2)^o$	3593.88
		$4f5d^2\,(J = 7/2)^o$	3703.59

Table 8.1 (continued)

Atomic number Z	Positive ion (ground-state term); ionization limit [cm^{-1}]	Excited state	Energy level [cm^{-1}]
1	2	3	4
		$4f5d^2\,(^4D^o_{3/2})$	3745.48
		$4f5d^2\,(^4I^o_{13/2})$	3793.63
		$4f^2\,(^3H)\,6s\,(^4H_{7/2})$	3854.01
		$4f5d\,(^3H^o)\,6s\,(^4H^o_{7/2})$	3995.46
		$4f^2\,(^3H)\,6s\,(^4H_{9/2})$	4165.55
		$4f5d^2\,(^4F^o_{3/2})$	4201.89
		$4f5d^2\,(^4H^o_{13/2})$	4203.93
59	Pr^+ $(4f^3\,6s\,(9/2,\,1/2)^o_4)$ 85000	$4f^3\,(^4I^o_{9/2})\,6s\,(9/2,\,1/2)^o_5$	441.9
		$4f^3\,(^4I^o_{11/2})\,6s\,(11/2,\,1/2)^o_6$	1649.0
		$4f^3\,6s\,(11/2,\,1/2)^o_5$	1743.7
		$4f^3\,(^4I^o_{13/2})\,6s\,(13/2,\,1/2)^o_7$	2998.4
		$4f^3\,6s\,(13/2,\,1/2)^o_6$	3403.2
		$4f^3\,(^4I^o)\,5d\,(^5L^o_6)$	3893.5
		$4f^3\,5d\,(^5K^o_5)$	4097.6
		$4f^3\,(^4I^o_{15/2})\,6s\,(15/2,\,1/2)^o_8$	4437.1
		$4f^3\,6s\,(15/2,\,1/2)^o_7$	5079.3
		$4f^3\,(^4I^o)\,5d\,(^5L^o_7)$	5108.4
		$4f^3\,5d\,(^5K^o_6)$	5226.5
		$4f^2\,(^3H)\,5d^2\,(^3F)\,(^5L_6)$	5854.6
		$4f^3\,5d\,(^5K^o_7)$	6413.9
		$4f^3\,5d\,(^5L^o_8)$	6417.8
		$4f^2\,(^3H)\,5d^2\,(^3F)\,(^5I_4)$	7228.0
		$4f^3\,5d\,(^3I^o_5)$	7438.2
		$4f^3\,5d\,(^5I^o_4)$	7446.4
		$4f^3\,5d\,(^5K^o_8)$	7659.8
		$4f^3\,5d\,(^5H^o_3)$	7744.3
60	Nd^+ $(4f^4\,6s - {}^6I_{7/2})$ 86000	$4f^4\,(^5I)\,6s\,(^6I_{9/2})$	513.33
		$4f^4\,6s\,(^6I_{11/2})$	1470.10
		$4f^4\,(^5I)\,6s\,(^4I_{9/2})$	1650.20
		$4f^4\,6s\,(^6I_{13/2})$	2585.46
		$4f^4\,6s\,(^4I_{11/2})$	3066.75
		$4f^4\,6s\,(^6I_{15/2})$	3801.93
		$4f^4\,(^5I)\,5d\,(^6L_{11/2})$	4437.56
		$4f^4\,6s\,(^4I_{13/2})$	4512.49
		$4f^4\,6s\,(^6I_{17/2})$	5085.64
		$4f^4\,5d\,(^6L_{13/2})$	5487.66
		$4f^4\,6s\,(^4I_{15/2})$	5985.58
		$4f^4\,5d\,(^6K_{9/2})$	6005.27
		$4f^4\,5d\,(^6L_{15/2})$	6637.43
		$4f^4\,5d\,(^6K_{11/2})$	6931.80
		$4f^4\,5d\,(^6I_{7/2})$	7524.74
		$4f^4\,5d\,(^6L_{17/2})$	7868.91
		$4f^4\,5d\,(^6K_{13/2})$	7950.07
		$4f^3\,(^4I^o)\,5d^2\,(^3F)\,(^6M^o_{13/2})$	8009.81
		$4f^4\,5d\,(^6I_{9/2})$	8420.32

Table 8.1 (continued)

Atomic number Z	Positive ion (ground-state term); ionization limit [cm^{-1}]	Excited state	Energy level [cm^{-1}]
1	2	3	4
		$4f^4 5d\,(^6G_{3/2})$	8716.45
		$4f^4 5d\,(^6G_{5/2})$	8796.36
		$4f^4 5d\,(^6K_{15/2})$	9042.76
		$4f^4 5d\,(^6L_{19/2})$	9166.21
		$4f^4 5d\,(^6G_{7/2})$	9198.40
		$4f^4 5d\,(^6I_{11/2})$	9357.91
		$4f^3\,(^4I^\circ)\,5d^2\,(^3F)\,(^6M^\circ_{15/2})$	9448.18
		$4f^4 5d\,(^6H_{5/2})$	9674.84
61	$Pm^+\,(4f^5 6s - {}^7H^\circ_2)$ 88000	$4f^5\,(^6H^\circ)\,6s\,(^7H^\circ_3)$	446.4
		$4f^5 6s\,(^7H^\circ_4)$	1133.4
		$4f^5 6s\,(^5H^\circ_3)$	1603.0
		$4f^5 6s\,(^7H^\circ_5)$	1983.5
		$4f^5 6s\,(^5H^\circ_4)$	2666.8
		$4f^5 6s\,(^7H^\circ_6)$	2950.3
		$4f^5 6s\,(^5H^\circ_5)$	3812.3
		$4f^5 6s\,(^7H^\circ_7)$	4000.1
		$4f^5 6s\,(^5H^\circ_6)$	5017.8
		$4f^5\,(^6F^\circ)\,6s\,(^7F^\circ_0)$	5280.9
		$4f^5\,(^6H^\circ)\,5d\,(^7K^\circ_4)$	5332.4
		$4f^5\,(^6F^\circ)\,6s\,(^7F^\circ_1)$	5391.5
		$4f^5 6s\,(^7F^\circ_2)$	5632.4
		$4f^5 6s\,(^7F^\circ_3)$	6048.5
		$4f^5\,(^6F^\circ)\,6s\,(^5F^\circ_1)$	6629.4
		$4f^5 6s\,(^7F^\circ_4)$	6705.2
		$4f^5 6s\,(^5F^\circ_2)$	7012.9
		$4f^5 6s\,(^5F^\circ_3)$	7701.1
62	$Sm^+\,(4f^6 6s - {}^8F_{1/2})$ 89000	$4f^6\,(^7F)\,6s\,(^8F_{3/2})$	326.6
		$4f^6 6s\,(^8F_{5/2})$	838.2
		$4f^6 6s\,(^8F_{7/2})$	1489.2
		$4f^6\,(^7F)\,6s\,(^6F_{1/2})$	1518.3
		$4f^6 6s\,(^6F_{3/2})$	2003.2
		$4f^6 6s\,(^8F_{9/2})$	2238.0
		$4f^6 6s\,(^6F_{5/2})$	2688.7
		$4f^6 6s\,(^8F_{11/2})$	3052.6
		$4f^6 6s\,(^6F_{7/2})$	3499.1
		$4f^6 6s\,(^8F_{13/2})$	3909.6
		$4f^6 6s\,(^6F_{9/2})$	4386.0
		$4f^6 6s\,(^6F_{11/2})$	5317.6
		$4f^6\,(^7F)\,5d\,(^8H_{3/2})$	7135.1
		$4f^6 5d\,(^8H_{5/2})$	7524.9
		$4f^6 5d\,(^8H_{7/2})$	8046.0
		$4f^6 5d\,(^8D_{3/2})$	8578.7
		$4f^6 5d\,(^8H_{9/2})$	8679.2
		$4f^6 5d\,(^8H_{11/2})$	9406.6
		$4f^6 5d\,(^8D_{5/2})$	9410.0

Table 8.1 (continued)

Atomic number Z	Positive ion (ground-state term); ionization limit [cm^{-1}]	Excited state	Energy level [cm^{-1}]
1	2	3	4
63	$Eu^+ (4f^7 6s - {}^9S^o_4)$	$4f^7 ({}^8S^o) 6s ({}^7S^o_3)$	1669.2
	90660	$4f^7 ({}^8S^o) 5d ({}^9D^o_2)$	9923.0
		$4f^7 5d ({}^9D^o_3)$	10081.6
		$4f^7 5d ({}^9D^o_4)$	10312.8
		$4f^7 5d ({}^9D^o_5)$	10643.5
		$4f^7 5d ({}^9D^o_6)$	11128.2
		$4f^7 ({}^8S^o) 5d ({}^7D^o_5)$	16860.7
		$4f^7 5d ({}^7D^o_4)$	17004.1
		$4f^7 5d ({}^7D^o_3)$	17140.9
		$4f^7 5d ({}^7D^o_2)$	17247.7
		$4f^7 5d ({}^7D^o_1)$	17324.7
64	$Gd^+ (4f^7 5d 6s - {}^{10}D^o_{5/2})$	$4f^7 ({}^8S^o) 5d ({}^9D^o) 6s ({}^{10}D^o_{7/2})$	261.84
	97500	$4f^7 5d 6s ({}^{10}D^o_{9/2})$	633.27
		$4f^7 5d 6s ({}^{10}D^o_{11/2})$	1158.94
		$4f^7 5d 6s ({}^{10}D^o_{13/2})$	1935.31
		$4f^7 5d 6s ({}^8D^o_{3/2})$	2856.68
		$4f^7 5d 6s ({}^8D^o_{5/2})$	3082.01
		$4f^7 5d 6s ({}^8D^o_{7/2})$	3427.27
		$4f^7 ({}^8S^o) 6s^2 ({}^8S^o_{7/2})$	3444.24
		$4f^7 5d 6s ({}^8D^o_{9/2})$	3972.17
		$4f^7 ({}^8S^o) 5d^2 ({}^3F) ({}^{10}F^o_{3/2})$	4027.16
		$4f^7 5d^2 ({}^{10}F^o_{5/2})$	4212.76
		$4f^7 5d^2 ({}^{10}F^o_{7/2})$	4483.85
		$4f^7 5d 6s ({}^8D^o_{11/2})$	4841.11
		$4f^7 5d^2 ({}^{10}F^o_{9/2})$	4852.30
		$4f^7 5d^2 ({}^{10}F^o_{11/2})$	5339.48
		$4f^7 5d^2 ({}^{10}F^o_{13/2})$	5897.26
		$4f^7 5d^2 ({}^{10}F^o_{15/2})$	6605.15
		$4f^8 ({}^7F) 6s ({}^8F_{13/2})$	7992.27
		$4f^7 ({}^8S^o) 5d ({}^7D^o) 6s ({}^8D^o_{11/2})$	8551.05
		$4f^7 5d 6s ({}^8D^o_{9/2})$	8884.81
		$4f^8 ({}^7F) 6s ({}^8F_{11/2})$	9092.49
		$4f^7 5d 6s ({}^8D^o_{7/2})$	9142.90
		$4f^7 5d 6s ({}^8D^o_{5/2})$	9328.86
		$4f^7 5d 6s ({}^8D^o_{3/2})$	9451.70
65	$Tb^+ (4f^9 6s - (15/2, 1/2)^o_8)$	$4f^9 ({}^6H^o_{15/2}) 6s (15/2, 1/2)^o_7$	1016.4
	93000	$4f^9 ({}^6H^o_{13/2}) 6s (13/2, 1/2)^o_7$	3010.0
		$4f^8 ({}^7F) 5d ({}^8G) 6s ({}^9G_7)$	3235.2
		$4f^8 5d 6s ({}^9G_8)$	3423.3
		$4f^8 5d 6s (J = 6)$	3440.8
		$4f^9 6s (13/2, 1/2)^o_6$	3542.4
		$4f^8 5d 6s (J = 5)$	4158.7
		$4f^8 ({}^7F) 5d ({}^8D) 6s (J = 6)$	5147.2
		$4f^9 ({}^6H^o_{11/2}) 6s (11/2, 1/2)^o_6$	5171.8
		$4f^9 6s (11/2, 1/2)^o_5$	5235.0

Table 8.1 (continued)

Atomic number Z	Positive ion (ground-state term); ionization limit [cm^{-1}]	Excited state	Energy level [cm^{-1}]
1	2	3	4
		$4f^8\,5d\,(^8G)\,6s\,(J = 5)$	5761.3
		$4f^8\,6s^2\,(^7F_6)$	5898.3
		$4f^8\,5d\,(^8G)\,6s\,(J = 7)$	6223.4
		$4f^9\,(^6F^{\circ}_{11/2})\,6s\,(11/2,\ 1/2)^{\circ}_6$	6372.9
		$4f^8\,5d\,(^8G)\,6s\,(^7G_6)$	6428.7
		$4f^8\,5d\,(^8G)\,6s\,(J = 5)$	6582.9
		$4f^9\,(^6H^{\circ}_{9/2})\,6s\,(9/2,\ 1/2)^{\circ}_5$	6912.5
66	$Dy^+\,(4f^{10}\,6s - (8,\ 1/2)_{17/2})$ 94100	$4f^{10}\,(^5I_8)\,6s\,(8,\ 1/2)_{15/2}$	828.3
		$4f^{10}\,(^5I_7)\,6s\,(7,\ 1/2)_{15/2}$	4341.1
		$4f^{10}\,6s\,(7,\ 1/2)_{13/2}$	4755.7
		$4f^{10}\,(^5I_6)\,6s\,(6,\ 1/2)_{11/2}$	7463.9
		$4f^{10}\,6s\,(6,\ 1/2)_{13/2}$	7485.1
		$4f^{10}\,(^5I_5)\,6s\,(5,\ 1/2)_{9/2}$	9432.1
		$4f^{10}\,6s\,(5,\ 1/2)_{11/2}$	9871.0
		$4f^9\,(^6H^{\circ})\,5d\,(^7H^{\circ})\,6s\,(^8H^{\circ}_{17/2})$	10594.2
		$4f^{10}\,(^5I_4)\,6s\,(4,\ 1/2)_{7/2}$	10953.9
		$4f^9\,(^6H^{\circ})\,5d\,(^7H^{\circ})\,6s\,(^8H^{\circ}_{15/2})$	11394.9
		$4f^{10}\,(^5I_4)\,6s\,(4,\ 1/2)_{9/2}$	11801.0
		$4f^9\,6s^2\,(^6H^{\circ}_{15/2})$	12336.3
		$4f^9\,5d\,6s\,(^8H^{\circ}_{13/2})$	12674.7
67	$Ho^+\,(4f^{11}\,6s - (15/2,\ 1/2)^{\circ}_8)$ 95200	$4f^{11}\,(^4I^{\circ}_{15/2})\,6s\,(15/2,\ 1/2)^{\circ}_7$	637.4
		$4f^{11}\,(^4I^{\circ}_{13/2})\,6s\,(13/2,\ 1/2)^{\circ}_7$	5617.0
		$4f^{11}\,6s\,(13/2,\ 1/2)^{\circ}_6$	5849.7
		$4f^{11}\,(^4I^{\circ}_{11/2})\,6s\,(11/2,\ 1/2)^{\circ}_5$	8850.5
		$4f^{11}\,6s\,(11/2,\ 1/2)^{\circ}_6$	9001.6
		$4f^{11}\,(^4I^{\circ}_{9/2})\,6s\,(9/2,\ 1/2)^{\circ}_4$	10838.8
		$4f^{11}\,6s\,(9/2,\ 1/2)^{\circ}_5$	11204.5
68	$Er^+\,(4f^{12}\,6s - (6,\ 1/2)_{13/2})$ 96200	$4f^{12}\,(^3H_6)\,6s\,(6,\ 1/2)_{11/2}$	440.43
		$4f^{12}\,(^3F_4)\,6s\,(4,\ 1/2)_{9/2}$	5132.61
		$4f^{12}\,6s\,(4,\ 1/2)_{7/2}$	5403.69
		$4f^{11}\,6s^2\,(^4I^{\circ}_{15/2})$	6824.77
		$4f^{12}\,(^3H_5)\,6s\,(5,\ 1/2)_{11/2}$	7149.63
		$4f^{12}\,6s\,(5,\ 1/2)_{9/2}$	7195.36
		$4f^{11}\,(^4I^{\circ})\,5d\,6s\,(J = 13/2)^{\circ}$	10667.19
		$4f^{12}\,(^3H_4)\,6s\,(4,\ 1/2)_{7/2}$	10893.94
		$4f^{12}\,6s\,(4,\ 1/2)_{9/2}$	11042.64
		$4f^{11}\,(^4I^{\circ})\,5d\,6s\,(J = 15/2)^{\circ}$	11309.18
		$4f^{11}\,5d\,6s\,(J = 11/2)^{\circ}$	12388.09
		$4f^{12}\,(^3F_3)\,6s\,(3,\ 1/2)_{7/2}$	12588.00
		$4f^{12}\,6s\,(3,\ 1/2)_{5/2}$	12600.09
		$4f^{11}\,5d\,6s\,(J = 19/2)^{\circ}$	12815.07
69	$Tm^+\,(4f^{13}\,6s - (7/2,\ 1/2)^{\circ}_4)$ 97200	$4f^{13}\,(^2F^{\circ}_{7/2})\,6s\,(7/2,\ 1/2)^{\circ}_3$	236.9
		$4f^{13}\,(^2F^{\circ}_{5/2})\,6s\,(5/2,\ 1/2)^{\circ}_2$	8769.7
		$4f^{13}\,6s\,(5/2,\ 1/2)^{\circ}_3$	8957.5
		$4f^{12}\,(^3H)\,6s^2\,(^3H_6)$	12457.3

Table 8.1 (continued)

Atomic number Z	Positive ion (ground-state term); ionization limit [cm^{-1}]	Excited state	Energy level [cm^{-1}]
1	2	3	4
		$4f^{12}\,(^3H_6)\,5d\,6s\,(^3D_1)\,(6,\,1)_5$	16567.5
		$4f^{13}\,(^2F^o_{7/2})\,5d\,(7/2,\,3/2)^o_2$	17624.6
		$4f^{12}\,(^3F)\,6s^2\,(^3F_4)$	17974.3
		$4f^{12}\,(^3H_6)\,5d\,6s\,(^3D_2)\,(6,\,2)_4$	18291.4
70	$Yb^+\,(4f^{14}\,6s - {}^2S_{1/2})$ 98300	$4f^{13}\,(^2F^o)\,6s^2\,(^2F^o_{7/2})$	21418.7
		$4f^{14}\,5d\,(^2D_{3/2})$	22960.8
		$4f^{14}\,5d\,(^2D_{5/2})$	24332.7
		$4f^{13}\,(^2F^o_{7/2})\,5d\,6s\,(^3D)\,^3[3/2]^o_{5/2}$	26759.0
		$4f^{14}\,6p\,(^2P^o_{1/2})$	27061.8
		$4f^{13}\,(^2F^o_{7/2})\,5d\,6s\,(^3D)\,^3[3/2]^o_{3/2}$	28758.0
71	$Lu^+\,(4f^{14}\,6s^2 - {}^1S_0)$ 112000	$5d\,6s\,(^3D_1)$	11796.2
		$5d\,6s\,(^3D_2)$	12435.3
		$5d\,6s\,(^3D_3)$	14199.1
		$5d\,6s\,(^1D_2)$	17332.6
		$6s\,6p\,(^3P^o_0)$	27264.4
		$6s\,6p\,(^3P^o_1)$	28503.2
		$5d^2\,(^3F_2)$	29406.7
		$5d^2\,(^3F_3)$	30889.1
		$6s\,6p\,(^3P^o_2)$	32453.3
		$5d^2\,(^3F_4)$	32503.6
72	$Hf^+\,(5d\,6s^2 - {}^2D_{3/2})$ 120000	$5d\,6s^2\,(^2D_{5/2})$	3050.9
		$5d^2\,(^3F)\,6s\,(^4F_{3/2})$	3644.6
		$(^4F_{5/2})$	4904.8
		$(^4F_{7/2})$	6344.3
		$(^4F_{9/2})$	8361.8
73	$Ta^+\,(5d^3\,6s - {}^5F_1)$	$5d^3\,(^4F)\,6s\,(^5F_2)$	1031.3
		$6s\,(^5F_3)$	2642.2
		$5d^2\,6s^2\,(^3F_2)$	3180.0
		$5d^2\,6s^2\,(^3P_0)$	4124.8
		$6s\,(^5F_4)$	4415.7
		$5d^2\,6s^2\,(^3P_1)$	5330.7
		$(^3P_2)$	5658.0
		$6s\,(^5F_5)$	6186.7
		$5d^2\,6s^2\,(^3F_3)$	6831.3
74	$W^+\,(5d^4\,6s - {}^6D_{1/2})$	$5d^4\,(^5D)\,6s\,(^6D_{3/2})$	1518.8
		$(^6D_{5/2})$	3172.5
		$(^6D_{7/2})$	4716.3
		$(^6D_{9/2})$	6147.2
		$5d^5\,(^6S_{5/2})$	7420.4
75	$Re^+\,(5d^5\,6s - {}^7S_3)$	$5d^4\,6s^2\,(^5D_0)$	13777
		$(^5D_2)$	14352
		$(^5D_1)$	14824
		$(^5D_4)$	14883
		$(^5D_3)$	14930
		$5d^5\,(^6S)\,6s\,(^5S_2)$	17223

Table 8.1 (continued)

Atomic number Z	Positive ion (ground-state term); ionization limit [cm^{-1}]	Excited state	Energy level [cm^{-1}]
1	2	3	4
76	$Os^+ (5d^6 6s - {}^6D_{9/2})$	$5d^6 ({}^5D) 6s ({}^6D_{7/2})$	3593.1
		$({}^6D_{5/2})$	3928.9
		$({}^6D_{3/2})$	5592.0
		$({}^6D_{1/2})$	6636.6
		$5d^7 ({}^4F_{9/2})$	7401.2
77	$Ir^+ (5d^7 6s - {}^5F_5)$	$5d^8 ({}^3F_4)$	2265.7
		$5d^8 ({}^3P_2)$	3090.2
		$5d^7 ({}^4F) 6s ({}^5F_4)$	4787.9
		$({}^5F_3)$	8187.0
		$({}^5F_2)$	8975.0
		$5d^8 ({}^3P_1)$	9062.1
		$5d^8 ({}^3F_3)$	9927.8
		$5d^8 ({}^3P_0)$	11211.9
		$5d^8 ({}^1D_2)$	11307.3
		$5d^7 ({}^4F) 6s ({}^3F_4)$	11719.1
		$6s ({}^5F_1)$	11957.7
		$5d^7 ({}^4P) 6s ({}^5P_3)$	12714.6
		$6s ({}^5P_2)$	15676.3
		$5d^8 ({}^1G_4)$	17210.1
78	$Pt^+ (5d^9 - {}^2D_{5/2})$ 149720	$5d^8 ({}^3F) 6s ({}^4F_{9/2})$	4787
		$5d^9 ({}^2D_{3/2})$	8420
		$6s ({}^4F_{7/2})$	9356
		$6s ({}^4F_{5/2})$	13329
		$6s ({}^4F_{3/2})$	15791
		$5d^8 ({}^3P) 6s' ({}^4P_{5/2})$	16821
		$5d^8 ({}^3F) 6s ({}^2F_{7/2})$	18098
79	$Au^+ (5d^{10} - {}^1S_0)$ 165000	$5d^9 ({}^2D_{5/2}) 6s ({}^3D_3)$	15039
		$6s ({}^3D_2)$	17639
		$5d^9 ({}^2D_{3/2}) 6s ({}^3D_1)$	27764
80	$Hg^+ (6s - {}^2S_{1/2})$ 151280	$5d^9 6s^2 ({}^2D_{5/2})$	35510
		$6s^2 ({}^2D_{3/2})$	50550
		$6p ({}^2P^o_{1/2})$	51480
		$6p ({}^2P^o_{3/2})$	60610
81	$Tl^+ (6s^2 - {}^1S_0)$ 164760	$6p ({}^3P^o_0)$	49448
		$6p ({}^3P^o_1)$	52390
		$6p ({}^3P^o_2)$	61722
		$6p ({}^1P^o_1)$	75660
		$7s ({}^3S_1)$	105220
82	$Pb^+ (6p - {}^2P^o_{1/2})$ 121245.1	$6p ({}^2P^o_{3/2})$	14081.07
		$6s 6p^2 ({}^4P_{1/2})$	57910.48
		$7s ({}^2S_{1/2})$	59448.56
		$6s 6p^2 ({}^4P_{3/2})$	66124.53
		$6d ({}^2D_{5/2})$	68964.31
		$6d ({}^2D_{3/2})$	69739.60

Table 8.1 (continued)

Atomic number Z	Positive ion (ground-state term); ionization limit [cm^{-1}]	Excited state	Energy level [cm^{-1}]
1	2	3	4
		$6s\,6p^2\,(^4P_{5/2})$	73905.71
		$7p\,(^2P^o_{1/2})$	74459.0
		$7p\,(^2P^o_{3/2})$	77272.6
		$6s\,6p^2\,(^2D_{5/2})$	88972.2
		$8s\,(^2S_{1/2})$	89180.2
83	$Bi^+\,(6p^2 - {}^3P_0)$ 135000	$6p^2\,(^3P_1)$	13320
		$6p^2\,(^3P_2)$	17030
		$6p^2\,(^1D_2)$	33940
		$6p^2\,(^1S_0)$	44170
88	$Ra^+\,(6p^6 7s - {}^2S_{1/2})$ 81842.3	$6d\,(^2D_{3/2})$	12084.4
		$6d\,(^2D_{5/2})$	13743.1
		$7p\,(^2P^o_{1/2})$	21351.2
		$7p\,(^2P^o_{3/2})$	26208.9
		$8s\,(^2S_{1/2})$	43405.0
89	$Ac^+\,(7s^2 - {}^1S_0)$ 95000	$6d\,(^2D)\,7s\,(^3D_1)$	4739.6
		$6d\,7s\,(^3D_2)$	5267.2
		$6d\,7s\,(^3D_3)$	7426.5
		$6d\,(^2D)\,7s\,(^1D_2)$	9087.5
		$6d^2\,(^3F_2)$	13236.5
		$6d^2\,(^3F_3)$	14949.2
		$6d^2\,(^3F_4)$	16756.9
		$6d^2\,(^3P_0)$	17737.1
		$6d^2\,(^3P_1)$	19015.3
		$6d^2\,(^1D_2)$	19203.0
		$6d^2\,(^1G_4)$	20848.2
		$7s\,(^2S)\,7p\,(^3P^o_0)$	20956.4
		$7s\,7p\,(^3P^o_1)$	22180.5
		$6d^2\,(^3P_2)$	22199.5
90	$Th^+\,(6d^2 7s\,(^4F_{3/2}) + 6d\,7s^2\,(^2D_{3/2}))$ 96000	$6d^2\,(^3F)\,7s\,(^4F_{5/2})$	1521.89
		$6d^2\,(^3F)\,7s\,(^4F_{3/2})$	1859.94
		$6d\,(^2D)\,7s^2\,(^2D_{5/2}) + 6d^2\,(^3F)\,7s\,(^4F_{5/2})$	4113.36
		$6d^2\,(^3F)\,7s\,(^4F_{7/2})$	4146.58
		$5f\,(^2F^o)\,7s^2\,(^2F^o_{5/2})$	4490.26
		$5f6d\,(^3H^o)\,7s\,(^4H^o_{5/2})$	6168.35
		$6d^2\,(^3F)\,7s\,(^4F_{9/2})$	6213.49
		$6d^2\,(^3P)\,7s\,(^4P_{1/2})$	6244.29
		$5f6d\,(^3F^o)\,7s\,(^4F^o_{3/2})$	6691.39
		$5f6d\,(^3H^o)\,7s\,(^4H^o_{9/2}) + 5f\,(^1G^o)\,6d^2\,(^2G^o_{9/2})$	6700.18
		$6d^3\,(^4F_{3/2}) + 6d\,(^2D)\,7s^2\,(^2D_{3/2})$	7001.43
		$5f6d\,(^3F^o)\,7s\,(^4F^o_{5/2})$	7331.49
		$6d^2\,(^1P)\,7s\,(^2P_{1/2}) + 6d^3\,(^2P_{1/2})$	7828.56
		$6d^2\,(^3P)\,7s\,(^4P_{3/2})$	8018.19
		$5f\,(^2F^o)\,7s^2\,(^2F^o_{7/2})$	8378.85
		$6d^3\,(^4F_{3/2}) + 6d\,(^2P)\,7s^2\,(^2P_{3/2})$	8460.3
		$6d^2\,(^1F)\,7s\,(^2F_{5/2})$	8605.8

Table 8.1 (continued)

Atomic number Z	Positive ion (ground-state term); ionization limit [cm^{-1}]	Excited state	Energy level [cm^{-1}]
1	2	3	4
		$6d^2(^3P)7s(^4P_{5/2})$	9061.1
		$5f6d(^1G^o)7s(^2G^o_{7/2})$	9202.3
		$5f6d(^1G^o)7s(^2G^o_{9/2}) + 5f(^3H^o)6d^2(^4H^o_{9/2})$	9238.0
		$6d^3(^4F_{5/2})$	9401.0
		$5f6d(^3G^o)7s(^4G^o_{5/2})$	9585.4
		$6d^2(^1G)7s(^2G_{7/2})$	9712.0
		$5f6d(^3F^o)7s(^4F^o_{7/2})$	9720.3
92	$U^+(5f^37s^2 - {}^4I^o_{9/2})$ 96000	$5f^36d7s(^6L^o_{11/2})$	289.04
		$5f^36d7s(^6K^o_{9/2})$	914.76
		$5f^36d7s(^6L^o_{13/2})$	1749.12
		$5f^36d7s(^6K^o_{11/2})$	2294.70
		$5f^37s^2(^4I^o_{11/2})$	4420.87
		$(5f^36d^2 + 5f^36d7s)(^6M^o_{13/2})$	4585.43
		$5f^47s(^6I_{7/2})$	4663.80
		$5f^36d7s(^6H^o_{5/2})$	4706.27
		$5f^36d7s(^6L^o_{15/2})$	5259.65
		$5f^36d7s(J = 7/2)^o$	5401.50
		$5f^36d7s(^6K^o_{13/2})$	5526.75
		$5f^36d7s(^6I^o_{7/2})$	5667.33
		$5f^47s(^6I_{9/2})$	5716.45
		$5f^36d7s(^6K^o_{11/2})$	5790.64
		$(5f^36d7s + 5f^36d^2)(J = 13/2)^o$	6283.43
		$5f^36d7s(^6I^o_{9/2})$	6445.03
		$(5f^37s^2 + 5f^36d7s)(^4F^o_{3/2})$	7017.17
		$5f^36d7s(J = 9/2)^o$	7166.63
		$(5f^36d7s + 5f^36d^2)(^4H^o_{7/2})$	7547.37
		$5f^36d7s(J = 11/2)^o$	7598.35
		$(5f^37s^2 + 5f^36d7s)(^4I^o_{13/2})$	8276.73
93	$Np^+(5f^36d7s^2 - {}^5L^o_6)$	$5f^57s(J = 2)$	83.5
		$5f^57s(J = 3)$	1053.1
94	$Pu^+(5f^67s - {}^8F_{1/2})$	$5f^67s(1, 1/2)_{3/2}$	2014.97
		$5f^67s(1, 1/2)_{1/2}$	3235.8
		$5f^67s(2, 1/2)_{5/2}$	3969.85
		$5f^67s(2, 1/2)_{3/2}$	5502.1
		$5f^67s(3, 1/2)_{7/2}$	5717.98
		$5f^67s(4, 1/2)_{9/2}$	7278.86
		$5f^67s(3, 1/2)_{5/2}$	7498.36
		$5f^67s(5, 1/2)_{11/2}$	8638.23
		$5f^67s(4, 1/2)_{7/2}$	9242.36
		$5f^67s(6, 1/2)_{13/2}$	9707.98
		$5f^67s(J = 1/2)$	10188.46
		$5f^67s(5, 1/2)_{9/2}$	10726.32
		$5f^67s(6, 1/2)_{11/2}$	11799.24
		$5f^67s(J = 3/2)$	13990.95
		$5f^67s(J = 1/2)$	14693.1

Table 8.1 (continued)

Atomic number Z	Positive ion (ground-state term); ionization limit [cm^{-1}]	Excited state	Energy level [cm^{-1}]
1	2	3	4
96	Cm^+ $(5f^7 7s^2 - {}^8S^o_{7/2})$	$5f^8 7s\ (J = 13/2)$	2093.88
		$5f^8 7s\ (J = 11/2)$	3941.46
		$5f^8 7s\ (J = 9/2)$	5919.28
		$5f^8 7s\ (J = 11/2)$	6347.92
		$5f^8 7s\ (J = 7/2)$	7067.13
		$5f^8 7s\ (J = 9/2)$	8144.32
		$5f^8 7s\ (J = 5/2)$	8436.11
		$5f^8 7s\ (J = 7/2)$	9073.60
		$5f^8 7s\ (J = 3/2)$	9127.87
		$5f^8 7s\ (J = 1/2)$	9801.32
97	Bk^+ $(5f^9 7s - {}^7H^o_8)$	$5f^9 7s\ ({}^5H^o_7)$	1487.5
		$5f^9 7s\ ({}^7F^o_6)$	5598.1
		$5f^9 7s\ ({}^5F^o_5)$	6051.2
		$5f^9 7s\ ({}^7H^o_7)$	6809.5
		$5f^9 7s\ ({}^7F^o_5)$	6906.1
		$5f^9 7s\ ({}^5F^o_4)$	7038.5
		$5f^9 7s\ ({}^5H^o_6)$	7786.7
		$5f^9 7s\ (J = 6)$	10034.0
		$5f^9 6d\ (J = 6)$	10282.4
		$5f^9 7s\ (J = 5)$	10711.2
98	Cf^+ $(5f^{10} 7s - {}^6I_{17/2})$	$5f^{10} 7s\ ({}^4I_{15/2})$	1180.5
		$5f^{10} 7s\ (J = 11/2)$	8852.0
		$5f^{10} 7s\ (J = 15/2)$	9350.2
		$5f^{10} 7s\ (J = 9/2)$	9633.1
		$5f^{10} 7s\ (J = 13/2)$	9922.26
		$5f^{10} 7s\ (J = 5/2)$	11114.47
		$5f^{10} 7s\ (J = 13/2)$	11647.9
		$5f^{10} 7s\ (J = 11/2)$	12029.5
99	Es^+ $(5f^{11} 7s - {}^5I^o_8)$	$5f^{11}\ ({}^4I^o_{15/2})\ 7s\ ({}^5I^o_7)$	938.2
		$5f^{11}\ ({}^2H^o_{11/2})\ 7s\ ({}^3I^o_6)$	9085.3
		$5f^{11}\ ({}^2H^o_{11/2})\ 7s\ ({}^3I^o_5)$	9580.2
		$5f^{11} 7p\ ({}^5I_7)$	27751.1
		$5f^{11} 7p\ (J = 8)$	28178.8

8.2 Lifetimes of Resonant Excited States in Atomic Ions

Included in Table 8.2 are values of the radiative lifetimes τ_k of low-lying resonant levels of positive atomic ions with a charge of $+1$. The fourth column of the table gives the value of the excitation energy of the multiplet centre of gravity (unless otherwise specified, see Sect. 7.1) of a given level with respect to the ionic ground state.

We selected the data for Table 8.2 with reference to [7.4.1–16].

Table 8.2. Radiative lifetimes of low-lying states in singly ionized atoms

Atomic number Z	Positive ion (ground-state term)	Excited state	Excitation energy [eV] for the centre of gravity of the multiplet	Lifetime τ [ns]
1	2	3	4	5
2	$He^+ (1s - {}^2S_{1/2})$	$2p\,({}^2P^o)$	40.814	0.10 (*A*)
		$3s\,({}^2S)$	48.3717	9.94 (*A*)
		$3p\,({}^2P^o)$	48.3718	0.33 (*A*)
		$3d\,({}^2D)$	48.3719	0.97 (*A*)
3	$Li^+ (1s^2 - {}^1S_0)$	$2p\,({}^3P^o)$	61.28	37 (*B*)
		$2p\,({}^1P^o_1)$	62.22	0.039 (*B*)
		$3p\,({}^1P^o_1)$	69.65	0.13 (*B*)
4	$Be^+ (2s - {}^2S_{1/2})$	$2p\,({}^2P^o)$	3.959	8.7 (*B*)
5	$B^+ (2s^2 - {}^1S_0)$	$2p\,({}^1P^o_1)$	9.100	0.9 (*D*)
		$2p^2\,({}^3P)$	12.266	1.0 (*D*)
6	$C^+ (2p - {}^2P^o_{1/2})$	$2s\,2p^2\,({}^2D)$	9.290	3.9 (*C*)
7	$N^+ (2p^2 - {}^3P_0)$	$2s\,2p^3\,({}^3P^o)$	13.54	2.7 (*C*)
		$2p\,3s\,({}^3P^o)$	18.48	0.9 (*C*)
		$3s\,({}^1P^o_1)$	18.50	0.21 (*C*)
		$2s\,2p^3\,({}^1P^o_1)$	20.68	0.24 (*C*)
		$3d\,({}^1P^o_1)$	23.57	0.47 (*C*)
8	$O^+ (2p^3 - {}^4S^o_{3/2})$	$2s\,2p^4\,({}^4P)$	14.87	1.2 (*C*)
		$2s\,2p^4\,({}^2D)$	20.58	0.45 (*C*)
		$3s\,({}^4P)$	22.99	0.9 (*C*)
		$3s\,({}^2P)$	23.43	0.27 (*C*)
		$2s\,2p^4\,({}^2S_{1/2})$	24.27	0.20 (*C*)
		$2s\,2p^4\,({}^2P)$	26.37	0.14 (*C*)
		$2p^2\,({}^1D)\,3p'\,({}^2P^o)$	28.83	4.9 (*C*)
		$3d\,({}^2P)$	28.95	0.58 (*C*)
		$5p\,({}^4P^o)$	32.40	1.4 (*C*)
9	$F^+ (2p^4 - {}^3P_2)$	$2s\,2p^5\,({}^3P^o)$	2.045	0.32 (*C*)
10	$Ne^+ (2p^5 - {}^2P^o_{3/2})$	$2p^4\,({}^3P)\,3p\,({}^4P^o_{5/2})$	30.52	7.8 (*C*)
		$({}^4P^o_{3/2})$	30.55	10.4 (*C*)
		$3p\,({}^4D^o_{7/2})$	30.89	6.1 (*D*)
		$({}^4D^o_{5/2})$	30.93	6.3 (*D*)
		$3p\,({}^2D^o_{5/2})$	31.12	7.5 (*D*)
		$({}^2D^o_{3/2})$	31.18	8.4 (*C*)
		$3p\,({}^2S^o_{1/2})$	31.34	4.9 (*C*)
		$3p\,({}^2P^o_{3/2})$	31.51	5.4 (*C*)
		$({}^2P^o_{1/2})$	31.53	5.7 (*C*)
		$2p^4\,({}^1D)\,3p'\,({}^2F^o_{7/2})$	34.02	8.4 (*C*)
		$3p'\,({}^2D^o_{5/2})$	34.39	6.5 (*D*)
11	$Na^+ (2p^6 - {}^1S_0)$	$3p\,({}^3S_1)\,(2p_{10})$	36.35	9.7 (*C*)
		$3p\,({}^3D_3)\,(2p_9)$	36.85	8.0 (*C*)
		$({}^3D_2)\,(2p_8)$	36.90	7.8 (*C*)
		$({}^3D_1)\,(2p_7)$	36.97	–

Table 8.2 (continued)

Atomic number Z	Positive ion (ground-state term)	Excited state	Excitation energy [eV] for the centre of gravity of the multiplet	Lifetime τ [ns]
1	2	3	4	5
		$3p\,(^1D_2)\,(2p_6)$	37.10	7.1 (*C*)
		$3p\,(^1P_1)\,(2p_5)$	37.18	6.5 (*C*)
		$3p\,(^3P_2)\,(2p_4)$	37.21	–
		$(^3P_0)\,(2p_3)$	37.24	5.4 (*C*)
		$(^3P_1)\,(2p_2)$	37.26	6.5 (*C*)
		$3p\,(^1S_0)\,(2p_1)$	38.29	2.9 (*C*)
12	$Mg^+\,(3s - {}^2S_{1/2})$	$3p\,(^2P^o)$	4.430	3.7 (*C*)
		$4s\,(^2S_{1/2})$	8.655	2.7 (*D*)
		$3d\,(^2D)$	8.864	2.1 (*D*)
		$4p\,(^2P^o)$	9.998	21 (*C*)
		$4d\,(^2D)$	11.57	12 (*C*)
		$4f\,(^2F^o)$	11.63	4.5 (*D*)
13	$Al^+\,(3s^2 - {}^1S_0)$	$3p\,(^1P^o_1)$	7.421	0.68 (*D*)
14	$Si^+\,(3p - {}^2P^o_{1/2})$	$3p\,(^2S_{1/2})$	9.505	0.6 (*D*)
		$3d\,(^2D)$	9.838	0.5 (*E*)
		$4p\,(^2P^o)$	10.07	9.6 (*C*)
		$4f\,(^2F^o)$	12.84	3.7 (*C*)
15	$P^+\,(3p^2 - {}^3P_0)$	$3s\,3p^3\,(^3P^o)$	9.521	5.4 (*D*)
		$4s\,(^3P^o)$	10.78	0.8 (*D*)
		$4s\,(^1P^o_1)$	11.02	0.7 (*D*)
		$3d\,(^3D^o)$	12.88	0.2 (*D*)
16	$S^+\,(3p^3 - {}^4S^o_{3/2})$	$3s\,3p^4\,(^4P)$	9.871	22 (*C*)
17	$Cl^+\,(3p^4 - {}^3P_2)$	$3p^3\,(^4S^o)\,4p\,(^3P)$	16.337	12 (*C*)
		$3p^3\,(^2D^o)\,4p'\,(^3D_3)$	18.160	9 (*C*)
		$4p'\,(^1F_3)$	18.301	13 (*C*)
		$4p'\,(^3P_2)$	18.573	5.0 (*D*)
		$3p^3\,(^2P^o)\,4p''\,(^3P_2)$	20.040	7 (*D*)
18	$Ar^+\,(3p^5 - {}^2P^o_{3/2})$	$3s\,3p^6\,(^2S_{1/2})$	13.48	4.8 (*C*)
		$3p^4\,(^3P)\,4s\,(^2P)$	17.18	0.32 (*C*)
		$4p\,(^4P^o_{5/2})$	19.223	7.4 (*B*)
		$(^4P^o_{3/2})$	19.261	7.0 (*B*)
		$4p\,(^4D^o_{7/2})$	19.495	6.9 (*B*)
		$(^4D^o_{5/2})$	19.549	7.5 (*C*)
		$(^4D^o_{3/2})$	19.610	7.4 (*C*)
		$4p\,(^2D^o_{5/2})$	19.680	9.6 (*B*)
		$(^2D^o_{3/2})$	19.762	8.9 (*C*)
		$4p\,(^2P^o_{1/2})$	19.801	8.7 (*C*)
		$(^2P^o_{3/2})$	19.867	9.4 (*C*)
		$4p\,(^4S^o_{3/2})$	19.968	7.2 (*C*)
		$4p\,(^2S^o_{1/2})$	19.973	8.8 (*C*)
		$3p^4\,(^1D)\,4p'\,(^2F^o_{5/2})$	21.13	8.0 (*B*)
		$4p'\,(^2F^o_{7/2})$	21.14	8.8 (*B*)

Table 8.2 (continued)

Atomic number Z	Positive ion (ground-state term)	Excited state	Excitation energy [eV] for the centre of gravity of the multiplet	Lifetime τ [ns]
1	2	3	4	5
		$4p'\,(^2P^o)$	21.38	4.3 (C)
		$4p'\,(^2D^o)$	21.50	7.6 (C)
19	$K^+\,(3p^6 - {}^1S_0)$	$4p\,(^3D_2)\,(2p_8)$	23.15	8.4 (C)
		$(^3D_1)\,(2p_7)$	23.25	8.3 (C)
		$(^1D_2)\,(2p_6)$	23.33	8.6 (C)
		$(^1P_1)\,(2p_4)$	23.46	9.6 (C)
		$(^3P_2)\,(2p_3)$	23.515	8.3 (C)
		$(^3P_0)\,(2p_5)$	23.529	6.7 (C)
		$(^3P_1)\,(2p_2)$	23.57	7.2 (C)
		$3p^5\,(^2P^o_{3/2})\,5s'\,[3/2]^o_2\,(2s_5)$	26.36	3.3 (C)
20	$Ca^+\,(4s - {}^2S_{1/2})$	$4p\,(^2P^o)$	3.142	6.8 (C)
		$5s\,(^2S_{1/2})$	6.468	4.0 (D)
		$4d\,(^2D)$	7.049	3.1 (D)
21	$Sc^+\,(3d\,4s - {}^3D_1)$	$4p\,(^3F^o_2)$	3.403	6.2 (C)
		$4p\,(^1P^o_1)$	3.821	9.2 (C)
24	$Cr^+\,(3d^5 - {}^6S_{5/2})$	$4p - z\,^6P$	6.017	3.3 (D)
25	$Mn^+\,(3d^5\,4s - {}^7S_3)$	$3d^5\,(^6S)\,4p - z\,^7P^o_2$	4.757	3.7 (C)
		$z\,^7P^o_3$	4.779	3.8 (C)
		$4p - z\,^5P^o$	5.387	4.0 (C)
		$3d^5\,(^4G)\,4p - z\,^5G^o_4$	7.996	4.8 (D)
		$z\,^5G^o_6$	8.003	3.9 (D)
		$4p - z\,^5H^o_3$	8.119	4.3 (D)
		$z\,^5H^o_4$	8.129	4.0 (D)
		$z\,^5H^o_5$	8.141	3.2 (D)
		$4p - z\,^5F^o_5$	8.250	4.2 (C)
26	$Fe^+\,(3d^6\,4s - {}^6D_0)$	$3d^6\,(^5D)\,4p - z\,^6D^o$	4.799	4.0 (C)
		$4p - z\,^6F^o_{9/2}$	5.222	3.4 (C)
		$4p - z\,^6P^o_{3/2}$	5.408	3.9 (C)
		$4p - z\,^4F^o$	5.544	4.0 (C)
		$4p - z\,^4D^o_{7/2}$	5.511	3.7 (C)
		$z\,^4D^o_{5/2}$	5.553	3.4 (C)
		$z\,^4D^o_{3/2}$	5.585	3.4 (C)
		$4p - z\,^4P^o_{3/2}$	5.876	3.8 (C)
29	$Cu^+\,(3d^{10} - {}^1S_0)$	$3d^9\,4p\,(^3F^o)$	8.543	2.5 (C)
		$4p\,(^1F^o_3)$	8.917	2.1 (D)
		$4p\,(^1D^o_2)$	9.095	13 (C)
		$4p\,(^1P^o_1)$	9.125	1.8 (D)
		$5s\,(^1D_2)$	13.68	2.3 (D)
		$5p\,(^3P^o_1)$	14.99	12 (C)
		$5f\,(^1F^o_3)$	18.36	16 (C)
30	$Zn^+\,(3d^{10}\,4s - {}^2S_{1/2})$	$4p\,(^2P^o)$	6.083	2.75 (D)
		$3d^9\,4s^2\,(^2D_{5/2})$	7.777	$1.6 \cdot 10^3$ (C)

Table 8.2 (continued)

Atomic number Z	Positive ion (ground-state term)	Excited state	Excitation energy [eV] for the centre of gravity of the multiplet	Lifetime τ [ns]
1	2	3	4	5
		$(^2D_{3/2})$	8.114	$2.2 \cdot 10^3\,(C)$
		$5s\,(^2S_{1/2})$	10.96	2.5 (D)
		$4d\,(^2D)$	12.02	2.3 (D)
		$5p\,(^2P)$	12.59	17 (D)
		$4f\,(^2F^o_{5/2})$	14.54	6.4 (D)
31	$Ga^+\,(4s^2 - {}^1S_0)$	$4p\,(^1P^o_1)$	8.77	0.5 (C)
		$4p^2\,(^1D_2)$	13.36	55 (C)
		$4d\,(^1D_2)$	15.65	0.75 (C)
36	$Kr^+\,(4p^5 - {}^2P^o_{1/2})$	$4s\,4p^6\,(^2S_{1/2})$	13.515	0.33 (D)
		$4p^4\,(^3P)\,5p\,(^4P^o_{5/2})$	16.60	7.9 (B)
		$(^4P^o_{3/2})$	16.65	8.1 (B)
		$(^4P^o_{1/2})$	16.84	8.5 (C)
		$5p\,(^4D^o_{7/2})$	16.84	7.2 (C)
		$(^4D^o_{5/2})$	16.87	9.6 (B)
		$(^4D^o_{3/2})$	17.16	8.4 (B)
		$(^4D^o_{1/2})$	17.38	8.1 (B)
		$5p\,(^2P^o_{1/2})$	17.25	9.1 (B)
		$(^2P^o_{3/2})$	17.37	8.5 (B)
		$5p\,(^2D^o_{5/2})$	17.37	8.2 (B)
		$(^2D^o_{3/2})$	17.61	8.7 (B)
		$5p\,(^4S^o_{3/2})$	17.57	8.3 (B)
		$5p\,(^2S^o_{1/2})$	17.65	9.4 (B)
		$4p^4\,(^1D)\,5p'\,(^2F^o_{5/2})$	18.50	8.6 (B)
		$(^2F^o_{7/2})$	18.56	8.3 (B)
		$5p'\,(^2P^o_{3/2})$	18.62	6.3 (B)
		$(^2P^o_{1/2})$	18.88	6.0 (B)
		$5p'\,(^2D^o_{3/2})$	18.87	7.7 (B)
		$(^2D^o_{5/2})$	18.88	7.8 (B)
37	$Rb^+\,(4p^6 - {}^1S_0)$	$4p^5\,(^2P^o_{3/2})\,4d\,[1/2]^o_1$	15.68	2.0 (D)
		$5s\,[3/2]^o_1$	16.72	1.2 (C)
		$4d\,[3/2]^o_1$	17.43	1.1 (D)
		$4p^5\,(^2P^o_{1/2})\,5s'\,[1/2]^o_1$	17.79	0.9 (C)
		$4d'\,[3/2]^o_1$	18.06	0.8 (E)
		$4p^5\,(^2P^o_{3/2})\,5p\,[1/2]_1\,(5p_1)$	19.13	8.0 (B)
		$5p\,[5/2]_2\,(5p_2)$	19.43	8.5 (D)
		$[5/2]_3\,(5p_3)$	19.45	7 (D)
		$5p\,[3/2]_1\,(5p_4)$	19.61	7.5 (D)
		$[3/2]_2\,(5p_5)$	19.68	–
		$5p\,[1/2]_0\,(5p_6)$	19.99	6 (D)
		$4p^5\,(^2P^o_{1/2})\,5p'\,[3/2]_1\,(5p_7)$	20.32	9.5 (D)
		$[3/2]_2\,(5p_8)$	20.45	8 (D)
		$5p'\,[1/2]_1\,(5p_9)$	20.47	7.5 (D)
		$[1/2]_0\,(5p_{10})$	20.78	6 (D)

Table 8.2 (continued)

Atomic number Z	Positive ion (ground-state term)	Excited state	Excitation energy [eV] for the centre of gravity of the multiplet	Lifetime τ [ns]
1	2	3	4	5
38	$Sr^+ (4p^6 5s - {}^2S_{1/2})$	$5p\,({}^2P^o_{1/2})$	2.940	7.5 (*B*)
		$5p\,({}^2P^o_{3/2})$	3.040	6.7 (*B*)
39	$Y^+ (5s^2 - {}^1S_0)$	$4d\,({}^2D)\,5p - z\,{}^1D^o_2$	3.242	6.3 (*C*)
		$5p - z\,{}^3F^o$	3.450	6 (*C*)
		$5p - z\,{}^3D^o$	3.587	4.3 (*C*)
40	$Zr^+ (4d^2 5s - {}^4F_{3/2})$	$4d^2 5s - z\,{}^4G^o_{5/2}$	3.450	7.1 (*C*)
41	$Nb^+ (4d^4 - {}^5D_0)$	$4d^3\,({}^4F)\,5p - z\,{}^5G^o$	4.353	5.5 (*C*)
		$5p - z\,{}^3D^o$	4.448	5.5 (*C*)
42	$Mo^+ (4d^5 - {}^6S_{5/2})$	$4d^4\,({}^5D)\,5p - z\,{}^6F^o$	5.908	4.9 (*C*)
		$z\,{}^6P^o$	6.128	3.0 (*D*)
		$z\,{}^6D^o$	6.255	4.3 (*C*)
48	$Cd^+ (4d^{10} 5s - {}^2S_{1/2})$	$5p\,({}^2P^o_{1/2})$	5.472	3.5 (*C*)
		$5p\,({}^2P^o_{3/2})$	5.780	3.5 (*C*)
		$4d^9 5s^2\,({}^2D_{5/2})$	8.587	830 (*C*)
		$({}^2D_{3/2})$	9.286	300 (*C*)
		$6p\,({}^2P^o_{3/2})$	11.83	19 (*C*)
		$4f\,({}^2F^o)$	13.44	6.0 (*C*)
		$6d\,({}^2D)$	13.67	6.5 (*D*)
		$5f\,({}^2F_{7/2})$	14.69	15 (*C*)
53	$I^+ (5p^4 - {}^3P_2)$	$5p^3\,({}^4S^o)\,6p\,({}^5P_3)$	12.45	13 (*C*)
		$6p\,({}^3P_2)$	12.72	13.5 (*C*)
		$6p\,({}^3P_1)$	12.60	14 (*C*)
		$5p^3\,({}^2D^o)\,6p'\,({}^3D)$	13.96	15 (*C*)
		$6p'\,({}^3F)$	14.18	16 (*C*)
		$6p'\,({}^3P_2)$	14.35	11 (*C*)
		$6p'\,({}^1D_2)$	14.77	12 (*C*)
54	$Xe^+ (5p^5 - {}^2P^o_{3/2})$	$5s\,5p^6\,({}^2S_{1/2})$	11.267	34 (*B*)
		$5p^4\,({}^3P)\,6p\,({}^4P^o_{3/2})$	13.86	9 (*C*)
		$({}^4P^o_{5/2})$	13.88	7.9 (*B*)
		$({}^4P^o_{1/2})$	14.09	11 (*D*)
		$6p\,({}^4D^o_{5/2})$	14.07	7.3 (*B*)
		$({}^4D^o_{7/2})$	14.10	6.9 (*B*)
		$({}^4D^o_{3/2})$	14.48	9 (*C*)
		$({}^4D^o_{1/2})$	14.93	7.4 (*C*)
		$6p\,({}^2S^o_{1/2})$	15.02	8.4 (*C*)
		$6p\,({}^2D^o_{5/2})$	15.26	7.5 (*C*)
		$({}^2D^o_{3/2})$	15.41	9 (*C*)
		$6p\,({}^2P^o_{3/2})$	15.28	10 (*C*)
		$({}^2P^o_{1/2})$	15.45	8.5 (*C*)
		$5p^4\,({}^1D)\,6p'\,({}^2P^o_{3/2})$	16.08	7.6 (*C*)
		$({}^2P^o_{1/2})$	16.46	7.8 (*C*)
		$6p'\,({}^2D^o_{3/2})$	16.36	9 (*C*)
		$({}^2D^o_{5/2})$	16.39	10 (*C*)

Table 8.2 (continued)

Atomic number Z	Positive ion (ground-state term)	Excited state	Excitation energy [eV] for the centre of gravity of the multiplet	Lifetime τ [ns]
1	2	3	4	5
55	$Cs^+ (5p^6 - {}^1S_0)$	$6p\,[1/2]_1$	15.69	9.2 (*C*)
		$6p\,[5/2]_3$	16.009	6.6 (*C*)
		$6p\,[3/2]_1$	16.118	7.8 (*C*)
		$6p\,[3/2]_2$	16.214	9.6 (*C*)
56	$Ba^+ (6s - {}^2S_{1/2})$	$6p\,({}^2P^o_{1/2})$	2.512	7.9 (*B*)
		$6p\,({}^2P^o_{3/2})$	2.722	6.31 (*A*)
		$6d\,({}^2D_{3/2})$	5.697	60 (*C*)
57	$La^+ (5d^2 - {}^3F_2)$	$5d\,({}^2D)\,4f - y\,{}^3F^o_2$	2.134	510 (*B*)
		$y\,{}^3F^o_3$	2.261	430 (*B*)
		$y\,{}^3F^o_4$	2.382	450 (*B*)
		$4f - z\,{}^1D^o_2$	2.343	570 (*C*)
		$4f - z\,{}^3D^o_1$	2.658	44 (*B*)
		$z\,{}^3D^o_2$	2.741	51 (*C*)
		$z\,{}^3D^o_3$	2.794	68 (*B*)
73	$Ta^+ (5d^3\,6s - {}^5F_1)$	$({}^5G^o_2)$	4.616	5.5 (*C*)
		$({}^5F^o_2)$	4.775	13 (*C*)
		$({}^5F^o_1)$	4.778	7 (*C*)
		$({}^5F^o_3)$	4.872	15 (*C*)
74	$W^+ (5d^4\,6s - {}^6D_{1/2})$	$5d^4\,({}^5D)\,6p\,({}^6F^o_{1/2})$	4.48	14 (*C*)
		$5d^4\,({}^3P)\,6p\,({}^2S^o_{1/2})$	4.78	11 (*C*)
80	$Hg^+ (5d^{10}\,6s - {}^2S_{1/2})$	$6p\,({}^2P^o_{1/2})$	6.383	2.3 (*D*)
		$({}^2P^o_{3/2})$	7.514	1.9 (*C*)
		$5d^9\,6s\,6p\,({}^2D_{5/2})$	9.88	250 (*C*)
		$({}^4D_{7/2})$	10.44	39 (*C*)
		$({}^4F_{5/2})$	10.52	150 (*C*)
		$6d\,({}^2D_{3/2})$	13.02	1.7 (*D*)
		$({}^2D_{5/2})$	13.09	1.9 (*D*)

8.3 Energy Levels and Lifetimes for Metastable States in Singly Ionized Atoms

Table 8.3 gives values of the excitation energy T for low-lying metastable states of positive ions with a charge of $+1$, the transition wavelengths λ (in a number of cases, values of T and λ were referred to the positions of multiplet centres of gravity) and the spontaneous emission lifetimes τ_m of such states, obtained mainly by calculation [7.1.1–10, 7.5.1–15].

Table 8.3. Excitation energies and spontaneous emission lifetimes of metastable singly ionized atoms

Atomic number Z	Element (ground-state term)	Metastable level	Excitation energy $[cm^{-1}]$	Excitation energy [eV]	Lower state for radiative decay of metastable level	Wavelength of radiative transition to lower state λ [Å]	Radiative lifetime τ_m [s] (and accuracy)
2	$He^+ (1^2S_{1/2})$	$2^2S_{1/2}$	329179.8	40.81	$1^2S_{1/2}$	$(2E1 + M1)$	0.00190 (A)
3	$Li^+ (1^1S_0)$	2^3S_1	476034.6	59.02	1^1S_0	210.1	49 (B)
		2^1S_0	491374.6	60.92	1^1S_0	203.5	$0.51 \cdot 10^{-3}$ (C)
6	$C^+ (2^2P^o_{1/2})$	$2^2P^o_{1/2}$	63.4	0.0079	$2^2P^o_{1/2}$	$1.6 \cdot 10^6$	$4 \cdot 10^5$ (E)
7	$N^+ (2^3P_0)$	2^3P_1	48.7	0.0060	2^3P_0	$2.1 \cdot 10^6$	$5 \cdot 10^5$ (E)
		2^3P_2	130.8	0.016	2^3P_1	$1.2 \cdot 10^6$	$1 \cdot 10^5$ (E)
		2^1S_0	32688.8	4.053	2^1D_2	5755	0.92 (C)
8	$O^+ (2^4S^o_{3/2})$	$2^2P^o_{3/2}$	40467	5.017	2^2D^o	7320–7330	7.7 (C)
		$2^2P^o_{1/2}$	40468	5.017	2^2D^o	7320–7330	7.6 (C)
10	$Ne^+ (2^2P^o_{3/2})$	$2^2P^o_{1/2}$	780.3	0.097	$2^2P^o_{3/2}$	$1.3 \cdot 10^5$	120 (D)
14	$Si^+ (3^2P^o_{1/2})$	$3^2P^o_{3/2}$	287.4	0.036	$3^2P^o_{1/2}$	$3.5 \cdot 10^5$	$5 \cdot 10^3$ (E)
15	$P^+ (3^3P_0)$	3^3P_1	165	0.0204	3^3P_0	$6.1 \cdot 10^5$	$1.2 \cdot 10^4$ (D)
		3^3P_2	469	0.0581	3^3P_1	$3.3 \cdot 10^5$	$2.7 \cdot 10^3$ (D)
		3^1D_2	8883	1.101	$3^3P_2 (3^3P_1)$	11882 (11468)	43 (D)
		3^1S_0	21576	2.675	$3^1D_2 (3^3P_1)$	7876 (4669)	0.46 (D)
16	$S^+ (3^4S^o_{3/2})$	$3^2D^o_{3/2}$	14852	1.841	$3^4S^o_{3/2}$	6731	$3.4 \cdot 10^3$ (D)
		$3^2D^o_{5/2}$	14883	1.845	$3^4S^o_{3/2}$	6717	$5.2 \cdot 10^3$ (D)
		$3^2P^o_{1/2}$	24524	3.041	$3^2D^o_{3/2} (3^4S^o_{3/2})$	10336 (4076)	5.1 (D)
		$3^2P^o_{3/2}$	24573	3.047	$3^2D^o_{5/2} (3^4S^o_{3/2})$	10318 (4068)	3.4 (D)
17	$Cl^+ (3^3P_2)$	3^3P_1	696.0	0.0863	3^3P_2	$1.4 \cdot 10^5$	720 (D)
		3^3P_0	996.5	0.124	3^3P_1	$3.3 \cdot 10^5$	$3.4 \cdot 10^3$ (D)
		3^1D_2	11653.6	1.445	$3^3P_2 (3^3P_1)$	8579 (9124)	28 (D)
		3^1S_0	27878.0	3.456	$3^1D_2 (3^3P_1)$	6162 (3678)	0.47 (D)
18	$Ar^+ (3^2P^o_{3/2})$	$3^2P^o_{1/2}$	1432	0.178	$3^2P^o_{3/2}$	$7 \cdot 10^4$	20 (D)
26	$Fe^+ (4^6D_{9/2})$	$3d^6 4s - a^6D_{7/2}$	384.79	0.0477	$a^6D_{9/2}$	$2.6 \cdot 10^5$	470 (D)
		$a^6D_{5/2}$	667.68	0.0828	$a^6D_{7/2}$	$3.5 \cdot 10^5$	630 (D)

Table 8.3 (continued)

		$a\,^6D_{3/2}$	862.61	0.107	$a\,^6D_{5/2}$	$5.1 \cdot 10^5$	$1.4 \cdot 10^3\,(D)$
		$a\,^6D_{1/2}$	977.05	0.121	$a\,^6D_{3/2}$	$8.7 \cdot 10^5$	$5.3 \cdot 10^3\,(D)$
		$3d^7 - a\,^4F_{9/2}$	1872.6	0.232	$a\,^6D_{9/2}$	$5.3 \cdot 10^4$	$3.6 \cdot 10^3\,(D)$
		$a\,^4F_{7/2}$	2439.1	0.302	$a\,^4F_{9/2}\,(a\,^6D_{7/2})$	$1.8 \cdot 10^5\,(4.9 \cdot 10^4)$	150 (D)
		$a\,^4F_{5/2}$	2837.9	0.352	$a\,^4F_{7/2}\,(a\,^6D_{5/2})$	$2.5 \cdot 10^5\,(4.6 \cdot 10^4)$	260 (D)
		$a\,^4F_{3/2}$	3117.5	0.387	$a\,^4F_{5/2}\,(a\,^6D_{3/2})$	$3.6 \cdot 10^5\,(4.4 \cdot 10^4)$	670 (D)
		$3d^6 4s - a\,^4D_{7/2}$	7955.3	0.986	$a\,^6D_{9/2}\,(a\,^4F_{9/2})$	12567 (16436)	96 (D)
		$a\,^4D_{5/2}$	8391.9	1.040	$a\,^4D_{7/2}\,(a\,^6D_{5/2})$	$2.3 \cdot 10^5$ (12943)	108 (D)
		$a\,^4D_{3/2}$	8680.5	1.076	$a\,^6D_{3/2}\,(a\,^4F_{7/2},\ a\,^4D_{5/2})$	12788 (16018, $3.5 \cdot 10^5$)	124 (D)
		$a\,^4D_{1/2}$	8846.8	1.097	$a\,^6D_{1/2}\,(a\,^4F_{5/2})$	12703 (16637)	133 (D)
		$3d^7 - a\,^4P_{5/2}$	13474.4	1.671	$a\,^4F_{9/2}\,(a\,^4F_{7/2})$	8617 (9059)	38 (D)
		$a\,^4P_{3/2}$	13673.2	1.695	$a\,^4F_{7/2}\,(a\,^4F_{5/2})$	8899 (9227)	43 (D)
		$a\,^4P_{1/2}$	13904.8	1.724	$a\,^4F_{5/2}\,(a\,^4F_{3/2})$	9033 (9268)	43 (D)
29	$Cu^+\,(3\,^1S_0)$	$4\,^3D_2$	22847.0	2.833	$3\,^1S_0$	4376	8.3 (D)
		$4\,^1D_2$	26264.5	3.256	$3\,^1S_0$	3806	0.53 (D)
32	$Ge^+\,(4\,^2P^o_{1/2})$	$4\,^2P^o_{3/2}$	1767.4	0.219	$4\,^2P^o_{1/2}$	$5.7 \cdot 10^4$	20 (D)
35	$Br^+\,(4\,^3P_2)$	$4\,^3P_1$	3140	0.389	$4\,^3P_2$	$3.2 \cdot 10^4$	1.5 (D)
		$4\,^3P_0$	3840	0.476	$4\,^3P_1$	$1.4 \cdot 10^5$	56 (D)
		$4\,^1D_2$	11410	1.415	$4\,^3P_2\,(4\,^3P_1)$	8762	0.46 (D)
		$4\,^1S_0$	–	–	$4\,^3P_1\,(4\,^1D_2)$	–	0.04 (E)
36	$Kr^+\,(4\,^2P^o_{3/2})$	$4\,^2P^o_{1/2}$	5370.1	0.666	$4\,^2P^o_{3/2}$	$1.9 \cdot 10^4$	0.36 (D)
50	$Sn^+\,(5\,^2P^o_{1/2})$	$5\,^2P^o_{3/2}$	4251.5	0.527	$5\,^2P^o_{1/2}$	$2.4 \cdot 10^4$	1.4 (D)
53	$I^+\,(5\,^3P_2)$	$5\,^3P_0$	6448	0.799	$5\,^3P_2$	$1.6 \cdot 10^4$	45 (D)
		$5\,^3P_1$	7086.9	0.879	$5\,^3P_2$	$1.4 \cdot 10^4$	0.14 (D)
		$5\,^1D_2$	13727	1.702	$5\,^3P_2\,(5\,^3P_1)$	7283	0.11 (D)
		$5\,^1S_0$	29501	3.658	$5\,^3P_1\,(5\,^1D_2)$	3389	0.011 (D)
54	$Xe^+\,(5\,^2P^o_{3/2})$	$5\,^2P^o_{1/2}$	10537	1.306	$5\,^2P^o_{1/2}$	9487	0.047 (D)
82	$Pb^+\,(6\,^2P^o_{1/2})$	$6\,^2P^o_{3/2}$	14081.1	1.746	$6\,^2P^o_{1/2}$	7100	0.038 (D)
83	$Bi^+\,(6\,^3P_0)$	$6\,^3P_1$	13320	1.651	$6\,^3P_0$	7505	0.03 (D)
		$6\,^3P_2$	17030	2.111	$6\,^3P_0\,(6\,^3P_1)$	5870	0.51 (D)
		$6\,^1D_2$	33940	4.208	$6\,^3P_1\,(6\,^3P_2)$	4848 (5912)	0.009 (D)

8.4 Optical Parameters of Multiply Charged Atomic Ions

Tables 8.4, 5 give quantitative information about the spectroscopic constants of low-lying ($n \leqslant 2$) states of atomic ions, isoelectronic to hydrogen and helium, in a wide range of nuclear charge variation ($Z \lesssim 90$). A characteristic feature of spectra of multiply charged atomic ions is the existence of intense forbidden radiative transitions between the states, which are governed by the contributions of strong relativistic effects [8.4.1, 2].

Figure 8.1 shows a schematic diagram of the energy levels of one-electron multiply charged ions, illustrating their relative positions and the character of the dominating (multipole) radiative transitions between them. Table 8.4 includes data for the transition probability A_{ki} and radiative lifetime τ_k of the upper level based on calculations [8.4.3–5].

In Fig. 8.2 there is a schematic diagram of the relative positions (not to scale!) of the energy levels of helium-like ions with nuclear charge $Z = 10$ and $Z = 80$. The dominant decay modes of $n = 2$ levels are marked on the diagram. Table 8.5 presents the calculated data for A_{ki} and τ_k [8.4.6–13].

In addition, the transition wavelengths λ for $n \leqslant 2$ levels in hydrogen- and helium-like ions are included in Tables 8.4, 5. These data are based on the energy differences between the relevant levels of the multiply charged ions (Sect. 6.3).

All the values given are truncated to the point where the uncertainty is at most ± 1 or 2 in the last significant figure quoted.

References

8.4.1 R.Marrus, P.J.Mohr: "Forbidden transitions in one- and two-electron atoms", Adv. At. Mol. Phys. **14**, 181 (1978)

8.4.2 I.A.Sellin: "Highly ionized ions", Adv. At. Mol. Phys. **12**, 215 (1976)

8.4.3 F.A.Parpia, W.R.Johnson: Phys. Rev. **A 26**, 1142 (1982)

8.4.4 S.P.Goldman, G.W.F.Drake: Phys. Rev. **A 24**, 183 (1981)

8.4.5 D.S.Viktorov, S.A.Zapryagaev, W.G.Pal'chikov: *Radiative Transition Probabilities of One-Electron Atoms with Various Nuclear Charge*, Preprint N 13 of the Institute of Spectroscopy, USSR Academy of Science (ISAS Print. Office, Troitsk 1977)

8.4.6 C.D.Lin, W.R.Johnson, A.Dalgarno: Phys. Rev. **A 15**, 154 (1977)

8.4.7 B.Schiff, C.L.Pekeris, Y.Accad: Phys. Rev. **A 4**, 885 (1971)

8.4.8 G.W.F.Drake: Phys. Rev. **A 3**, 908 (1971); Phys. Rev. **A 19**, 1387 (1979)

8.4.9 J.Hata, I.P.Grant: J. Phys. **B 14**, 2111 (1981); J. Phys. **B 17**, 931 (1984)

8.4.10 L.A.Vainstein, U.I.Safronova: At. Data Nucl. Data Tables **21**, 49 (1978)

8.4.11 J.Sucher: Rep. Prog. Phys. **41**, 1781 (1978)

8.4.12 H.Gould, R.Marrus, P.J.Mohr: Phys. Rev. Lett. **33**, 676 (1974)

8.4.13 P.J.Mohr: "Hyperfine Quenching of the $2\,^3P_0$ State in Heliumlike Ions", in *Beam-Foil Spectroscopy*, Vol. 1, ed. by I. A. Sellin, D. J. Pegg (Plenum, New York 1976) pp. 97–103

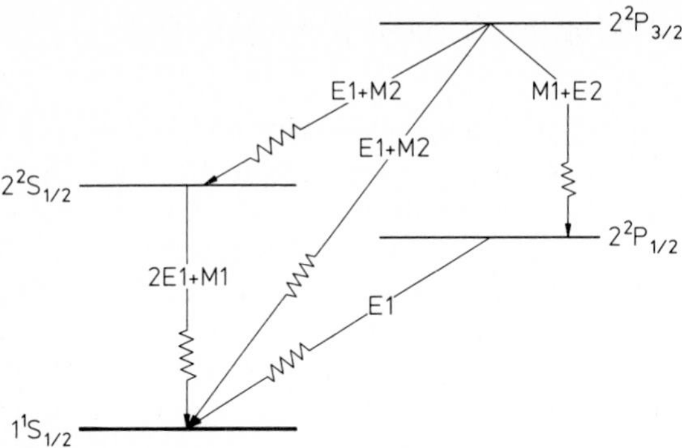

Fig. 8.1. Energy-level diagram for hydrogen isoelectronic sequence showing examples of allowed and forbidden radiative transitions

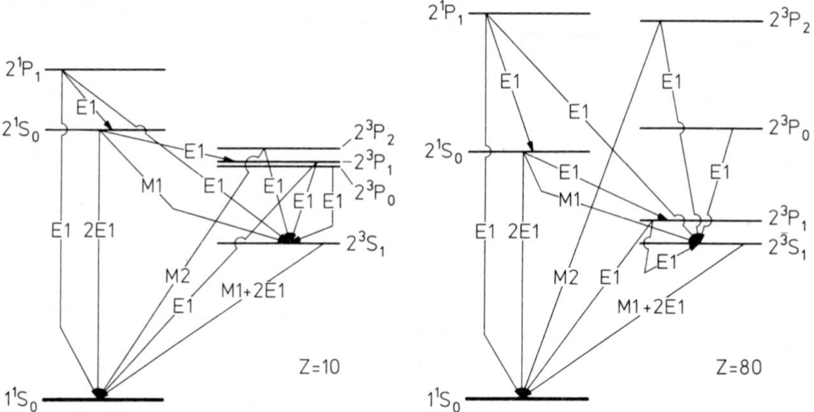

Fig. 8.2. Energy-level diagram for helium isoelectronic sequence showing examples of allowed and forbidden radiative transitions

Table 8.4. Wavelengths λ [Å] (*upper line* in each entry), transition probabilities A_{ki} [$10^{12}\,s^{-1}$] (*middle line*) and radiative lifetimes τ_k [ns] (*lower line*) for H(nl)-like isoelectronic sequence: $n = 1, 2$; $\Delta n = 0, 1$; $2 \leqslant Z \leqslant 94$ [$X(Y)$ means $X \cdot 10^Y$]

Atomic number Z	Ion	$2p\,^2P_{1/2} \rightarrow 1s\,^2S_{1/2}$	$2s\,^2S_{1/2} \rightarrow 1s\,^2S_{1/2}$	$2p\,^2P_{3/2} \rightarrow 1s\,^2S_{1/2}$	$2p\,^2P_{3/2} \rightarrow 2p\,^2P_{1/2}$	$2p\,^2P_{3/2} \rightarrow 2s\,^2S_{1/2}$
1	2	3	4	5	6	7
2	^{4}He II	303.7858 1.003(−2) 9.970(−2)	303.7854 5.265(−10) 1.899(+6)	303.7804 1.003(−2) 9.970(−2)	1.7(+7) 3.6(−21)	1.9(+7) 9.1(−16)
3	^{7}Li III	135.0012 5.077(−2) 1.970(−2)	135.0008 5.997(−9) 1.668(+5)	134.9958 5.077(−2) 1.970(−2)	3.4(+6) 4.7(−19)	3.6(+6) 5.2(−14)
4	^{9}Be IV	75.9313 0.1605 6.231(−3)	75.9310 3.369(−8) 2.968(+4)	75.9259 0.1604 6.234(−3)	1.1(+6) 1.5(−17)	1.1(+6) 9.3(−13)
5	^{11}B V	48.5909 0.392 2.55(−3)	48.5906 1.285(−7) 7.783(+3)	48.5855 0.392 2.55(−3)	4.4(+5) 2.1(−16)	4.7(+5) 8.7(−12)
6	^{12}C VI	33.7396 0.813 1.23(−3)	33.7393 3.836(−7) 2.607(+3)	33.7342 0.812 1.23(−3)	2.1(+5) 1.9(−15)	2.2(+5) 5.4(−11)
7	^{14}N VII	24.7846 1.51 6.62(−4)	24.7843 9.672(−7) 1.03(+3)	24.7792 1.50 6.67(−4)	1.1(+5) 1.2(−14)	1.2(+5) 2.5(−10)
8	^{16}O VIII	18.9725 2.57 3.89(−4)	18.9723 2.155(−6) 464	18.9671 2.57 3.89(−4)	6.7(+4) 6.1(−14)	7.0(+4) 9.6(−10)
9	^{19}F IX	14.9877 4.11 2.43(−4)	14.9875 4.370(−6) 229	14.9823 4.11 2.43(−4)	4.2(+4) 2.5(−13)	4.4(+4) 3.1(−9)

Table 8.4 (continued)

Atomic number Z	Ion	$2p\,^2P_{1/2} \rightarrow 1s\,^2S_{1/2}$	$2s\,^2S_{1/2} \rightarrow 1s\,^2S_{1/2}$	$2p\,^2P_{3/2} \rightarrow 1s\,^2S_{1/2}$	$2p\,^2P_{3/2} \rightarrow 2p\,^2P_{1/2}$	$2p\,^2P_{3/2} \rightarrow 2s\,^2S_{1/2}$
1	2	3	4	5	6	7
10	^{20}Ne X	12.1375 6.27 1.59 (− 4)	12.1372 8.226 (− 6) 122	12.1321 6.26 1.60 (− 4)	2.7 (+ 4) 8.8 (− 13)	2.8 (+ 4) 9.0 (− 9)
11	^{23}Na XI	10.0286 9.18 1.09 (− 4)	10.0284 1.46 (− 5) 68.6	10.0232 9.16 1.09 (− 4)	1.86 (+ 4) 2.8 (− 12)	1.94 (+ 4) 2.3 (− 8)
12	^{24}Mg XII	8.4246 13.0 7.69 (− 5)	8.4244 2.461 (− 5) 40.6	8.4192 13.0 7.69 (− 5)	1.311 (+ 4) 7.9 (− 12)	1.366 (+ 4) 5.6 (− 8)
13	^{27}Al XIII	7.1763 17.9 5.59 (− 5)	7.1761 3.98 (− 5) 25.1	7.1709 17.9 5.59 (− 5)	9509 2.1 (− 11)	9894 1.2 (− 7)
14	^{28}Si XIV	6.1858 24.1 4.15 (− 5)	6.1856 6.228 (− 5) 16.1	6.1804 24.0 4.17 (− 5)	7063 5.1 (− 11)	7338 2.6 (− 7)
15	^{31}P XV	5.3868 31.8 3.14 (− 5)	5.3866 9.43 (− 5) 10.6	5.3813 31.6 3.16 (− 5)	5355 1.2 (− 10)	5556 5.2 (− 7)
16	^{32}S XVI	4.7328 41.1 2.43 (− 5)	4.7326 1.397 (− 4) 7.16	4.7274 40.9 2.44 (− 5)	4132 2.5 (− 10)	4282 1.0 (− 6)
17	^{35}Cl XVII	4.1908 52.4 1.91 (− 5)	4.1906 2.01 (− 4) 4.98	4.1853 52.2 1.92 (− 5)	3239 5.2 (− 10)	3352 1.8 (− 6)

18	^{40}Ar XVIII	3.7365 65.9 1.52 (−5)	3.7363 2.859 (−4) 3.50	3.7311 65.5 1.53 (−5)	2574 1.0 (−9)	2661 3.3 (−6)
19	^{39}K XIX	3.3521 81.9 1.22 (−5)	3.3519 3.96 (−4) 2.53	3.3467 81.3 1.23 (−5)	2071 2.0 (−9)	2139 5.6 (−6)
20	^{40}Ca XX	3.0239 101 9.90 (−6)	3.0237 5.458 (−4) 1.83	3.0185 99.8 1.00 (−5)	1684.8 3.7 (−9)	1738.7 9.4 (−6)
21	^{45}Sc XXI	2.7414 122 8.20 (−6)	2.7413 7.35 (−4) 1.36	2.7360 121 8.26 (−6)	1384.2 6.7 (−9)	1427.2 1.5 (−5)
22	^{48}Ti XXII	2.4966 147 6.80 (−6)	2.4965 9.862 (−4) 1.01	2.4912 146 6.85 (−6)	1147.5 1.2 (−8)	1182.2 2.5 (−5)
23	^{51}V XXIII	2.2830 176 5.68 (−6)	2.2829 1.30 (−3) 0.77	2.2776 174 5.75 (−6)	959.14 2.0 (−8)	987.36 3.8 (−5)
24	^{52}Cr XXIV	2.0956 209 4.78 (−6)	2.0954 1.706 (−3) 0.586	2.0901 206 4.85 (−6)	807.7 3.4 (−8)	830.9 5.9 (−5)
25	^{55}Mn XXV	1.9301 246 4.07 (−6)	1.9300 2.20 (−3) 0.45	1.9247 243 4.12 (−6)	684.9 5.5 (−8)	704.0 8.9 (−5)
26	^{56}Fe XXVI	1.7834 288 3.47 (−6)	1.7833 2.852 (−3) 0.351	1.7780 284 3.52 (−6)	584.4 8.9 (−8)	600.4 1.3 (−4)
27	^{59}Co XXVII	1.6527 335 2.99 (−6)	1.6526 3.62 (−3) 0.28	1.6473 330 3.03 (−6)	501.6 1.4 (−7)	515.0 1.9 (−4)

Table 8.4 (continued)

Atomic number Z	Ion	$2p\,^2P_{1/2} \rightarrow 1s\,^2S_{1/2}$	$2s\,^2S_{1/2} \rightarrow 1s\,^2S_{1/2}$	$2p\,^2P_{3/2} \rightarrow 1s\,^2S_{1/2}$	$2p\,^2P_{3/2} \rightarrow 2p\,^2P_{1/2}$	$2p\,^2P_{3/2} \rightarrow 2s\,^2S_{1/2}$
1	2	3	4	5	6	7
28	^{58}Ni XXVIII	1.5358 390 2.56(−6)	1.5356 4.637(−3) 0.216	1.5303 380 2.63(−6)	432.9 2.2(−7)	444.2 2.8(−4)
29	^{63}Cu XXIX	1.4307 450 2.2(−6)	1.4306 5.82(−3) 0.17	1.4253 440 2.3(−6)	375.5 3.4(−7)	385.1 4.0(−4)
30	^{64}Zn XXX	1.3359 510 2.0(−6)	1.3358 7.376(−3) 0.136	1.3305 500 2.0(−6)	327.2 5.1(−7)	335.4 5.7(−4)
31	$^{69}Ga^{30+}$	1.2502 580 1.7(−6)	1.2501 9.16(−3) 0.11	1.2448 570 1.8(−6)	286.4 7.6(−7)	293.4 7.9(−4)
32	$^{74}Ge^{31+}$	1.1724 660 1.5(−6)	1.1723 1.14(−2) 0.088	1.1670 650 1.5(−6)	251.7 1.1(−6)	257.7 1.1(−3)
33	$^{75}As^{32+}$	1.1015 750 1.3(−6)	1.1014 1.42(−2) 0.070	1.0961 730 1.4(−6)	222.0 1.6(−6)	227.3 1.5(−3)
34	$^{80}Se^{33+}$	1.0368 840 1.2(−6)	1.0367 1.775(−2) 0.056	1.0314 830 1.2(−6)	196.6 2.3(−6)	201.1 2.0(−3)
35	$^{79}Br^{34+}$	0.9776 950 1.1(−6)	0.9775 2.17(−2) 0.046	0.9722 930 1.1(−6)	174.7 3.3(−6)	178.6 2.7(−3)

36	$^{84}Kr^{35+}$	0.9232 1060 9.4(−7)	0.9231 2.67(−2) 0.037	0.9178 1040 9.6(−7)	155.7 5.0(−6)	159.1 3.6(−3)
37	$^{85}Rb^{36+}$	0.8732 1200 8.3(−7)	0.8731 3.29(−2) 0.030	0.8678 1150 8.7(−7)	139.1 6.6(−6)	142.2 4.8(−3)
38	$^{88}Sr^{37+}$	0.8271 1300 7.7(−7)	0.8270 4.068(−2) 0.025	0.8216 1300 7.7(−7)	124.7 9.1(−6)	127.4 6.3(−3)
39	$^{89}Y^{38+}$	0.7845 1500 6.7(−7)	0.7843 4.93(−2) 0.020	0.7790 1400 7.1(−7)	112.1 1.3(−5)	114.5 8.2(−3)
40	$^{90}Zr^{39+}$	0.7450 1600 6.2(−7)	0.7449 6.02(−2) 0.017	0.7395 1600 6.2(−7)	101.0 1.7(−5)	103.1 1.1(−2)
41	$^{93}Nb^{40+}$	0.7084 1800 5.6(−7)	0.7083 7.33(−2) 0.014	0.7029 1700 5.9(−7)	90.9 2.3(−5)	93.5 1.4(−2)
42	$^{98}Mo^{41+}$	0.6743 2000 5.0(−7)	0.6742 8.99(−2) 0.011	0.6689 1900 5.3(−7)	82.6 3.1(−5)	84.7 1.8(−2)
47	$^{107}Ag^{46+}$	0.5354 3100 3.2(−7)	0.5353 0.230 4.3(−3)	0.5300 3000 3.3(−7)	51.8 1.3(−4)	52.9 5.6(−2)
48	$^{114}Cd^{47+}$	0.5127 3400 2.9(−7)	0.5126 0.277 3.6(−3)	0.5073 3200 3.1(−7)	47.6 1.6(−4)	48.5 7.0(−2)
49	$^{115}In^{48+}$	0.4914 3700 2.7(−7)	0.4913 0.332 3.0(−3)	0.4859 3500 2.9(−7)	43.5 2.1(−4)	44.4 8.7(−2)

Table 8.4 (continued)

Atomic number Z	Ion	$2p\,^2P_{1/2} \rightarrow 1s\,^2S_{1/2}$	$2s\,^2S_{1/2} \rightarrow 1s\,^2S_{1/2}$	$2p\,^2P_{3/2} \rightarrow 1s\,^2S_{1/2}$	$2p\,^2P_{3/2} \rightarrow 2p\,^2P_{1/2}$	$2p\,^2P_{3/2} \rightarrow 2s\,^2S_{1/2}$
1	2	3	4	5	6	7
50	$^{120}Sn^{49+}$	0.4713 4000 2.5(−7)	0.4712 0.402 2.5(−3)	0.4658 3800 2.6(−7)	40.0 2.7(−4)	40.8 0.11
51	$^{121}Sb^{50+}$	0.4524 4400 2.3(−7)	0.4523 0.50 2.0(−3)	0.4469 4100 2.4(−7)	37.0 4.0(−4)	38.5 0.14
52	$^{130}Te^{51+}$	0.4346 4700 2.1(−7)	0.4345 0.60 1.7(−3)	0.4291 4500 2.2(−7)	33.3 5.2(−4)	34.5 0.18
53	$^{127}I^{52+}$	0.4177 5100 2.0(−7)	0.4176 0.70 1.4(−3)	0.4123 4800 2.1(−7)	31.2 6.5(−4)	31.2 0.22
54	$^{132}Xe^{53+}$	0.4018 5500 1.8(−7)	0.4017 0.813 1.2(−3)	0.3963 5200 1.9(−7)	29.4 7.8(−4)	29.4 0.26
55	$^{133}Cs^{54+}$	0.3868 5800 1.7(−7)	0.3867 1.00 1.0(−3)	0.3813 5500 1.8(−7)	26.3 9.0(−4)	27.0 0.29
56	$^{138}Ba^{55+}$	0.3726 6300 1.6(−7)	0.3724 1.20 8.3(−4)	0.3671 6000 1.7(−7)	25.0 1.3(−3)	25.6 0.38
72	$^{180}Hf^{71+}$	0.2188 1.8(+4) 5.6(−8)	0.2187 14.4 6.9(−5)	0.2132 1.6(+4) 6.2(−8)	8.33 3.2(−2)	8.47 5.7

73	$^{181}Ta^{72+}$	0.2124	0.2123	0.2068	7.94	8.00
		1.9(+4)	16.5	1.7(+4)	3.7(−2)	6.6
		5.3(−8)	6.1(−5)	5.9(−8)		
74	$^{184}W^{73+}$	0.2062	0.2061	0.2007	7.46	7.58
		2.0(+4)	18.5	1.7(+4)	4.3(−2)	7.5
		5.0(−8)	5.4(−5)	5.9(−8)		
75	$^{187}Re^{74+}$	0.2003	0.2002	0.1947	6.99	7.09
		2.1(+4)	22.0	1.8(+4)	4.9(−2)	8.3
		4.8(−8)	4.5(−5)	5.6(−8)		
76	$^{192}Os^{75+}$	0.195	0.195	0.189	6.67	6.67
		2.2(+4)	25.4	1.9(+4)	6.3(−2)	10
		4.5(−8)	3.9(−5)	5.3(−8)		
77	$^{193}Ir^{76+}$	0.189	0.189	0.184	6.25	6.25
		2.3(+4)	28.9	2.0(+4)	7.6(−2)	12
		4.3(−8)	3.5(−5)	5.0(−8)	5.0(−8)	
78	$^{195}Pt^{77+}$	0.184	0.184	0.178	5.88	5.88
		2.4(+4)	32.3	2.1(+4)	9.0(−2)	14
		4.2(−8)	3.1(−5)	4.8(−8)		
79	$^{197}Au^{78+}$	0.179	0.179	0.173	5.56	5.88
		2.5(+4)	38.1	2.2(+4)	0.10	15
		4.0(−8)	2.6(−5)	4.5(−8)		
80	$^{202}Hg^{79+}$	0.174	0.174	0.168	5.3	5.3
		2.7(+4)	43.8	2.3(+4)	0.12	17
		3.7(−8)	2.3(−5)	4.3(−8)		
81	$^{205}Tl^{80+}$	0.169	0.169	0.164	5.0	5.0
		2.8(+4)	49.6	2.5(+4)	0.15	21
		3.6(−8)	2.0(−5)	4.0(−8)		
82	$^{208}Pb^{81+}$	0.165	0.165	0.159	4.5	4.8
		3.0(+4)	55.2	2.6(+4)	0.18	24
		3.3(−8)	1.8(−5)	3.8(−8)		

Table 8.4 (continued)

Atomic number Z	Ion	$2p\,^2P_{1/2} \rightarrow 1s\,^2S_{1/2}$	$2s\,^2S_{1/2} \rightarrow 1s\,^2S_{1/2}$	$2p\,^2P_{3/2} \rightarrow 1s\,^2S_{1/2}$	$2p\,^2P_{3/2} \rightarrow 2p\,^2P_{1/2}$	$2p\,^2P_{3/2} \rightarrow 2s\,^2S_{1/2}$
1	2	3	4	5	6	7
83	$^{209}Bi^{82+}$	0.160	0.160	0.155	4.3	4.3
		3.1(+4)	64.8	2.7(+4)	0.21	28
		3.2(−8)	1.5(−5)	3.7(−8)		
90	$^{232}Th^{89+}$	0.133	0.133	0.128	2.9	3.0
		4.3(+4)	153	3.6(+4)	0.59	67
		2.3(−8)	6.5(−6)	2.8(−8)		
92	$^{238}U^{91+}$	0.127	0.127	0.121	2.7	2.8
		4.7(+4)	196	4.0(+4)	0.86	92
		2.1(−8)	5.1(−6)	2.5(−8)		
94	Pu^{93+}	0.121	0.121	0.115	2.4	2.5
		5.2(+4)	240	4.3(+4)	1.1	120
		1.9(−8)	4.2(−6)	2.3(−8)		

Table 8.5. Wavelengths λ [Å] (*upper line* in each entry), transition probabilities A_{ki} [ps^{-1}] (*middle line*) and radiative lifetimes τ_k [ns] (*lower line*) for He (nl)-like isoelectronic sequence: $n = 1, 2$; $\Delta n = 0, 1$; $2 \leqslant Z < 100$ [$X(Y)$ means $X \cdot 10^Y$]

Atomic number Z	Ion	$2\,^3S_1 \to 1\,^1S_0$	$2\,^1S_0 \to 1\,^1S_0$	$2\,^3P_2 \to 1\,^1S_0$	$2\,^3P_2 \to 2\,^3S_1$	$2\,^3P_1 \to 1\,^1S_0$	$2\,^3P_1 \to 2\,^3S_1$	$2\,^3P_0 \to 2\,^3S_1$	$2\,^1P_1 \to 1\,^1S_0$
1	2	3	4	5		6		7	8
2	He I	625.56 1.2(−16) 9.0(+12)	601.40 5.10(−11) 2.0(+7)	591.41 3.3(−13)	10830.34 1.022(−5) 98	591.412 1.76(−10)	10830.25 1.022(−5) 98	10829.09 1.022(−5) 98	584.334 1.80(−3) 0.556
3	Li II	210.07 2.0(−14) 4.9(+10)	203.51 1.9(−9) 5.1(+5)	202.32 3.5(−11)	5484.50 2.28(−5) 44	202.322 1.80(−8)	5485.09 2.28(−5) 44	5483.50 2.28(−5) 44	199.278 2.56(−2) 3.91(−2)
4	Be III	104.55 5.6(−13) 1.8(+9)	101.92 1.8(−8) 5.5(+4)	101.69 6.2(−10)	3720.85 3.42(−5) 29	101.691 4.01(−7)	3722.91 3.42(−5) 29	3721.31 3.42(−5) 29	100.255 0.122 8.20(−3)
5	B IV	62.45 6.7(−12) 1.5(+8)	61.13 9.3(−8) 1.1(+4)	61.10 5.0(−9)	2821.66 4.54(−5) 22	61.105 4.23(−6)	2825.87 4.53(−5) 20	2824.56 4.53(−5) 22	60.315 0.372 2.69(−3)
6	C V	41.47 4.9(−11) 2.1(+7)	40.73 3.3(−7) 3.0(+3)	40.73 2.6(−8)	2270.91 5.67(−5) 18	40.731 2.83(−5)	2277.92 5.63(−5) 12	2277.25 5.63(−5) 18	40.268 0.887 1.13(−3)
7	N VI	29.53 2.5(−10) 3.9(+6)	29.08 9.4(−7) 1.1(+3)	29.08 1.0(−7)	1896.8 6.8(−5) 15	29.084 1.40(−4)	1907.3 6.7(−5) 4.8	1907.8 6.7(−5) 15	28.787 1.81 5.52(−4)
8	O VII	22.10 1.0(−9) 9.6(+5)	21.79 2.3(−6) 430	21.80 3.3(−7)	1623.5 8.0(−5) 12.5	21.807 5.52(−4)	1638.3 7.9(−5) 1.6	1639.9 7.8(−5) 13	21.602 3.31 3.02(−4)

Table 8.5 (continued)

Atomic number Z	Ion	$2\,^3S_1 \to 1\,^1S_0$	$2\,^1S_0 \to 1\,^1S_0$	$2\,^3P_2 \to 1\,^1S_0$	$2\,^3P_2 \to 2\,^3S_1$	$2\,^3P_1 \to 1\,^1S_0$	$2\,^3P_1 \to 2\,^3S_1$	$2\,^3P_0 \to 2\,^3S_1$	$2\,^1P_1 \to 1\,^1S_0$
1	2	3	4	5		6		7	8
9	F VIII	17.15	16.94	16.95	1414.4	16.949	1433.8	1437.0	16.807
		3.6(−9)	5.1(−6)	9.2(−7)	9.3(−5)	1.84(−3)	9.0(−5)	9.0(−5)	5.59
		2.8(+5)	200	11		0.52		10	1.79(−4)
10	Ne IX	13.69	13.55	13.55	1248.1	13.549	1272.8	1277.7	13.447
		1.1(−8)	1.0(−5)	2.3(−6)	1.1(−4)	5.4(−3)	1.0(−4)	1.0(−4)	8.88
		9.2(+4)	100	9.2		0.18		9.9	1.12(−4)
11	Na X	11.19	11.08	11.08	1112	11.083	1142	1149	11.003
		2.9(−8)	1.8(−5)	5.1(−6)	1.3(−4)	1.4(−2)	1.2(−4)	1.2(−4)	13.4
		3.4(+4)	54	7.5		7.0(−2)		8.3	7.5(−5)
12	Mg XI	9.314	9.226	9.228	997.4	9.231	1034	1043	9.169
		7.3(−8)	3.2(−5)	1.1(−5)	1.5(−4)	3.4(−2)	1.4(−4)	1.3(−4)	19.5
		1.4(+4)	31	6.2		2.9(−2)		7.7	5.1(−5)
13	Al XII	7.872	7.803	7.804	900	7.807	943	954	7.757
		1.7(−7)	5.3(−5)	2.1(−5)	1.7(−4)	7.5(−2)	1.5(−4)	1.4(−4)	27.5
		6.0(+3)	19	5.2		1.3(−2)		4.8	3.6(−5)
14	Si XIII	6.740	6.685	6.685	815	6.688	865	879	6.648
		3.6(−7)	8.5(−5)	3.9(−5)	2.0(−4)	0.16	1.7(−4)	1.5(−4)	37.7
		2.8(+3)	12	4.2		6.3(−3)		6.7	2.6(−5)
15	P XIV	5.836	5.791	5.790	740	5.793	800	810	5.760
		7.3(−7)	1.3(−4)	6.9(−5)	2.3(−4)	0.31	1.8(−4)	1.7(−4)	50.4
		1.4(+3)	7.6	3.3		3.2(−3)		4.8	2.0(−5)
16	S XV	5.102	5.064	5.063	670	5.066	740	760	5.039
		1.4(−6)	2.0(−4)	1.2(−4)	2.6(−4)	0.58	2.0(−4)	1.8(−4)	66.1
		700	5.1	2.6		1.7(−3)		5.6	1.5(−5)

17	Cl XVI	4.497 2.7(−6) 360	4.466 2.8(−4) 3.5	4.465 2.0(−4)	610 3.0(−4) 2.0	4.468 1.04	680 2.2(−4) 9.6(−4)	700 2.0(−4) 4.3	4.444 85.0 1.2(−5)
18	Ar XVII	3.994 4.8(−6) 205	3.968 4.0(−4) 2.5	3.966 3.15(−4)	560 3.5(−4) 1.5	3.969 1.80	640 2.3(−4) 5.6(−4)	660 2.1(−4) 4.8	3.948 107 9.3(−6)
19	K XVIII	3.571 8.3(−6) 120	3.548 6(−4) 2	3.546 4.9(−4)	510 4.1(−4) 1.1	3.549 2.99	595 2.6(−4) 3.3(−4)	620 2.3(−4) 4.0	3.531 134 7.5(−6)
20	Ca XIX	3.211 1.4(−5) 70	3.192 8(−4) 1	3.189 7.6(−4)	470 4.9(−4) 0.80	3.192 4.82	560 2.8(−4) 2.1(−4)	580 2.4(−4) 4.2	3.176 165 6.0(−6)
21	Sc XX	2.903 2.3(−5) 43	2.886 1(−3) 0.9	2.884 1.1(−3)	430 5.8(−4) 0.60	2.887 7.2	530 3.1(−4) 1.4(−4)	560 2.6(−4) 0.23	2.872 200 5.0(−6)
22	Ti XXI	2.636 3.8(−5) 26	2.622 1.5(−3) 0.7	2.619 1.7(−3)	390 6.9(−4) 0.42	2.623 11	500 3.3(−4) 9.1(−5)	530 2.8(−4) 3.6	2.609 240 4.1(−6)
23	V XXII	2.406 5.9(−5) 17	2.393 2(−3) 0.5	2.389 2.4(−3)	360 8.3(−4) 0.30	2.393 16	480 3.6(−4) 6.3(−5)	500 3.0(−4) 0.09	2.381 290 3.4(−6)
24	Cr XXIII	2.204 9.2(−5) 11	2.193 2.5(−3) 0.4	2.189 3.4(−3)	325 9.9(−4) 0.23	2.192 23	440 3.8(−4) 4.3(−5)	450 3.2(−4) 3.1	2.182 340 2.9(−6)
25	Mn XXIV	2.026 1.4(−4) 7.2	2.016 3(−3) 0.3	2.012 4.8(−3)	290 1.2(−3) 0.17	2.016 32	420 4.1(−4) 3.1(−5)	450 3.4(−4) 0.09	2.006 400 2.5(−6)
26	Fe XXV	1.868 2.1(−4) 4.8	1.859 4(−3) 0.2	1.855 6.6(−3)	270 1.4(−3) 0.13	1.859 43	400 4.3(−4) 2.3(−5)	430 3.6(−4) 2.8	1.850 460 2.2(−6)

Table 8.5 (continued)

Atomic number Z	Ion	$2\,^3S_1 \to 1\,^1S_0$	$2\,^1S_0 \to 1\,^1S_0$	$2\,^3P_2 \to 1\,^1S_0$	$2\,^3P_2 \to 2\,^3S_1$	$2\,^3P_1 \to 1\,^1S_0$	$2\,^3P_1 \to 2\,^3S_1$	$2\,^3P_0 \to 2\,^3S_1$	$2\,^1P_1 \to 1\,^1S_0$
1	2	3	4	5		6		7	8
27	Co XXVI	1.728	1.721	1.716	250	1.721	380	420	1.712
		3.1(−4)	5(−3)	8.9(−3)	1.75(−3)	58	4.6(−4)	3.8(−4)	530
		3.3	0.2		9.4(−2)		1.7(−5)	0.03	1.9(−6)
28	Ni XXVII	1.604	1.596	1.592	230	1.596	360	380	1.588
		4.4(−4)	7(−3)	1.2(−2)	2.1(−3)	76	4.8(−4)	4.1(−4)	610
		2.3	0.15		7.1(−2)		1.3(−5)	2.4	1.6(−6)
29	Cu XXVIII	1.492	1.485	1.481	210	1.485	340	370	1.478
		6.4(−4)	8(−3)	1.6(−2)	2.6(−3)	98	5.1(−4)	4.3(−4)	700
		1.6	0.1		5.4(−2)		1.0(−5)	0.05	1.4(−6)
30	Zn XXIX	1.391	1.385	1.381	190	1.385	330	360	1.378
		9.0(−4)	1(−2)	2.1(−2)	3.2(−3)	123	5.3(−4)	4.6(−4)	790
		1.1	0.1		4.1(−2)		8.1(−6)	2.2	1.3(−6)
31	Ga XXX	1.30	1.30	1.30	200	1.30	300	300	1.29
		1.3(−3)	1(−2)	2.8(−2)	3.9(−3)	155	5.6(−4)	4.9(−4)	900
		0.79	0.08		3.1(−2)		6.5(−6)	–	1.1(−6)
32	Ge^{30+}	1.22	1.21	1.21	200	1.21	300	300	1.21
		1.7(−3)	1.5(−2)	3.6(−2)	4.8(−3)	190	5.8(−4)	5.2(−4)	990
		0.57	0.07		2.5(−2)		5.3(−6)	1.9	1.0(−6)
33	As^{31+}	1.14	1.14	1.13	100	1.14	300	300	1.13
		2.4(−3)	2(−2)	4.6(−2)	5.8(−3)	230	6.1(−4)	5.5(−4)	1.1(+3)
		0.42	0.05		1.9(−2)		4.3(−6)	–	9(−7)
34	Se^{32+}	1.08	1.07	1.07	100	1.07	300	300	1.06
		3.2(−3)	2(−2)	5.9(−2)	7.1(−3)	280	6.3(−4)	5.8(−4)	1.3(+3)
		0.31	0.05		1.5(−2)		3.6(−6)	1.7	8(−7)

35	Br^{33+}	1.01 4.4(−3) 0.23	1.01 2.5(−2) 0.04	1.00 7.5(−2)	100 8.7(−3) 1.2(−2)	1.01 330	300 6.6(−4) 3.0(−6)	300 6.1(−4) –	1.00 1.4(+3) 7(−7)
36	Kr^{34+}	0.96 5.8(−3) 0.17	0.95 3(−2) 0.03	0.95 9.4(−2)	110 1.1(−2) 1.1(−2)	0.95 390	200 6.9(−4) 2.6(−6)	280 6.5(−4) 1.5	0.95 1.5(+3) 7(−7)
37	Rb^{35+}	0.90 7.7(−3) 0.13	0.90 4(−2) 0.03	0.90 0.12	100 1.3(−2) 7.5(−3)	0.90 460	200 7.1(−4) 2.2(−6)	200 6.8(−4) –	0.89 1.6(+3) 6(−7)
38	Sr^{36+}	0.85 1.0(−2) 0.10	0.85 4.5(−2) 0.02	0.85 0.15	90 1.6(−2) 6.0(−3)	0.85 530	200 7.4(−4) 1.9(−6)	200 7.2(−4) 1.4	0.85 1.8(+3) 6(−7)
39	Y^{37+}	0.81 1.3(−2) 0.08	0.81 5(−2) 0.02	0.80 0.18	80 1.9(−2) 5.0(−3)	0.81 610	200 7.7(−4) 1.6(−6)	200 7.6(−4) –	0.80 2.0(+3) 5(−7)
40	Zr^{38+}	0.76 1.7(−2) 0.06	0.76 6(−2) 0.02	0.76 0.22	80 2.4(−2) 4.0(−3)	0.77 700	200 7.9(−4) 1.4(−6)	200 8.0(−4) 1.3	0.76 2.2(+3) 5(−7)
41	Nb^{39+}	0.73 2.2(−2) 0.04	0.73 7(−2) 0.01	0.72 0.27	70 2.9(−2) 3.3(−3)	0.73 800	200 8.2(−4) 1.2(−6)	200 8.5(−4) –	0.72 2.4(+3) 4(−7)
42	Mo^{40+}	0.69 2.9(−2) 0.03	0.69 8(−2) 0.01	0.69 0.33	70 3.5(−2) 2.7(−3)	0.69 910	200 8.5(−4) 1.1(−6)	200 9.0(−4) 1.1	0.69 2.6(+3) 4(−7)
50	Sn^{48+}	0.48 0.17 6(−3)	0.48 0.2 4(−3)	0.48 1.4	40 0.16 6.4(−4)	0.48 2.1(+3)	100 1.1(−3) 5(−7)	200 1.4(−3) 0.7	0.48 5.1(+3) 2(−7)
60	Nd^{58+}	0.33 1 9(−4)	0.33 0.7 1.5(−3)	0.32 6.0	20 1.0 1.5(−4)	0.33 4.9(+3)	100 1.6(−3) 2(−7)	100 6.4(−3) 0.2	0.32 1.0(+4) 1(−7)

Table 8.5 (continued)

Atomic number Z	Ion	$2\,^3S_1 \to 1\,^1S_0$	$2\,^1S_0 \to 1\,^1S_0$	$2\,^3P_2 \to 1\,^1S_0$	$2\,^3P_2 \to 2\,^3S_1$	$2\,^3P_1 \to 1\,^1S_0$	$2\,^3P_1 \to 2\,^3S_1$	$2\,^3P_0 \to 2\,^3S_1$	$2\,^1P_1 \to 1\,^1S_0$
1	2	3	4	5		6		7	8
70	Yb^{68+}	0.24	0.24	0.23	10	0.24	100	80	0.23
		6	2	20	5	9.5(+3)	2.2(−3)	1.4(−2)	1.8(+4)
		2(−4)	5(−4)		4(−5)		1(−7)	0.07	6(−8)
80	Hg^{78+}	0.18	0.18	0.17	5	0.18	100	50	0.17
		25	4	65	20	1.7(+4)	3(−3)	2.5(−2)	3.0(+4)
		4(−5)	2.5(−4)		1(−5)		6(−8)	0.04	3(−8)
90	Th^{88+}	0.14	0.13	0.13	3	0.14	100	50	0.13
		90	8	150	60	2.7(+4)	5(−3)	0.06	5(+4)
		1(−5)	1(−4)		5(−6)		4(−8)	0.02	2(−8)
100	Fm^{98+}	0.11	0.11	0.10	2	0.11	50	30	0.10
		300	15	400	240	4.3(+4)	7(−3)	0.09	7(+4)
		3(−6)	6(−5)		2(−6)		2(−8)	0.01	1(−8)

Part II Molecules and Molecular Ions

All molecules may be divided into two groups according to their type of bonding. In the first group are molecules with chemical bonding which results from the overlapping of the atomic electron shells of the interacting partners (covalent bonding) or the partial transfer of the valence electron from one atom to another (ionic bonding). The dissociation energies of these stable molecules are within the range $\sim$ 1–10 eV. Quantitative information about structural, energetic, spectroscopic and electromagnetic parameters of diatomic and polyatomic molecules and molecular ions makes up the bulk of this part of the reference guide.

The second group of molecules is made up of the so-called van der Waals molecules. Electron exchange interaction in these molecules results in repulsion, and the bonding here is provided by the weak long-range attractive forces. These van der Waals forces result in shallow potential wells on the curves of molecular electron terms. Hence, dissociation energies of these molecules are rather small, approximately two orders of magnitude lower than those of molecules with chemical bonds. Quantitative data concerning van der Waals molecules are given in Chap. 12.

Finally, the data in Chap. 9 concerns the electrostatic and exchange interactions of atomic particles at distances between their centres large compared to the sizes of the particles. Parameters for the short-range repulsive interaction between atomic and molecular species are also provided.

9. Interaction Potentials Between Atomic and Molecular Species

The numerical values of the van der Waals coefficients C_{2l+2} describing the multipole interactions of atomic and molecular species are given. The calculated parameters of the exponential exchange interaction of two like atoms are presented in the limit of large internuclear separations. Numerical data on the exponential repulsive interaction of atomic and molecular particles are also compiled for short-range internuclear separations.

9.1 Van der Waals Coefficients for Interatomic Multipole Interactions

At distances R between the centres of atomic particles large compared to their sizes, the electrostatic interaction potential of particles may be reduced to the interaction of their multipole moments and be represented by an expansion in inverse powers of R.

In the case of the ion-atom interaction, the long-range electrostatic potential may be written as

$$U(R) = -C_4/R^4 - C_6/R^6 - \dots ,$$

where $C_4 = \alpha Ze^2/2$, and α is the polarizability of an atom (Sect. 5.6). The coefficient C_6 describes the interaction of the ionic charge Ze with the quadrupole moment of an atom, the dipole-dipole interaction of particles in second-order perturbation theory and so on. The potential $U(R)$ corresponds to an interaction averaged over all possible orientations of the electronic angular momenta of the particles.

The long-range electrostatic interaction potential for two neutral atoms, also averaged over the possible orientations of their electronic angular momenta, has the following representation:

$$U(R) = -C_6/R^6 - C_8/R^8 - C_{10}/R^{10} - \dots ,$$

where the van der Waals coefficients C_{2l+2} $(l = 2, 3, 4, \dots)$ describe the interactions between induced atomic dipole moments ($C_6 > 0$), dipole and quadrupole moments ($C_8 > 0$), two induced quadrupole moments, and an induced dipole moment with an induced octupole moment ($C_{10} > 0$), etc.

In Tables 9.1–7 the calculated values of the dispersion energy coefficients C_6, C_8 and C_{10} for the long-range atomic interaction potential are given; coefficient C_6 is also given for some molecular particles [9.1.1–8].

References

9.1.1 A.Dalgarno: "New Methods for Calculating Long-Range Intermolecular Forces", in Adv. Chem. Phys., Vol. 12, ed. by J.O.Hirschfelder (Wiley, New York 1967) Chap. 3

9.1.2 A.D.Buckingham: "Basic Theory of Intermolecular Forces: Applications to Small Molecules", in *Intermolecular Interactions: From Diatomics to Biopolymers*, ed. by B.Pullman (Wiley, New York 1978) Chap. 1, pp. 1–67

9.1.3 K.T.Tang, J.M.Norbeck, P.R.Certain: J. Chem. Phys. **64**, 3063 (1976)

9.1.4 F.Maeder, W.Kutzelnigg: Chem. Phys. **42**, 95 (1979)

9.1.5 H.A.Human: J. Chem. Phys. **61**, 4063 (1974)

9.1.6 R.T.Pack: J. Chem. Phys. **61**, 2091 (1974); J. Phys. Chem. **86**, 2794 (1982)

9.1.7 R.Ahlberg, O.Goscinski: J. Phys. **B 7**, 1194 (1974)

9.1.8 P.Huxley, D.B.Knowles, J.N.Murrell, J.D.Watts: J. Chem. Soc., Faraday Trans. II, **80**, 1349 (1984)

Table 9.1. Dispersion energy coefficient C_6 [a.u.] for the diatomic system (hydrogen, alkali atom – hydrogen, alkali atom)

Atom	H	Li	Na	K	Rb	Cs
H	6.50 (*A*)	66.5 (*A*)	72 (*B*)	105 (*B*)	119 (*C*)	146 (*C*)
Li	–	1390 (*A*)	1460 (*A*)	2300 (*A*)	2550 (*B*)	3150 (*A*)
Na	–	–	1540 (*B*)	2420 (*A*)	2680 (*A*)	3310 (*B*)
K	–	–	–	3880 (*A*)	4270 (*B*)	5300 (*A*)
Rb	–	–	–	–	4700 (*B*)	5880 (*B*)
Cs	–	–	–	–	–	7020 (*A*)

Table 9.2. Dispersion energy coefficient C_6 [a.u.] for the diatomic system (rare gas atom – hydrogen, alkali, rare gas atom)

Atom	He (^1S)	Ne	Ar	Kr	Xe
H	2.83 (*A*)	5.7 (*B*)	20 (*B*)	29 (*B*)	41 (*B*)
Li	22.3 (*B*)	43 (*B*)	178 (*B*)	260 (*B*)	407 (*B*)
Na	24.7 (*B*)	48 (*B*)	190 (*B*)	280 (*B*)	445 (*B*)
K	36 (*C*)	70 (*C*)	280 (*C*)	420 (*C*)	650 (*B*)
Rb	41 (*D*)	82 (*D*)	330 (*C*)	470 (*C*)	740 (*C*)
Cs	50 (*D*)	99 (*D*)	380 (*D*)	570 (*C*)	890 (*C*)
He (^1S)	1.46 (*A*)	3.06 (*B*)	9.7 (*B*)	13.3 (*B*)	18.6 (*B*)
Ne	–	6.6 (*C*)	20.4 (*B*)	28 (*B*)	38 (*B*)
Ar	–	–	58 (*B*)	95 (*B*)	140 (*B*)
Kr	–	–	–	130 (*B*)	187 (*C*)
Xe	–	–	–	–	296 (*B*)

Table 9.3. Dispersion energy coefficient C_6 [a.u.] for the diatomic system (rare gas atom – atom)

Atom	He	Ne	Ar	Kr	Xe	He* (2^3S)	He* (2^1S)
He* (2^3S)	29.0 (*B*)	58 (*C*)	240 (*C*)	360 (*C*)	590 (*C*)	3300 (*B*)	5800 (*B*)
He* (2^1S)	41.5 (*C*)	82 (*C*)	340 (*C*)	510 (*C*)	840 (*C*)	–	11000 (*B*)
Be	14.7 (*A*)	29.4 (*B*)	111 (*B*)	165 (*B*)	259 (*B*)	–	–
C	7.9 (*D*)	16 (*D*)	54 (*D*)	75 (*D*)	110 (*D*)	–	–
N	6.7 (*C*)	14 (*C*)	44 (*C*)	59 (*C*)	84 (*C*)	–	–
O	4.9 (*B*)	12 (*C*)	35 (*C*)	47 (*C*)	65 (*C*)	–	–
F	4.7 (*C*)	10 (*C*)	28 (*C*)	38 (*C*)	52 (*C*)	–	–
Mg	22 (*B*)	43 (*B*)	170 (*B*)	250 (*B*)	385 (*C*)	–	–
P	16 (*D*)	32 (*D*)	110 (*D*)	160 (*D*)	230 (*C*)	–	–
S	16 (*D*)	33 (*D*)	110 (*C*)	160 (*C*)	230 (*C*)	–	–
Cl	14 (*D*)	28 (*C*)	90 (*C*)	120 (*C*)	180 (*C*)	–	–
Ca	32 (*B*)	63 (*C*)	250 (*C*)	370 (*C*)	590 (*C*)	–	–
Br	18 (*D*)	37 (*D*)	120 (*D*)	170 (*D*)	240 (*D*)	–	–
Sr	36.7 (*B*)	72 (*C*)	290 (*C*)	430 (*C*)	680 (*C*)	–	–
I	24 (*D*)	49 (*D*)	170 (*D*)	240 (*D*)	350 (*D*)	–	–
Ba	39.5 (*B*)	77 (*C*)	310 (*C*)	460 (*C*)	740 (*C*)	–	–
Hg	14 (*B*)	29 (*B*)	100 (*C*)	150 (*C*)	220 (*C*)	–	–

Table 9.4. Dispersion energy coefficient C_6 [a.u.] for the system (atom – diatomic, polyatomic molecule)

Atom	Molecule H_2	N_2	O_2	CO_2	N_2O	SF_6
H	8.7 (*C*)	21 (*C*)	–	33.0 (*B*)	–	65 (*C*)
Li	83 (*C*)	180 (*C*)	–	278 (*C*)	–	520 (*D*)
Na	91 (*C*)	200 (*C*)	–	340 (*C*)	–	640 (*D*)
K	130 (*C*)	280 (*C*)	–	503 (*C*)	–	950 (*D*)
Rb	140 (*C*)	310 (*C*)	–	574 (*C*)	–	1100 (*D*)
Cs	170 (*C*)	370 (*C*)	–	490 (*C*)	–	900 (*D*)
He (^1S)	4.0 (*B*)	10 (*C*)	10.8 (*B*)	16.7 (*B*)	17 (*C*)	35 (*C*)
Ne	8.1 (*B*)	21 (*C*)	23 (*C*)	36 (*C*)	35 (*D*)	77 (*D*)
Ar	28 (*B*)	69 (*C*)	73 (*C*)	114 (*C*)	115 (*D*)	240 (*D*)
Kr	40 (*B*)	96 (*C*)	101 (*B*)	162 (*C*)	160 (*C*)	340 (*D*)
Xe	58 (*C*)	140 (*C*)	150 (*C*)	282 (*C*)	240 (*D*)	590 (*D*)

Table 9.5. Dispersion energy coefficient C_6 [a.u.] for the system (diatomic molecule – diatomic, polyatomic molecule)

Molecule	H_2	N_2	O_2	CO	NO	SF_6
H_2	12.1 (*B*)	31 (*B*)	30 (*C*)	34 (*C*)	32 (*C*)	94 (*C*)
N_2	–	73 (*B*)	73 (*C*)	80 (*C*)	76 (*C*)	250 (*D*)
O_2	–	–	73 (*B*)	79 (*C*)	75 (*C*)	230 (*D*)
CO	–	–	–	87 (*D*)	82 (*D*)	270 (*D*)
NO	–	–	–	–	77 (*D*)	250 (*D*)
CO_2	46.4 (*B*)	118 (*C*)	91 (*C*)	–	–	400 (*D*)
N_2O	48 (*C*)	–	120 (*D*)	–	–	390 (*D*)

Table 9.6. Dispersion energy coefficient C_8 [a.u.]

Atom	H	Li	Na	K	Rb	Cs	He	Ne	Ar	Kr	Xe
H	124.4 (*A*)	3260 (*B*)	4100 (*C*)	$8.3 \cdot 10^3$ (*D*)	$1 \cdot 10^4$ (*D*)	$1.5 \cdot 10^4$ (*D*)	41.75 (*A*)	97 (*C*)	422 (*C*)	590 (*C*)	945 (*B*)
Li	–	$8.4 \cdot 10^4$ (*C*)	$9.9 \cdot 10^4$ (*C*)	$2 \cdot 10^5$ (*C*)	$2.4 \cdot 10^5$ (*C*)	$3.3 \cdot 10^5$ (*D*)	1090 (*B*)	2270 (*C*)	9430 (*C*)	$1.3 \cdot 10^4$ (*C*)	$2.2 \cdot 10^4$ (*C*)
Na	–	–	$1.2 \cdot 10^5$ (*C*)	$2.3 \cdot 10^5$ (*C*)	$2.7 \cdot 10^5$ (*C*)	$3.8 \cdot 10^5$ (*D*)	1320 (*C*)	2900 (*C*)	$1.27 \cdot 10^4$ (*C*)	$1.6 \cdot 10^4$ (*D*)	$2.65 \cdot 10^4$ (*C*)
K	–	–	–	$4.2 \cdot 10^5$ (*C*)	$4.9 \cdot 10^5$ (*D*)	$6.7 \cdot 10^5$ (*D*)	2800 (*C*)	5820 (*C*)	$2.4 \cdot 10^4$ (*C*)	$3.3 \cdot 10^4$ (*D*)	$5.3 \cdot 10^4$ (*C*)
Rb	–	–	–	–	$5.7 \cdot 10^5$ (*D*)	$7.8 \cdot 10^5$ (*D*)	3600 (*C*)	7350 (*C*)	$2.9 \cdot 10^4$ (*C*)	$3.9 \cdot 10^4$ (*D*)	$6.2 \cdot 10^4$ (*C*)
Cs	–	–	–	–	–	$1.1 \cdot 10^6$ (*D*)	5080 (*C*)	9580 (*C*)	$4.1 \cdot 10^4$ (*D*)	$5.3 \cdot 10^4$ (*D*)	$8.5 \cdot 10^4$ (*C*)
He	–	–	–	–	–	–	14.2 (*B*)	34 (*C*)	146 (*D*)	206 (*C*)	340 (*D*)
Ne	–	–	–	–	–	–	–	82 (*D*)	330 (*D*)	440 (*C*)	760 (*D*)
Ar	–	–	–	–	–	–	–	–	1345 (*D*)	1900 (*C*)	3150 (*D*)
Kr	–	–	–	–	–	–	–	–	–	2700 (*C*)	4600 (*D*)
Xe	–	–	–	–	–	–	–	–	–	–	7000 (*D*)

Table 9.7. Dispersion energy coefficient C_{10} [a.u.]

Atom	H	Li	Na	K	Rb	Cs	He	Ne	Ar	Kr	Xe
H	$3.29 \cdot 10^3$ (*A*)	$1.92 \cdot 10^5$ (*B*)	$2.5 \cdot 10^5$ (*C*)	$6.4 \cdot 10^5$ (*D*)	$7.9 \cdot 10^5$ (*C*)	$1.2 \cdot 10^6$ (*C*)	865 (*B*)	1940 (*C*)	$1.0 \cdot 10^4$ (*C*)	$1.6 \cdot 10^4$ (*C*)	$2.7 \cdot 10^4$ (*C*)
Li	–	$7.2 \cdot 10^6$ (*B*)	$8.8 \cdot 10^6$ (*D*)	$2.1 \cdot 10^7$ (*C*)	$2.5 \cdot 10^7$ (*C*)	$3.8 \cdot 10^7$ (*D*)	$6.7 \cdot 10^4$ (*C*)	$1.5 \cdot 10^5$ (*C*)	$6.4 \cdot 10^5$ (*C*)	$9.8 \cdot 10^5$ (*C*)	$1.6 \cdot 10^6$ (*D*)
Na	–	–	$1.1 \cdot 10^7$ (*C*)	$2.5 \cdot 10^7$ (*C*)	$3 \cdot 10^7$ (*D*)	$4.4 \cdot 10^7$ (*D*)	$8.7 \cdot 10^4$ (*C*)	$1.9 \cdot 10^5$ (*C*)	$8.2 \cdot 10^5$ (*C*)	$1.2 \cdot 10^6$ (*C*)	$2.1 \cdot 10^6$ (*D*)
K	–	–	–	$5.5 \cdot 10^7$ (*C*)	$6.6 \cdot 10^7$ (*D*)	$9.7 \cdot 10^7$ (*D*)	$2.2 \cdot 10^5$ (*D*)	$4.5 \cdot 10^5$ (*D*)	$2.0 \cdot 10^6$ (*D*)	$3.0 \cdot 10^6$ (*D*)	$4.9 \cdot 10^6$ (*D*)
Rb	–	–	–	–	$8 \cdot 10^7$ (*D*)	$1.2 \cdot 10^8$ (*D*)	$2.7 \cdot 10^5$ (*D*)	$5.6 \cdot 10^5$ (*D*)	$2.4 \cdot 10^6$ (*D*)	$3.6 \cdot 10^6$ (*D*)	$5.9 \cdot 10^6$ (*D*)
Cs	–	–	–	–	–	$1.7 \cdot 10^8$ (*D*)	$3.9 \cdot 10^5$ (*D*)	$8.1 \cdot 10^5$ (*D*)	$3.5 \cdot 10^6$ (*D*)	$5.3 \cdot 10^6$ (*D*)	$8.6 \cdot 10^6$ (*D*)
He	–	–	–	–	–	–	180 (*C*)	405 (*C*)	2660 (*D*)	$4.1 \cdot 10^3$ (*D*)	$8.2 \cdot 10^3$ (*D*)
Ne	–	–	–	–	–	–	–	880 (*D*)	$5.9 \cdot 10^3$ (*D*)	$9 \cdot 10^3$ (*D*)	$1.9 \cdot 10^4$ (*D*)
Ar	–	–	–	–	–	–	–	–	$3 \cdot 10^4$ (*D*)	$4.8 \cdot 10^4$ (*D*)	$9 \cdot 10^4$ (*D*)
Kr	–	–	–	–	–	–	–	–	–	$7.0 \cdot 10^4$ (*D*)	$1.3 \cdot 10^5$ (*D*)
Xe	–	–	–	–	–	–	–	–	–	–	$2.2 \cdot 10^5$ (*D*)

9.2 Long-Range Exchange Interactions of Atoms

Table 9.8 gives the parameters of the exchange interaction between two identical atoms at large distances R. The interaction potential of two atoms with closed electron shells at large internuclear distances has the form

$$U(R) = U_L(R) + \Delta(R) ,$$

where $U_L(R) = -C_6/R^6 - C_8/R^8 - \ldots$ is the long-range multipole interaction potential and $\Delta(R)$ is the exchange interaction potential, determined by the overlapping of electron shells. In the limit of large separations between the nuclei the exchange interaction potential takes the form

$$\Delta(R) = BR^{\alpha} \exp(-\beta R) ,$$

where B, α, β are constant parameters for a given particle. The interaction potential of two atoms, each having spin 1/2, includes the exchange interaction potential as

$$U(R) = U_L(R) \pm 0.5\Delta(R) ,$$

where a minus sign corresponds to zero total spin for the system, and the plus sign to a total spin of one. In Table 9.8 the calculated parameters of the exchange interaction potential for like atoms, expressed in atomic units, are presented [9.2.1]. The estimated accuracy in determining the potential $\Delta(R)$ is of order 20%.

Reference

9.2.1 B.M.Smirnov: *The Asymptotic Methods in the Theory of Atomic Collisions* (Atomizdat, Moscow 1973) Chap. 2 (in Russian)

Table 9.8. Parameters of long-range exchange potential for identical atoms [a.u.]

Parameters	H	He	Li	Be	Ne	Na	Mg	Ar	K	Ca
α	2.5	1.60	4.56	3.22	1.78	4.59	3.63	2.24	5.17	4.16
β	2	2.79	1.26	1.658	2.52	1.252	1.512	2.16	1.134	1.356
B	1.65	7.0	0.044	0.64	5.1	0.024	0.27	7.6	0.0056	0.067

Parameters	Zn	Kr	Rb	Sr	Cd	Xe	Cs	Ba
α	3.22	2.30	5.29	4.37	3.30	2.71	5.53	4.64
β	1.66	2.06	1.112	1.304	1.626	1.888	1.072	1.24
B	0.66	6.1	0.0039	0.044	0.60	3.8	0.0016	0.021

9.3 Short-Range Repulsive Interactions Between Atomic and Molecular Species

The short-range repulsive interaction in atomic species is related to the distance R between their nuclei when overlapping of electron shells of the particles takes place. The resulting intermolecular forces are of an electrostatic and exchange nature, the interaction potential increasing sharply with descreasing distance between them [9.3.1]. This allows one to use a simple approximation formula for the repulsive interaction potential of the particles with an exponential-type dependence:

$$U(R) = A \exp(-\beta R) ,$$

where the parameters A and β vary insignificantly within the range of internuclear distances ΔR considered. (To be more exact, this representation corresponds to the assumption that the logarithmic derivative of the potential $\beta = d\ln U/dR$ is constant within a given range of R.)

Tables 9.9–15 give numerical values of the parameters A and β for interacting atomic and molecular species as well as the range of internuclear separations for which the approximation is correct. The data are derived from measurements of cross sections for elastic scattering of the partners, and from measurements of ionic mobility in gases in strong electric fields [9.3.2–6]. The repulsive interaction potentials of particles based on the above parameters are characterized by an error of 20–40% within the range of internuclear separations quoted.

References

9.3.1 H.Margenau, N.R.Kestner: *Theory of Intermolecular Forces*, 2nd ed. (Pergamon, Oxford 1971)

9.3.2 V.B.Leonas: "Intermolecular Interactions and Collisions of Atoms and Molecules", in *Review of Science and Technique: Atomic and Molecular Physics, Optics, and Magnetic Resonance, Vol. 1*, ed. by S.A.Losev (All-Union Institute of Scientific and Technical Information, Moscow 1980) pp. 206 (in Russian)

9.3.3 E.A.Mason, J.T.Vanderslice: "High-Energy Elastic Scattering of Atoms, Molecules, and Ions", in *Atomic and Molecular Processes*, ed. by D.R.Bates (Academic, New York 1962) Chap. 17, pp. 663–695

9.3.4 I.Amdur, J.E.Jordan: "Elastic Scattering of High-Energy Beams: Repulsive Forces", in *Molecular Beams*, ed. by J. Ross, Adv. Chem. Phys., Vol. 10 (Wiley, New York 1966) Chap. 2, p. 29

9.3.5 L.A.Viehland, E.A.Mason: J. Chem. Phys. **80**, 416 (1984); ibid **81**, 903 (1984)

9.3.6 H.Inouye, K.Noda, S.Kita: J. Chem. Phys. **71**, 2135 (1979)

Table 9.9. Parameters A, β of short-range repulsive potential $U(r) = A\exp(-\beta r)$ for collision system (rare gas atom – rare gas atom) in the interaction energy range of about 0.3–2 eV (A [keV] – upper line in each entry, β [Å^{-1}] – lower line)

	He $(1\,^1S)$	Ne	Ar	Kr	Xe
He $(2\,^3S)$	0.054 2.69	0.030 2.50	–	–	–
He $(1\,^1S)$	0.2 4.20	0.32 3.80	0.34 3.25	0.38 2.98	0.19 2.65
Ne	–	1.0 3.94	0.39 2.98	1.2 3.21	0.69 2.96
Ar	–	–	0.61 2.83	0.79 2.52	0.21 2.05
Kr	–	–	–	0.31 2.28	0.87 2.61
Xe	–	–	–	–	1.6 2.52

Table 9.10. Parameters A, β of short-range repulsive potential $U(r) = A\exp(-\beta r)$ for collision system (H, O, F – rare gas atom) in the interaction energy range of about 0.1–10 eV (A [keV] – upper line in each entry, β [Å^{-1}] – lower line)

	He	Ne	Ar	Kr	Xe
H	0.087 3.59	0.24 4.22	0.98 4.53	0.87 4.31	3.3 4.63
O	0.68 4.51	1.1 4.08	1.5 3.72	3.9 4.22	0.53 3.44
F	0.63 4.91	2.7 5.18	9.8 4.82	12 5.23	27 5.38

Table 9.11. Parameters A, β of short-range repulsive potential $U(r) = A\exp(-\beta r)$ for collision system (alkali atom, atomic mercury – rare gas atom) in the interaction energy range of about 0.01–0.5 eV (A [keV] – upper line in each entry, β [Å^{-1}] – lower line)

	He	Ne	Ar	Kr	Xe
Na	–	–	0.020 1.69	–	0.039 1.75
K	–	0.0069 1.46	0.034 1.79	0.019 1.65	0.032 1.68
Cs	0.0088 1.42	0.034 1.88	0.224 2.26	0.087 2.02	0.0814 1.97
Hg	0.153 2.79	0.509 3.09	0.717 2.90	1.247 2.88	4.195 3.12

Table 9.12. Parameters A, β of short-range repulsive potential $U(r) = A\exp(-\beta r)$ for collision system (positive ion – rare gas atom) in the interaction energy range of about 0.5–10 eV (A [keV] – upper line in each entry, β [Å^{-1}] – lower line)

	He	Ne	Ar	Kr	Xe
Li^+	0.33 4.89	1.6 5.3	1.8 4.24	2.2 4.08	–
Na^+	1.2 4.92	5.4 5.12	11.3 4.68	9.6 4.33	–
K^+	1.1 4.15	4.9 4.4	2.9 3.46	6.4 3.69	–
Rb^+	1.1 3.69	3.4 3.87	4.7 3.48	4.6 3.23	7.5 3.23
Cs^+	1.4 3.60	5.6 3.86	12.3 3.65	7.2 3.24	10.4 3.25
O^+	–	–	0.02 1.11	–	–
Al^+	0.33 3.50	1.2 3.93	3.2 3.86	–	–

Table 9.13. Parameters A, β of short-range repulsive potential $U(r) = A\exp(-\beta r)$ for collision system (negative ion – rare gas atom) in the interaction energy range of about 0.5–10 eV (A [keV] – upper line in each entry, β [Å^{-1}] – lower line)

	H^-	Cl^-	Br^-
He	0.018 2.04	0.26 2.88	0.37 2.92
Ne	0.035 2.20	0.82 3.05	0.99 3.05
Ar	0.060 2.23	1.50 3.01	1.4 2.83

Table 9.14. Parameters A, β of short-range repulsive potential $U(r) = A\exp(-\beta r)$ for collision system (atom – molecule) in the interaction energy range of about 0.1–1 eV (A [keV] – upper line in each entry, β [Å^{-1}] – lower line)

	H_2	N_2	O_2	CO	NO	CO_2	N_2O
He	0.21 3.54	0.29 3.14	0.16 2.95	–	–	1.4 3.42	0.55 2.98
Ne	–	–	–	–	–	33 4.38	5.7 3.63
Ar	–	5.9 3.26	9.6 3.94	1.4 2.93	–	120 4.40	11 3.53
Kr	–	–	–	–	–	270 4.55	23 3.63
Xe	–	–	–	–	–	10^3 4.82	68 3.85
H	–	1.7 4.52	1.2 4.60	–	–	–	–
N	–	0.62 3.31	3.9 4.13	1.6 3.72	5.3 4.21	–	–
O	0.29 3.74	2.8 4.12	5.0 4.28	1.2 3.56	1.6 3.72	7 3.89	–

Table 9.15. Parameters A, β of short-range repulsive potential $U(r) = A\exp(-\beta r)$ for collision system (molecule – molecule) in the interaction energy range of about 0.1–10 eV (A [keV] – upper line in each entry, β [Å^{-1}] – lower line)

	H_2	N_2	O_2	CO	NO	CO_2	N_2O
H_2	0.25 3.22	–	–	–	–	–	–
N_2	–	2.3 3.16	1.4 3.02	7.1 3.66	5.8 3.64	30 3.78	6.8 3.25
O_2	–	–	0.82 2.85	2.7 3.15	7.6 3.78	8.7 3.33	–
CO	–	–	–	4.7 3.47	4.3 3.49	19 3.55	–
NO	–	–	–	–	2.2 3.26	–	–
CO_2	–	–	–	–	–	45 3.43	–
N_2O	–	–	–	–	–	–	11 3.07

10. Diatomic Molecules

The systematics of electron quantum states in diatomic molecules is briefly reviewed along with the systematics of molecular terms. The normal electronic configurations of molecular species and asymptotic parameters of valence electron wavefunctions are presented. Numerical data are compiled for spectroscopic constants, dissociation energies, ionization potentials and radiative lifetimes of electronically excited diatomic molecules. The potential energy curves of some diatomic molecules are reproduced and the diagrams of ir-spectra for one- and two-quantum vibrational transitions in a few diatomic molecules are supplied with the calculated values of Einstein coefficients for spontaneous emission from vibrationally excited states.

10.1 Electron Configurations of Diatomic Molecules

The state of an individual electron in a diatomic molecule can be characterized by four quantum numbers:

a) The component $\pm\lambda$ of the electron angular momentum (measured in units of $\hbar$) along the molecular axis passing through the two nuclei. This quantum number takes the values $\lambda = 0, 1, 2, \ldots$ and the corresponding electrons are called the $\sigma, \pi, \delta, \ldots$ electrons. The electron energy depends significantly only on the absolute value of the projected orbital angular momentum $|\lambda|$, as the directions of electron rotation in an electric field are equivalent.

b) The component $m_s = \pm 1/2$ of the electron spin along the internuclear axis.

c) The principal quantum number n and the angular momentum l of the electron state, either in one of the atomic dissociation products, when the internuclear distance R approaches infinity ($R \to \infty$) (these states are denoted as λnl and, for example, we have $\sigma 2s$, $\pi 3p$ orbitals, etc.) or in the combined atom, when $R \to 0$ (these states are denoted as $nl\lambda$ and, for example, we deal with $1s\sigma$, $2p\pi$ orbitals, etc.). If the electrons have the same n and l they are said to be equivalent, and according to the Pauli principle, the number of such σ electrons is $\leqslant 2$; when $\lambda \neq 0$, the number of corresponding electrons is $\leqslant 4$.

For homonuclear molecules (consisting of two like atoms) the states of individual electrons are distinguished by the additional symmetry characteristic of their wavefunctions, which reflects the result of coordinate inversion in the midpoint between the nuclei: the wavefunctions of even (g) states (for example σ_g, $\pi_g, \ldots$) are unchanged when the coordinates of the electrons change sign, while those of odd (u) states change sign (for example σ_u, $\pi_u, \ldots$). The former correspond to even l ($s\sigma$, $d\sigma \to \sigma_g$), and the latter to odd l ($p\sigma \to \sigma_u$, $p\pi \to \pi_u$). Note that the above-mentioned classification of electron states is rigorous only for diatomic particles with one electron; otherwise one can consider only approximately the motion of an individual electron in the "axially symmetric field" of the nuclei and all the other electrons.

The states of the electron subsystem as a whole in the diatomic molecule (neglecting the molecular rotation) can be characterized by the following quantum numbers:

a) The component M_L of the total orbital angular momentum of the electrons (measured in units of $\hbar$) along the internuclear axis, which is conserved in the axially symmetric field of the nuclei. This quantum number takes the values $M_L = 0, \pm 1, \pm 2, \pm 3, \ldots$ and the corresponding molecular states are designated Σ, Π, Δ, $\Phi, \ldots$ states. The energy of the electrons in the electric field of the nuclei depends on the absolute value of the component $\Lambda = |M_L|$ only and hence Σ states are not degenerate, but Π, Δ, etc. states are doubly degenerate, since M_L can have the two values $+\Lambda$ and $-\Lambda$.

b) The constant component M_S of the resultant electron spin $\boldsymbol{S}$ which has an integer or half-integer value according to whether there is an even or odd number of electrons in the molecule. In the case of $\Lambda \neq 0$, the non-conservation of total electron spin $\boldsymbol{S}$ is due to the internal magnetic field in the direction of the internuclear axis resulting from the orbital motion of the electrons. This field causes the precession of the $\boldsymbol{S}$ vector about the field direction, with a constant component $M_S = \Sigma$ along the internuclear axis. The values of Σ are $S, S-1, \ldots, -S$ and can be positive or negative; if $\Lambda = 0$, the quantum number Σ is not defined. The number $2S+1$ is called the multiplicity of the electronic term and is added to the term symbol as a left superscript.

c) The component Ω of the total electronic angular momentum about the internuclear axis, which is formed by an algebraic addition of the Λ and Σ components ($\Omega = \Lambda + \Sigma$). The number Ω is added to the term symbol as a subscript and in specifying an electronic term of a diatomic molecule the notation

$$^{2S+1}\Lambda_{\Omega}$$

is used. Thus for the singlet terms, $S = 0$ and $\Omega = \Lambda$, as, for example is the case for $^1\Pi_1$, $^1\Delta_2$ terms. If $\Lambda \neq 0$, there are $2S+1$ different values of Ω and due to the interaction of the S vector with the internal magnetic field produced by the orbital motion of the electrons, the electronic term with a given

Λ splits into a multiplet of $2S+1$ components. As a consequence each multiplet component possesses a slightly different energy. In the first approximation the electron energy of a multiplet term is $E^{\text{el}} = E_0^{\text{el}} + A\Lambda \cdot \Sigma$, where E_0^{el} is the energy of the unperturbed term and A is a constant for a given multiplet term which increases rapidly with the number of electrons in the molecule. (For example, the splitting of the first excited $A^2\Pi$ state in BeH is 2 cm^{-1}, while for BaH the analogous value is $\sim$ 480 cm^{-1}.) When the coupling constant A is positive the term is called "normal"; if $A < 0$, we have an "inverted" term and multiplet components lie in the inverse order to their Ω values. The inverted terms are marked with the additional suffix (i). If $\Lambda = 0$, there is no internal magnetic field in the molecule and consequently Σ terms have no splitting, but the value $2S+1$ is called the multiplicity as before.

The spin does not alter the double degeneracy of the electronic terms with $\Lambda \neq 0$: to any order of approximation each multiplet component is two-fold degenerate if we neglect the effect of molecular rotation on the electron state. If only the Ω quantum number is zero, the degeneracy occurs to first approximation, but in higher order approximations the spin-orbit interaction of electrons results in a small splitting of the energy level with $\Omega = 0$. Thus the $^3\Pi_0$ terms, as is the case for halogen molecules, Br_2, I_2, etc. have two sublevels with slightly different energies, which are designated $^3\Pi_{0^+}$ and $^3\Pi_{0^-}$. For heavy molecules the spin-orbit interaction is no longer small and independent quantization of the L and S vectors does not take place. In this case Ω, the component of the precessing total angular momentum of the electrons along the internuclear axis, is conserved, but the values Λ and Σ are not defined themselves. Such states are classified as 0, 1/2, 1 . . ., etc. according to the values of Ω only.

Finally, the molecular electronic states are also classified taking into account the symmetry properties of the electron wavefunctions. Besides rotations through any angle about the axis, the symmetry of a diatomic molecule allows a reflection in any plane passing through the axis. Thus we have doubly degenerate states for $\Lambda \neq 0$ which differ in the direction of the projection of the orbital angular momentum on the molecular axis, and Σ^+, Σ^- terms for $\Lambda = 0$, whose wavefunctions are unaltered on reflection and change sign, respectively. In addition, the homonuclear diatomic molecule has a centre of symmetry at the midpoint of the internuclear axis. Therefore we can classify the terms with a given value of Λ according to their parity: the wavefunction of the even (g) state is unchanged and the wavefunction of the odd (u) state changes sign on reflection of the nuclei at this centre of symmetry. This symmetry property is indicated by adding a right subscript g or u, respectively, to the symbol of the electronic term.

The reader is reminded that molecular electron states of the same type are distinguished by letters as follows: the ground state is referred to as X, the excited states of the same multiplicity as A, B, C, . . ., those of different multiplicity as a, b, c, . . . (for example, $X^1\Sigma_g^+$ relates to a ground singlet term,

$b^3\Pi_{1u}$ relates to an excited triplet term, etc.). A detailed description of the above-mentioned problems is given in the monographs [10.1.1–3].

Table 10.1 contains information about the electron configurations of some diatomic molecules. These configurations relate to well-known electronic terms of molecules. We have shown the molecular orbital (MO) configurations with the help of "filling numbers"; thus the third column of Table 10.1 is not overloaded with reiterated MO symbols. The dissociation products of molecules and their atomic states are included in the fourth column of Table 10.1.

References

10.1.1 G.Herzberg: *Molecular Spectra and Molecular Structure. 1. Spectra of Diatomic Molecules*, 2nd ed. (Van Nostrand, Princeton 1950)

10.1.2 J.C.Slater: *Quantum Theory of Molecules and Solids. 1. Electronic Structure of Molecules* (McGraw-Hill, New York 1963)

10.1.3 G.Herzberg: *The Spectra and Structure of Simple Free Radicals* (Cornell University Press, Ithaca 1971)

Table 10.1. Electronic configurations and terms of diatomic molecules

Molecule	Electronic term	MO configuration	Dissociation products and their states
1	2	3	4
B_2	$X^3\Sigma_g^-$	$1\sigma_g^2 1\sigma_u^2 2\sigma_g^2 2\sigma_u^2 1\pi_u^2$	$^2P + {}^2P$
BF	$X^1\Sigma^+$	$1\sigma^2 2\sigma^2 3\sigma^2 4\sigma^2 1\pi^4 5\sigma^2$	$B(^2P) + F(^2P)$
BH	$X^1\Sigma^+$	$1\sigma^2 2\sigma^2 3\sigma^2$	$B(^2P) + H(^2S)$
BO	$X^2\Sigma^+$	$1\sigma^2 2\sigma^2 3\sigma^2 4\sigma^2 1\pi^4 5\sigma$	$B(^2P) + O(^3P)$
C_2	$A^1\Pi_u$	$2\,2\,2\,2\,(\pi_u 2p)^3 (\sigma_g 2p)$	$^3P + {}^3P$
	$b^3\Sigma_g^-$	$2\,2\,2\,2\,(\pi_u 2p)^2 (\sigma_g 2p)^2$	$^3P + {}^3P$
	$a^3\Pi_u$	$2\,2\,2\,2\,(\pi_u 2p)^3 (\sigma_g 2p)$	$^3P + {}^3P$
	$X^1\Sigma_g^+$	$(\sigma_g 1s)^2 (\sigma_u 1s)^2 (\sigma_g 2s)^2 (\sigma_u 2s)^2 (\pi_u 2p)^4$	$^3P + {}^3P$
CF	$X^2\Pi_r$	$1\sigma^2 2\sigma^2 3\sigma^2 4\sigma^2 1\pi^4 5\sigma^2 2\pi$	$C(^3P) + F(^2P)$
CH	$X^2\Pi_r$	$1\sigma^2 2\sigma^2 3\sigma^2 1\pi$	$C(^3P) + H(^2S)$
CN	$A^2\Pi_i$	$2\,2\,2\,2\,1\pi^3 5\sigma^2$	$C(^3P) + N(^4S)$
	$X^2\Sigma^+$	$1\sigma^2 2\sigma^2 3\sigma^2 4\sigma^2 1\pi^4 5\sigma$	$C(^3P) + N(^4S)$
CO	$I^1\Sigma^-$	$2\,2\,2\,2\,1\pi^3 5\sigma^2 2\pi$	$C(^3P) + O(^3P)$
	$A^1\Pi$	$2\,2\,2\,2\,1\pi^4 5\sigma\ 2\pi$	$C(^3P) + O(^3P)$
	$e^3\Sigma^-$	$2\,2\,2\,2\,1\pi^3 5\sigma^2 2\pi$	$C(^3P) + O(^3P)$
	$d^3\Delta_i$	$2\,2\,2\,2\,1\pi^3 5\sigma^2 2\pi$	$C(^3P) + O(^3P)$
	$a'^3\Sigma^+$	$2\,2\,2\,2\,1\pi^3 5\sigma^2 2\pi$	$C(^3P) + O(^3P)$
	$a^3\Pi_r$	$2\,2\,2\,2\,4\quad 5\sigma\ 2\pi$	$C(^3P) + O(^3P)$
	$X^1\Sigma^+$	$1\sigma^2 2\sigma^2 3\sigma^2 4\sigma^2 1\pi^4 5\sigma^2$	$C(^3P) + O(^3P)$

Table 10.1 (continued)

Mole-cule	Electronic term	MO configuration	Dissociation products and their states
1	2	3	4
F_2	$X^1\Sigma_g^+$	$1\sigma_g^2 1\sigma_u^2 2\sigma_g^2 2\sigma_u^2 3\sigma_g^2 1\pi_u^4 1\pi_g^4$	${}^2P + {}^2P$
H_2	$E, F^1\Sigma_g^+$	$(1s\sigma)\,(2s\sigma) + (2p\sigma)^2$	${}^2S + 2\,{}^2S$
	$C^1\Pi_u$	$(1s\sigma)\,(2p\pi)$	${}^2S + 2\,{}^2P$
	$a^3\Sigma_g^+$	$(1s\sigma)\,(2s\sigma)$	${}^2S + 2\,{}^2S$
	$c^3\Pi_u$	$(1s\sigma)\,(2p\pi)$	${}^2S + 2\,{}^2P$
	$B^1\Sigma_u^+$	$(1s\sigma)\,(2p\sigma)$	${}^2S + 2\,{}^2P$
	$X^1\Sigma_g^+$	$(1s\sigma)^2$	${}^2S + 1\,{}^2S$
HCl	$X^1\Sigma^+$	$1\sigma^2 2\sigma^2 3\sigma^2 1\pi^4 4\sigma^2 5\sigma^2 2\pi^4$	$H({}^2S) + Cl({}^2P)$
HF	$X^1\Sigma^+$	$1\sigma^2 2\sigma^2 3\sigma^2 1\pi^4$	$H({}^2S) + F({}^2P)$
He_2	$B^1\Pi_g$	$(1s\sigma)^3\,(2p\pi)$	${}^1S + 2^1P$
	$b^3\Pi_g$	$(1s\sigma)^3\,(2p\pi)$	${}^1S + 2^3P$
	$A^1\Sigma_u^+$	$(1s\sigma)^3\,(2s\sigma)$	${}^1S + 2^1S$
	$a^3\Sigma_u^+$	$(1s\sigma)^3\,(2s\sigma)$	${}^1S + 2^3S$
Li_2	$X^1\Sigma_g^+$	$1\sigma_g^2 1\sigma_u^2 2\sigma_g^2$	${}^2S + {}^2S$
LiF	$X^1\Sigma^+$	$1\sigma^2 2\sigma^2 3\sigma^2 4\sigma^2 1\pi^4$	$Li({}^2S) + F({}^2P)$
LiH	$X^1\Sigma^+$	$1\sigma^2 2\sigma^2$	$Li({}^2S) + H({}^2S)$
MgH	$X^2\Sigma^+$	$1\sigma^2 2\sigma^2 1\pi^4 3\sigma^2 4\sigma^2 5\sigma$	$Mg({}^1S) + H({}^2S)$
N_2	$w^1\Delta_u$	$2\,2\,2\,2\;1\pi_u^3 3\sigma_g^2 1\pi_g$	${}^2D^o + {}^2D^o$
	$a^1\Pi_g$	$2\,2\,2\,2\;1\pi_u^4 3\sigma_g 1\pi_g$	${}^2D^o + {}^2D^o$
	$a'^1\Sigma_u^-$	$2\,2\,2\,2\;1\pi_u^3 3\sigma_g^2 1\pi_g$	${}^2D^o + {}^2D^o$
	$B'^3\Sigma_u^-$	$2\,2\,2\,2\;1\pi_u^3 3\sigma_g^2 1\pi_g$	${}^4S^o + {}^2P^o$
	$W^3\Delta_u$	$2\,2\,2\,2\;1\pi_u^3 3\sigma_g^2 1\pi_g$	${}^4S^o + {}^2D^o$
	$B^3\Pi_g$	$2\,2\,2\,2\;1\pi_u^4 3\sigma_g 1\pi_g$	${}^4S^o + {}^2D^o$
	$A^3\Sigma_u^+$	$2\,2\,2\,2\;1\pi_u^3 3\sigma_g^2 1\pi_g$	${}^4S^o + {}^4S^o$
	$X^1\Sigma_g^+$	$1\sigma_g^2 1\sigma_u^2 2\sigma_g^2 2\sigma_u^2 1\pi_u^4 3\sigma_g^2$	${}^4S^o + {}^4S^o$
NF	$X^3\Sigma^-$	$1\sigma^2 2\sigma^2 3\sigma^2 4\sigma^2 1\pi^4 5\sigma^2 2\pi^2$	$N({}^4S^o) + F({}^2P)$
NH	$X^3\Sigma^-$	$1\sigma^2 2\sigma^2 3\sigma^2 1\pi^2$	$N({}^4S^o) + H({}^2S)$
NO	$X^2\Pi_r$	$(1s\sigma)^2\,(2s\sigma)^2\,(2p\sigma)^2\,(2p\pi)^4\,(3s\sigma)^2\,(3p\sigma)^2\,(3p\pi)$	$N({}^4S^o) + O({}^3P)$
NaH	$X^1\Sigma^+$	$1\sigma^2 2\sigma^2 1\pi^4 3\sigma^2 4\sigma^2$	$Na({}^2S) + H({}^2S)$
O_2	$B^3\Sigma_u^-$	$2\,2\,2\,2\,2\;1\pi_u^3 1\pi_g^3$	$({}^3P + {}^1D)_{pr}$
	$A^3\Sigma_u^+$	$2\,2\,2\,2\,2\;1\pi_u^3 1\pi_g^3$	${}^3P + P^3$
	$c^1\Sigma_u^-$	$2\,2\,2\,2\,2\;1\pi_u^3 1\pi_g^3$	${}^3P + P^3$
	$b^1\Sigma_g^+$	$2\,2\,2\,2\,2\,4\,2$	${}^3P + P^3$
	$a^1\Delta_g$	$2\,2\,2\,2\,2\,4\,2$	${}^3P + P^3$
	$X^3\Sigma_g^-$	$1\sigma_g^2 1\sigma_u^2 2\sigma_g^2 2\sigma_u^2 3\sigma_g^2 1\pi_u^4 1\pi_g^2$	${}^3P + P^3$
OH	$A^2\Sigma^+$	$1\sigma^2 2\sigma^2 3\sigma\; 1\pi^4$	$O({}^3P) + H({}^2S)$
	$X^2\Pi_i$	$1\sigma^2 2\sigma^2 3\sigma^2 1\pi^3$	$O({}^3P) + H({}^2S)$
SO	$X^3\Sigma^-$	$(1s\sigma)^2\,(2s\sigma)^2\,(2p\sigma)^2\,(2p\pi)^4\,(3s\sigma)^2\,(3p\sigma)^2\,(3p\pi)^4 \cdot (4s\sigma)^2\,(4p\sigma)^2\,(4p\pi)^2$	$S({}^3P) + O({}^3P)$

10.2 Asymptotic Parameters of Wavefunctions for Valence Electrons in Diatomic Molecules

Tables 10.2–4 give the asymptotic coefficients of wavefunctions for valence electrons in diatomic molecules [10.2.1]. These parameters characterize the one-electron distribution (molecular orbital) within a relatively large range of distances r from the centre of a molecule, compared to the mean size of the molecule itself. Similarly to the atomic case (Sect. 4.3), in this range of the electronic coordinates the wavefunction may be approximated by a dependence

$$\psi_{\mathrm{as}}(r,\ \theta,\ \phi) = A(\theta) r^{\frac{Z_c}{\gamma}-1} \exp(-\gamma r) \exp(im\phi) ,$$

where r is the distance from the centre of charge of the molecule, θ is the angle between the radius vector of the electron and the molecular axis, ϕ is the azimuthal angle, $(-\gamma^2/2)$ is the binding energy of the valence electron in the molecule, m is the projection of the electronic angular momentum on the molecular axis, and Z_c is the net charge of the molecular core, equal to 1 and 0 for a molecule and a negative molecular ion, respectively.

Values of the asymptotic coefficient $A(\theta)$ were determined in [10.2.1] by matching the wavefunction ψ_{as} with Hartree-Fock (self-consistent field) one-electron wavefunctions of the valence electrons [10.2.2, 3]. Numerical values of the parameters a, b and c of the asymptotic coefficient expansion were rounded off taking into account the errors in the determination of the HF wavefunctions and in matching the electron wavefunctions, so that possible corrections might involve the last significant figure within the range ± 1.

References

10.2.1 A.V.Evseev, A.A.Radzig, B.M.Smirnov: Sov. Phys.-JETP **50**, 283–289 (1979)
10.2.2 P.E.Cade, W.Huo: At. Data Nucl. Data Tables **12**, 415 (1973); ibid **13**, 339 (1974)
10.2.3 P.E.Cade, A.C.Wahl: At. Data Nucl. Data Tables **15**, 1 (1975)

Table 10.2. Asymptotic parameters[a] of valence electron wavefunctions for homonuclear diatomic molecules

Molecule and ground-state term	Valence molecular orbital	Asymptotic parameters [a.u.]						
		γ	R_e	a	b	c	α	β
$H_2(X^1\Sigma_g^+)$	$1\sigma_g$	1.065	1.401	2.26	0.35	0.09	0	0
$C_2(X^1\Sigma_g^+)$	$1\pi_u$	0.935	2.353	2.0	0.5	0.3	1	0
$N_2(X^1\Sigma_g^+)$	$1\pi_u$	1.070	2.075	1.5	1.20	−0.03	0	0
$O_2(X^3\Sigma_g^-)$	$1\pi_g$	0.941	2.285	2.53	0.65	0.01	1	1
$F_2(X^1\Sigma_g^+)$	$1\pi_g$	1.074	2.840	2.7	0.4	0.4	0	0

[a] $A(\theta) = [a\,\mathrm{ch}(b\gamma R_e \cos\theta)\,(1 + c\cos^2\theta)] \sin^\alpha\theta \cos^\beta\theta$

Table 10.3. Asymptotic parameters[a] of valence electron wavefunctions for heteronuclear diatomic molecules

Molecule and ground-state term	Valence molecular orbital	Asymptotic parameters [a.u.]					
		γ	R_e	a	b	c	α
$NF\,(X^3\Sigma^-)$	2π	0.951	2.489	1.6	0.9	−0.6	1
$CO\,(X^1\Sigma^+)$	5σ	1.015	2.132	2.0	0.2	−0.4	0
$NO\,(X^2\Pi_r)$	2π	0.842	2.175	0.5	0.2	−0.2	1

[a] $A(\theta) = \left[\frac{a\,\mathrm{ch}(\gamma R_e \cos\theta)}{1 + b\cos\theta}(1 + c\cos^2\theta)\right] \sin^\alpha\theta \cos^\alpha\theta$

Table 10.4. Asymptotic parameters[a] of valence electron wavefunctions for diatomic hydrides

Molecule and ground-state term	Valence molecular orbital	Asymptotic parameters [a.u.]			
		γ	R_e	a	b
$OH\,(X^2\Pi_i)$	1π	0.985	1.862	1.9	0.2
$NH\,(X^3\Sigma^-)$	1π	0.981	1.999	1.8	0.2
$CH\,(X^2\Pi_r)$	1π	0.895	2.370	1.7	0.1
$PH\,(X^3\Sigma^-)$	2π	0.833	3.251	2.2	0.1
$SH\,(X^2\Pi_i)$	2π	0.874	2.919	2.15	0.07
$HF\,(X^1\Sigma^+)$	1π	1.077	1.609	1.4	0.2

[a] $A(\theta) = [1 + b\exp(\gamma R_e \cos\theta)]\sin\theta$

10.3 Spectroscopic Constants of Diatomic Molecules

Below, we give brief comments on the magnitudes and symbols for physical quantities relevant to diatomic molecules and used in preparing Table 10.5.

Notations and Formulae

1) The total energy of a state in the diatomic molecule is given by $T = T_e + G(v) + F_v(J)$, where T_e is the electronic energy, $G(v)$ the vibrational energy and $F_v(J)$ the rotational energy.

2) $G(v) \simeq \omega_e(v + 1/2) - \omega_e x_e(v + 1/2)^2$, where v is the vibrational quantum number, ω_e the vibrational frequency and $\omega_e x_e$ the anharmonic vibrational constant.

3) $F_v(J) \simeq B_v J(J + 1)$, where J is the rotational quantum number, $B_v \simeq B_e - \alpha_e(v + 1/2)$ and B_e is the rotational constant.

4) The subscripts e and 0 are added to values which relate to the equilibrium position and to the zero-vibrational level, respectively.

5) The wavenumber ν corresponding to a transition from an upper state T' to a lower state T'' is $\nu = T' - T'' = T_e' + G' + F' - (T_e'' + G'' + F'')$. For a given system of bands, $\nu_e = T_e' - T_e'' = \text{const}$ is equal to the electron transition energy. If we neglect rotation, the transition energy between the vibrational levels v' and v'' is

$$\nu_{v',v''} = \nu_e + \omega_e' (v' + 1/2) - \omega_e' x_e' (v' + 1/2)^2 + \ldots$$
$$- [\omega_e'' (v'' + 1/2) - \omega_e'' x_e'' (v'' + 1/2)^2 + \ldots],$$

and if $v' = v'' = 0$, it follows that

$$\nu_{00} = \nu_e + 1/2\,(\omega_e' - \omega_e'') - 1/4\,(\omega_e' x_e' - \omega_e'' x_e'') + \ldots ,$$

so that

$$\nu_{v',v''} \simeq \nu_{00} + \omega_0' v' - \omega_0' x_0' v'^2 - (\omega_0'' v'' - \omega_0'' x_0'' v''^2) ,$$

where $\omega_0 \simeq \omega_e - \omega_e x_e$. The transition energy between two adjacent vibrational levels $v + 1$ and v (for unaltered electronic states and neglecting rotation) is $\Delta G_{v+1/2} = G(v + 1) + F_{v+1}(J) - G(v) - F_v(J) \simeq \omega_e - 2\,\omega_e x_e (v + 1)$; thus for $v = 0$, it follows that $\Delta G_{1/2} \simeq \omega_e - 2\,\omega_e x_e \approx \omega_0 - \omega_0 x_0$ (see also Fig. 10.1).

6) The observed electronic transitions in emission are indicated by arrows $\rightarrow$, those in absorption by $\leftarrow$ and those which occur both in emission and absorption by $\leftrightarrow$.

7) The force constant of diatomic molecules $k_e = \mu\omega_e^2 \simeq 5.892 \cdot 10^{-2}\,\mu_A \omega_e^2$, where ω_e is in cm^{-1} and μ_A is the reduced mass in a.m.u.

8) The rotational constant is defined as $B_e = \hbar^2/2\mu_A r_e^2$, where r_e is the equilibrium internuclear distance and, consequently, one finds (in common units):

$$r_e\,[\text{Å}] = 4.10581/\sqrt{\mu_A\,[\text{a.m.u.}]\,B_e\,[\text{cm}^{-1}]} .$$

9) The additional asterisk (* or **) was used in the cases when the value in Table 10.5 differed from that indicated in the heading: * after T_e, r_e, ω_e, $\omega_e x_e$ and B_e label the values ν_{00}, r_0, $\Delta G_{1/2}$, $\omega_0 x_0$, and B_0, respectively; ** after T_e and ω_e label the values T_0 and ω_0, respectively.

Table 10.5, which presents spectroscopic constants of diatomic molecules, is based on quantitive information from reference books [10.3.1–6] and some later journal publications. It includes mainly data for well-known molecular states and neglects questionable values. The numerical values listed are estimated to be accurate to within an error of not more than one, or occasionally two, in the last significant figure quoted.

References

10.3.1 K.P.Huber, G.Herzberg: *Molecular Spectra and Molecular Structure. IV. Constants of Diatomic Molecules* (Van Nostrand Reinhold, New York 1979)

10.3.2 B.Rosen (ed.): *Spectroscopic Data Relative to Diatomic Molecules,* Inernational Tables of Selected Constants **17** (Pergamon, Oxford 1970)

10.3.3 S.N.Suchard (ed.): *Spectroscopic Data. I. Heteronuclear Diatomic Molecules*, Parts A, B (IFI/Plenum, New York 1975)

10.3.4 S.N.Suchard, J.E.Melzer (eds.): *Spectroscopic Data. II. Homonuclear Diatomic Molecules* (IFI/Plenum, New York 1976)

10.3.5 K.S.Krasnov (ed.): *Molecular Constants for Inorganic Compounds* (Chimia, Moscow 1979) pp. 10–72 (in Russian)

10.3.6 N.G.Rambidi, S.M.Tolmachev, G.I.Gurova, I.V.Solovjeva, N.N.Veniaminov, A.I.Dementjev: *Nuclear Configuration and Internuclear Distances in Molecules and Ions in Gas Phase. I. Diatomic Molecules and Ions in Ground and Excited Electronic States* (Gosstandart, Moscow 1978) (in Russian)

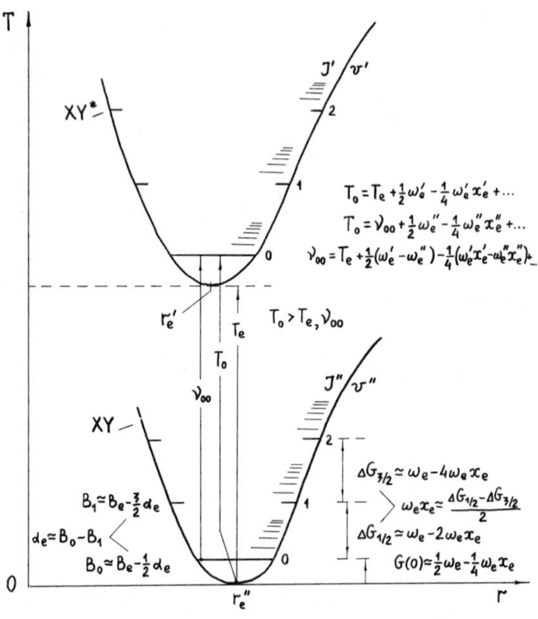

Fig. 10.1. Schematic diagram of electronic terms for molecular species XY and XY* explaining the notations used in Tables 10.5 and 11.3

Table 10.5. Spectroscopic constants of diatomic molecules

Molecule	Electronic term	Term energy T_e [cm^{-1}]	Equilibrium internuclear distance r_e [Å]	Vibrational frequency ω_e [cm^{-1}]	Anharmonic constant $\omega_e x_e$ [cm^{-1}]	Rotational constant B_e [cm^{-1}]	Rotation-vibration interaction constant α_e [10^{-3} cm^{-1}]	Force constant k_e [10^5 dyn/cm]	Observable electronic transition	Reduced mass μ_A [a.m.u.]
1	2	3	4	5	6	7	8	9	10	11
Ag_2	$X\,^1\Sigma_g^+$	0	2.7	192.4	0.64	0.496	0.19	1.18	–	53.93
AgBr	$X\,^1\Sigma^+$	0	2.393	247.7	0.68	0.0634	0.228	1.66	–	45.90
AgCl	$X\,^1\Sigma^+$	0	2.281	343.5	1.17	0.123	0.6	1.85	–	26.68
AgF	$X\,^1\Sigma^+$	0	1.983	513.45	2.59	0.266	1.92	2.51	–	16.15
AgH	$C\,^1\Pi$	41174**	1.6	1589	42	6.70	310	1.49	$C \leftarrow X$	–
	$A\,^1\Sigma^+$	29959	1.64	1663	87.0	6.265	348	1.63	$A \leftrightarrow X$	–
	$X\,^1\Sigma^+$	0	1.618	1760	34.1	6.449	201	1.82	–	0.999
AgI	$X\,^1\Sigma^+$	0	2.545	206.5	0.44	0.0449	0.15	1.46	–	58.31
AgO	$X\,^2\Pi_{1/2}$	0	2.003	490.2	3.1	0.3020	2.5	1.97	–	13.93
Al_2	$^3\Sigma_u^-(1)$	17700	2.56	278.8	0.83	0.1907	1.3	0.62	$^3\Sigma_u^- \rightarrow X$	–
	$A\,^3\Sigma_g^-$	320	2.47	350.0	2.02	0.2054	1.2	0.97	–	–
	$X\,^3\Pi_u$	0	2.8	270	1.6	0.16	–	0.58	–	13.49
AlBr	$A\,^1\Pi$	35879.5	2.32	297.2	6.40	0.1555	2.16	1.05	$A \leftrightarrow X$	–
	$a_2\,^3\Pi_1$	23779.3	2.26	410.3	1.75	0.1646	1	2.00	$a_2 \rightarrow X$	–
	$a_1\,^3\Pi_0$	23647	2.26	411.2	1.75	0.1643	–	2.01	$a_1 \rightarrow X$	–
	$X\,^1\Sigma^+$	0	2.295	378.2	1.33	0.1592	0.86	1.70	–	20.17
AlCl	$b\,^3\Sigma^+$	43527*	2.203	~ 350	–	0.2280	–	1.11	$b \rightarrow a$	–
	$A\,^1\Pi$	37997*	2.124	450.0	4.37	0.2454	2.52	1.83	$A \leftrightarrow X$	–
	$a\,^3\Pi_r$	{ 24658 24594 24528	2.094	524.4	2.17	0.2524	1.1	2.48	$a \rightarrow X$	–
	$X\,^1\Sigma^+$	0	2.130	481.3	1.95	0.2439	1.61	2.09	–	15.32

Table 10.5 (continued)

Molecule	Electronic term	Term energy T_e [cm^{-1}]	Equilibrium internuclear distance r_e [Å]	Vibrational frequency ω_e [cm^{-1}]	Anharmonic constant $\omega_e x_e$ [cm^{-1}]	Rotational constant B_e [cm^{-1}]	Rotation-vibration interaction constant α_e [10^{-3} cm^{-1}]	Force constant k_e [10^5 dyn/cm]	Observable electronic transition	Reduced mass μ_A [a.m.u.]
1	2	3	4	5	6	7	8	9	10	11
AlF	$H\,^1\Sigma^+$	67320	1.60*	958	7	0.592*	–	6.03	$H \to B, A;\ H \leftarrow X$	–
	$G\,^1\Sigma^+$	66334.0	1.581	931.5*	8	0.605	7.7	5.9	$G \to B, A;\ G \leftarrow X$	–
	$f\,^3\Pi$	65803	1.596	938.9*	6	0.594	4.8	5.9	$f \to c, b, a$	–
	$F\,^1\Pi$	65795.6	1.597	955.3	5.4	0.593	4.6	6.00	$F \to B, A;\ F \leftarrow X$	–
	$E\,^1\Pi$	63689.4	1.605	923.0	5.3	0.587	4.6	5.60	$E \to A;\ E \leftarrow X$	–
	$D\,^1\Delta$	61229.5	1.610	901.0	6.1	0.583	5.0	5.33	$D \to A$	–
	$C\,^1\Sigma^+$	57688.0	1.601	938.2	5.1	0.590	4.6	5.78	$C \to A;\ C \leftrightarrow X$	–
	$c\,^3\Sigma^+$	54957.7	1.603	933.7	4.8	0.589	4.6	5.73	$c \to b, a$	–
	$B\,^1\Sigma^+$	54251.0	1.615	866.6	7.4	0.580	5.6	4.93	$B \leftrightarrow X$	–
	$b\,^3\Sigma^+$	44813.2	1.639	786.4	7.6	0.563	6.5	4.06	$b \to a$	–
	$A\,^1\Pi$	43949.2	1.648	803.9	6.0	0.556	5.3	4.25	$A \leftrightarrow X$	–
	$a\,^3\Pi_r$	27241	1.648	827.8	3.9	0.557	4.5	4.50	$a \to X$	–
	$X\,^1\Sigma^+$	0	1.6544	802.3	4.8	0.552	5.0	4.23	–	11.15
AlH	$E\,^1\Pi$	52983*	1.76*	–	–	5.62*	–	–	$E \leftrightarrow X, \to A$	–
	$D\,^1\Sigma^+$	49288*	1.63*	–	–	6.56*	–	–	$D \leftrightarrow X$	–
	$C\,^1\Sigma^+$	44676	1.614	1575.3	125.5	6.664	544	1.42	$C \leftrightarrow X,\ C \to A$	–
	$A\,^1\Pi$	23470.93*	1.648	1083*	–	6.3869	732	–	$A \leftrightarrow X$	–
	$X\,^1\Sigma^+$	0	1.648	1683	29.1	6.3907	186	1.62	–	0.972
AlI	$a_2\,^3\Pi_1$	22089.5	–	333.4	2.0	0.123*	–	1.46	$a_2 \leftrightarrow X$	–
	$a_1\,^3\Pi_{0^+}$	21889.3	–	337.2	2.0	0.123*	–	1.49	$a_1 \leftrightarrow X$	–
	$X\,^1\Sigma^+$	0	2.537	316.1	1.0	0.1177	0.559	1.31	–	22.25

AlO	$E\,^2\Delta_i$	45431	1.84*	~ 500	–	0.49*	–	–	$E \leftrightarrow A$	–
	$D\,^2\Sigma^+$	40266.7	1.72	819.6	5.8	0.565	4.6	3.98	$D \rightarrow B,\ D \leftrightarrow X,\ A$	–
	$C\,^2\Pi_r$	33079	1.67	856	6	0.603	–	4.33	$C \leftrightarrow X,\ C \rightarrow B$	–
	$B\,^2\Sigma^+$	20688.95	1.667	870.0	3.5	0.6041	4.47	4.48	$B \leftrightarrow X$	–
	$A\,^2\Pi_i$	5341.7	1.771*	728.5	4.15	0.533*	–	3.14	$A \leftrightarrow X$	–
	$X\,^2\Sigma^+$	0	1.618	979.2	6.97	0.6414	5.8	5.67	–	10.04
AlS	$C\,^2\Sigma^+$	35714.9*	2.19	430*	~ 14	0.2402	3.6	1.8	$C \leftarrow X$	–
	$A\,^2\Sigma^+$	23433.8	2.164	510.9	1.45	0.2461	1.2	2.25	$A \leftrightarrow X$	–
	$X\,^2\Sigma^+$	0	2.029	617.1	3.33	0.2799	1.8	3.29	–	14.65
AlSe	$A\,^2\Sigma$	23183.5	–	389.8	1.23	–	–	1.31	$A \leftrightarrow X$	–
	$X\,^2\Sigma$	0	–	467.6	2.08	–	–	1.89	–	20.11
Ar_2	$X\,^1\Sigma_g^+$	0	3.76	25.7*	2.6	0.060	3.7	0.01	–	19.97
As_2	$A\,^1\Sigma_u^+$	40349	2.5*	260.3*	–	0.0720*	0.31	–	$A \leftrightarrow X$	–
	$d\,^3\Pi_g\,(1_g)$	30819	2.209	336.7	1.36	0.0922	0.33	2.50	$d \rightarrow X,\ c$	–
	$a\,^3\Sigma_u^-\,(0_u^+)$	24641	2.279	337.0	0.83	0.0866	0.30	2.51	$a \rightarrow X,\ c$	–
	$c\,^3\Sigma_u^+\,(0_u^-)$	14644	2.302	314.3	1.17	0.0849	0.35	2.18	$c \rightarrow X$	–
	$c\,^3\Sigma_u^+\,(1_u)$	14482	2.304	314.3	1.17	0.0847	0.35	2.18	$c \rightarrow X$	–
	$X\,^1\Sigma_g^+$	0	2.103	429.6	1.12	0.1018	0.333	4.07	–	37.46
AsF	$c\,^1\Pi$	48672.5	1.667	817.3	4.4	0.400	2.7	5.96	$c \rightarrow b;\ c \leftrightarrow X$	–
	$B\,^3\Pi$	48202.2*	–	815.5	4.0	–	–	5.94	$B \leftrightarrow X$	–
	$c'\,^1\Pi$	32479.5	1.95	399.4	1.34	0.293	1.8	1.42	$c' \rightarrow b,\ a$	–
	A_4 ($^3\Pi_r$, 2)	27152							$A_4 \rightarrow X_2$	
	A_3 ($^3\Pi_r$, 1)	26348	1.95	412.2	1.4	0.292	2	1.52	$A_3 \rightarrow X_1$	–
	A_2 ($^3\Pi_r$, 0^+)	25751							$A_2 \rightarrow X_2$	
	A_1 ($^3\Pi_r$, 0^-)	25719							$A_1 \rightarrow X_2$	
	$b\,^1\Sigma^+$	13648.6	1.729	697.3	3.1	0.372	2.8	4.34	$b \rightarrow X$	–
	$a\,^1\Delta$	7053.5	1.732	694.4	3.1	0.371	2.6	4.31	–	–
	X_2 ($^3\Sigma^-$)	138.7	1.736	685.5	3.0	0.369	2.8	4.20	–	15.155
	X_1 ($^3\Sigma^-$)	0		685.8	3.1	0.365	2.4	4.20		
AsN	$A\,^1\Pi$	35999.7	1.687	853.3*	8.24	0.502	9	5.16	$A \rightarrow X$	–
	$^1\Sigma^+$	29124.9*	1.689*	–	–	0.5011*	–	–	$^1\Sigma^+ \rightarrow X$	–
	$X\,^1\Sigma^+$	0	1.618	1068.5	5.41	0.5455	3.37	7.94	–	11.80

Table 10.5 (continued)

Molecule	Electronic term	Term energy T_e [cm^{-1}]	Equilibrium internuclear distance r_e [Å]	Vibrational frequency ω_e [cm^{-1}]	Anharmonic constant $\omega_e x_e$ [cm^{-1}]	Rotational constant B_e [cm^{-1}]	Rotation-vibration interaction constant α_e [10^{-3} cm^{-1}]	Force constant k_e [10^5 dyn/cm]	Observable electronic transition	Reduced mass μ_A [a.m.u.]
1	2	3	4	5	6	7	8	9	10	11
AsO	$B\,^2\Sigma^+$	39866.0	1.576	1098.3	6.1	0.5128*	3.6	9.36	$B \leftrightarrow X$	–
	$C\,^2\Delta_{5/2}$	38686	1.765	655.7	4.5	0.4164	4.0	3.34	$C \rightarrow X$	–
	$C\,^2\Delta_{3/2}$	38638		–	–	0.403*	–	–		
	$D\,^2\Sigma^-$	37555.4	1.794	629.9	3.79	0.3973	3.4	3.08	$D \rightarrow X$	–
	$H\,^2\Pi_{3/2}$	37053.7	1.871	606.9	4.91	0.3654	2.7	2.86	$H \rightarrow X$	–
	$A\,^2\Sigma^+$	31652.45	1.663	686.7	10.8	0.4624	7.1	3.66	$A \rightarrow X$	–
	$A'\,^2\Pi_{1/2}$	26485.2	1.855	630.3	3.01	0.3718	2.70	3.09	$A' \rightarrow X$	–
	$A'\,^2\Pi_{3/2}$	26168.4		633.2	2.89	0.3712	2.62	3.11		
	$X\,^2\Pi_{3/2}$	1025.97	1.624	965.9	4.91	0.4855	3.32	7.25	–	–
	$X\,^2\Pi_{1/2}$	0		967.1	4.85	0.4848	3.30	7.26	–	13.18
AsP	$A\,^1\Pi$	32417.05	2.10	475.5	2.1	0.174	0.9	2.92	$A \rightarrow X$	–
	$X\,^1\Sigma^+$	0	2.00	604.0	2.0	0.192	0.8	4.71	–	21.91
AsS	$A'_2\,^2\Pi_{1/2}$	20474.9	2.250	399.8*	–	0.1487	0.7	–	$A'_2 \rightarrow X_1$	–
	$A'_1\,^2\Pi_{3/2}$	19183.2*		405.6	1.11	0.1486	0.7	2.18	$A'_1 \rightarrow X_2$	–
	$X_2\,^2\Pi_{3/2}$	–	2.017	566.1	1.96	0.1849	0.8	4.24	–	–
	$X_1\,^2\Pi_{1/2}$	0		567.9	1.97	0.1848	0.8	4.27	–	22.45
Au_2	$B\,0_u^+$	25685.5	2.520*	179.8	0.680	0.027*	0.096	1.88	$B \leftrightarrow X$	–
	$A\,0_u^+$	19668.1	2.568	142.3	0.445	0.026	0.090	1.17	$A \leftrightarrow X$	–
	$X\,^1\Sigma_g^+$	0	2.472	190.9	0.420	0.028	0.072	2.11	–	98.48
AuAl	$C\,1$	24623	2.400*	250	2	0.1233*	–	0.87	$C \leftrightarrow X$	–
	$B\,1$	22490.3	2.326	291.8	3.0	0.1313	1.3	1.19	$B \leftrightarrow X$	–
	$A\,0^+$	16265.06	2.282	348.0	1.85	0.1365	0.85	1.69	$A \leftrightarrow X$	–
	$X\,^1\Sigma^+\,(0^+)$	0	2.338	333.0	1.16	0.1299	0.67	1.55	–	23.73

AuBe	$B\,^2\Pi_{1/2}$ ½	18946.0	2.020	628.9	3.22	0.4794	4.3	2.01	$B \leftrightarrow X$	–
	$A\,^2\Sigma^+$ ½	17171.0	1.993	655.4	3.59	0.4926	4.6	2.18	$A \leftrightarrow X$	–
	$X\,^2\Sigma^+$	0	2.060	607.7	3.5	0.4607	4.0	1.88	–	8.618
AuH	$B\,(0^+)$	38545	1.695	1544*	74	5.849	187	1.7	$B \leftrightarrow X$	–
	$A\,(0^+)$	27665.7	1.673	1669.5	55.1	6.007	249	1.65	$A \leftrightarrow X$	–
	$X\,^1\Sigma^+$	0	1.524	2305.0	43.1	7.240	214	3.14	–	1.003
AuMg	$B\,^2\Pi_{1/2}$ ½	19492.3	2.370*	338.5	1.5	0.1404*	–	1.46	$B \leftrightarrow X$	–
	$A\,^2\Sigma^+$ ½	18392.7	2.356*	341.7	3.3	0.1420*	–	1.49	$A \leftrightarrow X$	–
	$X\,^2\Sigma^+$	0	2.443	307.9	1.1	0.1321	0.7	1.21	–	21.64
B_2	$A\,^3\Sigma_u^-$	30573.4	1.625	937.4	2.6	1.160	11	2.80	$A \leftrightarrow X$	–
	$X\,^3\Sigma_g^-$	0	1.589	1051.3	9.4	1.212	14	3.52	–	5.41
BBr	$A\,^1\Pi$	33935.3	1.87	637.6	17.6	0.496*	9.0	2.28	$A \rightarrow X$	–
	a $\{^3\Pi_1$	18851.5	1.85	757.1	4.8	0.508	3.6	3.22	$a \rightarrow X$	–
	a $\{^3\Pi_{0+}$	18673.8		759.8	4.8	0.506	3.6	3.24		
	$X\,^1\Sigma^+$	0	1.89	684.3	3.52	0.489	3.5	2.63	–	9.52
BCl	$A\,^1\Pi$	36750.9	1.689	849.0	11.4	0.7054	8.20	3.52	$A \leftrightarrow X$	–
	$a\,^3\Pi_1$	20200	1.70	911	5.7	0.699	4.7	4.05	$a \rightarrow X$	–
	$X\,^1\Sigma^+$	0	1.715	840.29	5.49	0.6843	6.81	3.48	–	8.285
BF	$O\,^1\Sigma^+\,(5p\sigma)$	83680	1.219*	1676	9.5	1.627*	–	11.4	$O \leftarrow X$	–
	$J\,^1\Pi\,(4p\pi)$	80544	1.210	1673.1*	–	1.652	16.2	–	$J \leftarrow X$	–
	$h\,^3\Pi$	80230	1.212*	1679*	–	1.647*	–	–	$h \rightarrow b$	–
	$I\,^1\Sigma^+\,(4p\sigma)$	79631.4	1.215	1666.3	12.6	1.638	17.4	11.3	$I \leftarrow X$	–
	$F\,^1\Pi\,(3d\pi)$	77542.8*	1.203*	1670	–	1.672*	–	11	$F \rightarrow A$; $F \leftarrow X$	–
	$f\,^3\Pi$	77405	1.214*	1678.1*	–	1.642*	–	–	$f \rightarrow c, b$	–
	$G\,^1\Sigma^+\,(4s\sigma)$	76952	1.227	1685.6*	–	1.605	15	–	$G \rightarrow B$; $G \leftrightarrow X$	–
	$E\,^1\Delta\,(3d\delta)$	76290	1.221*	1581	–	1.621*	–	10.1	$E \rightarrow A$	–
	$e\,^3\Sigma^+$	75916	1.213	1654.3*	–	1.645	15	–	$e \rightarrow b$	–
	$D\,^1\Pi\,(3p\pi)$	72144.4	1.219	1662.0*	12	1.628	17	11.5	$D \leftrightarrow X$	–
	$d\,^3\Pi$	70710.4	1.210	1696.7	11.0	1.652	18	11.7	$d \rightarrow b$	–
	$C\,^1\Sigma^+\,(3p\sigma)$	69030.4	1.220	1613.1	14.5	1.624	19	10.6	$C \rightarrow A$; $C \leftrightarrow X$	–
	$c\,^3\Sigma^+$	67045	1.228*	1541*	–	1.603*	–	–	$c \rightarrow a$	–
	$B\,^1\Sigma^+\,(3s\sigma)$	65353.9	1.207	1693.5	12.6	1.659	18	11.6	$B \rightarrow A$; $B \leftrightarrow X$	–
	$b\,^3\Sigma^+$	61035.3	1.215	1629.3	22.25	1.638	20	10.8	$b \rightarrow a$	–

Table 10.5 (continued)

Molecule	Electronic term	Term energy T_e [cm^{-1}]	Equilibrium internuclear distance r_e [Å]	Vibrational frequency ω_e [cm^{-1}]	Anharmonic constant $\omega_e x_e$ [cm^{-1}]	Rotational constant B_e [cm^{-1}]	Rotation-vibration interaction constant α_e [10^{-3} cm^{-1}]	Force constant k_e [10^5 dyn/cm]	Observable electronic transition	Reduced mass μ_A [a.m.u.]
1	2	3	4	5	6	7	8	9	10	11
	$A^1\Pi$	51157.45	1.304	1265.0	12.5	1.423	18	6.50	$A \leftrightarrow X$	–
	$a^3\Pi_r$	29144.3	1.308	1323.9	9.2	1.413	16	7.11	$a \rightarrow X$	–
	$X^1\Sigma^+$	0	1.2626	1402.1	11.8	1.5072*	19.8	7.98	–	6.890
BH	$C^1\Sigma^+$	55281.1	1.213	2475	54.4	12.41	432	3.33	$C \leftarrow X$, $C \rightarrow A$	–
	$B^1\Sigma^+$	52335.8	1.216	2400	69.5	12.34	485	3.13	$B \leftarrow X$, $B \rightarrow A$	–
	$C'^1\Delta$	45981.0	1.196	2610	46.6	12.76	390	3.70	$C' \leftrightarrow A$	–
	$A^1\Pi$	23135.8	1.219	2251	56.7	12.30	835	2.75	$A \leftrightarrow X$	–
	$X^1\Sigma^+$	0	1.232	2367	49.4	12.02	412	3.04	–	0.922
BN	$A^3\Pi$	27875	1.326	1318	14.9	1.555	10	6.24	$A \leftrightarrow X$	–
	$X^3\Pi$	0	1.281	1515	12.3	1.666	25	8.25	–	6.10
BO	$C^2\Pi$	55346.1	1.320	1315	11.1	1.483	18	6.58	$C \rightarrow X$	–
	$B^2\Sigma^+$	43174.0	1.305	1282	10.7	1.517	21.0	6.24	$B \rightarrow X, A$	–
	$A^2\Pi_i$	23958.8 23833.7	1.353	1261	11.2	1.402*	19.6	6.04	$A \leftrightarrow X$	–
	$X^2\Sigma^+$	0	1.204	1886	11.8	1.782	16.6	13.5	–	6.45
BS	$D^2\Delta_i$	~48000	1.849*	676	–	0.60*	–	2.18	$D \rightarrow A$	–
	$C^2\Pi_r$	39041.2 38925.8	1.712*	892.6	6.74	0.7025*	–	3.80	$C \rightarrow X$	–
	$B^2\Sigma^+$	36223	1.806*	770	4.0	0.6311*	–	2.82	$B \rightarrow A$	–
	$A^2\Pi_i$	16209.7 15876.0	1.818	753.6	4.67	0.621* 0.619*	5.9	2.70	$A \rightarrow X$	–
	$X^2\Sigma$	0	1.609	1180.2	6.31	0.7949	6.0	6.63	–	8.085

BaF	$E\,^2\Sigma^+$	28139.7	2.10*	538.4	1.90	0.2290*	1.1	2.85	$E \leftrightarrow X$	–
	$D'\,^2\Sigma^+$	26227.0	2.1	504.9	1.54	$B_1 = 0.2269$	1.0	2.51	$D' \leftrightarrow X$	–
	$D\,^2\Sigma^+$	24156.8	2.107*	508.4	1.88	0.2273*	1.1	2.54	$D \leftrightarrow X$	–
	$C\,^2\Pi$	20197 19998.2	2.17	456	1.7	0.215 0.214	1.2	2.04	$C \leftrightarrow X$	–
	$B\,^2\Sigma^+$	14062.5	2.208*	424.4	1.88	0.2071*	1.2	1.77	$B \leftrightarrow X$	–
	$A\,^2\Pi_{3/2}$	12278.2	2.183*	436.7	1.82	0.2119*	1.2	1.88	$A \leftrightarrow X$	–
	$A\,^2\Pi_{1/2}$	11646.9		435.5	1.68	0.2118*	1.1	1.87		
	$X\,^2\Sigma^+$	0	2.163*	468.9	1.79	0.2158*	1.2	2.16	–	16.69
BaH	$F\,^2\Sigma^+$	30747.9*	2.16*	–	–	3.626*	–	–	$F \leftarrow X$	–
	$C\,^2\Sigma^+$	23675	2.17	1282	15	3.59	64	0.97	$C \leftrightarrow X$	–
	$D\,^2\Sigma^+$	21885	3.22	428	4.5	1.62	17	0.11	$D \leftrightarrow X$	–
	$E\,^2\Pi_{3/2}$	15055	2.187	1229	16.9	3.560	75	0.89	$E \rightarrow X$	–
	$E\,^2\Pi_{1/2}$	14605		1187*	–	3.486	72	–		
	$B\,^2\Sigma^+$	11092.6	2.270	1088.9	15.5	3.269	71	0.70	$B \rightarrow X$	–
	$A\,^2\Pi_{3/2}$	9939.82	2.249	1110.0	13.6	3.322	82	0.73	$A \leftrightarrow X$	–
	$A\,^2\Pi_{1/2}$	9457.45		1110.5	15.3	3.2789	73			
	$X\,^2\Sigma^+$	0	2.2319	1168.4	14.6	3.3823	66	0.80	–	1.001
BaO	$B\,(^1\Pi)$	32866.4	–	488	3.6	–	–	2.01	$B \leftrightarrow X$	–
	$A'\,^1\Pi$	17691	2.28	443	1.7	0.2252	1.3	1.66	$A' \leftrightarrow X$	–
	$a\,^3\Pi_i$	17483	2.29	448	2.4	0.224	1.4	1.7	–	–
	$A\,^1\Sigma^+$	16807	2.134	500	1.6	0.2583	1.07	2.11	$A \leftrightarrow X$	–
	$X\,^1\Sigma^+$	0	1.9397	669.8	2.03	0.3126	1.39	3.79	–	14.33
BaS	$B\,^1\Sigma^+$	27060.29	2.747	254.1	0.438	0.0860	0.44	0.99	$B \leftarrow X$	–
	$A\,^1\Sigma^+$	14493	2.635	294.3	3.1	0.0935	0.7	1.33	$A \leftarrow X$	–
	$X\,^1\Sigma^+$	0	2.507	379.4	0.884	0.1033	0.32	2.20	–	25.995
Be_2	$B\,^1\Sigma_u^+$	27860*	2.20	511	4.7	0.77	14	0.69	$B \leftrightarrow X$	–
	$A\,^1\Pi_u$	21678.4*	2.00	685.7	4.9	0.94	12	1.25	$A \leftrightarrow X$	–
	$X\,^1\Sigma_g^+$	0	2.45	276	26	0.62	28	0.20	–	4.506
BeBr	$A\,^2\Pi_r$	26554 26353	1.976*	695 702	5.2 4.4	0.533*	–	2.30 2.35	$A \rightarrow X$	–
	$X\,^2\Sigma^+$	0	1.953*	715	3.8	0.546*	–	2.44	–	8.099

Table 10.5 (continued)

Molecule	Electronic term	Term energy T_e [cm^{-1}]	Equilibrium internuclear distance r_e [Å]	Vibrational frequency ω_e [cm^{-1}]	Anharmonic constant $\omega_e x_e$ [cm^{-1}]	Rotational constant B_e [cm^{-1}]	Rotation-vibration interaction constant α_e [10^{-3} cm^{-1}]	Force constant k_e [10^5 dyn/cm]	Observable electronic transition	Reduced mass μ_A [a.m.u.]
1	2	3	4	5	6	7	8	9	10	11
BeCl	$B\,^2\Sigma^+$	48827.6*	1.742	952.5*	–	0.775	4.3	–	$B \to X$	–
	$A\,^2\Pi_r$	27992.0	1.821	822.1	5.24	0.7094	6.8	2.86	$A \leftrightarrow X$	–
	$X\,^2\Sigma^+$	0	1.797	846.7	4.8	0.7285	6.9	3.04	–	7.19
BeF	$C\,^2\Sigma^+$	50364.0	1.325	1420	9.9	1.570	14	7.26	$C \to X, A$	–
	$B\,^2\Sigma^+$	49563.9	1.335*	1351	12.6	1.55*	–	6.57	$B \to X$	–
	$A\,^2\Pi_r$	33233.7	1.394	1171	8.8	1.420	17.5	4.94	$A \leftrightarrow X$	–
	$X\,^2\Sigma^+$	0	1.361	1267	9.1	1.489	17.6	5.78	–	6.11
BeH	$G\,^2\Pi$	58711	1.92	405.3	22.7	5.02	–556	0.088	$G \leftarrow X$	–
	$F\,^2\Sigma^+$ $(4p\sigma)$	56661.2*	1.33*	2150	–	10.58*	–	2.5	$F \leftarrow X$	–
	$E\,^2\Sigma^+$ $(4s\sigma)$	54097.6*	1.33*	1970	–	10.58*	–	2.1	$E \leftarrow X$	–
	$B\,^2\Pi$ $(3p\pi)$	50882	1.309	2265.9	71.5	10.850	102	2.74	$B \leftrightarrow X$	–
	$C\,^2\Sigma^+$ $(3p\sigma)$	30953.9	2.301	1061.1	42.2	3.514	–22	0.69	$C \to X$	–
	$A\,^2\Pi_r$ $(2p\pi)$	20032.6	1.334	2085.2	37.3	10.46	322	2.32	$A \leftrightarrow X$	–
	$X\,^2\Sigma^+$ $(2p\sigma)$	0	1.345	2071.9	48.1	10.274	207	2.29	–	0.907
BeI	A $^2\Pi_{3/2}$	–	2.189*	–	–	0.418*	–	–	$A \to X$	–
	A $^2\Pi_{1/2}$	23544.7	2.180*	603.8	2.1	0.422*	–	1.81		–
	$X\,^2\Sigma^+$	0	2.179*	611.7	1.6	0.422*	–	1.86	–	8.415
BeO	$C\,^1\Sigma$	39120.2	1.49	1082	9.1	1.31	10	3.97	$C \to A$	–
	$B\,^1\Sigma^+$	21253.94	1.362	1370.8	7.75	1.576	15.4	6.38	$B \leftrightarrow X$, $B \to A$	–
	$A\,^1\Pi$	9407.6	1.463	1144	8.4	1.366	–16	4.45	$A \to X$	–
	$a\,^3\Pi$	8570	1.462	1130	8	1.367	–17	4.35	–	–
	$X\,^1\Sigma^+$	0	1.331	1486	11.6	1.652	–19	7.5	–	5.765

BeS	$A\,^1\Pi$	7961.6	1.909	762.1	4.1	0.658	5.8	2.41	$A \to X$	–
	$X\,^1\Sigma^+$	0	1.742	997.9	6.14	0.7906	6.64	4.13	–	7.035
Bi_2	$V\ 0_u^+$	30172.4	3.480	32.24	0.046	0.0133	0.057	0.064	$V \leftrightarrow X$; $V \to B'$	–
	$A\ 0_u^+$	17739.3	2.863	132.5	0.302	0.0197	0.053	1.08	$A \leftrightarrow X$	–
	$B'\ 0_g^+$	10826.4	3.108	106.28	0.24	0.0167	0.038	0.695	$B' \leftarrow X$	–
	$X\,^1\Sigma_g^+$	0	2.661	173.06	0.376	0.0228	0.042	1.84	–	104.5
BiF	$B\ 0^+$	26000.3	2.047	627	6.6	0.231	2	4.03	$B \leftrightarrow X_1$	–
	$A\ 0^+$	22960	–	381	3.0	–	–	1.49	$A \leftrightarrow X_1$	–
	$X_1\ 0^+$	0	2.051	513	2.3	0.230	1.5	2.70	–	17.42
Br_2	$F(0_u^+)$	53900	3.28	156	0.8	0.040	0.13	0.56	$B \to F$; $F \to X$	–
	$f(0_g^+)$	53102	3.17	153	0.4	0.043	0.15	0.54	$f \leftrightarrow B$	–
	$D\,(0_u^+)$	49928	3.18	134.5	0.09	0.042	0.11	0.42	$D \leftrightarrow B$; $D \to X$	–
	$E\,(0_g^+)$	49779	3.20	150.5	0.38	0.042	0.14	0.53	$E \leftrightarrow B$	–
	$D'\,(2_g)$	48900	3.17	151	0.4	0.043	–	0.53	$D \leftrightarrow B$	–
	$B\,^3\Pi_u\,(0_u^+)$	15902.5	2.678	167.61	1.636	0.0596	0.489	0.66	$B \leftrightarrow X$	–
	$A\,^3\Pi_u\,(1_u)$	13905	2.69	153	2.7	0.059	0.8	0.55	$A \leftrightarrow X$	–
	$X\,^1\Sigma_g^+\,(0_g^+)$	0	2.281	325.32	1.077	0.0821	0.318	2.49	–	39.95
BrCl	$B\,^3\Pi_{0^+}$	16879.9	2.541	222.7	2.88	0.1077	–	0.72	$B \leftrightarrow X$	–
	$X\,^1\Sigma^+$	0	2.136	444.28	1.84	0.1525	0.77	2.86	–	24.56
BrF	$B\,^3\Pi_{0^+}$	18272.0	–	372.2	3.5	–	–	1.25	$B \leftrightarrow X$	–
	$A\,^3\Pi_1$	17385	–	378	16	–	–	1.29	$A \leftarrow X$	–
	$X\,^1\Sigma^+$	0	1.759	670.75	4.05	0.3558	2.61	4.07	–	15.35
BrO	$X_1\,^2\Pi_{3/2}$	0	1.717	779	6.8	0.4296	3.64	4.76	–	13.33
C_2	$F\,^1\Pi_u$	75457**	1.31	1557.5*	–	1.64	19	–	$F \leftarrow X$	–
	$g\,^3\Delta_g$	73184**	1.358	1458.1*	–	1.524	17	–	$g \leftarrow a$	–
	$f\,^3\Sigma_g^-$	71045.8	1.39	1360.5	14.8	1.448	40	6.55	$f \leftarrow a$	–
	$E\,^1\Sigma_g^+$	55034.7	1.253	1671.5	40.0	1.790	39	9.89	$E \to A$	–
	$D\,^1\Sigma_u^+$	43239.4	1.238	1829.6	13.9	1.833	19.6	11.85	$D \leftrightarrow X$	–
	$e\,^3\Pi_g$	40796.6	1.535	1106.6	39.3	1.192	24.2	4.33	$e \to a$	–
	$C\,^1\Pi_g$	34261.3	1.255	1809	15.8	1.783	18.0	11.6	$C \to A$	–
	$d\,^3\Pi_g$	20022.50	1.266	1788.2	16.44	1.753	16.1	11.3	$d \leftrightarrow a$	–
	$A\,^1\Pi_u$	8391.0	1.318	1608.3	12.1	1.616	16.9	9.15	$A \leftrightarrow X$	–
	$b\,^3\Sigma_g^-$	6433.7	1.369	1470.37	11.14	1.499	16.3	7.65	$b \to a$	–

Table 10.5 (continued)

Molecule	Electronic term	Term energy T_e [cm^{-1}]	Equilibrium internuclear distance r_e [Å]	Vibrational frequency ω_e [cm^{-1}]	Anharmonic constant $\omega_e x_e$ [cm^{-1}]	Rotational constant B_e [cm^{-1}]	Rotation-vibration interaction constant α_e [10^{-3} cm^{-1}]	Force constant k_e [10^5 dyn/cm]	Observable electronic transition	Reduced mass μ_A [a.m.u.]
1	2	3	4	5	6	7	8	9	10	11
	$a\,^3\Pi_u$	716.24	1.312	1641.3	11.7	1.632	16.6	9.53	–	–
	$X\,^1\Sigma_g^+$	0	1.2425	1854.7	13.34	1.820	17.6	12.2	–	6.006
CCl	$A\,^2\Delta_r$	35870.3*	1.635*	848*	–	0.7062*	–	–	$A \leftrightarrow X$	–
	$X\ ^2\Pi_{3/2}$	–	1.645	865.5*	–	0.701	6.8	–	–	–
	$X\ ^2\Pi_{1/2}$	0		866.7*	6.2	0.694	6.7	4.1	–	8.972
CF	$D\,^2\Pi$	52272.5	1.151	1803.9	13.0	1.730	19.3	14.1	$D \leftarrow X$	–
	$B\,^2\Delta_r$	49399.6	1.317	1153.3*	19.4	1.321	23	6.2	$B \leftrightarrow X$	–
	$A\,^2\Sigma^+$	42692.9	1.153	1780.4	30.7	1.723	18.9	13.7	$A \leftrightarrow X$	–
	$X\,^2\Pi_r$	0	1.272	1308.1	11.1	1.417	18.4	7.42	–	7.36
CH	$C\,^2\Sigma^+$	31801.5	1.114	2840.2	126	14.60	718	4.42	$C \leftrightarrow X$	–
	$B\,^2\Sigma^-$	26044	1.197*	1794.9*	–	12.65*	–	–	$B \leftrightarrow X$	–
	$A\,^2\Delta$	23189.8	1.102	2930.7	96.6	14.93	700	4.71	$A \leftrightarrow X$	–
	$X\,^2\Pi_r$	0	1.120	2858.5	63.0	14.46	534	4.48	–	0.930
CN	$J\,^2\Delta_i$	65258.2	1.414	1121.8	14.20	1.305	21	4.80	$J \rightarrow A$	–
	$F\,^2\Delta_r$	60095.64	1.373	1240	12.8	1.383	18.7	5.85	$F \rightarrow A$	–
	$E\,^2\Sigma^+$	59151.18	1.324	1681.4	3.60	1.487	6.43	10.8	$E \leftrightarrow X,\ E \rightarrow A$	–
	$D\,^2\Pi_i$	54486.3	1.498	1005	8.8	1.162	13	3.85	$D \rightarrow A,\ X$	–
	$B\,^2\Sigma^+$	25752.0	1.149	2164	20.2	1.973	23	17.8	$B \rightarrow A,\ B \leftrightarrow X$	–
	$A\,^2\Pi_i$	9245.3	1.233	1812.6	12.6	1.715	17.1	12.5	$A \leftrightarrow X$	–
	$X\,^2\Sigma^+$	0	1.172	2068.6	13.1	1.8997	17.37	16.3	–	6.47
CO	$E\,^1\Pi\ (3p\pi)$	92930.0*	1.115	2154*	40	1.977	25.4	20	$E \rightarrow A,\ E \leftarrow X$	–
	$c\,^3\Pi\ (3p\pi)$	92076.9*	1.127*	–	–	1.935*	–	–	$c \rightarrow a,\ c \leftarrow X$	–
	$C\,^1\Sigma^+\ (3p\sigma)$	91916.5	1.122	2176	14.8	1.953	19.6	19.1	$C \leftrightarrow X,\ C \rightarrow A$	–

	$j^3\Sigma^+$ $(3p\sigma)$	90975	1.144*	2166*	15	1.878*	20	19	$j \leftarrow X$	–
	$D'^1\Sigma^+$	89438	1.584	650	20	0.9805	26.6	1.7	$D' \leftrightarrow X$	–
	$B^1\Sigma^+$ $(3s\sigma)$	86945.2	1.120	2113	15.2	1.961	26.1	18.0	$B \leftrightarrow X$, $B \rightarrow A$	–
	$b^3\Sigma^+$ $(3s\sigma)$	83832*	1.113	2199*	–	1.986	42	–	$b \leftarrow X$, $b \rightarrow a$	–
	$D^1\Delta$	65928	1.40	1094	10.2	1.257	17	4.84	$D \leftarrow X$	–
	$I^1\Sigma^-$	65084.4	1.391	1092.2	10.70	1.2705	18.5	4.82	$I \leftarrow X$	–
	$A^1\Pi$	65075.8	1.235	1518.2	19.4	1.6115	23.2	9.32	$A \leftrightarrow X$	–
	$e^3\Sigma^-$	64230.2	1.384	1117.7	10.69	1.284	17.5	5.05	$e \leftarrow X$, $e \rightarrow a$	–
	$d^3\Delta_i$	61120.1	1.370	1171.9	10.63	1.311	17.8	5.55	$d \leftarrow X$, $d \rightarrow a$	–
	$a'^3\Sigma^+$	55825.5	1.352	1228.6	10.47	1.345	18.9	6.10	$a' \leftarrow X$, $a' \rightarrow a$	–
	$a^3\Pi_r$	48686.7	1.2057	1743.4	14.4	1.6912	19.0	12.3	$a \leftrightarrow X$	–
	$X^1\Sigma^+$	0	1.1283	2169.81	13.29	1.93128	17.50	19.0	–	6.8606
CP	$B^2\Sigma^+$	29100.4	1.689	836.3	5.92	0.6829	6.28	3.57	$B \rightarrow X, A$	–
	$A^2\Pi_i$	7053.2 6894.9	1.65	1062.0	6.04	0.713	5.8	5.75	–	–
	$X^2\Sigma^+$	0	1.562	1239.7	6.86	0.7986	5.97	7.84	–	8.655
CS	$A'^1\Sigma^+$	56505	1.94	462	7.5	0.511	11	1.10	$A' \rightarrow X$	–
	$A^1\Pi$	38904.4	1.574	1073	10	0.780	6	5.93	$A \leftrightarrow X$	–
	$e^3\Sigma^-$	38680	1.77	753	5	0.619	4	2.92	$e \leftarrow X$	–
	$d^3\Delta_i$	35675.0	1.742	796	4.9	0.637	6	3.26	$d \leftarrow X$	–
	$a'^3\Sigma^+$	31331.4	1.725	831	5.0	0.649	6	3.55	$a' \leftarrow X$	–
	$a^3\Pi_r$	27661.0	1.569	1135	7.7	0.785	7	6.63	$a \rightarrow X$	–
	$X^1\Sigma^+$	0	1.535	1285.16	6.50	0.8200	5.92	8.50	–	8.738
CSe	$X^1\Sigma^+$	0	1.676	1035.4	4.9	0.575	3.8	6.58	–	10.425
Ca_2	$A^1\Sigma_u^+$	18960	3.807	137	0.73	0.05820	0.3	0.22	$A \leftarrow X$	–
	$X^1\Sigma_g^+$	0	4.278	64.93	1.065	0.04612	0.70	0.050	–	20.04
CaCl	$E^2\Sigma^+$	34268.3	2.352	413.7	1.7	0.1634	0.86	1.90	$E \rightarrow B$; $E \leftarrow X$	–
	$D^2\Sigma^+$	31111.0	2.355	423.2	1.6	0.1630	0.80	1.99	$D \leftarrow X$	–
	$C^2\Pi_r$	26574.6 26498.9	2.516	333.9* 336	1.4	0.143 0.142*	0.75	1.26 1.25	$C \leftrightarrow X$	–
	$B^2\Sigma^+$	16856.7	2.4172	367.2	1.52	0.1547	0.88	1.49	$B \leftrightarrow X$	–
	$A^2\Pi_r$	16130.8	2.4219	373.15	1.50	0.1541	0.84	1.54	$A \leftrightarrow X$	–
	$X^2\Sigma^+$	0	2.4367	370.20	1.373	0.1522	0.79	1.52	–	18.812

Table 10.5 (continued)

Molecule	Electronic term	Term energy T_e [cm^{-1}]	Equilibrium internuclear distance r_e [Å]	Vibrational frequency ω_e [cm^{-1}]	Anharmonic constant $\omega_e x_e$ [cm^{-1}]	Rotational constant B_e [cm^{-1}]	Rotation-vibration interaction constant α_e [10^{-3} cm^{-1}]	Force constant k_e [10^5 dyn/cm]	Observable electronic transition	Reduced mass μ_A [a.m.u.]
1	2	3	4	5	6	7	8	9	10	11
CaF	$B\,^2\Sigma^+$	18844	1.94	566.1	2.8	0.336*	–	2.43	$B \rightarrow X$	–
	$A_2\,^2\Pi_{3/2}$	16562.3	1.95	593.4	3.11	0.344	2.8	2.67	$A \leftrightarrow X$	–
	$A_1\,^2\Pi_{1/2}$	16489.8		587*	3.43			2.68		–
	$X\,^2\Sigma^+$	0	1.967	581*	2.74	0.338	2.6	2.61	–	12.89
CaH	$B\,^2\Sigma^+$	15762	1.974	1285	20	4.341*	116	0.96	$B \leftrightarrow X$	–
	$A\,^2\Pi_r$	14413	1.974	1333	20	4.348*	106	1.03	$A \leftrightarrow X$	–
	$X\,^2\Sigma^+$	0	2.002	1298.1	18.9	4.276	96	0.98	–	0.983
CaO	$C\,^1\Sigma^+$	28858	1.989	560.9	4	0.3731	3.2	2.12	$C \leftrightarrow X$	–
	$B\,^1\Pi$	25991	1.950	574*	–	0.388	5	–	$B \leftrightarrow X$	–
	$A\,^1\Sigma^+$	11555	1.907	718.9	2.1	0.4059	1.4	3.48	$A \leftrightarrow X$	–
	$A'\,^1\Pi$	8433	2.09	545.7	2.54	0.337	2	2.01	$A' \rightarrow X$	–
	$X\,^1\Sigma^+$	0	1.822	732.1	4.8	0.444	3.4	3.61	–	11.435
CaS	$A\,^1\Sigma^+$	15220.8	2.386	409.0	0.82	0.1667	0.6	1.76	$A \leftarrow X$	–
	$X\,^1\Sigma^+$	0	2.318	462.2	1.78	0.17667	0.84	2.24	–	17.81
CdH	$A\ ^2\Pi_{3/2}$	23116	1.657	1757.8	38.6	6.143	205	1.82	$A \rightarrow X$	–
	$A\ ^2\Pi_{1/2}$	22276.5*	1.669	1677*	–	6.061	193	–		–
	$X\,^2\Sigma^+$	0	1.781*	1337.1*	–	5.323*	–	–	–	0.999
Cl_2	$B\,^3\Pi\,(0_u^+)$	17809	2.435	259.5	5.3	0.1626	2.1	0.70	$B \leftrightarrow X$	–
	$X\,^1\Sigma_g^+$	0	1.988	559.7	2.67	0.2440	1.5	3.27	–	17.73
ClF	$B\,^3\Pi\,(0_u^+)$	18826	2.03	363.1	8.6	0.332	4.7	0.96	$B \leftarrow X$	–
	$X\,^1\Sigma^+$	0	1.6283	786.15	6.16	0.5165	4.36	4.50	–	12.37

[illegible]	$A^2\Pi_i$	[illegible] 31650	1.86	520	7.2	0.445	6	1.75	$A \leftrightarrow X$	–
	$X^2\Pi_i$	0	1.570	854	5.5	0.6234	6	4.74	–	11.02
Cr_2	$X^1\Sigma_g^+$	0	1.679	452.3*	≈9	0.230	4	3.4	–	25.97
CrH	$A^6\Sigma$	11552.3*	1.79	1479*	21	5.34	130	1.35	$A \to X$	–
	$X^6\Sigma$	0	1.656	1581*	30	6.22	180	1.6	–	0.989
CrO	$B^5\Pi$	16590	1.70	750	9.4	0.48	5	4.06	$B \to X$	–
	$X^5\Pi$	0	1.61	898	6.7	0.53	5	5.82	–	12.235
Cs_2	$E^1\Sigma_u^+$	20195.2	5.341	29.1	–	0.0089	–	0.033	$E \to X$, 1, 2, 3	–
	$D^1\Sigma_u^+$	16720	6.0	17	–	0.007	–	0.011	$D \leftrightarrow X$	–
	$(3)^1\Sigma_g^+$	15975.45	5.557	22.42	0.0058	0.0082	−0.0021	0.020	$E \to 3$	–
	$C^1\Pi_u$	15948.67	4.530	29.66	0.042	0.0124	0.067	0.034	$C \leftrightarrow X$	–
	$(1)^1\Pi_g$	13913.4	5.698	18.4	0.11	0.0078	–	0.013	$E \to 1$	–
	$B^1\Pi_u$	13043.9	–	34.33	0.080	–	–	0.046	$B \leftarrow X$	–
	$(2)^1\Sigma_g^+$	12114.09	5.832	23.35	−0.0072	0.0075	−0.0001	0.021	$D, E \to 2$	–
	$X^1\Sigma_g^+$	0	4.648	42.02	0.082	0.0117	0.022	0.069	–	66.45
CsBr	$X^1\Sigma^+$	0	3.072	149.7	0.37	0.036	0.12	0.66	–	49.90
CsCl	$X^1\Sigma^+$	0	2.906	214.2	0.73	0.072	0.34	0.76	–	27.99
CsF	$X^1\Sigma^+$	0	2.345	352.6	1.61	0.1844	1.18	1.22	–	16.62
CsH	$A^1\Sigma^+$	17841	3.96	166	−7.8	1.07	−22	0.02	$A \leftrightarrow X$	–
	$X^1\Sigma^+$	0	2.494	891.5	12.9	2.710	62	0.47	–	1.00
CsI	$X^1\Sigma^+$	0	3.315	119.2	0.254	0.0236	0.068	0.54	–	64.92
Cu_2	$B^1\Sigma_u^+$	21758.3	2.328	242.1*	2	0.0989	0.61	1.1	$B \leftrightarrow X$	–
	$A^1\Pi_u$	20433	2.558	192	0.35	0.082	0.6	0.69	$A \leftrightarrow X$	–
	$X^1\Sigma_g^+$	0	2.220	264.5*	1.02	0.1087	0.61	1.3	–	31.77
CuBr	$C^1\Sigma^+$	23461	2.26	295	1.1	0.094	0.4	1.81	$C \leftrightarrow X$	–
	$X^1\Sigma^+$	0	2.173	315	0.96	0.1019	0.45	2.07	–	35.40
CuCl	$F^1\Pi(1)$	25285.3	2.148	384.9	1.65	0.161	0.9	1.99	$F \leftrightarrow X$	–
	$E^1\Sigma^+(0^+)$	23074.2	2.112	403.3	1.6	0.166	1.1	2.18	$E \leftrightarrow X$	–
	$D^1\Pi(1)$	22969.7	2.103	392.9	1.74	0.1678	1.0	2.07	$D \leftrightarrow X$	–
	$C^1\Sigma^+(0^+)$	20630.9	2.09	396.9	1.48	0.169	0.9	2.11	$C \leftrightarrow X$	–
	$B^1\Pi(1)$	20484.1	2.10	399.3	1.61	0.168	0.9	2.14	$B \leftrightarrow X$	–
	$A^1\Pi(1)$	19001	–	407	1.7	–	–	2.22	$A \leftrightarrow X$	–
	$X^1\Sigma^+$	0	2.051	415.3	1.58	0.1763	1.00	2.31	–	22.76

Table 10.5 (continued)

Molecule	Electronic term	Term energy T_e [cm^{-1}]	Equilibrium internuclear distance r_e [Å]	Vibrational frequency ω_e [cm^{-1}]	Anharmonic constant $\omega_e x_e$ [cm^{-1}]	Rotational constant B_e [cm^{-1}]	Rotation-vibration interaction constant α_e [10^{-3} cm^{-1}]	Force constant k_e [10^5 dyn/cm]	Observable electronic transition	Reduced mass μ_A [a.m.u.]
1	2	3	4	5	6	7	8	9	10	11
CuF	D	22805.1	1.793	616.4	3.33	0.3594	2.8	3.27	–	–
	$C^1\Pi$	20258.7	1.754	643.7	3.7	0.3756	2.97	3.57	$C \leftrightarrow X$	–
	$B^1\Sigma^+$	19717.5	1.764	656.0	3.6	0.3713	2.85	3.71	$B \leftrightarrow X$	–
	$A(0^+, 1)$	17562.3	1.751	647.08	3.46	0.377	2.9	3.61	$A \leftrightarrow X$	–
	$a^3\Sigma^+$	14580.5	1.738	674.20	4.14	0.3825	2.98	3.92	–	–
	$X^1\Sigma^+$	0	1.745	621.55	3.49	0.3794	4.23	3.33	–	14.63
CuH	$C1$	27270	1.610	1627	86	6.55	350	1.55	$C \leftrightarrow X$	–
	$B^3\Pi_{0^+}$	26421	1.607	1670	51	6.58	290	1.63	$B \leftrightarrow X$	–
	$A^1\Sigma^+$	23434	1.572	1698	44	6.87	260	1.69	$A \leftrightarrow X$	–
	$X^1\Sigma^+$	0	1.463	1941.3	37.5	7.944	256	2.20	–	0.992
CuI	$E^1\Sigma^+(0^+)$	24001.70	2.471	229.5	0.99	0.0656	0.28	1.31	$E \leftrightarrow X$	–
	$C^1\Sigma^+(0^+)$	21867	2.43*	230	0.5	0.068*	–	1.32	$C \leftrightarrow X$	–
	$A^1\Pi(1)$	19734	2.43	213	2.2	0.0676	0.4	1.14	$A \leftrightarrow X$	–
	$X^1\Sigma^+$	0	2.338	264	0.6	0.0733	0.284	1.75	–	42.34
F_2	$H^1\Pi_u(3p\sigma)$	105520.1	1.32	1088.2	9.87	1.02	14	6.63	$H \leftarrow X$	–
	$F^1\Pi_g(3s\sigma)$	93100	1.30	1133.3	9.17	1.05	12	7.19	–	–
	$X^1\Sigma_g^+$	0	1.412	916.6	11.24	0.8902	13.85	4.70	–	9.50
FeO	a	21245	1.65*	820	1	0.50*	–	4.9	$a \leftrightarrow A$	–
	$A^5\Sigma^+$	3950	1.626	880.5	4.6	0.513	3.8	5.68	–	–
	$X^5\Delta$	0	1.6	970	–	–	–	6.8	–	12.44
GaBr	$X^1\Sigma^+$	0	2.352	263.0	0.81	0.08184	0.32	1.52	–	37.23
GaCl	$X^1\Sigma^+$	0	2.202	365.3	1.2	0.1499	0.794	1.84	–	23.50

GaF	$B\,^3\Pi_1$	33428	1.744	662.1	1.45	0.3720	3.0	3.86	$B \leftrightarrow X$	–
	$A\,^3\Pi_{0^+}$	33105	1.747	663.0	2.18	0.3710	3	3.87	$A \leftrightarrow X$	–
	$X\,^1\Sigma^+$	0	1.774	622.2	3.2	0.3595	2.86	3.41	–	14.93
GaH	$a\,^3\Pi$ 1	17622.0	1.59	1631.2	58.2	6.692	326	1.56	$1 \leftrightarrow X$	–
	$a\,^3\Pi$ 0^+	17337.1	1.63	1640.5	62.7	6.394	276	1.58	$0^+ \leftrightarrow X$	–
	$a\,^3\Pi$ 0^-	17333	1.63	1671	89	6.358	220	1.6	$0^- \rightarrow X$	–
	$X\,^1\Sigma^+$	0	1.66	1604.5	28.8	6.137	181	1.51	–	0.994
GaI	$X\,^1\Sigma^+$	0	2.575	216.6	0.5	0.0569	0.19	1.24	–	44.998
GeF	$G\,^2\Delta_r$	49412.9	1.704	716.0	2.8	0.3841	2.6	4.55	$G \rightarrow X$	–
	$D\,^2\Sigma^+$	48581.3	1.670	833.1	6.5	0.3997	2.1	6.16	$D \rightarrow A, X$	–
	$D'\,^2\Pi$	47920.7	1.668	804.0	3.4	0.4007	2.6	5.73	$D' \rightarrow A, X$	–
	$E\,^2\Sigma^+$	46645.4	1.673	760.1	2.97	0.3985	2.9	5.13	$E \rightarrow B$	–
	$C\,^2\Delta$	43977.5	1.695	702.6	9.3	0.3884	4.2	4.38	$C \rightarrow X$	–
	$C'\,^2\Pi$	43369.6	1.671	796.9	3.41	0.3996	2.6	5.63	$C' \rightarrow A, X$	–
	$a\,^4\Sigma^-$	35194.7	1.744	641.6	6.7	0.3668	3.7	3.65	$a \rightarrow X$	–
	$B\,^2\Sigma^+$	35010.8	1.682	797.0	3.61	0.3944	2.6	5.64	$B \rightarrow A, X$	–
	$A\,^2\Sigma^+$	23316.6	1.866	413.0	1.12	0.3204	3.1	1.51	$A \rightarrow X$	–
	X $^2\Pi_{3/2}$	934.3	1.745	667.3	3.15	0.3666	2.67	3.95	–	15.06
	X $^2\Pi_{1/2}$	0		665.7		0.3658		3.93		
GeH	$A\,^2\Delta$	25450	1.61	1185.1*	130	6.53	620	1.2	$A \leftrightarrow X$	–
	$X\,^2\Pi_r$	0	1.588	1833.8*	40	6.726	192	2.1	–	0.994
GeO	$E\,^1\Sigma^+$	46040	1.898	534	3.0	0.356	–	2.20	$E \leftarrow X$	–
	$A\,^1\Pi$	37763	1.759	649	4.0	0.4143	3.8	3.25	$A \leftrightarrow X$	–
	$D\,^1\Delta$	32710	1.88	590	3.1	0.362	2	2.7	–	–
	$C\,^1\Sigma^-$	32480	1.88	590	3.1	0.362	2	2.7	–	–
	$e\,^3\Sigma^-$	32210	1.88	600	3.1	0.362	2	2.8	–	–
	$b\,^3\Pi_1$	32130	1.71	735	5	0.44	–	4.2	$b \rightarrow X$	–
	$d\,^3\Delta_1$	30500	1.88	600	3.1	0.362	2	2.8	–	–
	$a\,^3\Sigma^+$	27730	1.81	633	3	0.40	–	3.10	$a \rightarrow X$	–
	$X\,^1\Sigma^+$	0	1.625	986.7	4.5	0.4857	3.08	7.52	–	13.11
GeS	$X\,^1\Sigma^+$	0	2.012	576	1.8	0.1866	0.749	4.36	–	22.32
GeSe	$X\,^1\Sigma^+$	0	2.135	409	1.4	0.0963	0.289	3.72	–	37.82
GeTe	$X\,^1\Sigma^+$	0	2.340	324	0.7	0.0653	0.172	2.86	–	46.27

Table 10.5 (continued)

Molecule	Electronic term	Term energy T_e [cm^{-1}]	Equilibrium internuclear distance r_e [Å]	Vibrational frequency ω_e [cm^{-1}]	Anharmonic constant $\omega_e x_e$ [cm^{-1}]	Rotational constant B_e [cm^{-1}]	Rotation-vibration interaction constant α_e [10^{-3} cm^{-1}]	Force constant k_e [10^5 dyn/cm]	Observable electronic transition	Reduced mass μ_A [a.m.u.]
1	2	3	4	5	6	7	8	9	10	11
H_2	Triplets									
	$u\,^3\Pi_u\,6p\pi$	123488**	1.07*	–	–	29.3*	–	–	$u \to a$	–
	$n\,^3\Pi_u\,5p\pi$	120953	1.06	2321	62.9	29.9	1240	1.60	$n \to a$	–
	$k\,^3\Pi_u\,4p\pi$	118366	1.055	2344.4	67.3	30.07	1460	1.63	$k \to a$	–
	$j\,^3\Delta_g\,3d\delta$	113530	1.054	2345.3	66.6	30.08	1700	1.63	$j \leftrightarrow c$	–
	$i\,^3\Pi_g\,3d\pi$	113130	1.070	2253.5	67.0	29.22	1510	1.51	$i \to e$; $i \leftrightarrow c$	–
	$d\,^3\Pi_u\,3p\pi$	112700	1.050	2371.6	66.3	30.36	1550	1.67	$d \to a$	–
	$e\,^3\Sigma_u^+\,3p\delta$	107775	1.11	2196.1	65.8	27.3	1510	1.43	$e \to a$	–
	$a\,^3\Sigma_g^+\,2s\sigma$	95936	0.989	2664.8	71.6	34.22	1670	2.11	$a \to b$	–
	$c\,^3\Pi_u\,2p\pi$	95838	1.038	2466.9	63.5	31.1	1420	1.81	–	–
	Singlets									
	$D''\,^1\Pi_u\,5p\pi$	121211	1.04	2319.9	63.04	30.8	1450	1.60	$D'' \leftarrow X$	–
	$D'\,^1\Pi_u\,4p\pi$	118865	1.06	2330.0	63.14	29.9	1100	1.61	$D' \leftarrow X$	–
	$B''\,^1\Sigma_u^+\,4p\sigma$	117984	1.120	2197.5	68.14	26.7	1200	1.43	$B'' \leftarrow X$	–
	$D\,^1\Pi_u\,3p\pi$	113876.9	1.049	2364.1	69.90	30.43	1680	1.66	$D \to E$; $D \leftrightarrow X$	–
	$J\,^1\Delta_g\,3d\delta$	113550	1.055	2341.1	63.2	30.08	1720	1.63	$J \to C, B$	–
	$I\,^1\Pi_g\,3d\pi$	113140	1.069	2259.1	78.4	29.26	1580	1.52	$I \to C, B$	–
	$B'\,^1\Sigma_u^+\,3p\sigma$	111643	1.119	2039.5	83.41	26.70	2780	1.24	$B' \to E, F$; $B' \leftarrow X$	–
	$E\,^1\Sigma_g^+\,2s\sigma$	100082	1.012	2589	130	32.7	1820	1.99	$E \to B$	–
	$C\,^1\Pi_u\,2p\pi$	100090	1.033	2443.8	69.52	31.36	1665	1.77	$C \leftrightarrow X$	–
	$B\,^1\Sigma_u^+\,2p\sigma$	91700	1.293	1358.1	20.89	20.02	1185	0.548	$B \leftrightarrow X$	–
	$X\,^1\Sigma_g^+\,(1s\sigma)^2$	0	0.7414	4401.21	121.34	60.85	3062	5.75	–	0.504

$^1H\,^2H$	$X\,^1\Sigma_g^+\,(1s\sigma)^2$	0	0.7414	3813.1	91.6	45.65	1990	5.75	–	0.672
$^1H\,^3H$	$X\,^1\Sigma_g^+\,(1s\sigma)^2$	0	0.7414	3597.0	81.68	40.60	1664	5.76	–	0.755
$^2H_2$	$X\,^1\Sigma_g^+\,(1s\sigma)^2$	0	0.7415	3115.5	61.8	30.44	1079	5.76	–	1.007
$^3H_2$	$X\,^1\Sigma_g^+\,(1s\sigma)^2$	0	0.7414	2546.5	41.2	20.33	589	5.76	–	1.508
HBr	$X\,^1\Sigma^+$	0	1.4144	2649.0	45.22	8.4649	233	4.11	–	0.995
^{1}HCl	$X\,^1\Sigma^+$	0	1.2746	2990.89	52.76	10.593	306.9	5.17	–	0.980
^{2}HCl	$X\,^1\Sigma^+$	0	1.2746	2145.16	27.18	5.4488	113.3	5.16	–	1.904
^{1}HF	$B\,^1\Sigma^+$	84776.6	2.091	1159.2	18.0	4.029	18	7.58	$B \leftrightarrow X$	–
	$X\,^1\Sigma^+$	0	0.9168	4138.3	89.9	20.96	798	9.66	–	0.957
^{2}HF	$X\,^1\Sigma^+$	0	0.9169	2998.2	45.76	11.010	302	9.65	–	1.821
^{1}HI	$b_0\,^3\Pi_0\,(0^+)$	60859	1.611	2315	54	6.49	118	3.16	$b_0 \leftarrow X$	–
	$X\,^1\Sigma^+$	0	1.6092	2309.01	39.64	6.4264*	169	3.14	–	1.000
^{2}HI	$X\,^1\Sigma^+$	0	1.6091	1639.65	19.9	3.2535*	60.8	3.14	–	1.983
He_2	Triplets									
	$v\,^3\Pi_g\,11p\pi$	33383.5*	1.08	1700	35	7.21	220	3.4	$v \rightarrow a$	–
	$u\,^3\Pi_g\,10p\pi$	33189.2*	1.08	1700	35	7.21	220	3.4	$u \rightarrow a$	–
	$t\,^3\Pi_g\,9p\pi$	32926.0*	1.08	1700	35	7.21	230	3.4	$t \rightarrow a$	–
	$s\,^3\Pi_g\,8p\pi$	32556.7*	1.081	1700	35	7.21	223	3.4	$s \rightarrow a$	–
	$r\,^3\Pi_g\,7p\pi$	32016.6*	1.09*	1700	35	7.104*	–	3.4	$r \rightarrow d,\ a$	–
	$p\,^3\Pi_g\,6p\pi$	31179.9*	1.080	1701.2	35.3	7.22	220	3.41	$p \rightarrow d,\ a$	–
	$n\,^3\Sigma_g^+\,6p\sigma$	30283.3*	1.061	1690	36	7.475	249	3.4	$n \rightarrow a$	–
	$l\,^3\Pi_g\,5p\pi$	29785.3*	1.080	1704	35	7.23	220	3.42	$l \rightarrow d,\ a$	–
	$k\,^3\Sigma_u^+\,5s\sigma$	29573.0*	1.08	1635*	–	7.23	230	3.4	$k \rightarrow c,\ b$	–
	$k'\,^3\Sigma_g^+\,5p\sigma$	28127.6*	1.068	1686.9	38.1	7.38	350	3.35	$k' \rightarrow a$	–
	j { $^3\Delta_u\,4d\delta$	27472.7*	1.081	1702.2	35.1	7.209	225	3.42	$j \rightarrow c,\ b$	–
	j { $^3\Pi_u\,4d\pi$	27290.2*	1.083	1680.9	40.8	7.186	230	3.33		–
	j { $^3\Sigma_u^+\,4d\sigma$	27206.0*	–	1669.8	39.1	–	–	3.29		–
	$i\,^3\Pi_g\,4p\pi$	27193.0*	1.078	1708	35	7.24	220	3.44	$i \rightarrow a$	–
	$h\,^3\Sigma_u^+\,4s\sigma$	26760.2*	1.08	1638*	–	7.26	230	–	$h \rightarrow c,\ b$	–
	$g\,^3\Sigma_g^+\,4p\sigma$	23597.0*	1.080	1672	41	7.221	248	3.30	$g \rightarrow a$	–
	f { $^3\Delta_u\,3d\delta$	22205.5*	1.079	1706.8	35.1	7.23	230	3.43	$f \rightarrow c,\ b$	–
	f { $^3\Pi_u\,3d\pi$	21754.0*	1.086	1661.5	44.8	7.14	235	3.25		–
	f { $^3\Sigma_u^+\,3d\sigma$	21548.8*	1.091	1635.8	44.4	7.07	250	3.15		–
	$e\,^3\Pi_g\,3p\pi$	21507.3*	1.075	1721.2	34.97	7.284	221	3.49	$e \leftrightarrow a$	–

Table 10.5 (continued)

Molecule	Electronic term	Term energy T_e [cm^{-1}]	Equilibrium internuclear distance r_e [Å]	Vibrational frequency ω_e [cm^{-1}]	Anharmonic constant $\omega_e x_e$ [cm^{-1}]	Rotational constant B_e [cm^{-1}]	Rotation-vibration interaction constant α_e [10^{-3} cm^{-1}]	Force constant k_e [10^5 dyn/cm]	Observable electronic transition	Reduced mass μ_A [a.m.u.]
1	2	3	4	5	6	7	8	9	10	11
	$d\,^3\Sigma_u^+\,3s\sigma$	20392.2*	1.071	1728.0	36.1	7.341	224	3.52	$d \to c,\ b$	–
	$c\,^3\Sigma_g^+\,3p\sigma$	10889.5*	1.097	1583.8	52.7	6.981	247	3.00	$c \to a$	–
	$b\,^3\Pi_g\,2p\pi$	4768.2*	1.063	1769.1	35.0	7.447	220	3.69	$b \to a$	–
	$a\,^3\Sigma_u^+\,2s\sigma$	0	1.046	1807.6	37.1	7.704	228	3.86	–	2.001
	Singlets									
	F $^1\Delta_u\,3d\delta$	19862*	1.079	1706.6	35.1	7.23	220	3.43	$F \to B$	–
	F $^1\Pi_u\,3d\pi$	19510*	1.085	1670.6	40.0	7.16	230	3.29	$F \to B$	–
	F $^1\Sigma_u^+\,3d\sigma$	19339*	1.089	1664	40	7.10	250	3.3	$F \to B$	–
	$E\,^1\Pi_g\,3p\pi$	19477*	1.076	1721.2	34.8	7.270	216	3.49	$E \to A$	–
	$D\,^1\Sigma_u^+\,3s\sigma$	18663*	1.069	1746.4	35.5	7.36	218	3.60	$D \to B,\ X$	–
	$C\,^1\Sigma_g^+\,3p\sigma$	10945*	1.092	1653.4	41.0	7.052	215	3.22	$C \to A$	–
	$B\,^1\Pi_g\,2p\pi$	3501*	1.067	1765.8	34.4	7.40	220	3.68	$B - A$	–
	$A\,^1\Sigma_u^+\,2s\sigma$	0	1.041	1861.0	34.9	7.779	217	4.08	$A \leftrightarrow X$	2.001
	$X\,^1\Sigma_g^+$	$-T_0$, where $T_0 \simeq 147279$ cm^{-1} is the energy of the $v = 0$, $J = 0$ level of $A\,^1\Sigma_u^+$ relative to $T(\mathrm{He}\,(^1S) + \mathrm{He}\,(^1S)) = \lim_{R\to\infty} T(X\,^1\Sigma_g^+)$ and $T_0(A\,^1\Sigma_u^+) - T_0(a\,^3\Sigma_u^+) = 2343.9$ cm^{-1}; term $X\,^1\Sigma_g^+$ characterizes the "repulsive" potential of the van der Waals molecule He_2 with well depth $\varepsilon = 7.4$ cm^{-1} and position of the potential minimum $r_m = 2.97$ Å								
HfO	$G\,^1\Sigma$	30090	1.761	860	3.7	0.3701	2.07	6.4	$G \leftrightarrow X$	–
	$F\,^1\Sigma$	27413.6	1.772	849.4	3.7	0.3656	1.9	6.24	$F \leftrightarrow X$	–
	$E\,^1\Pi$	25230.9	1.764	866.9	3.7	0.369	2.0	6.50	$E \leftrightarrow X$	–
	$D\,^1\Pi$	23554.4	1.764	872.6	3.3	0.369	1.8	6.59	$D \leftrightarrow X$	–
	$B\,^1\Pi$	17562.2	1.743	907.0	3.4	0.378	1.85	7.12	$B \leftrightarrow X$	–
	$A\,^1\Sigma$	16616.9	1.742	914.2	3.4	0.3780	1.83	7.23	$A \leftrightarrow X$	–
	$X\,^1\Sigma$	0	1.723	974.1	3.23	0.3865	1.72	8.21	–	14.68

Hg_2	$X\,^1\Sigma_g^+$	0	2.92	44	0.5	0.02	–	0.11	–	100.3
HgCl	$D\,^2\Pi_{3/2}$	39703.4	2.313	341.6	1.81	0.105	0.8	2.05	$D \leftrightarrow X$	–
	$X\,^2\Sigma^+$	0	2.395	292.6	1.63	0.098	0.5	1.50	–	30.13
HgH	$A_2\,^2\Pi_{3/2}$	28274	1.579	2068.2	43.0	6.741	230	2.53	$A_2 \rightarrow X$	–
	$A_1\,^2\Pi_{1/2}$	24933.1*	1.601*	1939.2*	64.8	6.561*	–	2.53	$A_1 \rightarrow X$	–
	$X\,^2\Sigma^+$	0	1.766*	1203.2*	120	5.389*	–	1.23	–	1.003
HoF	A	19152.8	2.006	539.4	4.4	0.2459	2.2	2.92	$A \leftarrow X$	–
	X	0	1.940	615.3	2.60	0.2630	1.4	3.80	–	17.04
I_2	$F\,^1\Sigma_u^+\,(0_u^+)$	47218	3.6	95.95	0.362	0.0205	–	0.34	$F \leftrightarrow X$	–
	$f\,^1\Sigma_g^+\,(0_g^+)$	47027.6	3.58	103.9	0.20	0.0208	0.055	0.40	$f \leftarrow B$	–
	$E\,^3\Pi_g\,(0_g^+)$	41411.4	3.64	101.35	0.197	0.0200	0.054	0.38	$E \rightarrow B$	–
	(1_g)	41031.8	3.56	103.81	0.206	0.0210	0.09	0.40	$1_g \rightarrow A,\ A'$	–
	$D\,(0_u^+)$	41026	3.58	95.7	0.13	0.0208	0.05	0.34	$D \leftrightarrow X$	–
	$B\,^3\Pi_u\,(0_u^+)$	15769.0	3.025	125.70	0.764	0.0290	0.158	0.59	$B \leftrightarrow X$	–
	$X\,^1\Sigma_g^+$	0	2.666	214.52	0.609	0.0374	0.114	1.72	–	63.45
IBr	$B\,^3\Pi_{0^+}$	16168	2.83	142	2.6	0.0432	0.5	0.59	$B \leftarrow X$	–
	$A\,^3\Pi_1$	12350	–	138	0.6	–	–	0.55	$A \leftrightarrow X$	–
	$X\,^1\Sigma^+$	0	2.469	268.6	0.814	0.0568	0.197	2.08	–	49.03
ICl	$B\,^3\Pi_{0^+}$	17363	2.66	221	9.6	0.087	1.7	0.80	$B \leftarrow X$	–
	$A\,^3\Pi_1$	13742	2.692	212	2.39	0.0848	0.4	0.74	$A \leftrightarrow X$	–
	$X\,^1\Sigma^+$	0	2.321	384.3	1.50	0.1142	0.535	2.41	–	27.71
IF	$B\,^3\Pi_{0^+}$	19052.28	2.117	411.28	2.86	0.2277	1.74	1.65	$B \leftrightarrow X$	–
	$X\,^1\Sigma^+$	0	1.910	610.23	3.13	0.2797	1.87	3.63	–	16.525
IO	$A\,^2\Pi_{3/2}$	21557.8	2.072	514.6	5.5	0.2764	2.7	2.22	$A \leftrightarrow X$	–
	$X\,^2\Pi_{3/2}$	0	1.868	681.5	4.3	0.3403	2.70	3.89	–	14.21
InBr	$X\,^1\Sigma^+$	0	2.543	221	0.65	0.0549	0.19	1.36	–	47.12
InCl	$B\,^3\Pi_1$	28560	2.34*	339	2	0.115*	–	1.84	$B \leftrightarrow X$	–
	$A\,^3\Pi_{0^+}$	27765	2.33	340	2	0.116	0.65	1.85	$A \leftrightarrow X$	–
	$X\,^1\Sigma^+$	0	2.401	317	1.0	0.1091	0.518	1.61	–	27.09
InF	$C\,^1\Pi$	42809	1.97	464	7.3	0.267	4.7	2.07	$C \leftrightarrow X$	–
	$B\,^3\Pi_1$	31255.7	1.944	572.2	2.6	0.2736	2.0	3.15	$B \leftrightarrow X$	–
	$A\,^3\Pi_{0^+}$	30445.9	1.945	575	3.68	0.273	2.0	3.18	$A \leftrightarrow X$	–
	$X\,^1\Sigma^+$	0	1.985	535.3	2.6	0.2623	1.88	2.75	–	16.30

Table 10.5 (continued)

Molecule	Electronic term	Term energy T_e [cm^{-1}]	Equilibrium internuclear distance r_e [Å]	Vibrational frequency ω_e [cm^{-1}]	Anharmonic constant $\omega_e x_e$ [cm^{-1}]	Rotational constant B_e [cm^{-1}]	Rotation-vibration interaction constant α_e [10^{-3} cm^{-1}]	Force constant k_e [10^5 dyn/cm]	Observable electronic transition	Reduced mass μ_A [a.m.u.]
1	2	3	4	5	6	7	8	9	10	11
InH	$A\,^1\Pi$	22017*	2.09*	142*	30	3.85*	–	0.02	$A \leftarrow X$	–
	$a\,^3\Pi$ 2	17781*	1.75	1301*	–	5.49	330	–	a $\rightarrow X$	–
	$a\,^3\Pi$ 1	16941.6	1.768	1415.1	43.5	5.400	236	1.18	a $\leftrightarrow X$	–
	$a\,^3\Pi$ 0^+	16278.1	1.779	1458.6	61.0	5.33	247	1.25	a $\leftrightarrow X$	–
	$a\,^3\Pi$ 0^-	16211.5*	1.78	1303.4*	–	5.35	330	–	a $\rightarrow X$	–
	$X\,^1\Sigma^+$	0	1.838	1476.0	25.6	4.994	143	1.28	–	0.999
InI	$X\,^1\Sigma^+$	0	2.754	177.1	0.4	0.0369	0.104	1.11	–	60.28
K_2	$C\,^1\Pi_u$	22969.4	4.43	61.48	0.13	0.0440	0.110	0.044	$C \leftarrow X$	–
	$B\,^1\Pi_u$	15377.16	4.236	74.88	0.327	0.0482	0.231	0.064	$B \leftrightarrow X$	–
	$X\,^1\Sigma_g^+$	0	3.925	92.40	0.328	0.0562	0.212	0.098	–	19.55
KBr	$X\,^1\Sigma^+$	0	2.821	213	0.8	0.0812	0.40	0.70	–	26.25
KCl	$X\,^1\Sigma^+$	0	2.667	281	1.3	0.1286	0.790	0.86	–	18.59
KF	$X\,^1\Sigma^+$	0	2.171	428	2.4	0.280	2.34	1.38	–	12.79
KH	$A\,^1\Sigma^+$	19053	3.7	228.2	−5.7	1.27	−37	0.030	$A \leftrightarrow X$	–
	$X\,^1\Sigma^+$	0	2.241	986.05	14.90	3.416	85.3	0.563	–	0.983
KI	$X\,^1\Sigma^+$	0	3.048	186.5	0.574	0.0609	0.268	0.61	–	29.89
Kr_2	$X\,^1\Sigma_g^+$ (0_g^+)	0	4.0	24.2	1.3	0.025	1.0	0.014	–	41.90
LaF	$X\,^1\Sigma^+$	0	2.026*	570*	–	0.2456*	–	–	–	16.71
LaO	$C\,^2\Pi_r$	22849	1.83	798	2.2	0.352	1.7	5.39	$C \rightarrow A'$; $C \leftrightarrow X$	–
		22631		792	2.2	0.350	1.6	5.31		
	$B\,^2\Sigma^+$	17879	1.856	734	2.0	0.341	1.7	4.6	$B \leftrightarrow X$	–
	$A\,^2\Pi_r$	13526	1.842*	761	2.0	0.346*	–	4.9	$A \leftrightarrow X$	–
		12663		762	2.1					

	$A'^2\Delta_r$	8191	1.848	772	2.3	0.344	1.7	5.05	–	–
		7493		768	2.2	0.344	1.6	4.99		
	$X^2\Sigma^+$	0	1.826	812.7*	2.2	0.353	1.4	5.65	–	14.35
LaS	$B^2\Sigma^+$	13790.2	2.417*	410.07	0.94	0.1110*	0.3	2.58	$B \leftrightarrow X$	–
	$X^2\Sigma^+$	0	2.355*	456.7	0.96	0.1169*	0.3	3.20	–	26.05
Li_2	$D^1\Pi_u$	34443.6*	3.22	201.7*	–	0.463	7	–	$D \leftarrow X$	–
	$C^1\Pi_u$	30551	3.08	238	3.3	0.507	9.7	0.12	$C \leftarrow X$	–
	$G^1\Pi_g$	31868.4	3.20	229.3	1.62	0.469	5.5	0.11	$G \leftarrow A$	–
	$F^1\Sigma_g^+$	29975.0	3.55	227.3	2.5	0.382	3.8	0.11	$F \leftarrow A$	–
	$E^1\Sigma_g^+$	27410.2	3.09	245.9	2.8	0.505	9.6	0.12	$E \leftarrow A$	–
	$B^1\Pi_u$	20436.3	2.936	270.66	2.920	0.5574	8.33	0.15	$B \leftrightarrow X$	–
	$A^1\Sigma_u^+$	14068.3	3.108	255.5	1.58	0.497	5.4	0.13	$A \leftarrow X$	–
	$X^1\Sigma_g^+$	0	2.673	351.43	2.595	0.6726	7.04	0.25	–	3.470
LiBr	$X^1\Sigma^+$	0	2.170	563.2	3.5	0.5554	5.64	1.19	–	6.385
LiCl	$X^1\Sigma^+$	0	2.021	643.3	4.50	0.7065	8.01	1.42	–	5.80
LiF	$X^1\Sigma^+$	0	1.564	910.3	7.9	1.345	20.3	2.48	–	5.08
LiH	$B^1\Pi$	34912	2.38	215.6	42	3.38	986	0.024	$B \leftarrow X$	–
	$A^1\Sigma^+$	26509.8	2.596	281.0*	–30	2.853*	–37	0.028	$A \leftrightarrow X$	–
	$X^1\Sigma^+$	0	1.596	1405.6	23.2	7.514	217	1.02	–	0.881
LiI	$X^1\Sigma^+$	0	2.392	498.2	3.4	0.4432	4.09	0.96	–	6.58
LiNa	$B^1\Pi$	20061.9	–	209.6	10.0	–	–	–	$B \rightarrow X$	–
	$A^1\Sigma^+$	14196	–	190	1.1	–	–	–	–	–
	$X^1\Sigma$	0	2.81	257.0	1.66	0.396	3.6	0.21	–	5.331
LiO	$X^2\Pi_i$	0	1.7	850	12	1.20	15	2.1	–	4.841
LuF	$F^1\Sigma$	25832	1.952	561	2.6	0.2581	17	3.2	$F \rightarrow X$	–
	$E^1\Pi$	24474.1	1.958	543.4	2.3	0.2565	16	2.98	$E \rightarrow X$	–
	$D^1\Pi$	20048	1.95*	570	2.5	0.259*	–	3.28	$D \rightarrow X$	–
	$B^1\Pi$	16800	1.936*	581	2.5	0.2624*	16	3.4	$B \rightarrow X$	–
	$A^1\Sigma$	16164.7	1.932	587.9	2.6	0.2636	16	3.49	$A \rightarrow X$	–
	$X^1\Sigma$	0	1.917	611.8	2.5	0.2676	16	3.78	–	17.14
LuH	$X^1\Sigma$	0	1.912	1500	20	4.602	100	1.4	–	1.002
LuO	$C^2\Sigma^+$	24440	1.828*	770	5	0.3441*	–	5.1	$C \rightarrow X$	–
	$B^2\Pi_{3/2}$	21470	1.806*	793	4	0.353*	–	5.43	$B \rightarrow X$	–
	$X^2\Sigma^+$	0	1.790	842	3.1	0.3588	1.6	6.13	–	14.66

Table 10.5 (continued)

Molecule	Electronic term	Term energy T_e [cm^{-1}]	Equilibrium internuclear distance r_e [Å]	Vibrational frequency ω_e [cm^{-1}]	Anharmonic constant $\omega_e x_e$ [cm^{-1}]	Rotational constant B_e [cm^{-1}]	Rotation-vibration interaction constant α_e [10^{-3} cm^{-1}]	Force constant k_e [10^5 dyn/cm]	Observable electronic transition	Reduced mass μ_A [a.m.u.]
1	2	3	4	5	6	7	8	9	10	11
Mg_2	$A\,^1\Sigma_u^+$	26068.8	3.082	190.61	1.15	0.1480	1.32	0.26	$A \leftrightarrow X$	–
	$X\,^1\Sigma_g^+$	0	3.889	51.12	1.64	0.0929	3.78	0.02	–	12.15
MgBr	$A\,^2\Pi$	{25877, 25767	2.33*	394	2.0	0.168*	–	1.70	$A \rightarrow X$	–
	$X\,^2\Sigma^+$	0	2.36*	374	1.3	0.164*	–	1.53	–	18.64
MgCl	$A\,^2\Pi_{1/2}$	26523	2.172*	492	2.5	0.2512*	1.8	2.05	$A \leftrightarrow X$	–
	$A\,^2\Pi_{3/2}$	26469	2.181*	–	–	0.2491*	1.8	–		–
	$X\,^2\Sigma^+$	0	2.199	462.1*	2.1	0.2450	1.6	1.85	–	14.42
MgF	$C\,^2\Sigma^+$	42539	1.700	823.2	5.0	0.5510	4.5	4.26	$C \leftarrow X$	–
	$B\,^2\Sigma^+$	37167	1.718	762.1	5.6	0.5384	5.1	3.65	$B \leftrightarrow X$	–
	$A\,^2\Pi_r$	{27851	–	746	4.0	–	–	3.50	$A \leftrightarrow X$	–
		27816	1.747	740.1*	–	0.5210	3.3	–		
	$X\,^2\Sigma^+$	0	1.750	721.6	4.9	0.5192	4.7	3.27	–	10.66
MgH	$C\,^2\Pi_r$	41160	1.68	1623*	–	6.16	140	–	$C \rightarrow A$; $C \leftrightarrow X$	–
	$E\,^2\Sigma^+$	35570	1.67	1445*	–	6.2	300	–	$E \leftarrow X$	–
	$B'\,^2\Sigma^+$	22410	2.60	828	12	2.59	–	0.39	$B' \rightarrow X$	–
	$A\,^2\Pi_r$	19227	1.678	1598.2	31.1	6.191	193	1.46	$A \leftrightarrow X$	–
	$X\,^2\Sigma^+$	0	1.730	1495.2	31.9	5.826	186	1.28	–	0.968
MgO	$F\,^1\Pi$	37920	1.773*	700*	–	0.559*	–	–	$F \rightarrow X$	–
	$E\,^1\Sigma^+$	37720	1.83*	705*	–	0.525*	–	–	$E \rightarrow A, X$	–
	$C\,^1\Sigma^-$	30081	1.873	632	5	0.501	5	3.40	$C \rightarrow A$	–
	$D\,^1\Delta$	29852	1.872	632	5	0.501	5	3.40	$D \rightarrow A$	–
	$d\,^3\Delta_i$	29300	1.9	650	–	0.5	–	3.6	$d \leftrightarrow a$	–

	$B\,^1\Sigma^+$	19984	1.737	824.1	4.8	0.582	4.5	5.77	$B \to A;\ B \leftrightarrow X$	–
	$A\,^1\Pi$	3563	1.864	664.4	3.9	0.506	5	3.75	–	–
	$X\,^1\Sigma^+$	0	1.749	785.1	5.2	0.574	5	5.24	–	14.42
MgS	$B\,^1\Sigma^+$	23052.6	2.196	497.3	2.33	0.2552	1.6	2.01	$B \leftrightarrow X$	–
	$X\,^1\Sigma^+$	0	2.142	528.7	2.70	0.2680	1.8	2.28	–	13.82
MnH	$A\,^7\Pi$	17600	1.63	1690	30	6.42	190	1.7	$A \leftrightarrow X$	–
	$X\,^7\Sigma$	0	1.731	1548	29	5.684	157	1.40	–	0.990
MnO	$X\,^6\Sigma$	0	1.77	840	4.8	0.43	–	5.15	–	12.39
N_2	$c_4\,^1\Pi_u$	115636	1.12	2220	19	1.926*	10	20.3	$c_4 \leftarrow a'',\ X$	–
	$z\,^1\Delta_g$	115435	1.17	1700	–	1.76	15	12	$z \to w$	–
	$y\,^1\Pi_g$	114305	1.18	1906.4	37.5	1.74	17	15.0	$y \to w,\ a'$	–
	$k\,^1\Pi_g$	113810	1.11	2182.3*	–	1.96	30	–	$k \to w,\ a'$	–
	$x\,^1\Sigma_g^-$	113438	1.17	1910	21	1.75	23	15	$x \to a'$	–
	$o_3\,^1\Pi_u$	105870	1.178	1987	16	1.734	9	16.3	$o_3 \to a;\ o_3 \leftarrow X$	–
	$H\,^3\Phi_u$	105720	1.488	924.2	12.3	1.087	19	3.52	$H \to G$	–
	$c_4'\,^1\Sigma_u^+$	104520	1.108	2201.8	25.20	1.961	44	20.0	$c_4' \to a;\ c_4' \leftrightarrow X$	–
	$c_3\,^1\Pi_u$	104480	1.116	2192.2	14.7	1.932	40	19.8	$c_3 \to a;\ c_3 \leftrightarrow X$	–
	$b'\,^1\Sigma_u^+$	104500	1.444	760.1	4.42	1.155	7.39	2.38	$b' \to a;\ b' \leftrightarrow X$	–
	$D\,^3\Sigma_u^+$	103571*	1.11*	–	–	1.96*	–	–	$D \to B$	–
	$b\,^1\Pi_u$	101675	1.284	635*	–	1.448*	–	–	$b \to a;\ b \leftrightarrow X$	–
	$C'\,^3\Pi_u$	98350	1.515*	790	33	1.050*	–	2.6	$C' \leftrightarrow B$	–
	$E\,^3\Sigma_g^+$	95860	1.118*	2185*	–	1.927*	–	–	$E \to B,\ A;\ E \leftarrow X$	–
	$C\,^3\Pi_u$	89136.9	1.149	2047.2	28.45	1.825	18.7	17.3	$C \to B;\ C \leftarrow X$	–
	$G\,^3\Delta_g$	87900	1.61	766	12	0.928	16	2.42	–	–
	$w\,^1\Delta_u$	72097	1.27	1559.3	11.6	1.50	17	10.0	$w \to a;\ w \leftarrow X$	–
	$a\,^1\Pi_g$	69283.1	1.220	1694.2	13.9	1.617	18	11.8	$a \to a';\ a \leftrightarrow X$	–
	$a'\,^1\Sigma_u^-$	68152.7	1.275	1530.3	12.07	1.480	16.6	9.66	$a' \leftrightarrow X$	–
	$B'\,^3\Sigma_u^-$	66271.3	1.278	1516.6	12.0	1.473	16.6	9.49	$B' \to B;\ B' \leftrightarrow X$	–
	$W\,^3\Delta_u$	59805.8	1.280	1506.49	12.55	1.4703	17.1	9.36	$W \leftrightarrow B;\ W \leftarrow X$	–
	$B\,^3\Pi_g$	59618.7	1.212	1733.98	14.39	1.6379	18.1	12.4	$B \leftrightarrow A;\ B \leftarrow X$	–
	$A\,^3\Sigma_u^+$	50203.6	1.287	1460.6	13.87	1.455	18	8.80	$A \leftrightarrow X$	–
	$X\,^1\Sigma_g^+$	0	1.0977	2358.6	14.32	1.998	17.3	23.0	–	7.003
NBr	$b\,^1\Sigma^+$	14787	1.731	785	4.36	0.4733	15	4.33	$b \to X$	–
	$X\,^3\Sigma^-\ (0^+)$	0	1.79	692	4.72	0.44	4	3.36	–	11.92

Table 10.5 (continued)

Molecule	Electronic term	Term energy T_e [cm^{-1}]	Equilibrium internuclear distance r_e [Å]	Vibrational frequency ω_e [cm^{-1}]	Anharmonic constant $\omega_e x_e$ [cm^{-1}]	Rotational constant B_e [cm^{-1}]	Rotation-vibration interaction constant α_e [10^{-3} cm^{-1}]	Force constant k_e [10^5 dyn/cm]	Observable electronic transition	Reduced mass μ_A [a.m.u.]
1	2	3	4	5	6	7	8	9	10	11
NCl	$b\,^1\Sigma^+$	14985	1.571*	936	5	0.6828*	–	5.18	$b \to X$	–
	$X\,^3\Sigma^-$	0	1.614*	827	5	0.6468*	–	4.05	–	10.04
NF	$b\,^1\Sigma^+$	18877.0	1.300	1197.5	8.6	1.238	14.5	6.81	$b \to X$	–
	$a\,^1\Delta$	11435.2*	1.308*	–	–	1.222*	–	–	$a \to X$	–
	$X\,^3\Sigma^-$	0	1.317	1141.4	9.0	1.206	14.9	6.19	–	8.063
NH	$d\,^1\Sigma^+$	83160	1.116	2673	71.2	14.39	621	3.96	$d \to c, b$	–
	$c\,^1\Pi$	43740	1.111	2551	214	14.54	593	3.61	$c \to b, a$	–
	$A\,^3\Pi_i$	29807	1.037	3231	99	16.674	745	5.78	$A \leftrightarrow X$	–
	$b\,^1\Sigma^+$	21200	1.036	3352.4	74.2	16.70	591	6.23	$b \to X$	–
	$a\,^1\Delta$	12570	1.034	3320	70	16.44*	660	6.1	–	–
	$X\,^3\Sigma^-$	0	1.036	3282	78.3	16.70	649	5.97	–	0.940
NO	$N\,^2\Delta\,4d\delta$	67374	1.07	2375	15	1.969	26	24.8	$N \to C$; $N \leftrightarrow X$	–
	$S\,^2\Sigma^+\,5s\sigma$	66900	1.07	2378	16	1.980	20	24.9	$S \leftarrow X$	–
	$M\,^2\Sigma^+\,4p\sigma$	64437	1.06	2352	19	2.022	18	24.3	$M \leftarrow X$	–
	$G\,^2\Sigma^-$	62913	1.343	1085.5	11.08	1.252	20	5.19	$G \leftarrow X$	–
	$H'\,^2\Pi\,3d\pi$	62485	1.058	2371	16.2	2.015	21	24.7	$H' \to D, C, A$; $H' \leftarrow X$	–
	$H\,^2\Sigma^+\,3d\sigma$	62473	1.062	2339*	–	2.003	18	–	$H \to D, C, A$; $H \leftarrow X$	–
	$F\,^2\Delta\,3d\delta$	61800	1.07	2394	20	1.982	23	25.2	$F \to C$; $F \leftrightarrow X$	–
	$E\,^2\Sigma^+\,4s\sigma$	60629	1.07	2375	16.4	1.986	18	24.8	$E \to D, A$; $E \leftarrow X$	–
	$B'\,^2\Delta_i$	60364	1.30	1217	15.6	1.332	21	6.52	$B' \to C, B$; $B' \leftrightarrow X$	–
	$D\,^2\Sigma^+\,3p\sigma$	53085	1.062	2323.9	22.9	2.003	21.7	23.8	$D \to A$; $D \leftrightarrow X$	–
	$C\,^2\Pi_r\,3p\pi$	52126	1.06	2395	15	2.00	30	25.2	$C \to A$; $C \leftrightarrow X$	–
	$B\,^2\Pi_r$	{ 45943 45914	1.417	1040 1037	8.3 7.7	1.152 1.092	12	4.76 4.73	$B \leftrightarrow X$	–

NS	$A\,^2\Sigma^+\,3s\sigma$		43966	1.063	2374.3	16.11	1.996	19.1	24.8	$A \leftrightarrow X$	–
	$X\,^2\Pi_r$	3/2	119.8	1.1508	1904.04	14.10	1.7202*	18	16.0	$3/2 \leftarrow 1/2$	–
		1/2	0		1904.20	14.07	1.6720*	17		–	7.468
	$H\,^2\Pi_i$		44050	1.70	–	–	0.597*	–	–	$H \rightarrow X$	–
			43876		768	5	0.591*	6	3.38		
	$C\,^2\Sigma^+$		43290	1.446*	1390*	–	0.827*	–	–	$C \leftrightarrow X$	–
	$G\,^2\Sigma^-$		43350	1.583*	880*	–	0.690*	13	–	$G \rightarrow X$	–
	$A\,^2\Delta_r$	5/2	40050	–	934*	–	–	–	–	$A \leftrightarrow X$	–
		3/2	40005	1.590*	944*	8	0.685*	–	5.3		
	$B'\,^2\Sigma^+$		36260	1.5	1060	15	0.8	–	6.5	$B' \rightarrow X$	–
	$B\,^2\Pi_r$	3/2	30384	1.70	798.8	3.6	0.601	5	3.66	$B \rightarrow X$	–
		1/2	30295		797.3	3.7	0.596	5	3.65		
	$X\,^2\Pi_r$	3/2	221	1.494	1219	7.3	0.7752*	6.3	8.53	–	9.748
		1/2	0				0.7696*				
NSe	$b\,^4\Sigma^-_{1/2}$		24840	1.792*	770*	–	0.441*	–	–	$b \rightarrow X_1$	–
	$A_2\,^2\Pi$	3/2	24800	1.84*	613*	–	0.417*	–	–	$A_2 \rightarrow X_2$	–
	A_1	1/2	24350	1.85*	659*	–	0.412*	–	–	$A_1 \rightarrow X_1$	–
	$a\,^4\Pi_i$		19700	1.98	711	10	0.36	2	3.5	–	–
	$X_2\,^2\Pi$	3/2	892	1.652	955.0	5.6	0.519	4	6.39	–	11.90
	X_1	1/2	0		956.8	5.6	0.518	4	6.42		
Na_2	$C\,^1\Pi_u$		29621.6	3.550	116.3	0.65	0.1163	0.86	0.092	$C \leftrightarrow$ X; $C \rightarrow 1, 2, 3$	
	$(3)\,^1\Sigma^+_g$		25691.5	3.563	112.7	1.05	0.1155	1.13	0.086	$3 \rightarrow B, A$	
	$(1)\,^1\Pi_g$		21795.54	4.560	42.73	0.45	0.0705	1.32	0.012	$C \rightarrow 1$	
	$B\,^1\Pi_u$		20320.0	3.423	124.1	0.700	0.1253	0.724	0.10	$B \leftrightarrow X$	–
	$(2)\,^1\Sigma^+_g$		19337.9	4.450	75.18	0.070	0.0740	–0.25	0.038	$C \rightarrow 2$	–
	$A\,^1\Sigma^+_u$		14680.69	3.638	117.27	0.353	0.1108	0.549	0.093	$A \leftrightarrow X$	–
	$X\,^1\Sigma^+_g$		0	3.079	159.18	0.760	0.1547	0.783	0.17	–	11.495
NaBr	$X\,^1\Sigma^+$		0	2.502	302	1.5	0.1513	0.941	0.96	–	17.85
NaCl	$X\,^1\Sigma^+$		0	2.361	366	2.1	0.2181	1.625	1.10	–	13.95
NaCs	$X\,^1\Sigma^+$		0	3.851	98.89	0.326	0.0580	0.230	0.11	–	19.60
NaF	$X\,^1\Sigma^+$		0	1.926	536	3.4	0.4369	4.559	1.76	–	10.40
NaH	$A\,^1\Sigma^+$		22713	3.193	317.6	–2.7	1.712	–91.5	0.057	$A \leftrightarrow X$	–
	$X\,^1\Sigma^+$		0	1.889	1176	21.2	4.89	131	0.79	–	0.966

Table 10.5 (continued)

Molecule	Electronic term		Term energy T_e [cm^{-1}]	Equilibrium internuclear distance r_e [Å]	Vibrational frequency ω_e [cm^{-1}]	Anharmonic constant $\omega_e x_e$ [cm^{-1}]	Rotational constant B_e [cm^{-1}]	Rotation-vibration interaction constant α_e [10^{-3} cm^{-1}]	Force constant k_e [10^5 dyn/cm]	Observable electronic transition	Reduced mass μ_A [a.m.u.]
1	2		3	4	5	6	7	8	9	10	11
NaI	$X^1\Sigma^+$		0	2.711	258	1.1	0.1178	0.648	0.76	–	19.46
NaK	$C^1\Pi$		16994	4.50	71.3	1.24	0.057	0.08	0.043	$C \leftarrow X$	–
	$X^1\Sigma^+$		0	3.59	124.13	0.511	0.0905	0.46	0.13	–	14.48
NbO	$G^4\Sigma^-$		21385	1.757	850	3.4	0.400	2	5.82	$G \leftrightarrow X$	–
	$X^4\Sigma^-$		0	1.691	990	3.8	0.432	2	7.87	–	13.65
Ne_2	$X^1\Sigma_g^+$		0	3.1*	14*	–	0.17*	60	–	–	10.09
NiH	$X_2\,{}^2\Delta$	3/2	980	1.479*	–	–	7.78*	–	–	–	–
	X_1	5/2	0	1.48	1927*	40	7.70*	230	–	–	0.991
O_2	$f^1\Sigma_u^+$		76090	1.11	1930	19	1.70	20	17.5	$f \leftarrow b, X$	–
	$D^3\Sigma_u^+$		75260	1.10	1960	20	1.7	25	18.1	$D \leftarrow X$	–
	$B^3\Sigma_u^-$		49793.3	1.604	709.3	10.6	0.819	12.1	2.37	$B \leftrightarrow X$	–
	$A^3\Sigma_u^+$		35398	1.522	799.1	12.2	0.911	14.2	3.01	$A \leftrightarrow X$	–
	$c^1\Sigma_u^-$		33057	1.517	794.3	12.74	0.915	13.9	2.97	$c \rightarrow a$; $c \leftrightarrow X$	–
	$b^1\Sigma_g^+$		13195	1.227	1432.8	14.0	1.400	18.2	9.68	$b \leftrightarrow X$; $b \rightarrow a$	–
	$a^1\Delta_g$		7918.1	1.216	1483.5*	13	1.4178*	17.1	10.7	$a \leftrightarrow X$	–
	$X^3\Sigma_g^-$		0	1.208	1580.2	12.0	1.446	15.9	11.8	–	8.000
OH	$C^2\Sigma^+$		89459	2.046	1233	19	4.25	80	0.85	$C \rightarrow A, X$	–
	$D^2\Sigma^-$		81759.8*	1.081*	2950	–	15.22*	–	4.9	$D \leftarrow X$	–
	$B^2\Sigma^+$		69774	1.870*	660*	100	5.09*	1.0	0.5	$B \rightarrow A$	–
	$A^2\Sigma^+$		32684	1.012	3178.9	92.9	17.36	0.79	5.64	$A \leftrightarrow X$	–
	$X^2\Pi_i$	1/2	126.2*	0.970	3737.8	84.9	18.91	0.724	7.80	–	0.948
		3/2	0								

P_2	$N^1\Sigma_u^+$		77287	1.910	701	30	0.298	5.0	4.5	$N \leftarrow X$	–
	$M^1\Sigma_u^+$		73168	1.97	684	3.0	0.280	1.6	4.27	$M \leftarrow X$	–
	$K^1\Pi_u$		72290	2.01*	702*	5	0.270*	–	4.6	$K \leftarrow X$	–
	$G^1\Sigma_u^+$		66313.4	1.913	694.1	4.18	0.2973	1.9	4.40	$G \leftrightarrow X$	–
	$E^1\Pi_u$		59446.2	1.969*	700.7	2.9	0.2807*	–	4.48	$E \leftrightarrow X$	–
	$B^1\Pi_u$		50846	2.19*	359.0*	3	0.227*	6	1.2	$B \rightarrow A$	–
		2	47177								
	$c^3\Pi_u$	1	47159	2.23	393.7	3.85	0.219	2.4	1.41	$c \rightarrow b$	–
		0	47139								
	$C^1\Sigma_u^+$		46941.3	2.120	473.9	2.34	0.2421	1.8	2.05	$C \leftrightarrow X$	–
	$A^1\Pi_g$		34515.2	1.989	618.9	3.0	0.2752	1.7	3.50	$A \rightarrow X$	–
		2	28330								
	$b^3\Pi_g$	1	28197	1.97	644.7	3.21	0.280	1.8	3.79	$b \rightarrow a$	–
		0	28069								
	$b'^3\Sigma_u^-$		28503	2.05	604.5	2	0.258	1.4	3.33	$b' \rightarrow X$	–
	$a^3\Sigma^+$		18790	$r_1 = 2.09$	565	3	$B_1 = 0.250$	–	2.9	–	–
	$X^1\Sigma_g^+$		0	1.893	780.8	2.83	0.3036	1.5	5.56	–	15.49
PF	$d^1\Pi$		36020	1.72	413.2*	–	0.485	6	–	$d \rightarrow b, a$	–
		2	29830		435.9*	–	0.469				
	$B^3\Pi$	1	29690	1.752	435.9*	–	0.466	4	–	$B \rightarrow X$	–
		0	29540		436.1*	–	0.463				
	$b^1\Sigma^+$		13353.9	1.581	866.1	4.5	0.572	5	5.21	$b \rightarrow X$	–
	$a^1\Delta$		7090.4	1.585	858.8	4.44	0.570	4.7	5.12	–	–
	$X^3\Sigma^-$		0	1.590	846.7	4.49	0.567	4.6	4.97	–	11.78
PH	$A^3\Pi_i$		29500	1.467*	1834*	100	8.022*	500	2.4	$A \leftrightarrow X$	–
	$a^1\Delta$		7700	1.430*	–	–	8.44*	120	–	–	–
	$X^3\Sigma^-$		0	1.422	2366.8	45	8.539	252	3.22	–	0.976
PN	$A^1\Pi$		39806	1.546	1103	7.2	0.731	6.4	6.91	$A \leftrightarrow X$	–
	$X^1\Sigma^+$		0	1.491	1337.0	6.9	0.7865	5.54	10.2	–	9.645
PO	$I^2\Sigma^+$		55458	1.432	1390	6	0.780	5	12.0	$I \rightarrow A, B, X$	–
	$E^2\Delta$		53092	1.451	1480.2	12.0	0.758	7	13.6	$E \rightarrow X$	–
	$G^2\Sigma^+$		52410	1.4	1380	10	0.78	5	12	$G \rightarrow A, B, X$	–
	$P^2\Pi$		51989	1.699	475.5	1.69	0.553	3	1.41	$P \rightarrow B; P \leftrightarrow X$	–
	$F^2\Sigma^+$		49880	1.6	850	7	0.61	4	4.5	$F \rightarrow A, B, X$	–

Table 10.5 (continued)

Molecule	Electronic term	Term energy T_e [cm^{-1}]	Equilibrium internuclear distance r_e [Å]	Vibrational frequency ω_e [cm^{-1}]	Anharmonic constant $\omega_e x_e$ [cm^{-1}]	Rotational constant B_e [cm^{-1}]	Rotation-vibration interaction constant α_e [10^{-3} cm^{-1}]	Force constant k_e [10^5 dyn/cm]	Observable electronic transition	Reduced mass μ_A [a.m.u.]
1	2	3	4	5	6	7	8	9	10	11
	$D\,^2\Pi_r$	48520	1.45	1370	7	0.75	7	11.7	$D \to B$; $D \leftrightarrow X$	–
	$C\,^2\Sigma^-$	44831.7	1.64	779.2	5.1	0.590	6	3.77	$C \to X$	–
	$C'\,^2\Delta$	43742.7	1.58	825.7	6.9	0.640	5	4.24	$C' \to X$	–
	$A\,^2\Sigma^+$	40407.0	1.431	1390.7	6.8	0.780	5.5	12.0	$A \to B$; $A \leftrightarrow X$	–
	$B'\,^2\Pi$	33121	1.717	759	3.8	0.542	5	3.58	$B' \to B$; $B' \leftrightarrow X$	–
	$B\,^2\Sigma^+$	30730.9	1.463	1164.5	13.5	0.746	8.9	8.43	$B \leftrightarrow X$	–
	$X\,^2\Pi_r$	0	1.476	1233.3	6.6	0.734	5	9.46	–	10.55
PS	$C\,^2\Sigma$	34686	2.013	535	3.3	0.2644	–	2.65	$C \to X$	–
	$X\,^2\Pi_r$	{ 321 0	1.901*	739	3.0	0.2967* 0.2963*	–	5.07	–	15.75
PbCl	$A\,^2\Sigma\,(1/2)$	21865	2.29	229	0.8	0.107	0.5	0.92	$A \to X_2$; $A \leftrightarrow X_1$	–
	$X_2\,^2\Pi_{3/2}$	8272	–	322	0.3	–	–	1.82	–	–
	$X_1\,^2\Pi_{1/2}$	0	2.18*	304	0.9	0.119*	–	1.63	–	30.27
PbF	$B\,^2\Sigma^+$	35644	1.976	612.6	3.4	0.2481	1.48	3.85	$B \leftrightarrow X_2, X_1$	–
	$A\,^2\Sigma^+\,(1/2)$	22556	2.160	398.3	1.8	0.2076	1.43	1.63	$A \to X_2$; $A \leftrightarrow X_1$	–
	$X_2\,^2\Pi_{3/2}$	8263	2.034	531.7	1.5	0.2340	1.45	2.90	–	–
	$X_1\,^2\Pi_{1/2}$	0	2.057	507.3	2.3	0.2287	1.47	2.64	–	17.40
PbH	$X\,^2\Pi_{1/2}$	0	1.839	1564	29.8	4.971	144	1.45	–	1.003
PbO	$D\,1$	30199	2.05	530	2.9	0.271	3	2.46	$D \leftrightarrow X$	–
	$C'\,1$	24950	2.1	490	3	0.25	2	2.1	$C' \leftarrow X$	–
	$C\,0^+$	23820	2.1	530	4	0.25	2	2.5	$C \leftarrow X$	–
	$B\,1$	22280	2.07	495	2.2	0.265	3	2.2	$B \leftrightarrow X$	–
	$A\,0^+$	19863	2.095	444	0.5	0.2587	1.4	1.73	$A \leftrightarrow X$	–
	$X\,^1\Sigma^+$	[illegible]	[illegible]	[illegible]	[illegible]	[illegible]	[illegible]	[illegible]		[illegible]

PbS	$D\,1$	29653	2.45	297.8	1.4	0.1016	0.6	1.45	$D \leftarrow X$	–
	$B\,1$	21847	2.47	282.2	0.86	0.100	0.6	1.30	$B \leftarrow X$	–
	$A\,0^+$	18853	2.515	260.99	0.456	0.0962	0.196	1.11	$A \leftarrow X$	–
	$a\,1$	14893	2.56	286	0.9	0.093	0.4	1.33	$a \leftarrow X$	–
	$X\,^1\Sigma^+$	0	2.287	429.17	1.266	0.1163	0.435	3.01	–	27.77
PbSe	$X\,^1\Sigma^+$	0	2.402	278	0.5	0.0506	0.13	2.62	–	57.17
PbTe	$X\,^1\Sigma^+$	0	2.595	212.0	0.4	0.0313	0.07	2.09	–	78.97
Pd^2H	$X\,^2\Sigma^+$	0	1.529	1446.0	19.6	3.649	81	2.44	–	1.977
PtC	$A\,^1\Pi$	18627.0	1.762	818.7	5.4	0.4802	4.1	4.46	$A \leftarrow X$	–
	$A'\,^1\Pi$	13263	1.717	918.1	6	0.5058	3.9	5.61	$A' \leftrightarrow X$	–
	$A''\,^1\Sigma$	12697.2	1.711	943.4	5.3	0.5096	3.7	5.93	$A'' \leftrightarrow X$	–
	$X\,^1\Sigma$	0	1.677	1051.1	4.9	0.5304	3.3	7.36	–	11.315
PtH	$X_1\,^2\Delta_{5/2}$	0	1.529	2390	50	7.196	200	3.4	–	1.003
PtO	$A\,^1\Sigma$	16995.1	1.795	727.1	5.4	0.3538	2.9	4.61	$A \rightarrow X$	–
	$X\,^1\Sigma$	0	1.727	851.1	5.0	0.3822	2.8	6.31	–	14.79
Rb_2	$B\,^1\Pi$	14665.4	4.46	47.43	0.15	0.0200	0.07	0.056	$B \leftrightarrow X$	–
	$X\,^1\Sigma_g^+$	0	4.17	57.75	0.16	0.0228	0.05	0.083	–	42.73
RbBr	$X\,^1\Sigma^+$	0	2.945	169.5	0.46	0.0475	0.186	0.70	–	41.30
RbCl	$X\,^1\Sigma^+$	0	2.787	230	0.9	0.0876	0.454	0.77	–	25.06
RbCs	$X\,^1\Sigma^+$	0	4.368	49.92	0.087	0.0170	0.094	0.076	–	51.81
RbF	$X\,^1\Sigma^+$	0	2.270	375	2	0.2107	1.52	1.29	–	15.54
RbH	$A\,^1\Sigma^+$	18220	3.864	212.5	−6.3	1.133	−43	0.026	$A \rightarrow X$	–
	$X\,^1\Sigma^+$	0	2.367	937.2	14.17	3.020	73	0.516	–	0.996
RbI	$X\,^1\Sigma^+$	0	3.177	138.5	0.33	0.0328	0.109	0.58	–	51.07
RhC	$C\,^2\Sigma$	21439	1.687	928	13.7	0.551	6	5.46	$C \leftarrow X$	–
	$A\,^2\Pi_r$ 3/2	10242.7	1.655	939.1	5.5	0.5715	4.3	5.59	$A \leftrightarrow X$	–
	$A\,^2\Pi_r$ 1/2	9462.9		949.4	5.36	0.5733	4.3	5.71		
	$X\,^2\Sigma$	0	1.613	1049.9	4.94	0.603	4.0	6.99	–	10.76
S_2	$D\,^3\Pi_{u,r}$	58979 58692 58518	1.855*	794	4.0	0.307*	–	5.95	$D \leftrightarrow X$	–
	$g\,^1\Delta_u$	57900	1.812*	816	2.7	0.321*	–	6.29	$g \leftrightarrow a$	–
	$C\,^3\Sigma_u^-$	55582	1.810	829.1	3.3	0.322	1.4	6.49	$C \leftrightarrow X$	–

Table 10.5 (continued)

Molecule	Electronic term	Term energy T_e [cm^{-1}]	Equilibrium internuclear distance r_e [Å]	Vibrational frequency ω_e [cm^{-1}]	Anharmonic constant $\omega_e x_e$ [cm^{-1}]	Rotational constant B_e [cm^{-1}]	Rotation-vibration interaction constant α_e [10^{-3} cm^{-1}]	Force constant k_e [10^5 dyn/cm]	Observable electronic transition	Reduced mass μ_A [a.m.u.]
1	2	3	4	5	6	7	8	9	10	11
	$f^1\Delta_u$	42590	2.155	438.3	2.7	0.2270	1.8	1.81	$f \leftrightarrow a$	–
	$B^3\Sigma_u^-$	31830	2.17	434	2.7	0.224	2	1.78	$B \leftrightarrow X$	–
	$B''^3\Pi_u$	31070	2.3	335	4	0.20	–	1.1	–	–
	$a^1\Delta_g$	5730*	1.898	702.3	3.1	0.2926	1.7	4.66	$a \leftrightarrow X$	–
	$X^3\Sigma_g^-$	0	1.889	725.6	2.84	0.2955	1.57	4.97	–	16.03
SF	$A_2\ {}^2\Pi$ 1/2	25606*	–	480	3	–	–	1.6	$A_2 \leftarrow X_2, X_1$	–
	$A_1\ {}^2\Pi$ 3/2	24995*	1.60	490	3	0.55	4	1.7	$A_1 \leftarrow X_1$	–
	$X_2\ {}^2\Pi$ 1/2	400	1.596	837.6	4.47	0.5552	0.45	4.93	–	11.93
	$X_1\ {}^2\Pi$ 3/2	0								
SH	$B^2\Sigma$	59640	1.401	2557.0*	57	8.79	260	4.11	$B \leftarrow X$	–
	$A^2\Sigma^+$	31040	1.423	1980	97.7	8.52	460	2.26	$A \leftrightarrow X$	–
	$X^2\Pi_i$	0	1.341	2712	60	9.461*	270	4.23	–	0.977
SO	$B^3\Sigma^-$	41630	1.77	630	4.8	0.502	6	2.50	$B \leftrightarrow X$	–
	$A^3\Pi$ 2	38620	1.602	413	1.7	0.616	20	1.07		
	$A^3\Pi$ 1	38460	1.609	413	1.6	0.611	19	1.07	$A \leftrightarrow X$	–
	$A^3\Pi$ 0	38310	1.614	415	1.6	0.607	19	1.08		
	$b^1\Sigma^+$	10510	1.500	1068.7	7.2	0.7026	6.3	7.18	$b \rightarrow X$	–
	$a^1\Delta$	5860*	1.492*	–	–	0.7103*	–	–	$a \leftrightarrow X$	–
	$X^3\Sigma^-$	0	1.481	1149.2	5.6	0.7208	5.74	8.31	–	10.67
SbF	$C_3 1$	44757	1.855	702.3	3	0.298	3	4.8	$C_3 \rightarrow b, a$; $C_3 \leftrightarrow X_1$	–
	$A_3 1$	28707	2.062	411	1.7	0.241	1.7	1.64	$A_3 \rightarrow b, a, X_2$	–
	$A_2 2$	24788	2.064	420	1.7	0.241	1.6	1.71	$A_2 \rightarrow a, X_2$	–
	$A 0^+$	21407	2.075*	416	2	0.2385*	–	1.7	$A \rightarrow X_2, X_1$	–
	$b^1\Sigma^+$	13651	1.910	615	3	0.281	2	3.67	$b \rightarrow X, X$	–

	$a\,{}^1\Delta$	6816	1.913	616	2.7	0.281	1.9	3.68	–	–
	$X_2\,{}^3\Sigma^-$ $\{1$	796	1.913	613	2.6	0.280	2	3.63	–	–
	X_1 $\{0^+$	0	1.918	610.2	2.6	0.279	2	3.61	–	16.43
SbO	$X\,{}^2\Pi_r$	2270 0	1.826	820	4	0.358	2	5.5	–	14.14
SbP	$B\,{}^1\Pi$	28140	2.31	394*	–	0.128	–	–	$B \rightarrow X$	–
	$X\,{}^1\Sigma^+$	0	2.21	500.1	1.63	0.141	0.5	3.64	–	24.69
ScCl	$D\,{}^1\Pi$	21521	2.34*	373	2	0.157*	–	1.63	$D \rightarrow X$	–
	$B\,{}^1\Pi$	17613	2.35*	374	2	0.155*	–	1.64	$B \rightarrow X$	–
	$A\,{}^1\Sigma^+$	12431	2.33*	374	1	0.157*	–	1.63	$A \rightarrow X$	–
	$X\,{}^1\Sigma^+$	0	2.23	447	2	0.173	1	2.34	–	19.82
ScF	$F\,{}^1\Pi$	26892	1.910	571	3	0.346	3	2.57	$F \leftarrow X$	–
	$E\,{}^1\Pi$	20383	1.865	622	4	0.363	3.0	3.05	$E \leftrightarrow X$	–
	$C\,{}^1\Sigma^+$	16165	1.906	590	2.6	0.347	2	2.74	$C \leftrightarrow X$	–
	$B\,{}^1\Pi$	10735.5	1.918	586.2	2.01	0.3431	2.6	2.70	$B \leftrightarrow X$	–
	$X\,{}^1\Sigma^+$	0	1.788	736	4	0.395	2.7	4.26	–	13.35
ScO	$B\,{}^2\Sigma^+$	20645	1.720*	825.5	4.2	0.4831*	3	4.74	$B \leftrightarrow X$	–
	$A\,{}^2\Pi_r$	16547	1.686*	876	5.0	0.5028*	4	5.34	$A \leftrightarrow X$	–
	$A'\,{}^2\Delta_r$	15136 15030	–	846	5	–	–	–	$A' \rightarrow X$	–
	$X\,{}^2\Sigma^+$	0	1.668*	973.3	4.2	0.5134*	3	6.59	–	11.80
Se_2	$E\,0_u^+$	54752	2.14	404	1	0.092	0.3	3.79	$E \leftarrow X_2, X_1$	–
	$D\,1_u$	53070	2.09*	426*	–	0.096*	–	–	$D \leftarrow X_1$	–
	$C_2\,{}^3\Sigma_u^-$ $\{0_u^+$	53320	–	414*	–	–	–	–	$C_2 \leftarrow X_1$	–
	C_1 $\{1_u$	53220	2.089	428	1.2	0.0966	0.3	4.26	$C_1 \leftarrow X_2$	–
	$B_2\,{}^3\Sigma_u^-$ $\{1_u$	26060.5	2.440	246.4	1.22	0.0709	0.55	1.41	$B_2 \rightarrow X_1$; $B_2 \leftrightarrow X_2$	–
	B_1 $\{0_u^+$	25980.3	2.447	246.3	1.02	0.0705	0.34	1.41	$B_1 \rightarrow X_2$; $B_1 \leftrightarrow X_1$	–
	$A_2\,{}^3\Pi_u$ $\{1_u$	24930	2.535	191.2	2.2	0.0657	0.6	0.86	–	–
	A_1 $\{0_u^+$	24158	2.511	189	0.8	0.067	0.5	0.84	–	–
	$a\,{}^3\Sigma_u^+$	16290	–	280	1.4	–	–	1.85	–	–
	$b\,{}^1\Sigma_g^+$	7957.2	2.193	355.07	1.095	0.0877	0.33	2.97	–	–
	$X_2\,{}^3\Sigma_g^-$ $\{1_g^+$	511.9	2.164	387.1	0.96	0.0901	0.28	3.53	–	–
	$\{1_g^-$		2.165	387.2	0.97	0.0900	0.28	3.53	–	–
	$X_1\,0_g^+$	0	2.166	385.37	0.98	0.0899	0.29	3.50	–	39.48

Table 10.5 (continued)

Molecule	Electronic term	Term energy T_e [cm^{-1}]	Equilibrium internuclear distance r_e [Å]	Vibrational frequency ω_e [cm^{-1}]	Anharmonic constant $\omega_e x_e$ [cm^{-1}]	Rotational constant B_e [cm^{-1}]	Rotation-vibration interaction constant α_e [10^{-3} cm^{-1}]	Force constant k_e [10^5 dyn/cm]	Observable electronic transition	Reduced mass μ_A [a.m.u.]
1	2	3	4	5	6	7	8	9	10	11
SeH	$X_2\ {}^2\Pi\ 1/2$	1815	1.47*	2400	–	7.8*	–	3.4	–	0.995
	$X_1\ {}^2\Pi\ 3/2$	0								
SeO	$B_2\ {}^3\Sigma^-\ 1$	34380	1.87	517*	–	0.342*	40	–	$B_2 \rightarrow X_2$	–
	$B_1\ {}^3\Sigma^-\ 0^+$	34278	1.91	522	4	0.333*	30	2.14	$B_1 \rightarrow X_1$	–
	$A_3\ {}^3\Pi_r\ 2$	16770		980*	–			–	$A_3 \rightarrow X_2$	–
	$A_2\ {}^3\Pi_r\ 1$	16460	1.6*	1000	7	0.47*	–	7.8	$A_2 \rightarrow X_2, X_1$	–
	$A_1\ {}^3\Pi_r\ 0$	16140		990	6			7.7	$A_1 \rightarrow X_2, X_1$	–
	$b\ {}^1\Sigma^+$	9723	1.665*	839	5.1	0.456*	3	5.52	$b \rightarrow X_2, X_1$	–
	$X_2\ {}^3\Sigma^-\ 1$	166	1.634	915.4	4.5	0.474	3.4	6.57	–	–
	$X_1\ {}^3\Sigma^-\ 0^+$	0	1.648	914.7	4.5	0.465	3.2	6.56	–	13.30
SeS	$B_2\ {}^3\Sigma^-\ 1$	28300	2.35*	327.6*	–	0.135*	4	–	$B_2 \leftarrow X_2$	–
	$B_1\ {}^3\Sigma^-\ 0^+$	28248	2.33	330.8*	3	0.137	1.2	1.52	$B_1 \leftarrow X_1$	–
	$A\ 0^+$	27328	2.50	332	2.7	0.119	1.0	1.48	$A \leftarrow X_1$	–
	$X_2\ {}^3\Sigma^-\ 1$	205	2.026	556.3	1.83	0.1812	0.9	4.16	–	–
	$X_1\ {}^3\Sigma^-\ 0^+$	0	2.037	555.6	1.85	0.1793	0.8	4.15	–	22.80
Si_2	$O\ {}^3\Sigma_u^-$	53395.6	2.33	404	3	0.222	3	1.35	$O \leftarrow X$	–
	$N\ {}^3\Sigma_u^-$	46789.1	2.34	459	5	0.219	2	1.74	$N \leftarrow X$	–
	$K\ {}^3\Sigma_u^-$	30794	2.35	463	6.0	0.219	3.2	1.77	$K \leftarrow X$	–
	$H\ {}^3\Sigma_u^-$	24429.1	2.65	275.3	2.0	0.171	1.3	0.63	$H \leftrightarrow X$	–
	$D\ {}^3\Pi_{u,i}$	34700*	2.16	548	2.4	0.260	1.5	2.48	$D \leftarrow X$	–
	$X\ {}^3\Sigma_g^-$	0	2.25	511.0	2.0	0.239	1.3	2.16	–	14.04

SiBr	$X^2\Pi_r$	423 0	2.26*	424	1	0.160*	–	2.20	–	20.78
SiCl	$C^2\Pi$	41177 41166	1.94	674	2.2	0.289*	1	4.20	$C \leftrightarrow X$	–
	$B'^2\Delta$	35631	2.035	511	6	0.2619	2.4	2.41	$B' \leftrightarrow B$	–
	$B^2\Sigma^+$	34109	1.97	707	4	0.278*	2	4.61	$B \leftrightarrow X$	–
	$A^2\Sigma$	23114	2.34	294.9	0.7	0.199	0.7	0.81	$A \rightarrow X$	–
	$X^2\Pi_r$	207 0	2.06	535.6	2.17	0.256	1.6	2.65	–	15.67
SiF	$D^2\Sigma^+$	47419	1.54	1003	5.6	0.62	5	6.72	$D \rightarrow X$	–
	$D'^2\Pi$	46612	1.534	1033	5.3	0.633	4	7.12	$D' \rightarrow B, A, X$	–
	$C'^2\Pi$	41965	1.529	1032	4.4	0.638	4	7.11	$C' \rightarrow A, X$	–
	$C^2\Delta$	39438	1.571	878.4*	6	0.6034	5.4	5.29	$C \rightarrow X$	–
	$B^2\Sigma^+$	34561	1.541	1011.2	4.82	0.6271	4.6	6.83	$B \rightarrow A, X$	–
	$a^4\Sigma^-$	29805.1	1.605	863.2	5.4	0.5786	5.0	4.97	$a \rightarrow X$	–
	$A^2\Sigma^+$	22858	1.605	718.6	10.17	0.5784	9.4	3.45	$A \leftrightarrow X$	–
	$X^2\Pi_r$	0	1.601	857.2	4.73	0.5812	4.9	4.91	–	11.33
SiH	$A^2\Delta$	24300	1.523	1858.9	99.2	7.466	340	1.98	$A \leftrightarrow X$	–
	$X^2\Pi_r$	0	1.520	2041.8	35.5	7.500	220	2.39	–	0.973
SiI	$X_1{}^2\Pi_{1/2}$	0	2.4	364	1.2	0.12*	–	1.79	–	23.00
SiN	$B^2\Sigma^+$	24299.2	1.580	1031.0	16.8	0.724	10.5	5.85	$B \leftrightarrow X$	–
	$X^2\Sigma^+$	0	1.572	1151.4	6.5	0.731	5.6	7.30	–	9.346
SiO	$f^3\Pi_i$	59283 59261 59237	1.68	488	3	0.59	1.4	1.43	$f \rightarrow b$	–
	$c^3\Sigma^+$	57551	1.556	949.1	17.3	0.684	8	5.41	$c \rightarrow b$	–
	$E^1\Sigma^+$	52861	1.740	675.5	4.20	0.5473	5.5	2.74	$E \leftrightarrow X$	–
	$A^1\Pi$	42835	1.621	853	6.4	0.631	6.6	4.37	$A \leftrightarrow X$	–
	$D^1\Delta$	38820	1.73	730	3.9	0.554	5.2	3.2	–	–
	$C^1\Sigma^-$	38620	1.73	740	4.3	0.555	5.2	3.3	–	–
	$e^3\Sigma^-$	38310	1.73	748	4.2	0.556	5.1	3.4	–	–
	$d^3\Delta_r$	36490	1.71	770	4.1	0.56	5.2	3.5	–	–

Table 10.5 (continued)

Molecule	Electronic term		Term energy T_e [cm^{-1}]	Equilibrium internuclear distance r_e [Å]	Vibrational frequency ω_e [cm^{-1}]	Anharmonic constant $\omega_e x_e$ [cm^{-1}]	Rotational constant B_e [cm^{-1}]	Rotation-vibration interaction constant α_e [10^{-3} cm^{-1}]	Force constant k_e [10^5 dyn/cm]	Observable electronic transition	Reduced mass μ_A [a.m.u.]
1	2		3	4	5	6	7	8	9	10	11
	$b\,^3\Pi_r$	2	34018				0.689*				
		1	33947	1.562	1014	7.6	0.676*	4.4	6.17	$b \to X$	–
		0	33875				0.664*				
	$a\,^3\Sigma^+$		33630	1.7	790	4	0.6	5	3.7	$a \to X$	–
	$X\,^1\Sigma^+$		0	1.510	1241.56	6.0	0.7268	5.04	9.26	–	10.19
SiS	$E\,^1\Sigma^+$		41916	2.259	406	1.6	0.2214	1.4	1.45	$E \leftrightarrow X$	–
	$C\,^1\Delta$		37270	2.23	440	4.0	0.227	3	1.71	–	–
	$D\,^1\Pi$		35026.9	2.059	513.1	2.9	0.2665	2.2	2.32	$D \leftrightarrow X$	–
	$X\,^1\Sigma^+$		0	1.929	749.6	2.58	0.3035	1.47	4.96	–	14.97
SiSe	$X\,^1\Sigma^+$		0	2.058	580	1.8	0.1920	0.78	4.11	–	20.72
SnCl	$B\,^2\Sigma^+$		33583	2.26	432	1.2	0.122	0.6	3.00	$B \leftrightarrow X$	–
	$A'\,^2\Sigma^+$		19418	2.62*	232	0.7	0.091*	–	0.87	$A' \to X_1$	–
	$X_2\,^2\Pi$	3/2	2357	2.36	354	1.0	0.112	0.4	2.02	–	–
	X_1	1/2	0	2.36	351	1.1	0.112	0.4	1.98	–	27.30
SnF	$C\,^2\Delta$		40830 40760	1.90*	600*	5	0.286*	–	3.60	$C \leftrightarrow X$	–
	$B\,^2\Sigma^+$		34109	1.89*	678	2.7	0.290*	–	4.43	$B \to A$; $B \leftrightarrow X$	–
	$A\,^2\Sigma^+$		20137	2.04	420	2.2	0.247	3	1.70	$A \to X$	–
	$X\,^2\Pi$	3/2	2317	1.94	588	2.8	0.274	1.4	3.34	–	–
		1/2	0	1.94	583	2.7	0.273	1.4	3.28	–	16.38

SnO	$E\,^1\Sigma^+$	36297.5	2.076	505.0	2.7	0.2770	1.86	2.12	$E \leftarrow X$	–
	$D\,^1\Pi$	29620	1.949	580	3.1	0.3146	2.5	2.79	$D \leftrightarrow X$	–
	$B\,1$	24890	1.99*	560	–	$B_1 = 0.301$	–	–	$B \leftarrow X$	–
	$A\,0^+$	24330	2.01*	550*	–	0.296*	–	–	$A \leftarrow X$	–
	$X\,^1\Sigma^+$	0	1.832	822.1	3.72	0.3557	2.14	5.62	–	14.10
SnS	$D\,^1\Pi$	28336.6	2.357	331.3	1.26	0.1202	0.7	1.63	$D \leftrightarrow X$	–
	$X\,^1\Sigma^+$	0	2.209	487.3	1.36	0.1369	0.506	3.53	–	25.24
SnSe	$X\,^1\Sigma^+$	0	2.326	331	0.74	0.0650	0.17	3.06	–	47.42
SnTe	$X\,^1\Sigma^+$	0	2.523	259	0.5	0.0425	0.10	2.44	–	61.49
Sr_2	$A\,^1\Sigma_u^+$	17300	4.0	40	–	0.024	–	0.04	–	–
	$X\,^1\Sigma_g^+$	0	4.5	83	–	0.019	–	0.18	–	43.81
SrCl	$B\,^2\Sigma^+$	15722.8	2.556	307.1	1.02	0.1031	0.47	1.39	$B \leftrightarrow X$	–
	$X\,^2\Sigma^+$	0	2.576	302.4	0.95	0.1016	0.45	1.35	–	25.02
SrF	$F\,^2\Sigma^+$	32824	2.000*	598	3.4	0.2697*	1.9	3.30	$F \leftrightarrow X$	–
	$B\,^2\Sigma^+$	17267.4	2.080	496	2.3	0.2494	1.59	2.26	$B \leftrightarrow X$	–
	$X\,^2\Sigma^+$	0	2.075	502	2.3	0.2505	1.55	2.32	–	15.61
SrH	$F\,^2\Sigma^+$	34096	2.056*	1337*	30	4.002*	90	1.16	$F \leftarrow X$	–
	$C\,^2\Sigma^+$	26230	2.05	1350	23	4.01	130	1.07	$C \leftrightarrow X$	–
	$D\,^2\Sigma^+$	20848	2.96	1014	15	1.920	24	0.60	$D \leftrightarrow X$	–
	$B\,^2\Sigma^+$	14340	2.088	1193*	20	3.879	93	0.89	$B \rightarrow X$	–
	$X\,^2\Sigma^+$	0	2.146	1206	17	3.675	81	0.85	–	0.996
SrO	$C\,^1\Sigma^+$	28633	2.13	480	3	0.274	2	1.84	$C \rightarrow X$	–
	$B\,^1\Pi$	24701	2.06	520.0	3.2	0.294	1.5	2.15	$B \rightarrow X$	–
	$A\,^1\Sigma^+$	10886.6	2.022	619.6	0.9	0.3047	1.1	3.06	$A \rightarrow X$	–
	$A'\,^1\Pi$	9400	2.20	473	2.1	0.256	2	1.78	$A' \rightarrow X$	–
	$a\,^3\Pi_i$	9150	2.20	464	1.6	0.258	2	1.71	–	–
	$X\,^1\Sigma^+$	0	1.920	653.5	4.0	0.3380	2.2	3.40	–	13.53
SrS	$B\,^1\Sigma^+$	39332	2.609	286.8	0.8	0.1057	0.32	1.14	$B \leftarrow X$	–
	$X\,^1\Sigma^+$	0	2.440	388.4	1.3	0.1207	0.44	2.09	–	23.47
TaO	$Q'\,^2\Delta\,(5/2)$	27353	1.733	904	4	0.3818	2.2	7.1	$Q' \rightarrow X_2$	–
	$P\,^2\Delta\,(3/2)$	26736.2	1.743	902.7	4.1	0.3775	1.8	7.06	$P \leftrightarrow X_1$	–
	$M\,^2\phi\,(5/2)$	24124	1.744	899	4	0.3771	1.8	7.0	$M \leftrightarrow X_1$	–
	$L\,^2\Pi\,(1/2)$	23408	1.743	896	4	0.3774	2.0	7.0	$L \leftrightarrow X_1$	–

Table 10.5 (continued)

Molecule	Electronic term	Term energy T_e [cm^{-1}]	Equilibrium internuclear distance r_e [Å]	Vibrational frequency ω_e [cm^{-1}]	Anharmonic constant $\omega_e x_e$ [cm^{-1}]	Rotational constant B_e [cm^{-1}]	Rotation-vibration interaction constant α_e [10^{-3} cm^{-1}]	Force constant k_e [10^5 dyn/cm]	Observable electronic transition	Reduced mass μ_A [a.m.u.]
1	2	3	4	5	6	7	8	9	10	11
	$K'\,^2\phi$ (7/2)	22981.6	1.736	903.1	3.6	0.3808	1.9	7.06	$K' \rightarrow X_2$	–
	$E\,^2\phi$ (5/2)	15930	1.723*	935	5	0.3862*	–	7.6	$E \leftrightarrow X_1$	–
	$C\,^2\Delta$ (3/2)	13610	1.720*	940	–	0.3875*	–	7.7	$C \leftrightarrow X_1$	–
	$B\,^2\phi$ (5/2)	12900	1.722*	930	–	0.3869*	–	7.5	$B \leftrightarrow X_1$	–
	$A''\,^2\Delta$ (3/2)	10910	1.721*	930	–	0.3873*	–	7.5	$A'' \rightarrow X_1$	–
	$X_2\,^2\Delta$ (5/2)	3504.4	1.686	1030.8	3.6	0.4036	1.9	9.20	–	–
	$X_1\,^2\Delta$ (3/2)	0	1.687	1028.7	3.5	0.4028	1.8	9.17	–	14.70
Te_2	$B_1\,^3\Sigma_u^-\,(0_u^+)$	22207	2.824	162.3	0.45	0.0325	0.12	1.01	$B_1 \rightarrow X_2$; $B_1 \leftrightarrow X_1$	–
	$A\,0_u^+$	19451	2.882	143.6	0.454	0.0312	0.13	0.79	$A \leftrightarrow X_1$	–
	$b\,^1\Sigma_g^+$	9600.0	2.587	229.0	0.602	0.0388	0.107	2.01	$b \rightarrow X_2$, X_1	–
	$X_2\,^3\Sigma_g^-$ 1_g	1974.8	2.552	250.03	0.513	0.03984	0.097	2.39	–	–
	$X_1\,^3\Sigma_g^-$ 0_g^+	0	2.558	247.07	0.522	0.03967	0.099	2.34	–	63.80
TeO	$A_1\,0^+$	28210	2.07	445.0*	18	0.276	5	1.94	$A_1 \leftrightarrow X_1$	–
	$X_2\,1$	680	1.822	798.1	4.0	0.356	2.4	5.34	–	–
	$X_1\,0^+$	0	1.825	797.1	4.0	0.355	2.4	5.33	–	14.22
TeS	$B\,0^+$	24530	2.53	250	3.4	0.103	1.2	0.95	$B \leftarrow X_1$	–
	$A\,0^+$	23550	2.59	204.2	1.02	0.098	0.7	0.63	$A \leftarrow X_1$	–
	$X_1\,0^+$	0	2.230	471.2	1.6	0.1322	0.5	3.35	–	25.62
TeSe	$X_2\,1$	1547	2.37	317.4	0.72	0.062	–	2.87	–	–
	$X_1\,0^+$	0	2.37	316.2	0.74	0.062	0.2	2.85	–	48.78
ThO	$X\,^1\Sigma$	0	1.840	895.8	2.4	0.3326	1.3	7.07	–	14.96

TiO	$e\,^1\Sigma^+$		30040	1.695	854	4	0.489	2	5.15	$e \leftrightarrow d$	–
	$f\,^1\Delta$		a + 19140.57	1.670	874.10	2.50	0.5038	3.1	5.40	$f \leftrightarrow a$	–
	$c\,^1\phi$		21330	1.639	918	4.2	0.523	3.1	5.95	$c \leftrightarrow a$	–
			19617								
	$C\,^3\Delta_r$		19525	1.694	838.3	4.8	0.4900	3.1	4.96	$C \rightarrow a$; $C \leftrightarrow X$	–
			19427								
			16331								
	$B\,^3\Pi_r$		16315	1.666*	875	5	0.5062*	–	5.4	$B \leftrightarrow X$	–
			16293								
	$b\,^1\Pi$		14760	1.655	919	3.7	0.5134	2.9	6.0	$b \leftrightarrow d, a$; $b \rightarrow X$	–
			14431								
	$A\,^3\phi_r$		14263	1.664	867.8	3.94	0.5074	3.1	5.32	$A \leftrightarrow X$	–
			14090								
	$d\,^1\Sigma^+$		5660	1.600	1024	4.6	0.5492	3.4	7.4	–	–
	$a\,^1\Delta$		3440	1.617	1018.27	4.52	0.5376	2.9	7.33	–	–
			197								
	$X\,^3\Delta_r$		96	1.620	1009.0	4.50	0.5354	3.0	7.19	–	11.99
			0								
TiS		3	11810		484.1		0.1890	1.0	2.65		–
	$C\,^3\Delta$	2	11720	2.161	484.3	2.5	0.1882	1.0	2.65	$C \leftrightarrow X$	–
		1	11624		–		0.1868*	–	–		–
		3	185		558.2*		0.2034			–	–
	$X\,^3\Delta$	2	90	2.083	558.3*	1.9	0.2027	0.9	3.6	–	–
		1	0		558.4*		0.2018			–	19.20
TlBr	$X\,^1\Sigma^+$		0	2.618	192.1	0.4	0.0424	0.13	1.25	–	57.45
TlCl	$A\,^3\Pi_{0^+}$		31049	2.473	223	11	0.0923	1.3	0.89	$A \leftrightarrow X$	–
	$X\,^1\Sigma^+$		0	2.485	283.7	0.82	0.0914	0.40	1.43	–	30.21
TlF	$B\,^3\Pi_1$		36863.1	2.076	366.6	10.2	0.225	3.1	1.38	$B \leftrightarrow X$	–
	$A\,^3\Pi_{0^+}$		35186.0	2.049	436	7	0.231	2.7	1.95	$A \leftrightarrow X$	–
	$X\,^1\Sigma^+$		0	2.084	477	2	0.22315	1.50	2.33	–	17.38
TlH	$C\,^1\Pi\,(1)$		23556.2*	2.88*	98*	20	2.03*	0.2	0.01	$C \leftarrow X$	–
	$A\,^3\Pi_{0^+}\,(0^+)$		17520*	1.91*	760*	140	4.62*	0.7	0.64	$A \leftrightarrow X$	–
	$X\,^1\Sigma^+$		0	1.87	1391	23	4.81	0.15	1.14	–	1.003

Table 10.5 (continued)

Molecule	Electronic term	Term energy T_e [cm^{-1}]	Equilibrium internuclear distance r_e [Å]	Vibrational frequency ω_e [cm^{-1}]	Anharmonic constant $\omega_e x_e$ [cm^{-1}]	Rotational constant B_e [cm^{-1}]	Rotation-vibration interaction constant α_e [10^{-3} cm^{-1}]	Force constant k_e [10^5 dyn/cm]	Observable electronic transition	Reduced mass μ_A [a.m.u.]
1	2	3	4	5	6	7	8	9	10	11
TlI	$X^1\Sigma^+$	0	2.814	143*	–	0.0272	0.07	1.0	–	78.29
VO	$C^4\Sigma^-$	17494	1.672	863	5.3	0.495	3	5.35	$C \leftrightarrow X$	–
	$B^4\Pi_{5/2}$	12761	1.63	911	5	0.525	4	5.95	$B \leftrightarrow X$	–
	$X^4\Sigma^-$	0	1.589	1011	4.9	0.5482	3.5	7.34	–	12.18
Xe_2	$X^1\Sigma_g^+$	0	4.36	21.1	0.6	0.014	–	0.017	–	65.65
XeCl	$B^2\Sigma^+$	32405	2.9	195	0.5	0.071	–	0.62	$B \leftrightarrow X$	–
	$X^2\Sigma^+$	0	3.2	26	–0.3	0.060	–	0.011	–	27.91
YCl	$C^1\Sigma$	14908	2.48	324	1.1	0.109	0.7	1.6	$C \rightarrow X$	–
	$X^1\Sigma$	0	2.41	381	1	0.116	0.3	2.1	–	25.35
YF	$G^1\Pi$	31254	1.973	541	2.1	0.2766	2.3	2.7	$G \leftrightarrow X$	–
	$F^1\Sigma^+$	28022	1.978*	553	2.7	0.2754*	–	2.82	$F \leftrightarrow X$	–
	$C^1\Sigma^+$	19242	2.010	532	2.4	0.2667	1.8	2.6	$C \leftrightarrow X$	–
	$B^1\Pi$	15934	2.008	539.4	2.3	0.2671	1.6	2.68	$B \leftrightarrow X$	–
	$X^1\Sigma^+$	0	1.926	636.3	2.5	0.2904	1.6	3.73	–	15.65
YO	$B^2\Sigma^+$	20793.3	1.825	758.7	4.0	0.3731	2.5	4.60	$B \leftrightarrow X$	–
	$A^2\Pi_r$	16529.7	1.793	820.34	3.38	0.3868	2.03	5.38	$A \leftrightarrow X$	–
	$X^2\Sigma^+$	0	1.788	861.46	2.87	0.3889	1.72	5.93	–	13.56
YbF	$B^2\Sigma^+$ 1/2	21074.2*	1.99*	512*	–	0.249*	–	–	$B \leftrightarrow X$	–
	A_2 $^2\Pi$ 3/2	19471.0*	1.99*	540*	–	0.249*	–	–	$A_2 \leftrightarrow X$	–
	A_1 $^2\Pi$ 1/2	18106.3*	2.00*	474*	–	0.247*	–	–	$A_1 \leftrightarrow X$	–
	$X^2\Sigma^+$	0	2.016	501.9*	2.2	0.2414*	1	2.6	–	17.12

YbH	$D\,^2\Pi$	3/2	16780	2.011	1360	19	4.161	90	1.1	$D \leftrightarrow X$	–
		1/2	15326	2.003	1350	21	4.195	90	1.1		
	$X\,^2\Sigma^+$		0	2.053	1249.5	21.1	3.993	96	0.92	–	1.002
ZnH	$C\,^2\Sigma^+$		41100	1.53*	1820	50	7.2*	–	1.94	$C \leftarrow X$	–
	$B\,^2\Sigma^+$		27588	2.27*	1021	16	3.29*	–	0.61	$B \rightarrow X$	–
	$A\,^2\Pi_r$		23277	1.512	1910	41	7.433	240	2.13	$A \rightarrow X$	–
	$X\,^2\Sigma^+$		0	1.595	1608	55	6.679	250	1.51	–	0.993
ZrO	$E\,^1\Sigma^+$		27212	1.773	843.3	3.0	0.395	2	5.69	$E \leftrightarrow X$	–
	$D\,^1\Delta$		25226	1.765	841	2.6	0.399	2	5.65	$D \rightarrow A$	–
	$d\,^3\Delta$		23414				0.395*				
			22993	1.776	821	3.3	0.393*	2	5.39	$d \leftrightarrow a$	–
			22693				0.390*				
	$c\,^3\Pi_r$		19237				0.406*				
			19178	1.76	845	3.6	0.403*	2	5.72	$c \leftrightarrow a$	–
			19140				0.396*				
	$b\,^3\phi_r$		17800				0.4044*				
			17169	1.751	854	3.1	0.4037*	2.0	5.83	$b \leftrightarrow a$	–
			16567				0.4031*				
	$B\,^1\Pi$		15383.4*	1.758*	860	3	0.4015*	–	5.9	$B \rightarrow A, a;\ B \leftrightarrow X$	–
	$A\,^1\Delta$		5904.2	1.726*	938	1.8	0.417*	1	7.04	–	–
	$a\,^3\Delta_r$		1725				0.4157*	1.7			
			1387.0	1.728	936	3.5	0.4147*	1.9	7.02	–	–
			1099.1				0.4133*	1.8			
	$X\,^1\Sigma^+$		0	1.712	969.8*	4.9	0.4226*	2	7.68	–	13.61

10.4 Potential Energy Curves

Figures 10.2–14 present potential energy curves and show the electron term energies for a number of diatomic molecules and molecular ions. In some figures the estimated uncertainties of particular potential energy curves are labelled. In selecting these curves, we made use of the data supplied by reference material [10.4.1–4], as well as some informative publications [10.4.5–13].

References

10.4.1 B.Rosen (ed.): *Spectroscopic Data Relative to Diatomic Molecules*, International Tables of Selected Constants **17** (Pergamon, Oxford 1970)

10.4.2 S.N.Suchard (ed.): *Spectroscopic Data. I. Heteronuclear Diatomic Molecules*, Parts A, B (IFI/Plenum, New York 1975)

10.4.3 S.N.Suchard, J.E.Melzer (eds.): *Spectroscopic Data. II. Homonuclear Diatomic Molecules* (IFI/Plenum, New York 1976)

10.4.4 K.P.Huber, G.Herzberg: *Molecular Spectra and Molecular Structure. IV. Constants of Diatomic Molecules* (Van Nostrand Reinhold, New York 1979) (see references therein on potential functions)

10.4.5 T.E.Sharp: At. Data **2**, 119 (1971) (H_2, H_2^+)

10.4.6 M.L.Ginter, R.Battino: J. Chem. Phys. **52**, 4469 (1970) (He_2, He_2^+)

10.4.7 G.Herzberg, A.Lagerquist, C.Malmberg: Can. J. Phys. **47**, 2735 (1969) (C_2)

10.4.8 P.H.Krupenie: "The Band Spectrum of Carbon Monoxide", Nat. Stand. Ref. Data Ser. Nat. Bur. Stand. **5** (1966) (CO)

10.4.9 A.Lofthus, P.H.Krupenie: J. Phys. Chem. Ref. Data **6**, 113 (1977) (N_2, N_2^+)

10.4.10 G.Krishnamurty, N.A.Narasimhan: J. Mol. Spectrosc. **29**, 410 (1969) (NH)

10.4.11 F.R.Gilmore: J. Quant. Spectrosc. Radiat. Transfer **5**, 369 (1965) (N_2, NO, O_2 and their ions)

10.4.12 P.H.Krupenie: J. Phys. Chem. Ref. Data **1**, 423 (1972) (O_2, O_2^+, O_2^-)

10.4.13 T.V.R.Rao, R.R.Reddy, P.S.Rao: Physica **106 C**, 445 (1981) (PO)

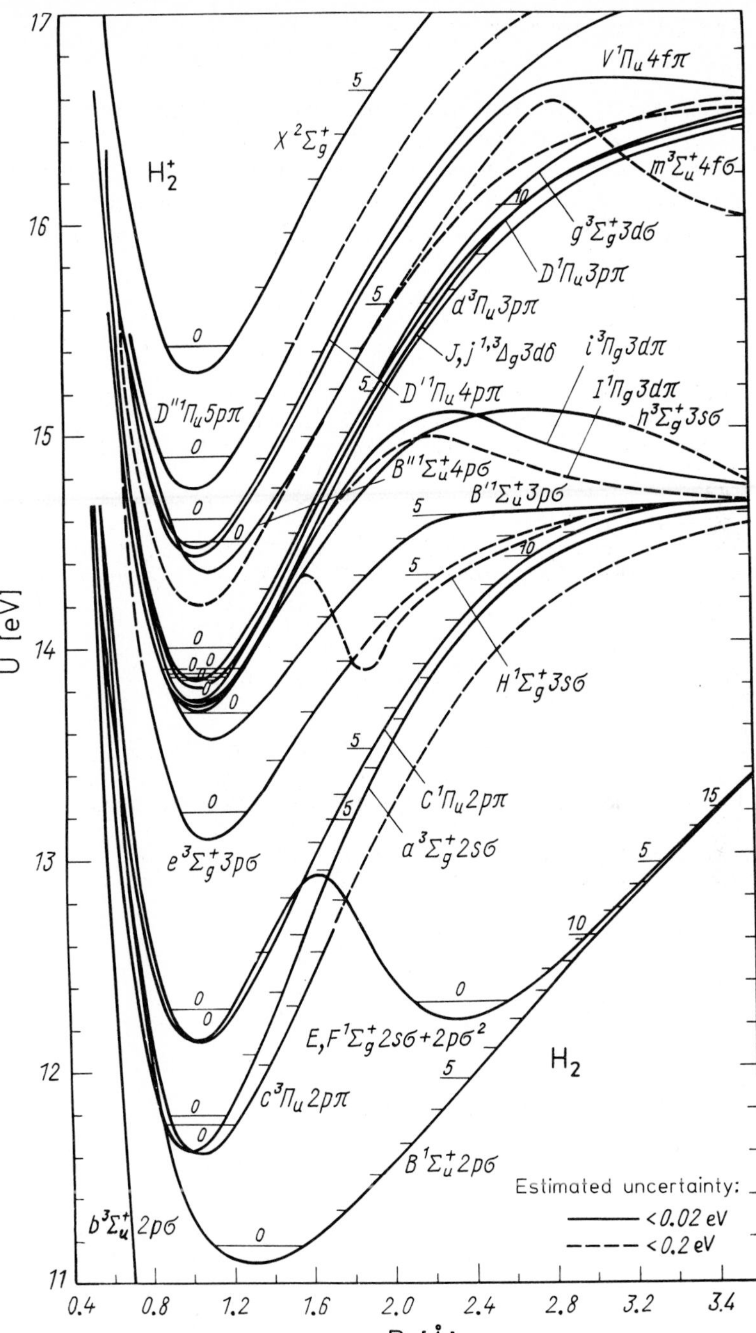

Fig. 10.2. Theoretical adiabatic terms of H_2 and H_2^+

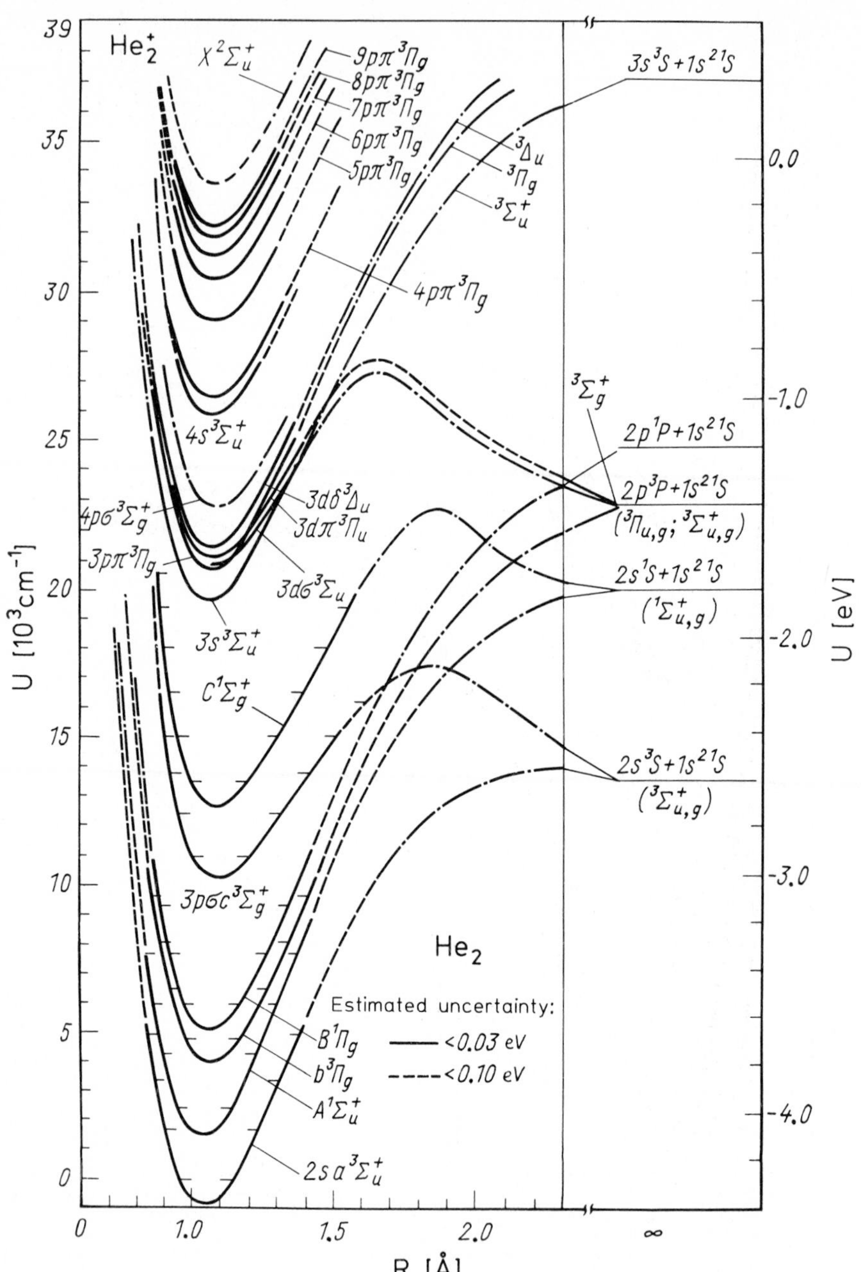

Fig. 10.3. Theoretical adiabatic terms of He_2 and He_2^+

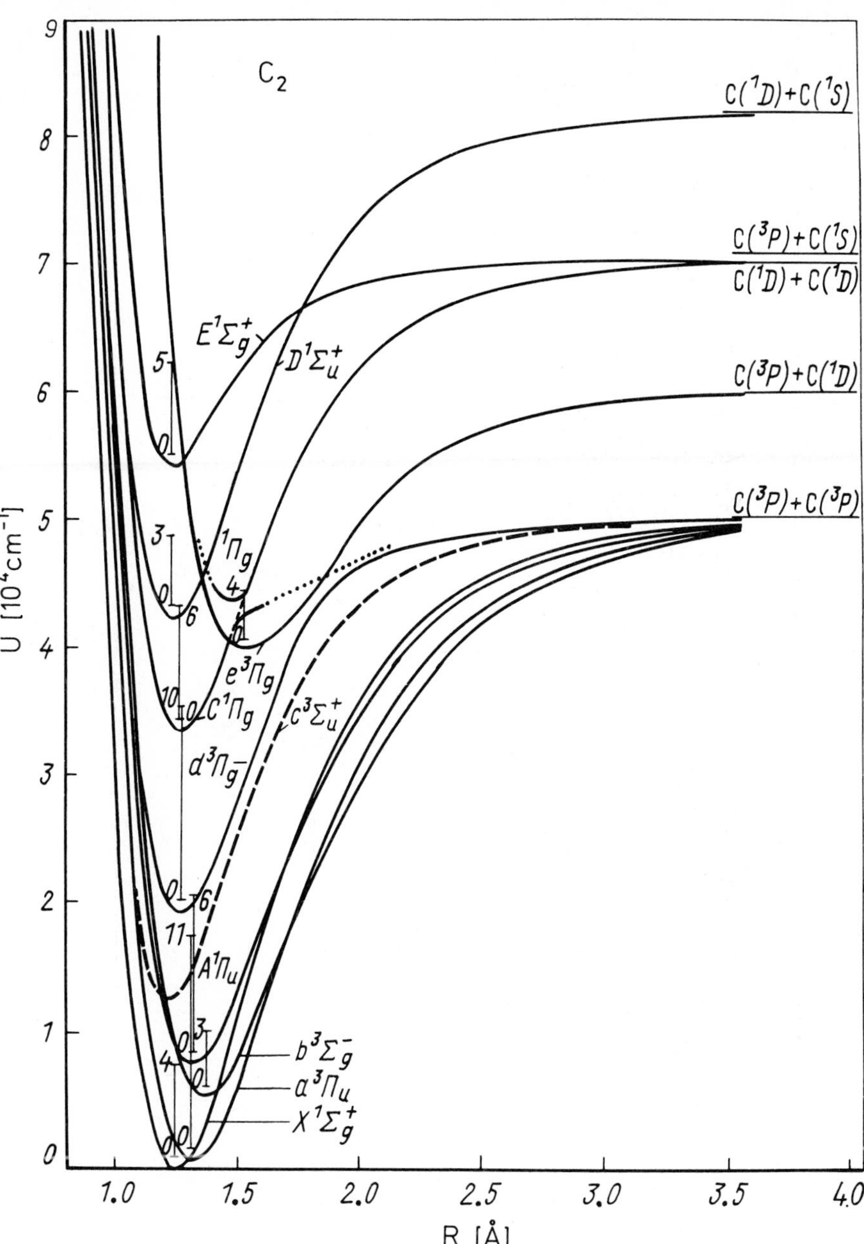

Fig. 10.4. Adiabatic terms of C_2 approximated by empirical Morse potential function

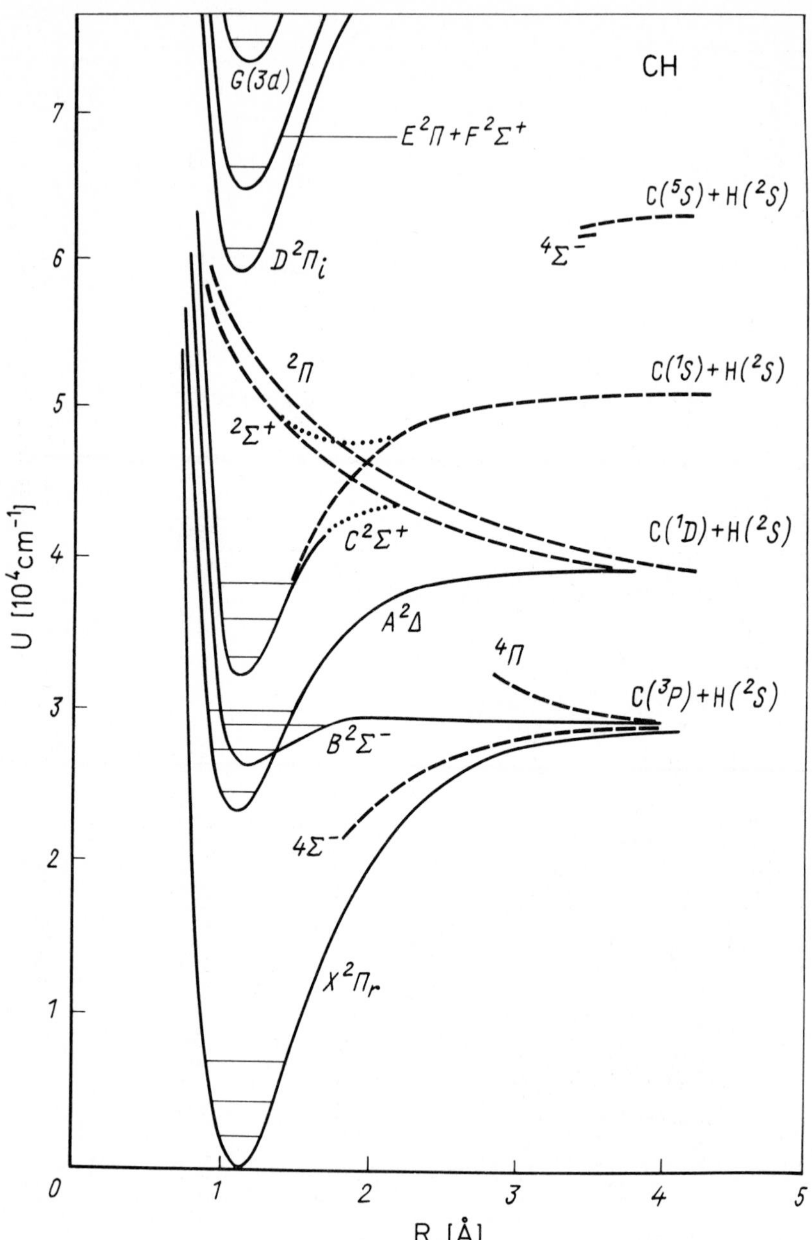

Fig. 10.5. Schematic diagram of CH adiabatic terms based on theoretical approach

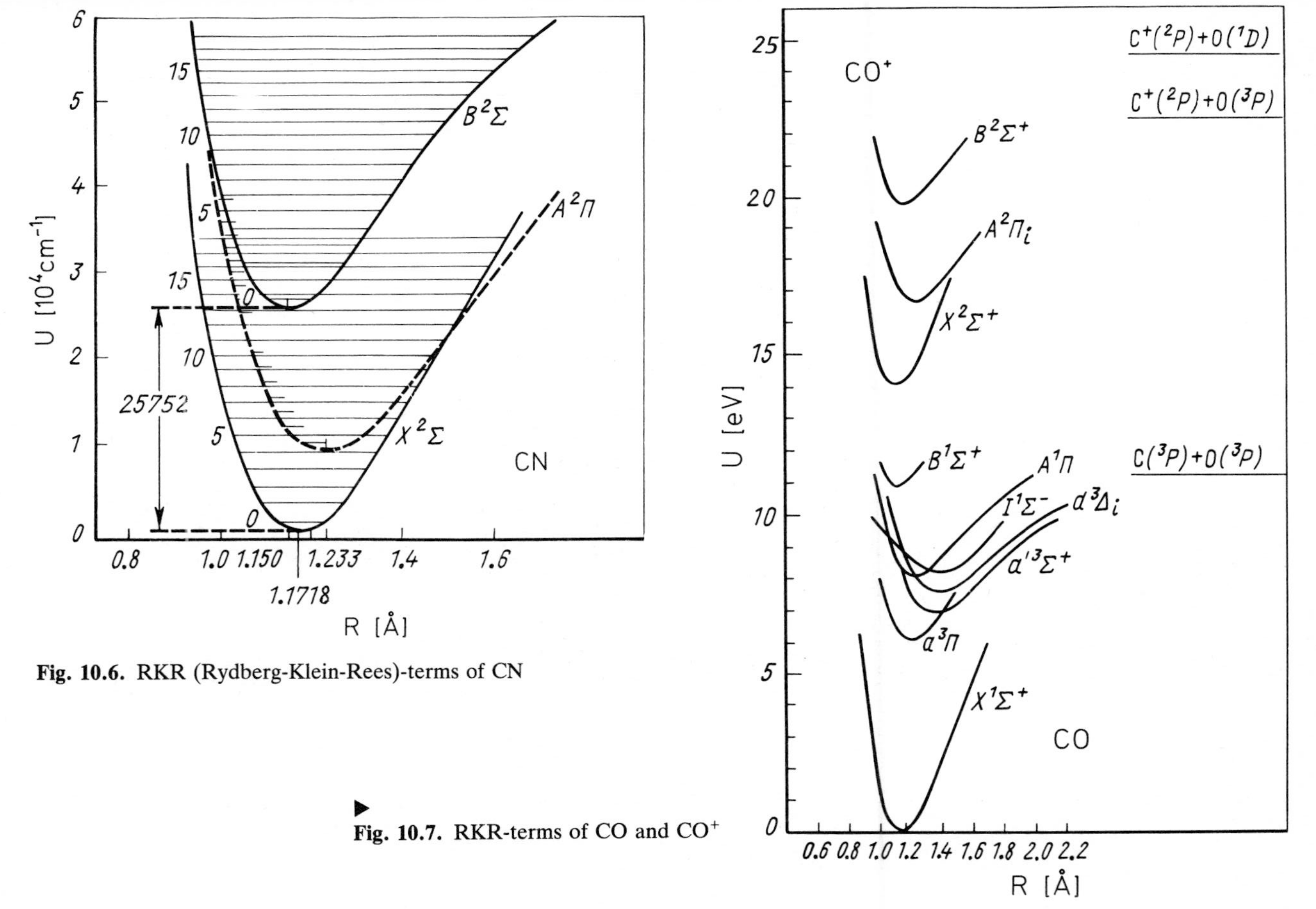

Fig. 10.6. RKR (Rydberg-Klein-Rees)-terms of CN

▶ **Fig. 10.7.** RKR-terms of CO and CO^+

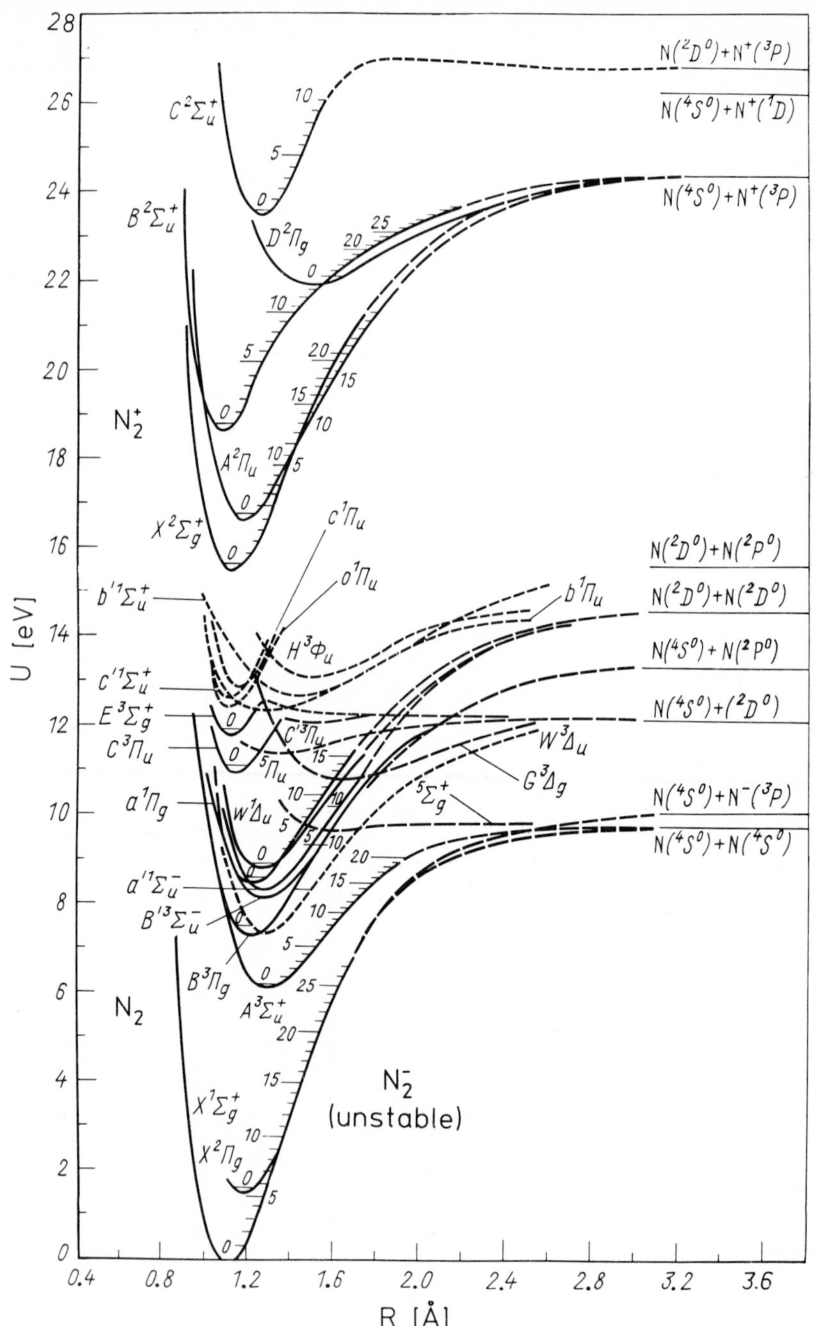

Fig. 10.8. Experimental potential energy curves for electronic states of N_2, N_2^+ and N_2^-

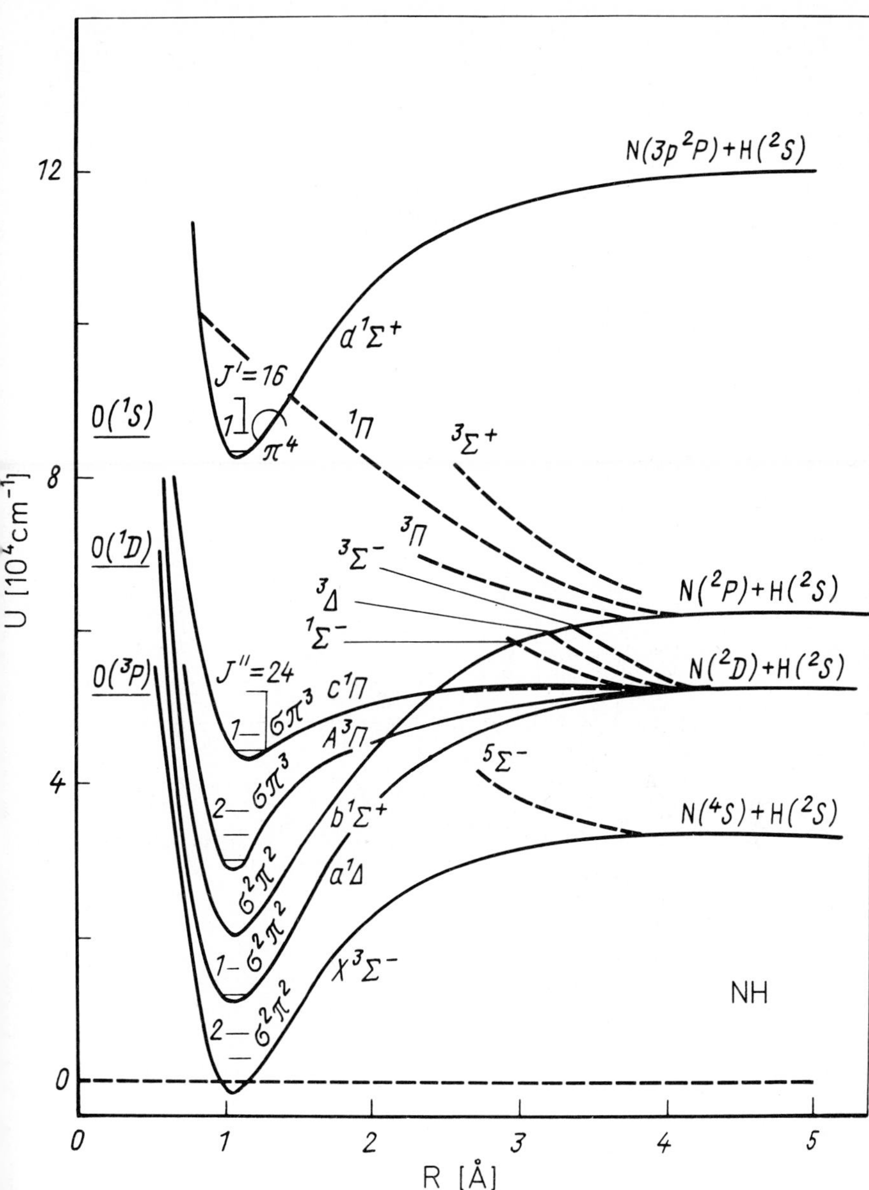

Fig. 10.9. Qualitative scheme of NH adiabatic terms

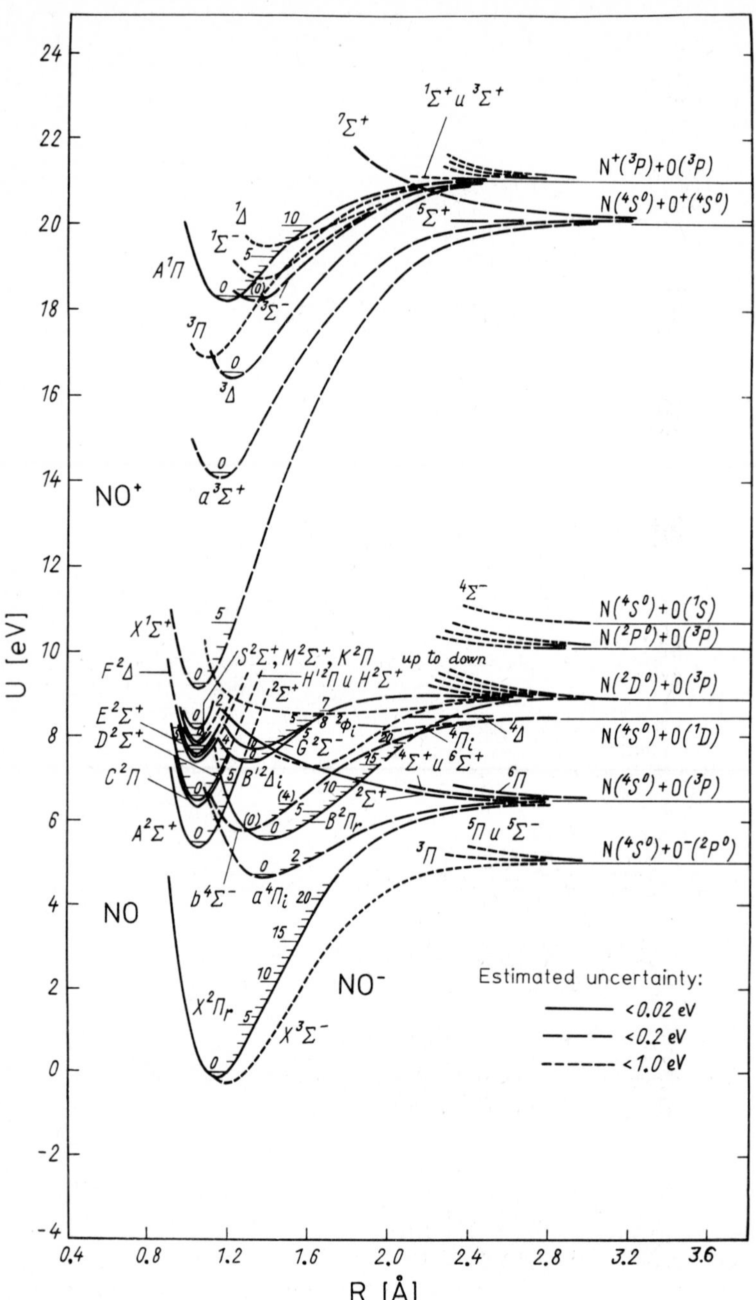

Fig. 10.10. Experimental potential energy curves for electronic states of NO, NO^+ and NO^-

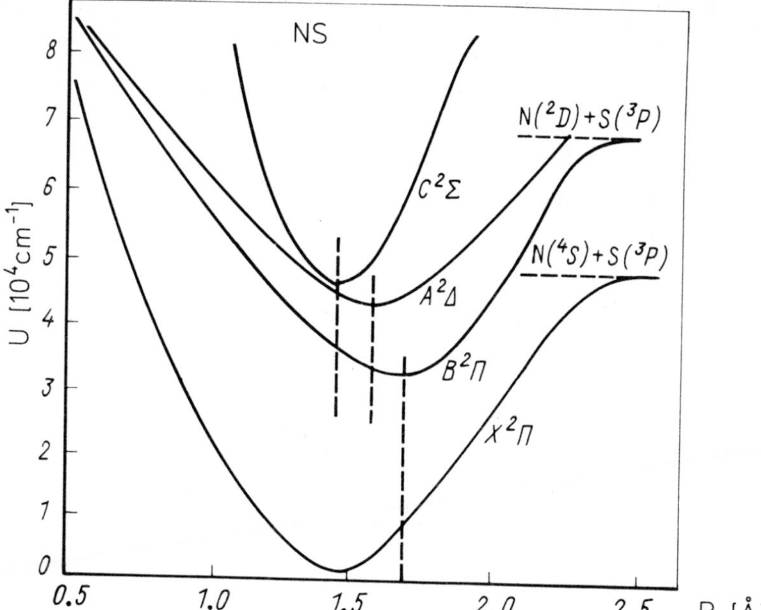

Fig. 10.11. Qualitative scheme of NS adiabatic terms

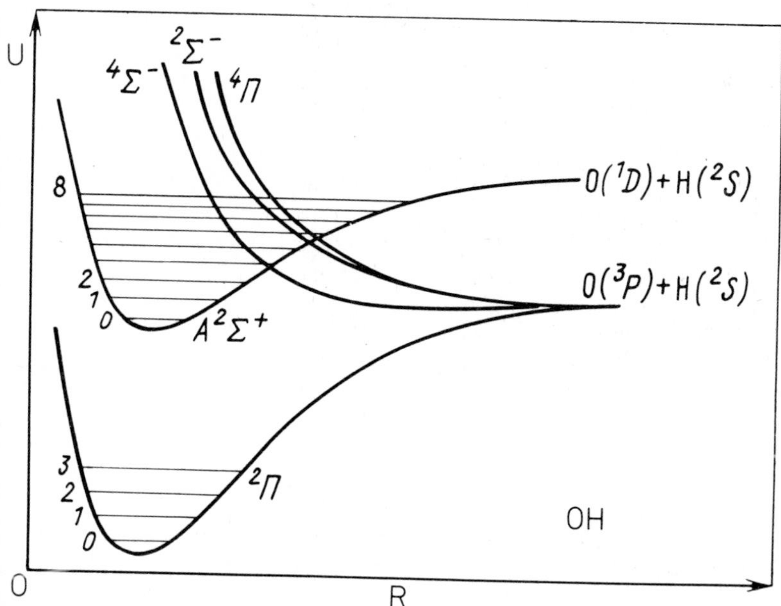

Fig. 10.12. Qualitative scheme of OH adiabatic terms

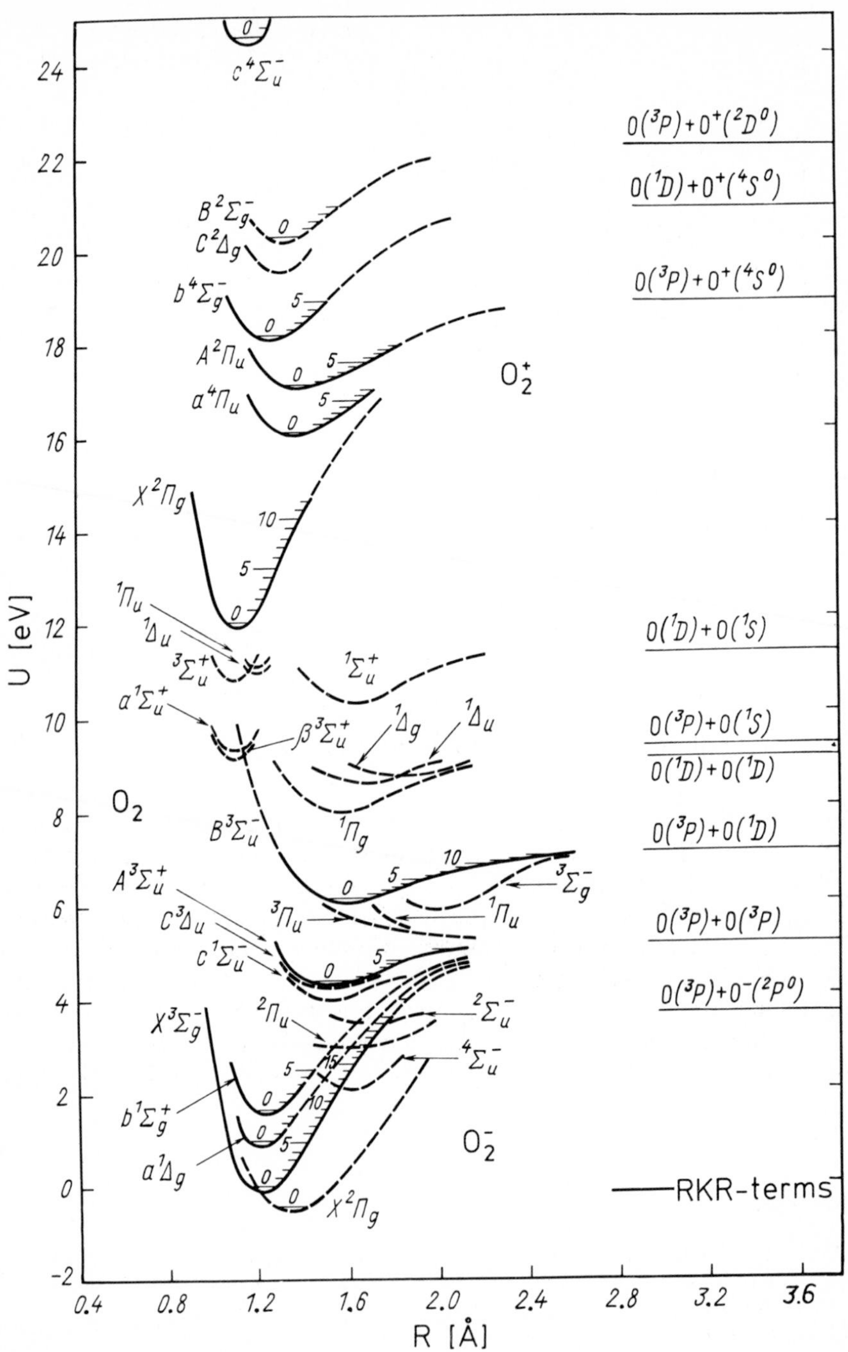

Fig. 10.13. Experimental potential energy curves for electronic states of O_2, O_2^+ and O_2^-

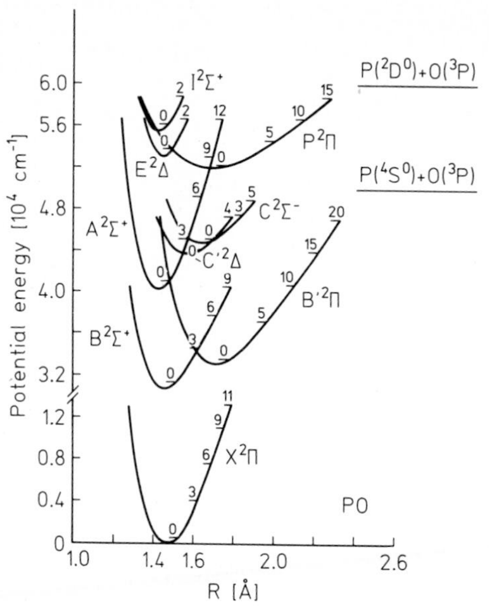

Fig. 10.14. Experimental potential energy curves for electronic states of PO molecule

10.5 Ionization Potentials of Diatomic Molecules

Table 10.6 incorporates the numerical values of the first ionization potentials (IP) for diatomic molecules, based on reference material from monographs [10.5.1–3] and some later publications. The energy values given apply to ground-state electronic terms of the neutral molecule and positive molecular ion; they also characterize the normal vibrational states. The basic techniques used for measuring molecular IPs are the spectroscopic method, and photoionization and electron impact experiments [10.5.1].

The values presented in Table 10.6 are grouped into accuracy classes (see the Introduction). The molecules are tabulated in alphabetical order.

References

10.5.1 L.V.Gurvich, G.V.Karachevtsev, V.N.Kondratjev, Y.A.Lebedev, V.A.Medvedev, V.K.Potapov, Y.S.Hodeev: *Bond Dissociation Energies, Ionization Potentials, and Electron Affinities*, 2nd ed. (Nauka, Moscow, 1974) (in Russian)

10.5.2 K.P.Huber, G.Herzberg: *Molecular Spectra and Molecular Structure. IV. Constants of Diatomic Molecules* (Van Nostrand Reinhold, New York 1979)

10.5.3 A.G.Gaydon: *Dissociation Energies and Spectra of Diatomic Molecules*, 3rd ed. (Chapman and Hall, London 1968)

Table 10.6. Ionization potentials (IP) of diatomic molecules

Molecule or radical	IP [eV]	Molecule or radical	IP [eV]
AgF	11.4 (*B*)	CsCl	8.3 (*B*)
AlCl	9.4 (*C*)	CsF	8.8 (*B*)
AlF	9.8 (*C*)	CsI	7.25 (*A*)
AlO	9.5 (*C*)	Cu_2	7.89 (*A*)
AlS	9.5 (*C*)	D_2	15.467 (*A*)
AlTe	9.0 (*C*)	DBr	11.67 (*A*)
Ar_2	14.5 (*A*)	DCl	12.76 (*A*)
ArKr	13.4 (*A*)	DF	16.06 (*A*)
ArXe	12.0 (*A*)	DI	10.39 (*A*)
As_2	12 (*C*)	DT	15.475 (*A*)
At_2	8.3 (*C*)	DyF	6.0 (*C*)
AuBr	9.2 (*C*)	ErF	6.3 (*C*)
AuSi	8.3 (*C*)	F_2	15.686 (*A*)
BC	10.5 (*C*)	FeO	8.7 (*B*)
BF	11.1 (*A*)	GaF	10.6 (*C*)
BH	9.8 (*B*)	GaO	9.4 (*C*)
BO	13.5 (*C*)	GaS	8.9 (*C*)
BSi	7.8 (*C*)	GaSe	8.8 (*C*)
BaBr	4.8 (*C*)	GaTe	8.4 (*C*)
BaCl	5.0 (*B*)	Ge_2	7.9 (*C*)
BaF	4.9 (*C*)	GeBr	7.3 (*C*)
BaI	4.7 (*C*)	GeC	10.3 (*C*)
BaO	6.9 (*C*)	GeCl	7.2 (*C*)
BeF	9.1 (*C*)	GeF	7.5 (*C*)
BeH	8.2 (*B*)	GeO	11.1 (*B*)
BeO	10.1 (*C*)	GeSi	8.2 (*C*)
Br_2	10.52 (*A*)	GeTe	10.1 (*C*)
BrCl	11.1 (*B*)	H_2	15.4259 (*A*)
BrF	11.8 (*B*)	HBr	11.67 (*A*)
BrO	10.3 (*A*)	HCl	12.75 (*A*)
C_2	12.15 (*A*)	HD	15.445 (*A*)
CF	9.20 (*B*)	HF	16.04 (*A*)
CH	10.64 (*A*)	HI	10.38 (*A*)
CN	14.2 (*A*)	HT	15.451 (*A*)
CO	14.014 (*A*)	He_2	22.22 (*A*)
CS	11.33 (*A*)	HfO	7.5 (*B*)
CaF	6.0 (*C*)	HoF	6.2 (*C*)
CaH	5.9 (*B*)	I_2	9.3 (*B*)
CaCl	6.0 (*B*)	IBr	9.85 (*A*)
CaO	6.5 (*C*)	ICl	10.08 (*A*)
CeO	4.9 (*C*)	IF	10.5 (*B*)
Cl_2	11.50 (*A*)	InBr	9.1 (*B*)
ClF	12.7 (*B*)	InCl	9.5 (*B*)
ClO	11.0 (*B*)	InF	9.6 (*C*)
CoO	9.0 (*C*)	InI	8.5 (*B*)
CrF	8.4 (*C*)	InS	7.0 (*C*)
CrO	8.4 (*C*)	InSe	7.1 (*C*)
Cs_2	3.64 (*B*)	InTe	7.6 (*C*)
CsBr	7.72 (*A*)	IrC	9.5 (*D*)

Table 10.6 (continued)

Molecule or radical	IP [eV]	Molecule or radical	IP [eV]
K_2	4.064 (*A*)	PdSi	8.4 (*C*)
KCl	8.4 (*B*)	PrO	4.9 (*C*)
KI	8.2 (*C*)	PtB	10 (*D*)
KLi	4.6 (*B*)	PtSi	7.9 (*C*)
KNa	4.416 (*A*)	Rb_2	3.45 (*B*)
Kr_2	12.87 (*A*)	RbBr	7.7 (*B*)
LaO	4.9 (*B*)	RbCl	8.3 (*B*)
Li_2	5.145 (*A*)	RbI	7.1 (*B*)
LiBr	10.0 (*C*)	RhO	9.3 (*C*)
LiCl	10.1 (*C*)	RuO	8.7 (*C*)
LiD	7.7 (*C*)	S_2	9.4 (*B*)
LiF	11.3 (*C*)	SH	10.4 (*A*)
LiH	7.85 (*B*)	SO	10.29 (*A*)
LiI	8.6 (*C*)	Sb_2	9.0 (*C*)
LiNa	5.0 (*B*)	Se_2	8.88 (*A*)
LiO	8.4 (*B*)	SeH	9.06 (*A*)
MgCl	7.5 (*B*)	Si_2	7.4 (*C*)
MgF	7.8 (*C*)	SiC	10.2 (*B*)
Mn_2	7.4 (*C*)	SiF	7.26 (*A*)
MnF	8.7 (*C*)	SiO	11.6 (*A*)
MoO	8.0 (*C*)	SnBr	7.4 (*C*)
N_2	15.581 (*A*)	SnCl	6.6 (*C*)
NF	12.3 (*B*)	SnF	7.4 (*C*)
NH	13.1 (*A*)	SnO	10.5 (*C*)
NO	9.264 (*A*)	SnS	9.7 (*C*)
NS	8.9 (*B*)	SnSe	9.7 (*C*)
Na_2	4.90 (*A*)	SnTe	9.1 (*C*)
NaBr	8.3 (*B*)	SrCl	5.6 (*B*)
NaCl	8.93 (*A*)	SrF	4.9 (*C*)
NaI	7.64 (*A*)	SrO	6.1 (*C*)
NaLi	5.0 (*B*)	T_2	15.487 (*A*)
NaK	4.416 (*A*)	TaO	6.0 (*C*)
NdO	5.0 (*D*)	Te_2	8.29 (*A*)
NiCl	11.4 (*B*)	TeO	8.7 (*B*)
NiO	9.5 (*C*)	Ti_2	6.3 (*C*)
O_2	12.071 (*A*)	TiO	6.4 (*C*)
OD	12.9 (*A*)	TiS	7.1 (*C*)
OH	12.9 (*A*)	TlBr	9.14 (*A*)
OT	12.9 (*A*)	TlCl	9.70 (*A*)
P_2	10.5 (*A*)	TlF	10.5 (*B*)
PC	10.5 (*C*)	TlI	8.47 (*A*)
PH	10.6 (*B*)	V_2	6.4 (*C*)
PO	8.2 (*B*)	VO	5.0 (*D*)
PbBr	7.8 (*C*)	WO	9.1 (*D*)
PbCl	7.5 (*C*)	UC	6.2 (*C*)
PbF	7.5 (*C*)	UN	7.0 (*C*)
PbO	9.0 (*C*)	UO	5.7 (*B*)
PbS	8.6 (*C*)	US	6.3 (*C*)
PbTe	8.2 (*C*)	Xe_2	11.13 (*A*)
Pd_2	7.7 (*C*)	ZrN	7.9 (*C*)
PdO	9.1 (*C*)	ZrO	6.1 (*C*)

10.6 Dissociation Energies of Diatomic Molecules

The dissociation energies D_0 of diatomic molecules are shown in Table 10.7. This energy is defined as the transition energy between the zero vibrational level of the ground electronic state of the molecule and the boundary of the continuum vibrational spectrum. The main part of the experimental data presented was derived from accurate spectroscopic measurements [10.6.1].

The values of dissociation energy D_0 are grouped into accuracy classes (see the Introduction). The molecules are tabulated in alphabetical order. The symbols of the ground-state molecular electronic terms are also included in the table. The primary information on D_0 values for diatomic molecules is contained in monographs [10.6.1–5].

References

10.6.1 L.V.Gurvich, G.V.Karachevtsev, V.N.Kondratjev, Y.A.Lebedev, V.A.Medvedev, V.K.Potapov, Y.S.Hodeev: *Bond Dissociation Energies, Ionization Potentials, and Electron Affinities*, 2nd ed. (Nauka, Moscow 1974) (in Russian)

10.6.2 K.P.Huber, G.Herzberg: *Molecular Spectra and Molecular Structure. IV. Constants of Diatomic Molecules* (Van Nostrand Reinhold, New York 1979)

10.6.3 K.S.Krasnov (ed.): *Molecular Constants for Inorganic Compounds* (Chimia, Moscow 1979) pp. 10–72 (in Russian)

10.6.4 A.G.Gaydon: *Dissociation Energies and Spectra of Diatomic Molecules*, 3rd ed. (Chapman and Hall, London 1968)

10.6.5 JANAF Thermochemical Tables, 2nd ed., Nat. Stand. Ref. Data Ser. Nat. Bur. Stand. **37** (1971);
1974 Supplement to JANAF Thermochemical Tables: J. Phys. Chem. Ref. Data **3**, 311–480 (1974);
1975 Supplement to JANAF Thermochemical Tables: ibid **4**, 1 (1975)
1978 Supplement to JANAF Thermochemical Tables: ibid **7**, 793 (1978)
1982 Supplement to JANAF Thermochemical Tables: ibid **11**, 659 (1982)

Table 10.7. Dissociation energies D_0 of diatomic molecules

Molecule or radical	D_0 [eV]	Molecule or radical	D_0 [eV]	Molecule or radical	D_0 [eV]
Ag_2 ($^1\Sigma_g^+$)	1.7 (*C*)	AuSe ($^2\Pi$)	2.6 (*C*)	CI ($^2\Pi$)	2.90 (*B*)
AgAl ($^1\Sigma^+$)	1.9 (*C*)	AuSi ($^2\Pi_{1/2}$)	3.2 (*C*)	CN ($^2\Sigma^+$)	7.8 (*B*)
AgBi (0^+)	2.0 (*D*)	AuSn ($^2\Pi_{1/2}$)	2.5 (*C*)	CO ($^1\Sigma^+$)	11.09 (*A*)
AgBr ($^1\Sigma^+$)	3.1 (*C*)	AuSr ($^2\Sigma$)	2.6 (*C*)	CP ($^2\Sigma^+$)	5.3 (*C*)
AgCu	1.8 (*C*)	AuTe ($^2\Pi$)	2.4 (*C*)	CS ($^1\Sigma^+$)	7.35 (*A*)
AgCl ($^1\Sigma^+$)	3.2 (*C*)	B_2 ($^3\Sigma_g^-$)	3.0 (*C*)	CSe ($^1\Sigma^+$)	6.0 (*C*)
AgF ($^1\Sigma^+$)	3.6 (*C*)	BBr ($^1\Sigma^+$)	4.5 (*C*)	Ca_2 ($^1\Sigma_g^+$)	0.132 (*A*)
AgGa ($^1\Sigma$)	1.9 (*C*)	BC ($^4\Sigma^-$)	4.6 (*C*)	CaBr ($^2\Sigma^+$)	3.2 (*C*)
AgH ($^1\Sigma^+$)	2.3 (*C*)	BCl ($^1\Sigma^+$)	5.6 (*C*)	CaCl ($^2\Sigma^+$)	4.3 (*B*)
AgI ($^1\Sigma^+$)	2.6 (*C*)	BF ($^1\Sigma^+$)	7.8 (*B*)	CaF ($^2\Sigma^+$)	5.5 (*B*)
AgIn ($^1\Sigma$)	1.7 (*C*)	BH ($^1\Sigma^+$)	3.4 (*B*)	CaH ($^2\Sigma^+$)	1.7 (*C*)
AgO ($^2\Pi_r$)	2.3 (*C*)	BN ($^3\Pi$)	4.0 (*D*)	CaI ($^2\Sigma^+$)	2.9 (*D*)
AgS ($^2\Pi$)	2.2 (*C*)	BO ($^2\Sigma^+$)	8.3 (*B*)	CaO ($^1\Sigma$)	4.7 (*C*)
AgSe ($^2\Pi$)	2.1 (*C*)	BP	3.6 (*C*)	CaS ($^1\Sigma^+$)	3.5 (*C*)
AgTe	2.0 (*C*)	BS ($^2\Sigma^+$)	6.0 (*B*)	Cd_2	0.09 (*D*)
Al_2 ($^3\Sigma_g^-$)	1.9 (*C*)	BSe ($^2\Sigma$)	4.7 (*C*)	CdCl ($^2\Sigma$)	2.11 (*B*)
AlBr ($^1\Sigma^+$)	4.4 (*C*)	BaBr ($^2\Sigma^+$)	3.7 (*B*)	CdF ($^2\Sigma$)	3.2 (*C*)
AlCl ($^1\Sigma^+$)	5.1 (*C*)	BaCl ($^2\Sigma^+$)	4.5 (*B*)	CdH ($^2\Sigma^+$)	0.68 (*C*)
AlF ($^1\Sigma^+$)	6.9 (*B*)	BaF ($^2\Sigma^+$)	6.1 (*B*)	CdIn ($^2\Sigma$)	1.4 (*D*)
AlH ($^1\Sigma^+$)	3.0 (*B*)	BaH ($^2\Sigma$)	1.8 (*C*)	Ce_2	2.83 (*A*)
AlI ($^1\Sigma^+$)	3.8 (*C*)	BaI ($^2\Sigma^+$)	4.4 (*B*)	CeB	3.1 (*C*)
AlN ($^1\Sigma^+$)	3.7 (*D*)	BaO ($^1\Sigma^+$)	5.81 (*A*)	CeN	5.3 (*C*)
AlO ($^2\Sigma^+$)	5.25 (*A*)	BaS ($^1\Sigma^+$)	4.4 (*B*)	CeO ($^3\Phi_2$)	8.2 (*B*)
AlP ($^1\Sigma^+$)	2.2 (*C*)	BeBr ($^2\Sigma^+$)	4.1 (*D*)	CeS	5.9 (*B*)
AlS ($^2\Sigma^+$)	3.8 (*B*)	BeCl ($^2\Sigma^+$)	4.5 (*C*)	Cl_2 ($^1\Sigma_g^+$)	2.479 (*A*)
AlSe ($^2\Sigma$)	3.5 (*C*)	BeF ($^2\Sigma^+$)	6.26 (*B*)	ClF ($^1\Sigma^+$)	2.62 (*B*)
AlTe ($^2\Sigma$)	2.7 (*C*)	BeH ($^2\Sigma^+$)	2.03 (*A*)	ClO ($^2\Pi_i$)	2.75 (*A*)
Ar_2 ($^1\Sigma_g^+$)	0.010 (*B*)	BeO ($^1\Sigma^+$)	4.6 (*C*)	Co_2	1.7 (*D*)
As_2 ($^1\Sigma_g^+$)	3.9 (*B*)	BeS ($^1\Sigma^+$)	3.9 (*D*)	CoBr	3.4 (*D*)
AsF ($^3\Sigma^-$)	4 (*D*)	Bi_2 ($^1\Sigma_g^+$)	2.0 (*C*)	CoCl ($^3\Sigma$)	4.1 (*B*)
AsH ($^3\Sigma^-$)	2.8 (*C*)	BiBr (0^+)	2.74 (*A*)	CoCu	1.6 (*C*)
AsN ($^1\Sigma^+$)	6.5 (*D*)	BiCl (0^+)	3.1 (*B*)	CoGe ($^3\Sigma$)	2.4 (*D*)
AsO ($^2\Pi_{1/2}$)	4.95 (*B*)	BiF (0^+)	3.3 (*C*)	CoO	3.8 (*C*)
AsP ($^1\Sigma^+$)	4.5 (*B*)	BiH ($^3\Sigma^-$)	2.5 (*D*)	CoS	3.4 (*C*)
AsS ($^2\Pi_{1/2}$)	3.7 (*C*)	BiI (0^+)	1.95 (*A*)	CoSi ($^2\Sigma$)	2.8 (*C*)
Au_2 ($^1\Sigma_g^+$)	2.30 (*B*)	BiO ($^2\Pi_{1/2}$)	3.5 (*B*)	Cr_2	1.6 (*D*)
AuAl ($^1\Sigma^+$)	3.3 (*B*)	BiS ($^2\Pi_{1/2}$)	3.2 (*B*)	CrBr	3.4 (*C*)
AuB ($^1\Sigma^+$)	3.7 (*C*)	BiSe ($\frac{1}{2}$)	2.8 (*C*)	CrCl ($^6\Sigma^+$)	3.8 (*C*)
AuBe ($^2\Sigma^+$)	2.9 (*C*)	BiTe ($\frac{1}{2}$)	2.4 (*C*)	CrCu	1.6 (*D*)
AuCa ($^2\Sigma$)	2.5 (*D*)	Br_2 ($^1\Sigma_g^+$)	1.971 (*A*)	CrF ($^6\Sigma^+$)	4.6 (*C*)
AuCl ($^1\Sigma^+$)	3.5 (*C*)	BrCl ($^1\Sigma^+$)	2.23 (*A*)	CrGe	1.7 (*D*)
AuCu	2.4 (*C*)	BrF ($^1\Sigma^+$)	2.55 (*A*)	CrH ($^6\Sigma^+$)	2.9 (*D*)
AuGa (0^+)	2.4 (*D*)	BrO ($^2\Pi_{3/2}$)	2.40 (*A*)	CrI	2.9 (*C*)
AuGe ($^2\Pi_{1/2}$)	2.8 (*C*)	C_2 ($^1\Sigma_g^+$)	6.2 (*B*)	CrN ($^4\Sigma$)	3.9 (*C*)
AuH ($^1\Sigma^+$)	3.2 (*C*)	CBr ($^2\Pi_{1/2}$)	3.8 (*C*)	CrO ($^5\Pi$)	4.0 (*C*)
AuMg ($^2\Sigma^+$)	2.5 (*D*)	CCl ($^2\Pi_{1/2}$)	4.1 (*C*)	CrS	3.4 (*C*)
AuO ($^2\Pi$)	2.3 (*C*)	CF ($^2\Pi_r$)	5.7 (*B*)	Cs_2 ($^1\Sigma_g^+$)	0.4499 (*A*)
AuS ($^2\Pi$)	2.6 (*C*)	CH ($^2\Pi_r$)	3.46 (*A*)	CsBr ($^1\Sigma^+$)	4.2 (*B*)

Table 10.7 (continued)

Molecule or radical	D_0 [eV]	Molecule or radical	D_0 [eV]	Molecule or radical	D_0 [eV]
$CsCl\,(^1\Sigma^+)$	4.6 (*B*)	$Ge_2\,(^3\Sigma)$	2.8 (*C*)	$InO\,(^2\Sigma)$	3.3 (*D*)
$CsF\,(^1\Sigma^+)$	5.2 (*B*)	$GeBr\,(^2\Pi_r)$	3.5 (*D*)	$InS\,(^2\Sigma)$	2.9 (*C*)
$CsH\,(^1\Sigma^+)$	1.781 (*A*)	$GeC\,(^3\Sigma)$	4.7 (*C*)	$InSe\,(^2\Sigma)$	2.5 (*C*)
$CsI\,(^1\Sigma^+)$	3.6 (*B*)	$GeCl\,(^2\Pi_r)$	4.2 (*C*)	$InTe\,(^2\Sigma)$	2.2 (*C*)
$CsO\,(^2\Sigma^+)$	3.0 (*C*)	$GeF\,(^2\Pi_{1/2})$	5.0 (*C*)	$IrBr\,(^1\Sigma)$	5.3 (*C*)
$Cu_2\,(^1\Sigma_g^+)$	2.0 (*B*)	$GeH\,(^2\Pi_r)$	3.2 (*B*)	$IrC\,(^2\Delta_{5/2})$	6.4 (*B*)
$CuAl\,(^1\Sigma^+)$	2.2 (*C*)	$GeI\,(^2\Pi_{1/2})$	3.0 (*C*)	$IrO\,(^2\Delta)$	3.6 (*C*)
$CuBr\,(^1\Sigma^+)$	2.3 (*B*)	$GeO\,(^1\Sigma^+)$	6.8 (*B*)	$K_2\,(^1\Sigma_g^+)$	0.51 (*B*)
$CuCl\,(^1\Sigma^+)$	3.9 (*B*)	$GeS\,(^1\Sigma^+)$	5.68 (*A*)	$KBr\,(^1\Sigma^+)$	3.9 (*B*)
$CuF\,(^1\Sigma^+)$	4.4 (*B*)	$GeSe\,(^1\Sigma^+)$	5.0 (*B*)	$KCl\,(^1\Sigma^+)$	4.3 (*B*)
$CuGe\,(^3\Sigma)$	2.0 (*C*)	$GeSi\,(^3\Sigma)$	3.1 (*C*)	$KCs\,(^1\Sigma^+)$	0.47 (*C*)
$CuH\,(^1\Sigma^+)$	2.8 (*B*)	$GeTe\,(^1\Sigma^+)$	4.2 (*B*)	$KF\,(^1\Sigma^+)$	5.1 (*B*)
$CuI\,(^1\Sigma^+)$	3.0 (*C*)	$H_2\,(^1\Sigma_g^+)$	4.478 (*A*)	$KH\,(^1\Sigma^+)$	1.8 (*C*)
$CuLi$	2.0 (*C*)	$HBr\,(^1\Sigma^+)$	3.76 (*B*)	$KI\,(^1\Sigma^+)$	3.3 (*B*)
$CuNa$	1.8 (*C*)	$HCl\,(^1\Sigma^+)$	4.434 (*A*)	$KO\,(^2\Sigma^+)$	3.1 (*C*)
$CuNi\,(^2\Sigma)$	2.1 (*C*)	$HD\,(^1\Sigma_g^+)$	4.514 (*A*)	$KRb\,(^1\Sigma^+)$	0.5 (*C*)
$CuO\,(^2\Pi_{3/2})$	2.8 (*C*)	$HF\,(^1\Sigma^+)$	5.87 (*A*)	$Kr_2\,(^1\Sigma_g^+)$	0.016 (*B*)
$CuS\,(^2\Pi_i)$	2.8 (*C*)	$HI\,(^1\Sigma^+)$	3.054 (*A*)	La_2	2.5 (*C*)
$CuSe\,(^2\Pi_i)$	2.6 (*C*)	$HS\,(^2\Pi_i)$	3.5 (*B*)	$LaF\,(^1\Sigma^+)$	6.2 (*C*)
$CuTe\,(^2\Pi)$	2.4 (*C*)	$HT\,(^1\Sigma_g^+)$	4.527 (*A*)	$LaO\,(^2\Sigma^+)$	8.2 (*B*)
$D_2\,(^1\Sigma_g^+)$	4.556 (*A*)	$He_2\,(a\,^3\Sigma_u^+)$	1.85 (*A*)	$LaS\,(^2\Sigma^+)$	5.9 (*B*)
$DT\,(^1\Sigma_g^+)$	4.573 (*A*)	$He_2\,(A\,^1\Sigma_u^+)$	2.35 (*A*)	$LaSe$	4.9 (*B*)
$Dy_2\,(^1\Sigma_g^+)$	0.7 (*D*)	$HfO\,(^1\Sigma)$	8.2 (*C*)	$LaTe$	3.9 (*C*)
Er_2	0.7 (*D*)	$Hg_2\,(^1\Sigma_g^+)$	0.11 (*C*)	$Li_2\,(^1\Sigma_g^+)$	1.034 (*A*)
Eu_2	0.5 (*D*)	$HgBr\,(^2\Sigma^+)$	0.74 (*B*)	$LiBr\,(^1\Sigma^+)$	4.3 (*B*)
$EuLi$	0.66 (*C*)	$HgCl\,(^2\Sigma^+)$	1.07 (*B*)	$LiCl\,(^1\Sigma^+)$	4.8 (*B*)
EuO	4.8 (*B*)	$HgF\,(^2\Sigma^+)$	2 (*D*)	$LiCs\,(^1\Sigma^+)$	0.7 (*C*)
$F_2\,(^1\Sigma_g^+)$	1.60 (*A*)	$HgH\,(^2\Sigma^+)$	0.374 (*A*)	$LiF\,(^1\Sigma^+)$	5.91 (*A*)
$FO\,(^2\Pi)$	2.39 (*B*)	$HgI\,(^2\Sigma^+)$	0.39 (*B*)	$LiH\,(^1\Sigma^+)$	2.429 (*A*)
Fe_2	1.1 (*C*)	HgS	2.1 (*C*)	$LiI\,(^1\Sigma^+)$	3.5 (*B*)
$FeBr$	2.5 (*E*)	$HgSe$	1.7 (*C*)	$LiK\,(^1\Sigma^+)$	0.75 (*B*)
$FeCl\,(^4\Sigma)$	3.6 (*D*)	$HgTe$	1.4 (*C*)	$LiNa\,(^1\Sigma^+)$	0.86 (*B*)
$FeGe\,(^3\Sigma)$	2.1 (*C*)	Ho_2	0.8 (*D*)	$LiO\,(^2\Pi_i)$	3.5 (*B*)
$FeO\,(^5\Delta)$	4.2 (*C*)	HoF	5.6 (*C*)	$LiRb\,(^1\Sigma^+)$	0.8 (*C*)
FeS	3.3 (*C*)	HoO	6.4 (*B*)	Lu_2	1.4 (*D*)
$FeSi$	3.0 (*C*)	HoS	4.4 (*C*)	$LuF\,(^1\Sigma^+)$	5.9 (*C*)
Ga_2	1.4 (*D*)	$HoSe$	3.4 (*C*)	$LuO\,(^2\Sigma)$	7.2 (*B*)
$GaBr\,(^1\Sigma^+)$	4.3 (*C*)	$I_2\,(^1\Sigma_g^+)$	1.542 (*A*)	LuS	5.2 (*B*)
$GaCl\,(^1\Sigma^+)$	4.9 (*C*)	$IBr\,(^1\Sigma^+)$	1.818 (*A*)	$LuSe$	4.3 (*C*)
$GaF\,(^1\Sigma^+)$	6.0 (*B*)	$ICl\,(^1\Sigma^+)$	2.153 (*A*)	$LuTe$	3.3 (*C*)
$GaH\,(^1\Sigma^+)$	2.8 (*B*)	$IF\,(^1\Sigma^+)$	2.88 (*A*)	$Mg_2\,(^1\Sigma_g^+)$	0.050 (*A*)
$GaI\,(^1\Sigma^+)$	3.5 (*B*)	$IO\,(^2\Pi_{3/2})$	2.3 (*C*)	$MgBr\,(^2\Sigma^+)$	3.3 (*C*)
$GaN\,(^3\Pi)$	5.4 (*C*)	In_2	1.0 (*C*)	$MgCl\,(^2\Sigma^+)$	3.3 (*C*)
$GaO\,(^2\Sigma^+)$	3.9 (*C*)	$InBr\,(^1\Sigma^+)$	4.0 (*C*)	$MgF\,(^2\Sigma^+)$	4.8 (*B*)
$GaP\,(^3\Sigma^-)$	2.4 (*C*)	$InCl\,(^1\Sigma^+)$	4.4 (*C*)	$MgH\,(^2\Sigma^+)$	1.3 (*C*)
$GaTe$	2.7 (*C*)	$InF\,(^1\Sigma^+)$	5.3 (*B*)	$MgI\,(^2\Sigma)$	2.9 (*C*)
Gd_2	1.8 (*D*)	$InH\,(^1\Sigma^+)$	2.5 (*C*)	$MgO\,(^1\Sigma^+)$	3.7 (*B*)
GdO	7.4 (*B*)	$InI\,(^1\Sigma^+)$	3.4 (*C*)	$MgS\,(^1\Sigma^+)$	2.4 (*D*)

Table 10.7 (continued)

Molecule or radical	D_0 [eV]	Molecule or radical	D_0 [eV]	Molecule or radical	D_0 [eV]
Mn_2	0.2 (D)	$O_2\,(^3\Sigma_g^-)$	5.116 (A)	$RhC\,(^2\Sigma^+)$	6.0 (B)
$MnBr\,(^7\Sigma)$	3.2 (B)	$OH\,(^2\Pi_i)$	4.39 (A)	RhO	4.2 (C)
$MnCl\,(^7\Sigma)$	3.7 (B)	$P_2\,(^1\Sigma_g^+)$	5.03 (A)	$RhSi\,(^2\Sigma)$	4.0 (C)
$MnF\,(^7\Sigma)$	4.3 (C)	$PF\,(^3\Sigma^-)$	4.6 (C)	$RhTi$	4.0 (C)
$MnH\,(^7\Sigma)$	2.4 (D)	$PH\,(^3\Sigma^-)$	3.5 (C)	$RuB\,(^2\Sigma)$	4.6 (C)
$MnI\,(^7\Sigma)$	2.9 (C)	$PN\,(^1\Sigma^+)$	6.4 (C)	RuC	6.7 (B)
$MnO\,(^6\Sigma)$	3.7 (C)	$PO\,(^2\Pi_r)$	6.10 (A)	RuO	5.3 (C)
MnS	2.8 (C)	$PS\,(^2\Pi_r)$	4.6 (B)	$RuSi\,(^1\Sigma)$	4.0 (C)
$Mo_2\,(^1\Sigma_g^+)$	4 (D)	PSe	3.7 (B)	$RuTh$	6.1 (C)
MoF	4.8 (C)	PTe	3.0 (C)	$S_2\,(^3\Sigma_g^-)$	4.369 (A)
MoO	6.3 (B)	$Pb_2\,(^3\Sigma_g^-)$	0.8 (D)	$SF\,(^2\Pi)$	3.3 (B)
$N_2\,(^1\Sigma_g^+)$	9.76 (A)	$PbBi$	1.4 (C)	$SH\,(^2\Pi_i)$	3.5 (C)
$NBr\,(^3\Sigma^-)$	2.9 (C)	$PbBr\,(^2\Pi_{1/2})$	2.5 (D)	$SO\,(^3\Sigma^-)$	3.36 (A)
$NCl\,(^3\Sigma^-)$	4.0 (C)	$PbCl\,(^2\Pi_{1/2})$	3.1 (C)	$Sb_2\,(^1\Sigma_g^+)$	3.1 (B)
$NF\,(^3\Sigma^-)$	3.5 (C)	$PbF\,(^2\Pi_{1/2})$	3.6 (C)	$SbBi\,(^1\Sigma^+)$	2.6 (B)
$NH\,(^3\Sigma^-)$	3.27 (A)	$PbH\,(^2\Pi_{1/2})$	1.6 (C)	$SbBr\,(^3\Sigma^-)$	3.2 (D)
$NI\,(^3\Sigma^-)$	1.6 (C)	$PbI\,(^2\Pi_{1/2})$	2.0 (D)	$SbCl\,(^3\Sigma^-)$	3.7 (D)
$NO\,(^2\Pi_r)$	6.497 (A)	$PbO\,(^1\Sigma^+)$	3.8 (B)	$SbF\,(^3\Sigma^-)$	4.5 (C)
$NS\,(^2\Pi_r)$	4.9 (C)	$PbS\,(^1\Sigma^+)$	3.51 (A)	$SbN\,(^1\Sigma^+)$	4.8 (D)
$NSe\,(^2\Pi_{1/2})$	3.9 (D)	$PbSe\,(^1\Sigma^+)$	3.1 (B)	$SbO\,(^2\Pi_r)$	4.0 (D)
$Na_2\,(^1\Sigma_g^+)$	0.737 (A)	$PbTe\,(^1\Sigma^+)$	2.5 (C)	$SbS\,(^2\Pi_r)$	3.9 (C)
$NaBr\,(^1\Sigma^+)$	3.7 (B)	$Pd_2\,(^1\Sigma)$	0.7 (D)	$SbTe$	2.8 (B)
$NaCl\,(^1\Sigma^+)$	4.2 (B)	$PdAl$	2.6 (C)	Sc_2	1.6 (D)
$NaCs\,(^1\Sigma^+)$	0.61 (B)	PdB	3.4 (C)	$ScCl\,(^1\Sigma^+)$	3.3 (C)
$NaF\,(^1\Sigma^+)$	5.3 (B)	$PdGe$	2.7 (C)	$ScF\,(^1\Sigma^+)$	6.2 (B)
$NaH\,(^1\Sigma^+)$	1.95 (B)	PdO	2.9 (D)	$ScO\,(^2\Sigma^+)$	6.9 (B)
$NaI\,(^1\Sigma^+)$	3.0 (B)	$PdSi\,(^1\Sigma)$	3.2 (C)	$ScS\,(^2\Sigma)$	4.9 (B)
$NaK\,(^1\Sigma^+)$	0.645 (A)	$Po_2\,(0_g^+)$	1.9 (C)	$ScSe$	3.9 (C)
$NaO\,(^2\Pi)$	2.6 (C)	Pr_2	1.3 (D)	$ScTe$	3.0 (C)
$NaRb\,(^1\Sigma^+)$	0.65 (B)	$PrO\,(^2\Pi_{3/2})$	7.7 (B)	$Se_2\,(^3\Sigma_g^-)$	3.41 (A)
$NbO\,(^4\Sigma^-)$	7.8 (B)	PrS	5.5 (B)	$SeBr\,(^2\Pi_i)$	3.0 (D)
Nd_2	0.8 (E)	$PrSe$	4.6 (C)	$SeCl\,(^2\Pi_i)$	3.3 (D)
$NdCl$	4.6 (C)	$PtB\,(^2\Sigma)$	4.9 (C)	$SeF\,(^2\Pi_i)$	3.2 (D)
NdF	5.9 (B)	$PtC\,(^1\Sigma)$	6.3 (B)	$SeH\,(^2\Pi_{3/2})$	3.2 (C)
NdO	7.3 (B)	$PtH\,(^2\Delta_{5/2})$	3.44 (A)	$SeO\,(^3\Sigma^-)$	4.4 (B)
NdS	4.8 (C)	$PtO\,(^1\Sigma)$	3.8 (C)	$SeS\,(^3\Sigma^-)$	3.7 (C)
$NdSe$	3.9 (C)	$PtSi\,(^1\Sigma)$	5.1 (C)	$Si_2\,(^3\Sigma_g^-)$	3.2 (C)
$NdTe$	3.1 (C)	$PtTh$	5.7 (C)	$SiBr\,(^2\Pi_r)$	3.9 (C)
$Ne_2\,(^1\Sigma_g^+)$	0.0020 (B)	PuF	5.5 (C)	$SiC\,(^3\Pi_i)$	4.7 (C)
Ni_2	2.07 (A)	$Rb_2\,(^1\Sigma_g^+)$	0.49 (C)	$SiCl\,(^2\Pi_r)$	4.7 (C)
$NiBr$	3.7 (B)	$RbBr\,(^1\Sigma^+)$	3.9 (B)	$SiF\,(^2\Pi_r)$	5.6 (B)
$NiCl\,(^2\Delta_{5/2})$	3.8 (C)	$RbCl\,(^1\Sigma^+)$	4.3 (B)	$SiH\,(^2\Pi_r)$	3.06 (A)
NiF	4.5 (D)	$RbCs\,(^1\Sigma^+)$	0.472 (A)	$SiI\,(^2\Pi_{1/2})$	3.0 (C)
$NiH\,(^2\Delta_{5/2})$	3.0 (C)	$RbF\,(^1\Sigma^+)$	5.0 (B)	$SiN\,(^2\Sigma^+)$	5.2 (C)
NiI	3.0 (C)	$RbH\,(^1\Sigma^+)$	1.8 (C)	$SiO\,(^1\Sigma^+)$	8.26 (A)
NiO	3.6 (C)	$RbI\,(^1\Sigma^+)$	3.3 (B)	$SiP\,(^2\Pi_i)$	3.7 (C)
NiS	3.5 (C)	Rh_2	2.9 (C)	$SiS\,(^1\Sigma^+)$	6.4 (B)
$NiSi$	3.3 (C)	$RhB\,(^1\Sigma)$	4.9 (C)	$SiSe\,(^1\Sigma^+)$	5.6 (B)

Table 10.7 (continued)

Molecule or radical	D_0 [eV]	Molecule or radical	D_0 [eV]	Molecule or radical	D_0 [eV]
SiTe ($^1\Sigma^+$)	4.6 (B)	TeS (0^+)	3.5 (D)	VN	4.9 (C)
$\mathrm{Sm_2}$	0.5 (E)	TeSe (0^+)	3.0 (C)	VO ($^4\Sigma^-$)	6.4 (B)
SmCl	4.3 (C)	$\mathrm{Th_2}$	3.0 (C)	VS	4.6 (C)
SmF	5.5 (C)	ThB	3.0 (D)	VSe	3.5 (C)
SmLi	0.47 (C)	ThN	5.9 (C)	WO ($^3\Sigma^-$)	6.8 (C)
SmO	5.9 (C)	ThO ($^1\Sigma^+$)	8.9 (B)	$\mathrm{Xe_2}$ ($^1\Sigma_g^+$)	0.023 (B)
$\mathrm{Sn_2}$	2.1 (B)	ThP	3.9 (C)	XeCl ($^2\Sigma^+$)	0.030 (B)
SnBr ($^2\Pi_{1/2}$)	3.5 (B)	$\mathrm{Ti_2}$ ($^7\Delta$)	1.3 (D)	XeF ($^2\Sigma$)	0.13 (C)
SnCl ($^2\Pi_{1/2}$)	4.3 (C)	TiF ($^4\Sigma$)	5.9 (C)	$\mathrm{Y_2}$	1.6 (D)
SnF ($^2\Pi_{1/2}$)	4.9 (B)	TiH	1.6 (D)	YCl ($^1\Sigma$)	3.5 (D)
SnH ($^2\Pi_r$)	2.7 (C)	TiN ($^2\Sigma$)	4.9 (C)	YF ($^1\Sigma^+$)	6.2 (C)
SnI ($^2\Pi_{1/2}$)	2.4 (D)	TiO ($^3\Delta_r$)	6.9 (B)	YO ($^2\Sigma^+$)	7.4 (B)
SnO ($^1\Sigma^+$)	5.47 (A)	TiS ($^3\Delta_1$)	4.7 (C)	YS ($^2\Sigma$)	5.5 (B)
SnS ($^1\Sigma^+$)	4.8 (B)	TiTe	2.9 (C)	YSe	4.5 (B)
SnSe ($^1\Sigma^+$)	4.2 (B)	TlAs ($^3\Sigma$)	2.0 (C)	YTe	3.5 (C)
SnTe ($^1\Sigma^+$)	3.7 (B)	TlBi ($^3\Sigma$)	1.2 (D)	YbLi	0.35 (C)
$\mathrm{Sr_2}$ ($^1\Sigma_g^+$)	0.13 (C)	TlBr ($^1\Sigma^+$)	3.42 (A)	ZnBr ($^2\Sigma$)	1.4 (D)
SrBr ($^2\Sigma^+$)	3.4 (C)	TlCl ($^1\Sigma^+$)	3.82 (A)	ZnCl ($^2\Sigma$)	2.1 (C)
SrCl ($^2\Sigma^+$)	4.2 (C)	TlF ($^1\Sigma^+$)	4.58 (A)	ZnF ($^2\Sigma$)	3.8 (C)
SrF ($^2\Sigma^+$)	5.6 (C)	TlH ($^1\Sigma^+$)	2.0 (B)	ZnH ($^2\Sigma^+$)	0.85 (B)
SrH ($^2\Sigma^+$)	1.7 (C)	TlI ($^1\Sigma^+$)	2.88 (A)	ZnI ($^2\Sigma$)	1.4 (C)
SrI ($^2\Sigma^+$)	2.8 (C)	$\mathrm{Tm_2}$	0.5 (E)	ZnO	2.8 (D)
SrO ($^1\Sigma^+$)	4.9 (B)	TmCl	3.9 (C)	ZnS	2.1 (C)
SrS ($^1\Sigma^+$)	3.5 (C)	TmLi	0.67 (C)	ZnSe	1.4 (C)
SrSe	2.9 (C)	TmO	5.8 (B)	ZnTe	0.9 (D)
$\mathrm{T_2}$ ($^1\Sigma_g^+$)	4.591 (A)	$\mathrm{U_2}$	2.3 (C)	ZrCl	5.3 (C)
TaO ($^2\Delta_{3/2}$)	8.2 (C)	UB	3.3 (D)	ZrI	3.7 (D)
$\mathrm{Tb_2}$	1.3 (D)	UC	4.8 (C)	ZrN ($^2\Sigma$)	5.8 (C)
TbCl	4.2 (C)	UN	5.5 (C)	ZrO ($^1\Sigma^+$)	7.8 (B)
TbO	7.3 (B)	UO	7.9 (C)	ZrS	5.9 (B)
$\mathrm{Te_2}$ ($^3\Sigma_g^-$)	2.68 (A)	US	5.4 (B)		
TeO (0^+)	3.9 (B)	$\mathrm{V_2}$ ($^3\Sigma_g^-$)	2.0 (C)		

10.7 Lifetimes of Excited Electron States in Diatomic Molecules

The radiative lifetime of a molecular excited state τ_n is related to the Einstein coefficient A_{nm} for spontaneous emission from level n to level m as follows:

$$\tau_n = 1 \Big/ \sum_m A_{nm} ,$$

where the summation is performed over all possible transitions. Dipole transitions are the most important for the electronic spectrum of a molecule. They are determined by the selection rules which, for the most common cases of molecular interactions (Hund's cases a and b, when electron states are characterized by quantum numbers Λ, S, where Λ is the projection of the electron orbital angular momentum on the molecular axis and S is the total spin of the electrons), are written as

$$S' - S = 0, \Lambda' - \Lambda = 0, \pm 1 ,$$

where the primed and unprimed quantum numbers refer to different electron states. For the case of two Σ terms one finds $\Sigma^+ \to \Sigma^+$ and $\Sigma^- \to \Sigma^-$. There is also a rigorous selection rule for homonuclear diatomic molecules:

$$\mathrm{g} \to \mathrm{u} \quad \text{and} \quad \mathrm{u} \to \mathrm{g} ,$$

where g, u are the parity symbols of the electron wavefunctions.

Table 10.8 provides the values of radiative lifetimes τ_{el} of excited electron states in a variety of diatomic molecules, their accuracy being indicated as usual. The most common procedures for determining radiative lifetimes of molecular states are based on measuring the duration of fluorescence intensity decay and the phase shift of radiation [10.7.1]. In separate columns of the table the symbols of the excited molecular electron states and the quantum of energy for the transition between the zero vibrational level of the upper electron state and the ground state of the molecule are given. The main information on diatomic molecular radiative lifetimes τ_{el} is presented in [10.7.1–7].

References

10.7.1 R.Anderson: At. Data **3**, 227 (1971)

10.7.2 D.K.Hsu, W.H.Smith: Spectrosc. Lett. **10**, 181 (1977)

10.7.3 N.E.Kuz'menko, L.A.Kuznetsova, A.P.Monyakin, Y.Y.Kuzyakov, Y.A.Plastinin: Usp. Fiz. Nauk **127**, 451 (1979)

10.7.4 L.A.Kuznetsova, N.E.Kuz'menko, Y.Y.Kuzyakov, Y.A.Plastinin: *Optical Transition Probabilities of Diatomic Molecules*, ed. by R.V.Khokhlov (Nauka, Moscow 1980) (in Russian)

10.7.5 K.P.Huber, G.Herzberg: *Molecular Spectra and Molecular Structure. IV. Constants of Diatomic Molecules* (Van Nostrand Reinhold, New York 1979)

10.7.6 R.W.Nicholls: "Transition Probability Data for Molecular Spectra of Astrophysical Interest", Annu. Rev. Astron. Astrophys. **15** (1978)

10.7.7 M.N.Dumont, F.Remy: Spectrosc. Lett. **15**, 699 (1982)

Table 10.8. Radiative lifetimes of electronically excited diatomic molecules

Molecule (ground state)	Excited state	Excitation energy T_{00} [eV] relative to zero vibrational level of the ground state	Radiative lifetime $\tau_{el}(v'=0)$ [μs] (upper vibrational level is indicated if $v' \neq 0$)
AlH ($^1\Sigma^+$)	$A\,^1\Pi$	2.91	0.066 (C)
AlO ($^2\Sigma^+$)	$B\,^2\Sigma^+$	2.56	0.11 (C)
Ar_2 ($^1\Sigma_g^+$)	$B\,^1\Sigma_u^+$	11.56	0.005 (D)
	$A\,^3\Sigma_u^+$	11.46	3.2 (C)
BBr ($^1\Sigma^+$)	$A\,^1\Pi$	4.20	0.026 (D)
BCl ($^1\Sigma^+$)	$A\,^1\Pi$	4.56	0.019 (D)
BF ($^1\Sigma^+$)	$A\,^1\Pi$	6.33	0.003 (D)
BH ($^1\Sigma^+$)	$A\,^1\Pi$	2.86	0.13 (C)
BaBr ($^2\Sigma^+$)	$C\,^2\Pi$	2.38–2.31	0.008 (E)
BaCl ($^2\Sigma^+$)	$C\,^2\Pi$	2.41–2.36	0.02 (D)
BaF ($^2\Sigma^+$)	$C\,^2\Pi$	2.50–2.48	0.024 (C)
BaI ($^2\Sigma^+$)	$C\,^2\Pi$	2.30–2.21	0.016 (C)
BaO ($^1\Sigma^+$)	$A\,^1\Sigma^+$	2.07	0.36 (C)
BeO ($^1\Sigma^+$)	$B\,^1\Sigma^+$	2.63	0.09 (C)
Br_2 ($^1\Sigma_g^+$)	$B\,^3\Pi_u\,(0_u^+)$	1.96	12 (E)
BrCl ($^1\Sigma^+$)	$B\,^3\Pi_{0^+}$	2.075	18 (D)
BrF ($^1\Sigma^+$)	$B\,^3\Pi_{0^+}$	2.25	25 (D)
C_2 ($^1\Sigma_g^+$)	$D\,^1\Sigma_u^+$	5.36	0.018 (C)
	$C\,^1\Pi_g$	4.24	0.031 (C)
	$d\,^3\Pi_g$	2.48	0.23 (C)
CF ($^2\Pi_r$)	$B\,^2\Delta_r$	6.12	0.019 (D)
	$A\,^2\Sigma^+$	5.32	0.019 ($v'=1$) (D)
CH ($^2\Pi_r$)	$B\,^2\Sigma^-$	3.19	0.36 (C)
	$A\,^2\Delta$	2.88	0.50 (C)
C^2H ($^2\Pi_r$)	$A\,^2\Delta$	2.88	0.47 (C)
CN ($^2\Sigma^+$)	$B\,^2\Sigma^+$	3.20	0.062 (C)
	$A\,^2\Pi_i$	1.13	8 (D)
CO ($^1\Sigma^+$)	$c\,^3\Pi$	11.42	0.016 (D)
	$C\,^1\Sigma^+$	11.40	0.002 (E)
	$B\,^1\Sigma^+$	10.78	0.024 (C)
	$b\,^3\Sigma^+$	10.39	0.056 (C)
	$D\,^1\Delta$	8.11	$100 \cdot 10^3$ (D)
	$A\,^1\Pi$	8.03	0.011 (C)
	$e\,^3\Sigma^-$	7.90	2 ($v'=4$) (D)
	$d\,^3\Delta_i$	7.52	7.3 ($v'=1$) (C)
	$a'\,^3\Sigma^+$	6.86	10 ($v'=4$) (C)
	$a\,^3\Pi_r$	6.01	$8 \cdot 10^3$ (D)
CS ($^1\Sigma^+$)	$A\,^1\Pi$	4.81	0.22 (D)
	$a\,^3\Pi_r$	3.42	$16 \cdot 10^3$ (D)
CaBr ($^2\Sigma^+$)	$C\,^2\Pi$	3.17–3.14	0.032 (D)
	$B\,^2\Sigma^+$	2.03	0.043 (C)
	$A\,^2\Pi$	1.98–1.97	0.034 ($v'=1$) (C)
CaCl ($^2\Sigma^+$)	$C\,^2\Pi_r$	3.29–3.28	0.025 (C)
	$B\,^2\Sigma^+$	2.09	0.038 (C)
	$A\,^2\Pi$	2.00	0.029 ($v'=2$) (C)

Table 10.8 (continued)

Molecule (ground state)	Excited state	Excitation energy T_{00} [eV] relative to zero vibrational level of the ground state	Radiative lifetime $\tau_{el}(v'=0)$ [μs] (upper vibrational level is indicated if $v' \neq 0$)
CaF ($^2\Sigma^+$)	$B\,^2\Sigma^+$	2.34	0.025 (*D*)
	$A\,^2\Pi_r$	2.05–2.04	0.020 (*D*)
CaH ($^2\Sigma^+$)	$B\,^2\Sigma^+$	1.95	0.057 (*C*)
CaI ($^2\Sigma^+$)	$B\,^2\Sigma^+$	1.95	0.051 (*C*)
	$A\,^2\Pi$	1.94–1.93	0.042 ($v'=3{,}5$) (*C*)
CdH ($^2\Sigma^+$)	$A\,^2\Pi$	2.89–2.76	0.07 (*C*)
Cl_2 ($^1\Sigma_g^+$)	$B\,^3\Pi_{0_u^+}$	2.19	< 10 (*D*)
Cu_2 ($^1\Sigma_g^+$)	C	2.71	1.0 (*D*)
	$B\,^1\Sigma_u^+$	2.70	0.035 (*D*)
	$A\,^1\Pi_u$	2.53	0.07 (*D*)
CuF ($^1\Sigma^+$)	$C\,^1\Pi$	2.51	0.6 (*E*)
	$B\,^1\Sigma^+$	2.45	1.2 (*C*)
	$A\,^1\Pi$	2.18	7.3 (*B*)
FeO ($^5\Delta$)	b	2.70–2.69	0.5 (*D*)
	a	2.63	0.5 (*D*)
H_2 ($^1\Sigma_g^+$)	$I\,^1\Pi_g$	13.90	0.040 (*D*)
	$G\,^1\Sigma_g^+$	13.86	0.030 (*D*)
	$d\,^3\Pi_u$	13.85	0.063 (*C*)
	$C\,^1\Pi_u$	12.29	0.0006 (*E*)
	$a\,^3\Sigma_g^+$	11.79	0.011 (*C*)
	$c\,^3\Pi_u$	11.76	$1.0\cdot10^3$ (*D*)
	$B\,^1\Sigma_u^+$	11.18	0.0005 (*D*)
$^2H_2$ ($^1\Sigma_g^+$)	$a\,^3\Sigma_g^+$	11.82	0.012 (*C*)
	$c\,^3\Pi_u$	11.80	$1.0\cdot10^3$ (*D*)
HgBr ($^2\Sigma^+$)	$B\,^2\Sigma^+$	2.91	0.023 (*C*)
HgH ($^2\Sigma^+$)	$A_1\,^2\Pi_{1/2}$	3.09	0.10 (*C*)
I_2 ($^1\Sigma_g^+$)	$E\,^3\Pi_g\,(0_g^+)$	5.13	0.028 (*C*)
	(1_g)	5.08	0.012 (*C*)
	$D\,^1\Sigma_u^+$	5.04	0.015 (*C*)
IBr ($^1\Sigma^+$)	$B\,^3\Pi_{0^+}$	2.00	0.54 ($v'=2$) (*C*)
ICl ($^1\Sigma^+$)	$B\,^3\Pi_{0^+}$	2.14	1.0 (*C*)
	$A\,^3\Pi_1$	1.69	100 (*D*)
IF ($^1\Sigma^+$)	$B\,^3\Pi_{0^+}$	2.35	$1\cdot10^3$ (*D*)
K_2 ($^1\Sigma_g^+$)	$B\,^1\Pi_u$	1.905	0.011 (*C*)
KH ($^1\Sigma^+$)	$A\,^1\Sigma^+$	2.32	< 0.01 (*D*)
Kr_2 ($^1\Sigma_g^+$)	$A\,^3\Sigma_u^+\,(1_u)$	9.87	0.35 (*D*)
KrF ($^2\Sigma$)	$B\,^2\Sigma\,(\frac{1}{2})$	4.99	0.009 (*C*)
LaO ($^2\Sigma^+$)	$C\,^2\Pi_r$	2.83–2.80	0.028 (*C*)
	$B\,^2\Sigma^+$	2.21	0.035 (*C*)
Li_2 ($^1\Sigma_g^+$)	$B\,^1\Pi_u$	2.53	0.008 (*C*)
	$A\,^1\Sigma_u^+$	1.74	0.018 (*C*)
LiH ($^1\Sigma^+$)	$A\,^1\Sigma^+$	3.22	0.030 (*C*)
MgH ($^2\Sigma^+$)	$A\,^2\Pi_r$	2.39	0.044 (*C*)
MgO ($^1\Sigma^+$)	$B\,^1\Sigma^+$	2.480	0.033 (*C*)

Table 10.8 (continued)

Molecule (ground state)	Excited state	Excitation energy T_{00} [eV] relative to zero vibrational level of the ground state	Radiative lifetime $\tau_{el}(v'=0)$ [μs] (upper vibrational level is indicated if $v' \neq 0$)
$N_2(^1\Sigma_g^+)$	$y\,^1\Pi_g$	14.15	0.020 (*C*)
	$x\,^1\Sigma_g^-$	14.04	0.023 (*C*)
	$c_4'\,^1\Sigma_u^+$	12.93	0.0009 (*D*)
	$D\,^3\Sigma_u^+$	12.84	0.014 (*C*)
	$E\,^3\Sigma_g^+$	11.87	190 (*D*)
	$C\,^3\Pi_u$	11.03	0.038 (*C*)
	$a\,^1\Pi_g$	8.55	110 (*D*)
	$a'\,^1\Sigma_u^-$	8.40	$13\cdot10^3$ (*D*)
	$B\,^3\Pi_g$	7.35	8.0 (*C*)
	$A\,^3\Sigma_u^+$	6.169	$13\cdot10^6$ (*E*)
$NF(^3\Sigma^-)$	$b\,^1\Sigma^+$	2.34	$23\cdot10^3$ (*C*)
$NH(^3\Sigma^-)$	$d\,^1\Sigma^+$	10.27	0.046 (*D*)
	$c\,^1\Pi$	5.37	0.44 (*D*)
	$A\,^3\Pi_i$	3.69	0.43 (*C*)
	$b\,^1\Sigma^+$	2.63	$20\cdot10^3$ (*D*)
$N^2H(^3\Sigma^-)$	$d\,^1\Sigma^+$	10.28	0.062 (*C*)
$NO(^2\Pi_r)$	$F\,^2\Delta$	7.68–7.69	0.09 (*D*)
	$B'\,^2\Delta_i$	7.43–7.44	0.11 (*C*)
	$D\,^2\Sigma^+$	6.59–6.61	0.018 (*C*)
	$C\,^2\Pi_r$	6.48–6.49	0.025 (*D*)
	$b\,^4\Sigma^-$	6.00	6.4 ($v'=1$) (*C*)
	$B\,^2\Pi_r$	5.63–5.64	3.1 (*C*)
	$A\,^2\Sigma^+$	5.47–5.48	0.20 (*C*)
	$a\,^4\Pi_i$	4.71	$160\cdot10^3$ (*E*)
$NS(^2\Pi_r)$	$C\,^2\Sigma^+$	5.35–5.38	0.0065 (*D*)
$Na_2(^1\Sigma_g^+)$	$B\,^1\Pi_u$	2.52	0.0075 (*B*)
	$A\,^1\Sigma_u^+$	1.82	0.015 (*C*)
$NaH(^1\Sigma^+)$	$A\,^1\Sigma^+$	2.76	0.024 ($v'=3$) (*D*)
$Ne_2(^1\Sigma_g^+)$	$a\,^3\Sigma_u^+$	–	6.6 (*C*)
$O_2(^3\Sigma_g^-)$	$b\,^1\Sigma_g^+$	1.63	$12\cdot10^6$ (*D*)
	$a\,^1\Delta_g$	0.98	$2.5\cdot10^9$ (*E*)
$OH(^2\Pi_i)$	$C\,^2\Sigma^+$	10.94	0.004 (*D*)
	$B\,^2\Sigma^+$	8.48	2 (*E*)
	$A\,^2\Sigma^+$	4.02	0.69 (*B*)
$O^2H(^2\Pi_i)$	$A\,^2\Sigma^+$	4.03	0.77 (*C*)
$PH(^3\Sigma^-)$	$A\,^3\Pi_i$	3.65–3.62	440 (*D*)
$PN(^1\Sigma^+)$	$A\,^1\Pi$	4.92	0.23 (*D*)
$PbO(^1\Sigma^+)$	$B\,1$	2.75	2.6 (*D*)
	$A\,0^+$	2.45	3.7 ($v'=2$) (*D*)
$Rb_2(^1\Sigma_g^+)$	D	2.82	0.061 (*D*)
	$C\,^1\Pi_u$	2.58	0.014 (*D*)
$S_2(^3\Sigma_g^-)$	$B\,^3\Sigma_u^-$	3.93	0.019 (*C*)
$SH(^2\Pi_i)$	$A\,^2\Sigma^+$	3.80	0.8 (*D*)
$S^2H(^2\Pi_i)$	$A\,^2\Sigma^+$	3.81	0.7 (*D*)

Table 10.8 (continued)

Molecule (ground state)	Excited state	Excitation energy T_{00} [eV] relative to zero vibrational level of the ground state	Radiative lifetime $\tau_{el}(v' = 0)$ [μs] (upper vibrational level is indicated if $v' \neq 0$)
SO ($^3\Sigma^-$)	$B\,^3\Sigma^-$	5.13	0.017 (D)
	$A\,^3\Pi$	4.74–4.70	0.012 (D)
	$b\,^1\Sigma^+$	1.298	$6 \cdot 10^3$ (E)
ScF ($^1\Sigma^+$)	$E\,^1\Pi$	2.52	110 (D)
	$c\,^3\Phi$	a + 1.9	100 (D)
ScO ($^2\Sigma^+$)	$B\,^2\Sigma^+$	2.55	0.033 (C)
	$A\,^2\Pi_r$	2.05	0.03 (D)
Se_2 ($^3\Sigma_g^-(0_g^+)$)	$B_2\,^3\Sigma_u^-\,(1_u)$	3.22	0.060 (D)
SiF ($^2\Pi_r$)	$A\,^2\Sigma^+$	2.83	0.23 (C)
SiH ($^2\Pi_r$)	$A\,^2\Delta$	3.00	0.53 (C)
SiO ($^1\Sigma^+$)	$E\,^1\Sigma^+$	6.52	0.010 ($v' = 1$) (C)
	$A\,^1\Pi$	5.29	0.010 (C)
	$b\,^3\Pi$	4.20–4.19	$48 \cdot 10^3$ (D)
SiS ($^1\Sigma^+$)	$a\,^3\Pi_1$	3.70	$30 \cdot 10^3$ (D)
SrBr ($^2\Sigma^+$)	$C\,^2\Pi$	3.06–3.02	0.029 ($v' = 1$) (C)
	$B\,^2\Sigma^+$	1.90	0.042 ($v' = 3$) (C)
	$A\,^2\Pi$	1.86–1.82	0.034 ($v' = 2$) (C)
SrCl ($^2\Sigma^+$)	$C\,^2\Pi$	3.15–3.13	0.026 ($v' = 1$) (C)
	$B\,^2\Sigma^+$	1.95	0.039 ($v' = 1$) (C)
	$A\,^2\Pi$	1.87–1.84	0.031 ($v' = 1$) (C)
SrF ($^2\Sigma^+$)	$A\,^2\Pi$	1.90–1.87	0.023 (C)
SrI ($^2\Sigma^+$)	$C\,^2\Pi$	2.88–2.81	0.036 ($v' = 8$) (C)
	$B\,^2\Sigma$	1.84	0.046 ($v' = 1$) (C)
	$A\,^2\Pi$	1.83–1.79	0.042 ($v' = 6$) (C)
Te_2 ($^3\Sigma_g^-(0_g^+)$)	$B_1\,^3\Sigma_u^-\,(0_u^+)$	2.75	0.075 (C)
	$A\,(0_u^+)$	2.41	0.67 (C)
TiO ($^3\Delta_r$)	$c\,^1\Phi$	2.64	0.017 (D)
	$C\,^3\Delta_r$	2.40	0.040 (D)
VO ($^4\Sigma^-$)	$C\,^4\Sigma^-$	2.16	0.41 (D)
XeF ($^2\Sigma$)	$B\,^2\Pi_{1/2}$	3.58	0.018 (C)
YO ($^2\Sigma^+$)	$B\,^2\Sigma^+$	2.57	0.030 (C)
	$A\,^2\Pi_r$	2.07–2.02	0.033 (C)
ZnH ($^2\Sigma^+$)	$A\,^2\Pi_r$	2.91	0.075 (C)
Zn^2H ($^2\Sigma^+$)	$A\,^2\Pi_r$	2.90	0.075 (C)

10.8 Parameters of Excimer Molecules

Molecules which contain an excited atom, but which do not form a stable chemical bond for the same ground-state atom, are referred to as excimer molecules. Most of the information available today concerns diatomic excimer molecules with excited rare gas atoms.

Tables 10.9, 10 supply quantitative data for the interaction potentials and radiative lifetimes of some excimer molecules. Table 10.9 (upper part) gives the parameters of the interaction potential for homonuclear excimer inert gas molecules in the region of its minimum, namely r_m, the equilibrium separation between nuclei, related to their minimum interaction energy and D_e, the potential well depth. The lower part of Table 10.9 includes values of the parameters r_m, D_e for excimer molecules containing an excited rare gas atom and a halogen atom (rare gas halides). States $B_{1/2}$, $C_{3/2}$, $D_{1/2}$ are the low-lying excited states of a molecule composed of an excited rare gas atom with an electron shell $np^5\ (n+1)s\ (n = 2-5)$ and a halogen atom in the ground state, the lower index indicating the total electron angular momentum for a given state. The interaction character of these states corresponds to the ionic bond between a positive rare gas ion and a negative halogen ion. Table 10.10 gives values of the radiative lifetime τ_{el} of excited electronic states for a variety of diatomic excimer molecules.

Primary information concerning these parameters can be found in monographs [10.8.1, 2] and the review [10.8.3]. All the numerical values in Tables 10.9, 10 are grouped into accuracy classes (see the Introduction).

References

10.8.1 C.K.Rhodes (ed.): *Excimer Lasers*, Topics Appl. Phys. Vol. 30, 2nd ed. (Springer, Berlin, Heidelberg, New York, Tokyo 1984)

10.8.2 B.M.Smirnov: *Excited Atoms and Molecules* (Wiley, New York 1985)

10.8.3 P.J.Hay, W.R.Wadt, T.H.Dunning, Jr.: "Theoretical Studies of Molecular Electronic Transition Lasers", Annu. Rev. Phys. Chem. **30**, 347–378 (1979)

Table 10.9. Potential parameters of diatomic excimers

Molecule (term)	Equilibrium internuclear distance r_m [Å]	Dissociation energy D_e [eV]
I) Homonuclear rare gas diatomic molecules		
$He_2\,(a\,^3\Sigma_u^+)$	1.05 (*B*)	2.0 (*C*)
$He_2\,(A\,^1\Sigma_g^+)$	1.06 (*B*)	2.47 (*B*)
$Ne_2\,(a\,^3\Sigma_u^+)$	1.79 (*B*)	0.47 (*C*)
$Ne_2\,(A\,^1\Sigma_u^+)$	1.79 (*B*)	0.40 (*C*)
$Ar_2\,(1_u,\,0_u^-)$	2.33 (*B*)	0.78 (*B*)
$Ar_2\,(0_u^+)$	2.32 (*B*)	0.74 (*B*)
$Xe_2\,(1_u,\,0_u^-)$	3.03 (*B*)	0.79 (*C*)
$Xe_2\,(0_u^+)$	3.02 (*B*)	0.77 (*C*)
II) Excited rare gas atom – halogen atom		
NeF $(B_{1/2})$	2.00 (*B*)	6.4 (*C*)
NeF $(C_{3/2})$	1.99 (*B*)	6.35 (*C*)
ArBr $(B_{1/2})$	2.81 (*C*)	4.7 (*D*)
KrF $(B_{1/2})$	2.51 (*B*)	5.30 (*B*)
KrF $(C_{3/2})$	2.44 (*B*)	5.24 (*B*)
KrF $(D_{1/2})$	2.47 (*B*)	5.26 (*C*)
XeF $(B_{1/2})$	2.63 (*B*)	5.30 (*B*)
XeF $(C_{3/2})$	2.56 (*B*)	5.03 (*C*)
XeF $(D_{1/2})$	2.51 (*B*)	5.46 (*C*)
XeCl $(B_{1/2})$	3.22 (*B*)	4.21 (*C*)
XeCl $(C_{3/2})$	3.14 (*C*)	4.14 (*C*)
XeCl $(D_{1/2})$	3.18 (*C*)	4.17 (*C*)
XeBr $(B_{1/2})$	3.38 (*B*)	4.30 (*C*)
XeBr $(C_{3/2})$	3.31 (*C*)	4.0 (*D*)
XeBr $(D_{1/2})$	3.34 (*C*)	4.0 (*D*)
XeI $(B_{1/2})$	3.62 (*C*)	4.08 (*C*)
XeI $(C_{3/2})$	3.57 (*C*)	3.7 (*D*)
XeI $(D_{1/2})$	3.59 (*C*)	3.7 (*D*)

Table 10.10. Radiative lifetimes of diatomic excimers

I) Homonuclear rare gas diatomic molecules

Excited molecule (term)	Radiative lifetime τ [ns]
$Ne_2\,(a\,^3\Sigma_u^+)$	8000 (*E*)
$Ne_2\,(A\,^1\Sigma_u^+)$	3 (*D*)
$Ar_2\,(1_u, 0_u^-)$	3600 (*D*)
$Ar_2\,(0_u^+)$	5 (*C*)
$Kr_2\,(1_u, 0_u^-)$	300 (*C*)
$Kr_2\,(0_u^+)$	6 (*D*)
$Xe_2\,(1_u, 0_u^-)$	110 (*D*)
$Xe_2\,(0_u^+)$	6 (*D*)

II) Mercury and mercury-halogen excimers

Excited molecule (term)	Emission wavelength for transition centre [μm]	Radiative lifetime τ [ns]
$Hg_2\,(A1_u)$	0.335 (*A*)	1200 (*D*)
$Hg_2\,(B0_u^+)$	0.225 (*A*)	2.4 (*C*)
$HgCl\,(B_{1/2})$	0.56 (*C*)	27 (*C*)
$HgBr\,(B_{1/2})$	0.50 (*C*)	24 (*C*)
$HgI\,(B_{1/2})$	0.44 (*C*)	27 (*D*)
Hg_3^*	0.485 (*B*)	$1.7 \cdot 10^4$ (*D*)

III) Diatomic rare gas halides

Excited molecule (term)	Emission wavelength for transition centre [μm]	Radiative lifetime τ [ns]
$NeF\,(B_{1/2})$	0.108 (*C*)	2.5 (*D*)
$ArF\,(B_{1/2})$	0.193 (*B*)	4 (*D*)
$ArCl\,(B_{1/2})$	0.175 (*C*)	9 (*D*)
$KrF\,(B_{1/2})$	0.248 (*B*)	8 (*C*)
$KrCl\,(B_{1/2})$	0.222 (*B*)	19 (*D*)
$XeF\,(B_{1/2})$	0.352 (*A*)	16 (*C*)
$XeF\,(C_{3/2})$	0.450 (*C*)	100 (*C*)
$XeF\,(D_{1/2})$	0.260 (*C*)	11 (*C*)
$XeCl\,(B_{1/2})$	0.308 (*C*)	11 (*C*)
$XeCl\,(C_{3/2})$	0.330 (*C*)	120 (*D*)
$XeCl\,(D_{1/2})$	0.236 (*C*)	10 (*D*)
$XeBr\,(B_{1/2})$	0.282 (*C*)	15 (*D*)
$XeBr\,(C_{3/2})$	0.302 (*C*)	120 (*D*)
$XeBr\,(D_{1/2})$	0.221 (*C*)	10 (*D*)
$XeI\,(B_{1/2})$	0.254 (*C*)	14 (*D*)
$XeI\,(C_{3/2})$	0.292 (*C*)	110 (*D*)
$XeI\,(D_{1/2})$	0.208 (*C*)	10 (*D*)
$Ar_2F\,(^2B_2)$	0.284 (*C*)	160 (*C*)
$Kr_2F\,(^2B_2)$	0.420 (*C*)	170 (*C*)
$Xe_2Cl\,(^2B_2)$	0.490 (*C*)	150 (*D*)

10.9 Einstein Coefficients for Spontaneous Emission from Vibrationally Excited Diatomic Molecules

In Figs. 10.15–21 detailed information is presented about the infrared spectra of the vibrationally excited diatomic molecules CO, NO, OH, HF, DF, HCl and DCl [10.9.1]. The energy levels, energy differences [cm^{-1}] and one-quantum and two-quantum transition (air) wavelengths [μm] are given. Values of the Einstein coefficient $A_{v''}^{v'}$ [s^{-1}] for spontaneous emission from the vibrationally excited state of a molecule in the ground electronic state are also included. They were calculated from the corresponding transition moments given in [10.9.2–6].

The numerical values listed on the figures are considered reliable to ± 1 in the last significant figure quoted.

References

10.9.1 K.P.Huber, G.Herzberg: *Molecular Spectra and Molecular Structure. IV. Constants of Diatomic Molecules* (Van Nostrand Reinhold, New York 1979)

10.9.2 G.Hancock, I.W.M.Smith: Appl. Optics **10**, 1827 (1971) (CO)

10.9.3 F.P.Billingsley II: J. Mol. Spectrosc. **61**, 53 (1976) (NO)

10.9.4 F.H.Mies: J. Mol. Spectrosc. **53**, 150 (1974) (OH)

10.9.5 G.Emanuel, N.Cohen, T.A.Jacobs: J. Quant. Spectrosc. Radiat. Transfer **13**, 1365 (1973) (HF)

10.9.6 J.M.Herbelin, G.Emanuel: J. Chem. Phys. **60**, 689 (1974) (HF, DF, HCl, DCl)

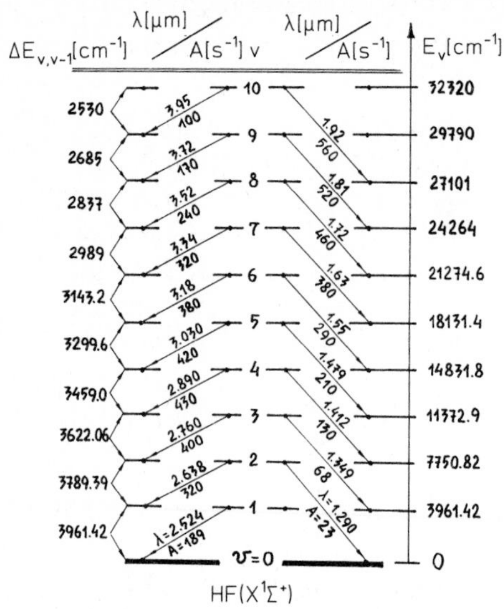

Fig. 10.15. Energy-level diagram of vibrationally excited ground-state HF molecule with one- and two-quantum transition wavelengths and Einstein coefficients

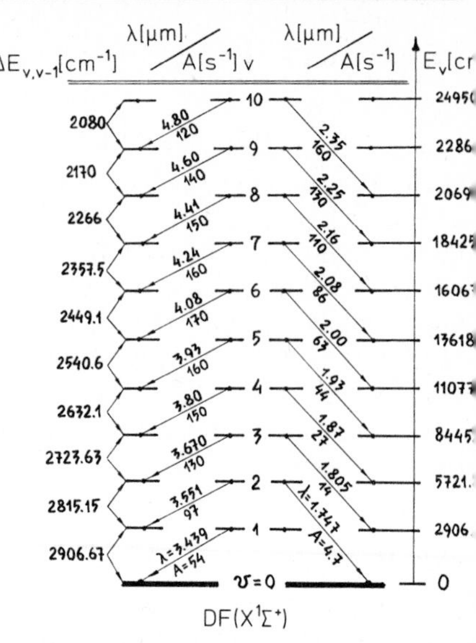

Fig. 10.16. Energy-level diagram of vibrationally excited ground state DF molecule with one- and two-quantum transition wavelengths and Einstein coefficients

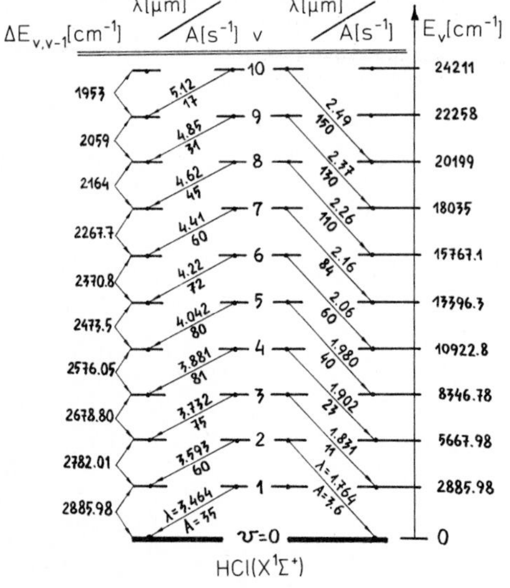

Fig. 10.17. Energy-level diagram of vibrationally excited ground state HCl molecule with one- and two-quantum transition wavelengths and Einstein coefficients

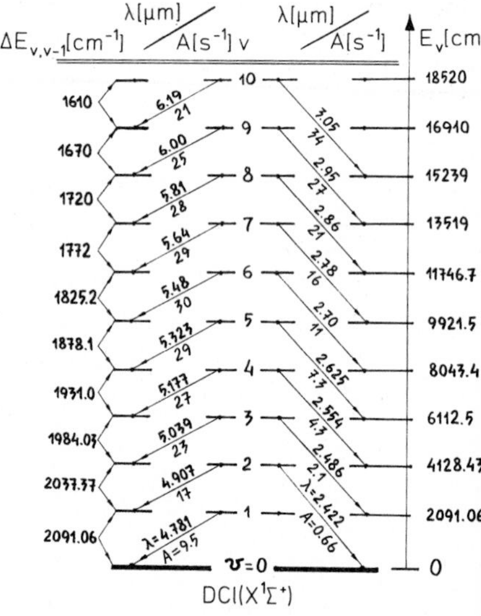

Fig. 10.18. Energy-level diagram of vibrationally excited ground state DCl molecule with one- and two-quantum transition wavelengths and Einstein coefficients

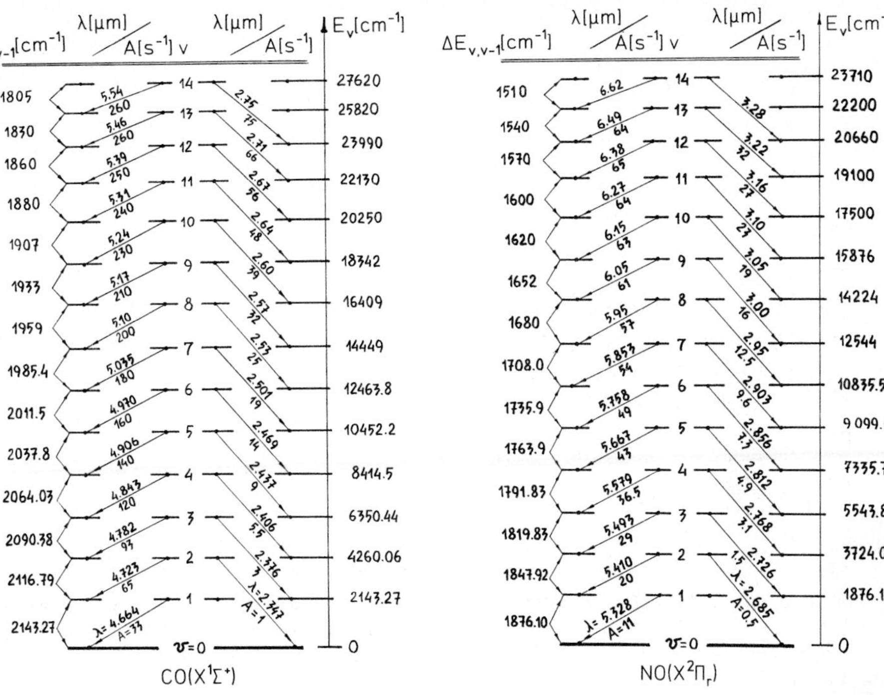

Fig. 10.19. Energy-level diagram of vibrationally excited ground state CO molecule with one- and two-quantum transition wavelengths and Einstein coefficients

Fig. 10.20. Energy-level diagram of vibrationally excited ground state NO molecule with one- and two-quantum transition wavelengths and Einstein coefficients

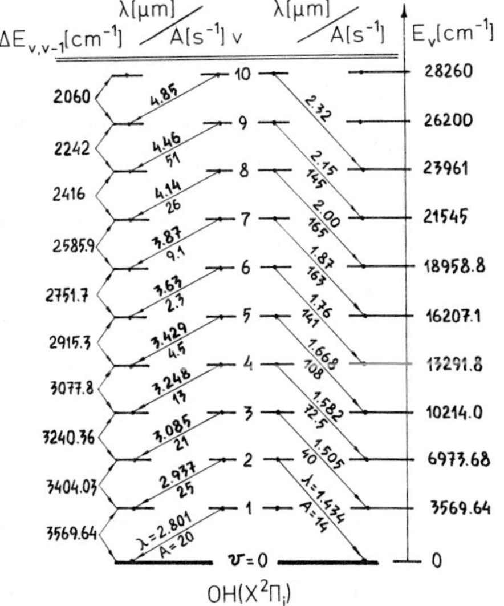

Fig. 10.21. Energy-level diagram of vibrationally excited ground state OH molecule with one- and two-quantum transition wavelengths and Einstein coefficients

11. Diatomic Molecular Ions

The normal electronic configurations and asymptotic parameters of valence electron wavefunctions in diatomic molecular ions are presented. Numerical data are compiled for spectroscopic constants, dissociation energies and radiative lifetimes of electronically excited diatomic molecular ions and, also, for the electron affinity of diatomic molecules and the proton affinity of atoms.

11.1 Electron Configurations and Asymptotic Parameters of Wavefunctions for Valence Electrons in Diatomic Molecular Ions

Table 11.1 provides information about the ground-state electron configurations of some diatomic molecular ions (see Sect. 10.1 for the details of state designations, etc.). The dissociation products of the molecular ions and their related atomic states are also indicated in a separate column.

The numerical values of asymptotic coefficients of wavefunctions for valence electrons in some negative diatomic molecular ions are presented in Table 11.2 (Sect. 10.2 gives details of wavefunction asymptotic expansions). The parameters $A(\theta)$ and γ characterize the one-electron distribution (valence orbital) in the range where the distance r from the centre of the molecular ion is large compared to the mean size of the ions themselves.

Most of the values have been rounded off to give uncertainties of a few units in the last place. The data mentioned above are based on [10.1.1–3, 10.2.1–3].

Table 11.1. Electronic configurations and terms of diatomic molecular ions

Molecule	Electronic term	MO configuration	Dissociation products and their states
C_2^-	$^2\Sigma_g^+$	$(\sigma_g 1s)^2(\sigma_u 1s)^2(\sigma_g 2s)^2(\sigma_u 2s)^2(\pi_u 2p)^4(\sigma_g 2p)$	$C(^3P) + C^-(^4S)$
CO^+	$X^2\Sigma^+$	$1\sigma^2 2\sigma^2 3\sigma^2 4\sigma^2 1\pi^4 5\sigma$	$C^+(^2P) + O(^3P)$
H_2^+	$X^2\Sigma_g^+$	$1s\sigma_g$	$H(^2S) + H^+$
HeH^+	$X^1\Sigma^+$	$(1\sigma)^2$	$He(^1S) + H^+$
N_2^+	$X^2\Sigma_g^+$	$1\sigma_g^2 1\sigma_u^2 2\sigma_g^2 2\sigma_u^2 1\pi_u^4 3\sigma_g$	$N(^4S^o) + N^+(^3P)$
NO^+	$X^1\Sigma^+$	$(1s\sigma)^2(2s\sigma)^2(2p\sigma)^2(2p\pi)^4(3s\sigma)^2(3p\sigma)^2$	$N(^4S^o) + O^+(^4S)$
NO^-	$X^3\Sigma^-$	$(1s\sigma)^2(2s\sigma)^2(2p\sigma)^2(2p\pi)^4(3s\sigma)^2(3p\sigma)^2(3p\pi)^2$	$N(^4S^o) + O^-(^2P)$
O_2^+	$X^2\Pi_{g,i}$	$1\sigma_g^2 1\sigma_u^2 2\sigma_g^2 2\sigma_u^2 3\sigma_g^2 1\pi_u^4 1\pi_g$	$O(^3P) + O^+(^4S)$
O_2^-	$X^2\Pi_{1/2,g}$	$1\sigma_g^2 1\sigma_u^2 2\sigma_g^2 2\sigma_u^2 3\sigma_g^2 1\pi_u^4 1\pi_g^3$	$O(^3P) + O^-(^2P)$
OH^+	$X^3\Sigma^-$	$1\sigma^2 2\sigma^2 3\sigma^2 1\pi^2$	$O^+(^4S) + H(^2S)$

Table 11.2. Asymptotic parameters of valence electron wavefunctions for diatomic negative ions

Molecule and ground-state term	Valence molecular orbital	Asymptotic parameters [a.u.]						
		γ	R_e	a	b	c	α	β
$C_2^-(X^2\Pi_u)^a$	$1\pi_u$	0.510	2.392	0.7	2.0	0.8	0	0
$O_2^-(X^2\Pi_g)^a$	$1\pi_g$	0.180	2.56	1.5	2.5	0.4	1	1
$OH^-(X^1\Sigma^+)^b$	1π	0.365	4.879	0.8	0.3	–	–	–
$NH^-(X^2\Pi_i)^b$	1π	0.167	1.151	0.3	0.1	–	–	–
$CH^-(X^3\Sigma^-)^b$	1π	0.233	0.953	0.6	0.2	–	–	–

[a] $A(\theta) = a\,\mathrm{ch}(b\gamma R_e \cos\theta)(1 + c\cos^2\theta)\sin^\alpha\theta\cos^\beta\theta$
[b] $A(\theta) = [1 + b\exp(\gamma R_e \cos\theta)]\sin\theta$

11.2 Spectroscopic Constants of Diatomic Molecular Ions

Table 11.3 presents spectroscopic constants of diatomic molecular ions. It includes mainly the numerical data for well-known low-lying electronic states and neglects questionable values. The information contained in each column of this table was briefly commented on in Sect. 10.3. We assume that the numerical values are accurate to one or two units in the last significant figure given. The data listed are based on [10.3.1–6].

Table 11.3. Spectroscopic constants of diatomic molecular ions

Ion	Electronic term	Term energy T_e [cm^{-1}]	Equilibrium internuclear distance r_e [Å]	Vibrational frequency ω_e [cm^{-1}]	Anharmonic constant $\omega_e x_e$ [cm^{-1}]	Rotational constant B_e [cm^{-1}]	Rotation-vibration interaction constant α_e [10^{-3} cm^{-1}]	Force constant k_e [10^5 dyn/cm]	Observable electronic transition	Reduced mass μ_A [a.m.u.]
1	2	3	4	5	6	7	8	9	10	11
AlH^+	$X^2\Sigma^+$	0	1.602	≈ 1620	–	6.76	400	1.5	–	0.972
ArH^+	$X^1\Sigma^+$	0	1.28	2589.28*	61	10.46	375	4.26	–	0.983
AsO^+	$A^1\Pi$	42594.2	1.69	780.8	7.4	0.449	4.7	4.73	$A \to X$	–
	$X^1\Sigma^+$	0	1.57	1091.3*	5.0	0.520	3.1	9.4	–	13.18
AsS^+	$A^1\Pi$	37359.7	2.08	441.2	3.4	0.173	1.2	2.57	$A \to X$	–
	$X^1\Sigma^+$	0	1.945	644.2	2.1	0.199	0.9	5.49	–	22.45
BeH^+	$A^1\Sigma^+$	39417.0	1.609	1476.1	14.8	7.184	125	1.16	$A \to X$	–
	$X^1\Sigma^+$	0	1.312	2221.7	39.8	10.80	294	2.64	–	0.907
C_2^-	$B^2\Sigma_u^+$	18390.88	1.223	1968.7	14.43	1.877	17.8	13.7	$B \leftrightarrow X$	–
	$X^2\Sigma_g^+$	0	1.2682	1781.0	11.58	1.747	16.7	11.2	–	6.006
CH^+	$B^1\Delta$	52530	1.232	2075.5	76.3	11.94	620	2.36	$B \to A$	–
	$b^3\Sigma^-$	38200	1.245	1940*	–	11.71	540	–	$b \to a$	–
	$A^1\Pi$	23596.9*	1.234	1865.3	115.8	11.90	941	1.91	$A \leftrightarrow X$	–
	$a^3\Pi$	≈ 9200	1.136	2810	–	14.05	603	4.3	–	–
	$X^1\Sigma^+$	0	1.131	2739.7*	64	14.178	492	4.5	–	0.930
CN^+	$f^1\Sigma$	$x + 45533.6$	1.171	2670.5	47	1.903	32	27.2	$f \to b, a$	–
	$c^1\Sigma$	$x + 31770$	1.36	1265	11	1.40	2	6.10	$c \to a$	–
	$b^1\Pi$	$x + 8313.6$	1.247	1688.3	15.1	1.677	19.1	10.9	–	–
	$a^1\Sigma$	x	1.173	2033.0	16.1	1.896	18.8	15.8	–	6.47
CO^+	$C^2\Delta_r$	63012	1.35	1144	33.3	1.357	24	5.29	$C \to A$	–
	$B^2\Sigma^+$	45876.7	1.169	1734.2	27.9	1.800	30.3	12.2	$B \leftrightarrow X$, $B \to A$	–
	$A^2\Pi_i$	20733.3	1.244	1562.1	13.5	1.589	19.4	9.86	$A \to X$	–
	$X^2\Sigma^+$	0	1.115	2214.2	15.2	1.977	19.0	19.8	–	6.8606

CS^+	$A\,^2\Pi_i$	11990	1.641	1013	6.5	0.718	6.2	5.28	$A \rightarrow X$	–
	$X\,^2\Sigma^+$	0	1.495*	138	–	0.864*	–	9.9	–	8.738
CdH^+	$A\,^1\Sigma^+$	42934.1	1.865	1252	8.6	4.85	82	0.92	$A \rightarrow X$	–
	$X\,^1\Sigma^+$	0	1.667	1772.5	35.4	6.07	190	1.85	–	0.999
Cl_2^+	$X_2\,^2\Pi_{1/2g}$	645	1.891	644.8	2.99	0.2697	1.67	4.34	–	–
	$X_1\,^2\Pi_{3/2g}$	0	1.892	645.6	3.02	0.2695	1.64	4.35	–	17.73
F_2^+	$X\,^2\Pi_{g,i}$	0	1.32	1073	9.1	1.01	10	6.45	–	9.50
H_2^+	$X\,^2\Sigma_g^+(1s\sigma)$	0	1.05	2322	66	30.2	1680	1.60	–	0.504
HBr^+	$A\,^2\Sigma^+$	28421	1.684	1404	37.7	5.970	248	1.16	$A \rightarrow X$	–
	$X\,^2\Pi_i$	0	1.448	2441.5	47.4	8.072	236	3.50	–	0.995
HCl^+	$A\,^2\Sigma^+$	28626	1.514	1606.5	40.3	7.505	331	2.90	$A \rightarrow X$	–
	$X\,^2\Pi_i$	0	1.3147	2673.7	52.54	9.9566	327	8.02	–	1.904
HF^+	$A\,^2\Sigma^+$	25449.8	1.224	1496.1	88.4	11.75	1026	1.26	$A \rightarrow X$	–
	$X\,^2\Pi_i$	0	1.001	3090.5	89.0	17.58	886	5.39	–	0.957
He_2^+	$X\,^2\Sigma_u^+$	0	1.081	1698	35	7.21	224	3.40	–	2.001
HeH^+	$X\,^1\Sigma^+$	0	0.774	3228	158	34.9	2640	4.94	–	0.805
$HgAr^+$	$X\,^2\Sigma^+$	0	2.87*	99	1.5	0.061*	–	0.2	–	33.31
HgH^+	$A\,^1\Sigma^+$	44317	1.693	1624	45.1	5.87	200	1.56	$A \rightarrow X$	–
	$X\,^1\Sigma^+$	0	1.594	2028	41	6.61	206	2.43	–	1.003
Li_2^+	$X\,^2\Sigma_g^+$	0	3.1	260	1.6	0.496	5	0.14	–	3.470
MgH^+	$B\,^1\Pi$	50480	2.27*	527*	–	3.38*	280	–	$B \rightarrow X$	–
	$A\,^1\Sigma^+$	35904.5	2.006	1136	8.18	4.330	68	0.74	$A \rightarrow X$	–
	$X\,^1\Sigma^+$	0	1.652	1699	31.9	6.387	182	1.65	–	0.968
N_2^+	$C\,^2\Sigma_u^+$	64609	1.263	2069	8	1.510	–1	17.7	$C \rightarrow X$	–
	$D\,^2\Pi_{g,i}$	52318	1.471	907.7	11.9	1.113	20	3.40	$D \rightarrow A$	–
	$B\,^2\Sigma_u^+$	25461.1	1.075	2421	24.1	2.085	21	24.2	$B \leftrightarrow X$	–
	$A\,^2\Pi_{u,i}$	9167.5	1.175	1903.46	15.02	1.7444	18.9	15.0	$A \rightarrow X$	–
	$X\,^2\Sigma_g^+$	0	1.1164	2207.27	16.26	1.9318	19.0	20.1	–	7.003
NH^+	$C\,^2\Sigma^+$	35000	1.163	2150.5	73.1	13.26	790	2.56	$C \rightarrow X$	–
	$A\,^2\Sigma^-$	22200	1.251	1707	61	11.46	690	1.6	$A \rightarrow X$	–
	$X\,^2\Pi_r$	0	1.07	2922*	–	15.3*	640	–	–	0.940

Table 11.3 (continued)

Ion	Electronic term	Term energy T_e [cm^{-1}]	Equilibrium internuclear distance r_e [Å]	Vibrational frequency ω_e [cm^{-1}]	Anharmonic constant $\omega_e x_e$ [cm^{-1}]	Rotational constant B_e [cm^{-1}]	Rotation-vibration interaction constant α_e [10^{-3} cm^{-1}]	Force constant k_e [10^5 dyn/cm]	Observable electronic transition	Reduced mass μ_A [a.m.u.]
1	2	3	4	5	6	7	8	9	10	11
NO^+	$A\,^1\Pi$	73471.8	1.194	1601.9	20.2	1.584	22	11.3	$A \to X$	–
	$W\,^1\Delta$	71650	1.30	1220	12	1.33	17	6.52	–	–
	$A'\,^1\Sigma^-$	69540	1.29	1280	13.2	1.36	18	7.21	–	–
	$b'\,^3\Sigma^-$	67780	1.28	1283	11	1.39	25	7.25	$b' \to X$	–
	$w\,^3\Delta$	61910	1.28	1320	11	1.38	16	7.63	–	–
	$b\,^3\Pi$	59180	1.176	1710	14.2	1.634	18.4	13	$b \to X$	–
	$a\,^3\Sigma^+$	52150	1.28	1300	15	1.37	20	7.47	–	–
	$X\,^1\Sigma^+$	0	1.063	2376.7	16.3	1.997	18.8	24.86	–	7.468
NO^-	$X\,^3\Sigma^-$	EA(NO)=240	1.26	1360	8	1.43	–	8.2	–	7.468
NS^+	$X\,^1\Sigma^+$	0	1.44	1410	15	0.83	10	11.5	–	9.748
O_2^+	$c\,^4\Sigma_u^-$	100910	1.16*	1545*	–	1.56*	–	–	$c \to b$	–
	$b\,^4\Sigma_g^-$	49550	1.280	1196.8	17.1	1.287	22.1	6.75	$b \to a$	–
	$A\,^2\Pi_u$	40669	1.409	898.2	13.6	1.062	19.4	3.80	$A \to X$	–
	$a\,^4\Pi_{u,i}$	32960	1.381	1035.7	10.4	1.105	15.8	5.06	–	–
	$X\,^2\Pi_g$	{197.3 0	1.116	1904.8	16.3	1.691	19.8	17.1	–	8.000
O_2^-	$X\,^2\Pi_{g,i}$ {3/2 1/2	160 0	1.35	1090	8	1.16	–	5.6	–	8.000
OH^+	$b\,^1\Sigma^+$	29050	1.03	2980*	–	16.32*	730	–	–	–
	$A\,^3\Pi_i(1)$	28439	1.135	2133.6	79.5	13.792	890	2.54	$A \to X$	–
	$X\,^3\Sigma^-$	0	1.029	3113.4	78.5	16.794	750	5.41	–	0.948
OH^-	$X\,^1\Sigma^+$	0	0.964	3700	–	19.13	770	7.6	–	0.948

P_2^+	$C_2\,{}^2\Pi_g$	3/2	28871	2.243	441.5	2.6	0.2163	1.4	1.78	$C_2 \to X_1$	–
	C_1	1/2	28687							$C_1 \to X_2$	
	$B\,{}^2\Sigma_u^+$		25570	2.12	410.5	3.2	0.242	2.1	1.54	$B \to A$	–
	$D_2\,{}^2\Pi_g$	1/2	18832	2.23	462	2.4	0.220	1.4	1.95	$D_2 \to X_2$	–
	D_1	3/2	18741							$D_1 \to X_1$	
	$A\,{}^2\Sigma_g^+$		2180	1.89	733*	–	0.304	2	–	–	–
	$X_2\,{}^2\Pi_u$	1/2	260	1.986	672.2	2.7	0.2760	1.5	4.12	–	15.49
	X_1	3/2	0								
PF^+	$A\,{}^2\Sigma$		35435	1.600	619.0	4.6	0.559	8	2.66	$A \to X$	–
	$X\,{}^2\Pi_r$		0	1.500	1053.2	5.1	0.636	5	7.70	–	11.78
PH^+	$A\,{}^2\Delta_r$		26220	1.549	1535	69	7.196	422	1.35	$A \to X$	–
	$X\,{}^2\Pi_r$		0	1.425	2382.7	41.7	8.509	244	3.27	–	0.976
SH^+	$A\,{}^3\Pi_i$		29912*	1.52*	–	–	7.475*	–	–	$A \to X$	–
	$X\,{}^3\Sigma^-$		0	1.374*	2560	50	9.134*	280	3.8	–	0.977
SO^+	$A\,{}^2\Pi_i$		31422.7	1.657	805.4	6.3	0.5759	5.8	4.07	$A \to X$	–
	$X\,{}^2\Pi_r$		{340, 0}	1.424	1307.1	7.74	0.7800	6.3	10.7	–	10.67
SiH^+	$A\,{}^1\Pi$		25846	1.878	390.2*	70	4.912	767	0.16	$A \to X$	–
	$X\,{}^1\Sigma^+$		0	1.504	2157.2	34.2	7.660	210	2.67	–	0.973
ZnH^+	$A\,{}^1\Sigma^+$		46700	1.72	1360	15	5.77	105	1.09	$A \to X$	–
	$X\,{}^1\Sigma^+$		0	1.51	1920	40	7.41	240	2.15	–	0.993

11.3 Dissociation Energies of Diatomic Molecular Ions

Tables 11.4, 5 give the values of ground-state dissociation energies (D_0) of diatomic molecular ions which correspond to the transition from the zero vibrational level to the continuum of atomic states. Dissociation energies of positive and negative molecular ions were calculated on the basis of thermochemical relations for the separate stages in the cycles (see also Fig. 11.1):

$$D_0(\mathrm{A-B^+}) = D_0(\mathrm{A-B}) + \mathrm{IP(B)} - \mathrm{IP(AB)}\ ,$$

$$D_0(\mathrm{A-B^-}) = D_0(\mathrm{A-B}) + \mathrm{EA(AB)} - \mathrm{EA(B)}\ ,$$

where IP is the ionization potential of a particle and EA the electron affinity.

In selecting the data for Tables 11.4, 5 we made use of [11.3.1–4] and some later publications. The numerical values listed are grouped into the accuracy classes defined in the Introduction.

References

11.3.1 H.M.Rosenstock, K.Draxl, B.W.Steiner, J.T.Herron: "Energetics of Gaseous Ions", J. Phys. Chem. Ref. Data **6**, Suppl. 1 (1977)

11.3.2 L.V.Gurvitch, G.V.Karachevtsev, V.N.Kondratjev, Y.A.Lebedev, V.A.Medvedev, V.K.Potapov, Y.S.Hodeev: *Bond Dissociation Energies, Ionization Potentials, and Electron Affinities*, 2nd ed. (Nauka, Moscow 1974) (in Russian)

11.3.3 K.P.Huber, G.Herzberg: *Molecular Spectra and Molecular Structure. IV. Constants of Diatomic Molecules* (Van Nostrand Reinhold, New York 1979)

11.3.4 K.S.Krasnov (ed.): *Molecular Constants for Inorganic Compounds* (Chimia, Moscow 1979) (in Russian)

Table 11.4. Dissociation energies D_0 [eV] of positive diatomic molecular ions

I) Homonuclear system: X_2^+

X =	Ag	Ar	As	Br	C	Cl	Cs
D_0	1.9 (*D*)	1.23 (*B*)	2.7 (*D*)	3.26 (*B*)	5.3 (*C*)	3.99 (*B*)	0.62 (*C*)
X =	D	F	Ge	H	He	Hg	I
D_0	2.692 (*A*)	3.34 (*B*)	2.8 (*D*)	2.650 (*A*)	2.36 (*A*)	0.9 (*D*)	2.68 (*A*)
X =	K	Kr	Li	N	Na	Ne	O
D_0	0.79 (*B*)	1.15 (*B*)	1.29 (*A*)	8.713 (*A*)	0.98 (*B*)	1.16 (*C*)	6.66 (*A*)
X =	P	Pb	Rb	S	Sb	Se	Sr
D_0	5.0 (*B*)	1.7 (*D*)	0.75 (*C*)	5.4 (*B*)	2.7 (*D*)	4.4 (*B*)	0.77 (*C*)
X =	Te	Ti	V	Xe			
D_0	3.4 (*B*)	1.8 (*D*)	2.8 (*D*)	1.03 (*C*)			

Table 11.4 (continued)

II) Heteronuclear system: XY^+

Ion	CH^+	CN^+	CP^+	CS^+	$CaCl^+$	CdH^+	$CoCl^+$
D_0	4.08 (*A*)	4.9 (*B*)	5.2 (*C*)	6.38 (*A*)	4.2 (*C*)	2.1 (*D*)	3.0 (*D*)
Ion	$GeBr^+$	GeC^+	$GeCl^+$	GeH^+	$GeSi^+$	$GeTe^+$	HBr^+
D_0	4.1 (*C*)	2.3 (*D*)	4.9 (*C*)	3.9 (*C*)	2.8 (*D*)	4.2 (*D*)	3.89 (*A*)
Ion	HCl^+	HI^+	OH^+	HS^+	InS^+	$InSe^+$	IrC^+
D_0	4.65 (*A*)	3.12 (*A*)	5.1 (*B*)	3.5 (*C*)	1.7 (*D*)	1.3 (*D*)	6.1 (*D*)
Ion	IBr^+	ICl^+	$MgCl^+$	NF^+	PF^+	PCl^+	PS^+
D_0	2.4 (*B*)	2.5 (*B*)	3.4 (*D*)	5.54 (*A*)	5.2 (*B*)	3.0 (*D*)	6.5 (*D*)
Ion	$PbCl^+$	PbS^+	$PbSe^+$	$PbTe^+$	$SiBr^+$	SiC^+	$SnBr^+$
D_0	3.0 (*D*)	2.4 (*D*)	2.0 (*D*)	1.7 (*D*)	4.4 (*C*)	3.5 (*D*)	3.5 (*D*)
Ion	SnC^+	$SnSe^+$	$SnTe^+$	$SrCl^+$	TiS^+		
D_0	2.4 (*D*)	1.7 (*D*)	1.6 (*D*)	4.2 (*C*)	4.1 (*D*)		

III) Rare gas system: $R\text{–}R^+$

	He	Ne	Ar	Kr	Xe
He^+	2.37 (*A*)	0.047 (*C*)	0.17 (*C*)	0.22 (*C*)	0.28 (*C*)
Ne^+	0.69 (*C*)	1.16 (*C*)	–	–	–
Ar^+	0.06 (*D*)	0.08 (*C*)	1.23 (*B*)	–	0.59 (*C*)
Kr^+	–	0.055 (*C*)	0.59 (*C*)	1.15 (*B*)	–
Xe^+	–	0.041 (*C*)	0.14 (*D*)	0.37 (*C*)	1.03 (*C*)

IV) Heteronuclear system: $X^+\text{–O}$

X = Al	As	B	Ba	Be	C	Ca	Ce	Cl
D_0 1.78 (*B*)	7.7 (*C*)	3.1 (*D*)	4.1 (*C*)	3.8 (*C*)	8.34 (*A*)	3.9 (*D*)	8.5 (*D*)	5.3 (*D*)
X = Co	Cr	Fe	Ga	Ge	Mn	N	Ni	O
D_0 2.8 (*C*)	3.4 (*B*)	3.0 (*C*)	0.5 (*D*)	4.6 (*C*)	2.5 (*C*)	10.85 (*A*)	1.9 (*C*)	6.66 (*A*)
X = P	Pb	Pd	Pr	Rh	S	Si	Sn	Sr
D_0 8.7 (*B*)	2.2 (*D*)	1.6 (*D*)	8.5 (*C*)	2.2 (*D*)	5.4 (*B*)	5.0 (*B*)	2.3 (*D*)	4.0 (*D*)
X = Ta	Te	Th	Ti	U	V	W	Zr	
D_0 10.3 (*C*)	4.2 (*C*)	8.9 (*C*)	6.8 (*C*)	8.1 (*C*)	7.7 (*D*)	5.6 (*D*)	8.0 (*C*)	

V) Heteronuclear system: $X^+\text{–F}$

X = Br	C	Ca	Cl	Cu	Dy	Er	Eu	Ga
D_0 2.6 (*C*)	8.0 (*C*)	5.6 (*C*)	2.9 (*C*)	2.8 (*D*)	5.2 (*C*)	5.5 (*D*)	5.1 (*D*)	1.3 (*D*)
X = Gd	Ge	H	He	Ho	I	In	Kr	La
D_0 5.9 (*D*)	5.6 (*C*)	3.42 (*A*)	1.4 (*D*)	5.6 (*D*)	2.8 (*C*)	1.5 (*D*)	1.6 (*C*)	5.5 (*D*)
X = Mg	N	Ne	Ni	P	Pb	S	Sc	Si
D_0 4.6 (*C*)	5.2 (*D*)	1.3 (*D*)	4.2 (*C*)	6.7 (*D*)	3.6 (*D*)	4.1 (*C*)	6.4 (*C*)	6.4 (*C*)
X = Sn	Sr	Y						
D_0 4.9 (*C*)	6.4 (*C*)	6.2 (*C*)						

Table 11.4 (continued)

VI) Potential parameters of alkali diatomic molecular ions (upper line: D_0 – dissociation energy in eV; lower line: r_m – equilibrium distance between nuclei in Å)

	Li	Na	K	Rb	Cs
Li^+	1.28 (*A*) 3.23 (*B*)	0.94 (*C*) 3.38 (*B*)	0.56 (*D*) 3.72 (*B*)	0.42 (*D*) 3.8 (*C*)	0.28 (*D*) 4.0 (*C*)
Na^+	–	1.0 (*C*) 3.43 (*B*)	0.61 (*C*) 3.9 (*C*)	0.49 (*D*) 4.0 (*C*)	0.36 (*D*) 4.3 (*C*)
K^+	–	–	0.80 (*C*) 4.14 (*B*)	0.74 (*C*) 4.3 (*C*)	0.57 (*D*) 4.6 (*C*)
Rb^+	–	–	–	0.74 (*B*) 4.52 (*B*)	0.60 (*C*) 4.6 (*C*)
Cs^+	–	–	–	–	0.65 (*C*) 4.78 (*B*)

Table 11.5. Dissociation energies D_0 [eV] of negative diatomic molecular ions

I) Homonuclear system: X_2^-

X =	Br	C	Cl	F	I	Li	O	S
D_0	1.15 (*B*)	8.5 (*B*)	1.3 (*B*)	1.3 (*C*)	1.0 (*C*)	0.9 (*D*)	4.09 (*A*)	3.95 (*A*)

II) Heteronuclear system: XY^-

Ion	BO^-	BeO^-	CH^-	CN^-	CS^-	$KrCl^-$	$LiCl^-$	LiF^-
D_0	9.6 (*C*)	4.9 (*C*)	3.43 (*A*)	10.3 (*A*)	6.3 (*B*)	0.13 (*D*)	1.8 (*C*)	2.9 (*C*)
Ion	LiH^-	NH^-	NO^-	NS^-	$NaCl^-$	OH^-	PH^-	PO^-
D_0	2.0 (*C*)	3.1 (*C*)	5.06 (*A*)	3.9 (*C*)	1.3 (*D*)	4.752 (*A*)	3.3 (*C*)	5.8 (*B*)
Ion	SF^-	SH^-	SO^-	SeF^-	SeH^-	SeS^-	SiH^-	$XeCl^-$
D_0	2.8 (*D*)	3.8 (*B*)	2.4 (*C*)	2.8 (*D*)	3.2 (*D*)	4.4 (*D*)	3.0 (*C*)	0.32 (*C*)

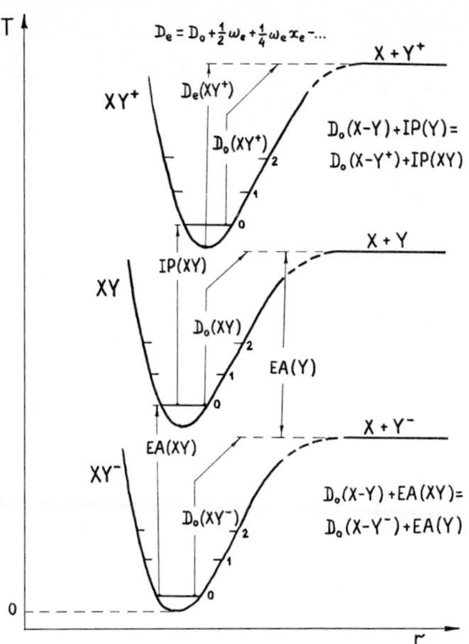

Fig. 11.1. Schematic diagram of electronic terms for molecular species XY, XY^+ and XY^- showing their relative positions and explaining some energy relations

11.4 Electron Affinities of Diatomic Molecules

Table 11.6 lists the values of the electron affinity (EA) of neutral diatomic molecules, characterizing the lowest energy required to remove an electron from the molecular negative ion. The numerical data on EAs were grouped into the accuracy classes defined in the Introduction. The main information about electron affinities of neutral molecules is collected in [11.4.1–7].

References

11.4.1 H.S.W.Massey: *Negative Ions*, 3rd ed. (Cambridge University Press, Cambridge 1976)

11.4.2 B.M.Smirnov: *Negative Ions* (McGraw-Hill, New York 1981)

11.4.3 B.K.Janousek, J.I.Brauman: "Electron Affinities", in *Gas Phase Ion Chemistry*, Vol. 2, ed. by M.T.Bowers (Academic, New York 1979) Chap. 10, pp. 53–86

11.4.4 R.R.Corderman, W.C.Lineberger: "Negative Ion Spectroscopy", Annu. Rev. Phys. Chem. **30**, 347–378 (1979)

11.4.5 H.M.Rosenstock, K.Draxl, B.W.Steiner, J.T.Herron: "Energetics of Gaseous Ions", J. Phys. Chem. Ref. Data **6**, Suppl. 1 (1977)

11.4.6 L.V.Gurvitch, G.V.Karachevtsev, V.N.Kondratjev, Y.A.Lebedev, V.A.Medvedev, V.K.Potapov, Y.S.Hodeev: *Bond Dissociation Energies, Ionization Potentials, and Electron Affinities*, 2nd ed. (Nauka, Moscow 1974) (in Russian)

11.4.7 P.S.Drzaic, J.Marks, J.I.Brauman: "Electron Photodetachment from Gas Phase Molecular Anions", in *Gas Phase Ion Chemistry*, Vol. 3, ed. by M.T.Bowers (Academic, New York 1984) Chap. 21, pp. 167–211

Table 11.6. Electron affinities (EA) of diatomic molecules

Molecule or radical	EA [eV]	Molecule or radical	EA [eV]
Al_2	2.42 (*C*)	Li_2	0.7 (*D*)
AlO	3.6 (*C*)	LiCl	0.6 (*D*)
AsBr	1.3 (*C*)	LiF	0.44 (*C*)
AsCl	1.3 (*C*)	LiH	0.3 (*D*)
AsF	1.3 (*C*)	LiN	0.4 (*D*)
AsH	1.0 (*D*)	MgH	1.05 (*C*)
BO	3.0 (*C*)	NH	0.38 (*B*)
Be_2	0.3 (*D*)	NO	0.03 (*D*)
BeH	0.7 (*D*)	NS	1.19 (*A*)
BeO	1.8 (*C*)	NaCl	0.66 (*C*)
Br_2	2.6 (*C*)	NaH	0.32 (*D*)
C_2	3.39 (*B*)	O_2	0.44 (*B*)
CBr	1.7 (*C*)	OH	1.8277 (*A*)
CF	3.3 (*C*)	OD	1.8255 (*A*)
CH	1.24 (*A*)	PH	1.03 (*A*)
CN	3.82 (*A*)	PO	1.11 (*B*)
CS	0.21 (*C*)	S_2	1.66 (*B*)
CaH	0.93 (*C*)	SF	2.5 (*C*)
Cl_2	2.44 (*B*)	SH	2.31 (*A*)
ClO	2.0 (*D*)	SO	1.09 (*C*)
F_2	2.96 (*B*)	SeH	2.21 (*C*)
FeO	1.49 (*B*)	SiH	1.28 (*A*)
I_2	2.51 (*B*)	ZnH	0.95 (*D*)
IBr	2.6 (*C*)		

11.5 Proton Affinities of Atoms

The proton affinity (PA) of an atom is defined as the energy released in the protonation reaction: $A + H^+ \rightarrow AH^+$. The most important methods for determining the atomic PAs are the collisional ionization technique (based on the measurement of the appearance potential in mass spectrometers) and thermochemical calculations [11.5.1].

The numerical values of atomic PAs in Table 11.7 are grouped according to accuracy classes (see the Introduction) and here we followed the authors' estimated errors. The basic information about the proton affinity of atoms may be found in [11.5.1–4].

References

11.5.1 L.V.Gurvitch, G.V.Karachevtsev, V.N.Kondratjev, Y.A.Lebedev, V.A.Medvedev, V.K.Potapov, Y.S.Hodeev: *Bond Dissociation Energies, Ionization Potentials, and Electron Affinities*, 2nd ed. (Nauka, Moscow 1974) (in Russian)

11.5.2 R.Walder, J.L.Franklin: Int. J. Mass Spectrom. Ion Phys. **36**, 85 (1980)

11.5.3 D.K.Bohme, G.I.Mackay, H.I.Schiff: J. Chem. Phys. **73**, 4976 (1980)

11.5.4 S.G.Lias, J.F.Liebman, R.D.Levin: J. Phys. Chem. Ref. Data **13**, 695 (1984)

Table 11.7. Proton affinities (PA) of atoms

Atom	PA [eV]	Atom	PA [eV]
Ar	3.87 (*A*)	N	3.4 (*C*)
Br	5.7 (*B*)	Ne	2.08 (*B*)
Cl	5.3 (*B*)	O	5.1 (*B*)
Cs	7.6 (*C*)	S	6.9 (*C*)
F	3.42 (*A*)	Xe	5.1 (*C*)
H	2.650 (*A*)	Zn	6.8 (*C*)
He	1.845 (*A*)		
I	6.3 (*B*)		
Kr	4.4 (*B*)		

11.6 Lifetimes of Excited Electron States in Diatomic Molecular Ions

Table 11.8 provides values of the radiative lifetimes of excited electron states in a number of diatomic molecular ions (see Sect. 10.7 for the details of notation and [10.7.1–7]). In the separate columns of the table one finds the symbols of excited electron states of the molecular ions and the values of the transition energy, referred to the zero vibrational levels of the upper electron state and of the ground state. The accuracy of the values of τ is indicated as usual (see the Introduction).

Table 11.8. Radiative lifetimes of electronically excited diatomic molecular ions

Ion (ground state)	Excited state	Excitation energy T_{00} [eV] relative to zero vibrational level of the ground state	Radiative lifetime τ_{el} ($v' = 0$) [µs]
CD^+ ($^1\Sigma^+$)	$A\,^1\Pi$	2.94	0.06 (D)
CH^+ ($^1\Sigma^+$)	$B\,^1\Delta$	6.46	0.23 (D)
	$b\,^3\Sigma^-$	4.6	0.48 (D)
	$A\,^1\Pi$	2.93	0.36 (D)
CO^+ ($^2\Sigma^+$)	$B\,^2\Sigma^+$	5.66	0.054 (C)
	$A\,^2\Pi_i$	2.53	3.8 (C)
CN^+ ($^1\Sigma$)	$d\,^1\Pi$	(a + 5.66)	0.024 (D)
DBr^+ ($^2\Pi_i$)	$A\,^2\Sigma^+$	3.48	3.6 (C)
F_2^+ ($^2\Pi_{g,i}$)	$A\,^2\Pi_{u,i}$	2.75	1.4 ($v' = 6$–9)(C)
HBr^+ ($^2\Pi_i$)	$A\,^2\Sigma^+$	3.46	4.4 (C)
HCl^+ ($^2\Pi_i$)	$A\,^2\Sigma^+$	3.48	2.6 (C)
N_2^+ ($^2\Sigma_g^+$)	$C\,^2\Sigma_u^+$	8.00	0.06 (D)
	$B\,^2\Sigma_u^+$	3.17	0.063 (B)
	$A\,^2\Pi_{u,i}$	1.12	17 (C)
NH^+ ($^2\Pi_r$)	$C\,^2\Sigma^+$	4.285	0.4 (D)
	$B\,^2\Delta_i$	2.85	1.0 (D)
	$A\,^2\Sigma^-$	2.67	1.1 (D)
NO^+ ($^1\Sigma^+$)	$A\,^1\Pi$	9.06	0.05 (D)
O_2^+ ($^2\Pi_g$)	$b\,^4\Sigma_g^-$	6.10	1.2 (C)
	$A\,^2\Pi_u$	4.97	0.8 (D)
OH^+ ($^3\Sigma^-$)	$A\,^3\Pi_i$	3.47	0.85 (C)
OD^+ ($^3\Sigma^-$)	$A\,^3\Pi_i$	3.48	1.0 (C)
SH^+ ($^3\Sigma^-$)	$A\,^3\Pi_i$	3.71	1.1 (C)

12. Van der Waals Molecules

The numerical parameters of potential wells are presented for loosely bound van der Waals molecules and molecular ions. The ionization potentials of van der Waals molecules are also given.

12.1 Potential Well Parameters of Van der Waals Molecules

Tables 12.1–8 incorporate the numerical values of well parameters characterizing the interaction potential $U(R)$ of van der Waals molecules with small binding energies in the vicinity of the potential minimum (see also the Abstract to Part II). These parameters are defined as follows: $U(R_{min}) = -\varepsilon$, where R_{min} is the atomic equilibrium separation in the molecule and ε the interaction energy in the potential minimum. For molecular species we give the parameters of the spherical potential only, and these equilibrium distances R_{min} are referred to the electrical centres of molecules.

All the data presented here were derived from the measured differential and total elastic cross-sections, measured bulk characteristics of a substance, spectroscopic measurements and, finally, extensive theoretical considerations [12.1.1–11]. In accordance with the established treatment of errors of measurement, the listed numerical values are grouped into the usual accuracy classes (see the Introduction).

References

12.1.1 R.B.Bernstein, J.T.Muckerman: "Determination of Intermolecular Forces via Low-Energy Molecular Beam Scattering", in *Intermolecular Forces*, ed. by J.O.Hirschfelder, Adv. Chem. Phys., Vol. 12 (Wiley, New York 1967) pp. 389–486

12.1.2 U.Buck: "Elastic Scattering", in *Molecular Scattering: Physical and Chemical Applications*, ed. by K.P.Lawley, Adv. Chem. Phys., Vol. 30 (Wiley, New York 1975) p. 313

12.1.3 U.Buck: Rev. Mod. Phys. **46**, 369 (1974)

12.1.4 G.Scoles: "Two-Body, Spherical, Atom-Atom, and Atom-Molecule Interaction Energies", Annu. Rev. Phys. Chem. **31**, 81–96 (1980)

12.1.5 H.Pauly: "Elastic Scattering Cross Sections I: Spherical Potentials", in *Atom-Molecule Collision Theory*, ed. by R.B.Bernstein (Plenum, New York 1979) pp. 111–200

12.1.6 Y.S.Kim, R.G.Gordon: J. Chem. Phys. **61**, 1 (1974)

12.1.7 J.L.Fraites, J.Bentley, D.H.Winicur: J. Phys. **B 10**, 127 (1977)

12.1.8 F.Pirani, F.Vecchiocattivi: J. Chem. Phys. **66**, 372 (1977)
12.1.9 K.T.Tang, J.P.Toennies: J. Chem. Phys. **66**, 1496 (1977)
12.1.10 J.P.Toennies, W.Welz, G.Wolf: J. Chem. Phys. **71**, 614 (1979)
12.1.11 J.T.Slankas, M.Keil, A.Kuppermann: J. Chem. Phys. **70**, 1482 (1979)

Table 12.1. Potential well parameters of diatomic van der Waals molecules: rare gas atom – rare gas atom (upper line – well depth ε [meV], lower line – position of the potential minimum r_m [Å])

	He	Ne	Ar	Kr	Xe
He	0.92 (B) 2.97 (B)	1.9 (*B*) 3.0 (*B*)	2.5 (*B*) 3.5 (*B*)	2.6 (*B*) 3.7 (*B*)	2.4 (*B*) 4.0 (*B*)
Ne	–	3.6 (*B*) 3.1 (*B*)	6.0 (*B*) 3.4 (*B*)	6.3 (*B*) 3.6 (*B*)	6.5 (*B*) 3.8 (*B*)
Ar	–	–	12.2 (*B*) 3.76 (*B*)	14 (*B*) 3.9 (*B*)	16 (*B*) 4.1 (*B*)
Kr	–	–	–	17.2 (*B*) 4.0 (*B*)	20 (*B*) 4.2 (*B*)
Xe	–	–	–	–	24 (*B*) 4.4 (*B*)

Table 12.2. Potential well parameters of diatomic van der Waals molecules: alkali atom – rare gas atom, mercury atom (upper line – well depth ε [meV], lower line – position of the potential minimum r_m [Å])

	Li	Na	K	Rb	Cs
Ne	1.2 (*B*) 5.0 (*B*)	1.0 (*B*) 5.3 (*B*)	5.6 (*D*) 5.0 (*D*)	5.4 (*D*) 5.0 (*D*)	5.0 (*D*) 5.0 (*D*)
Ar	5.3 (*B*) 4.8 (*B*)	5.3 (*B*) 5.0 (*B*)	5.6 (*C*) 5.0 (*C*)	5.6 (*C*) 5.2 (*C*)	5.8 (*D*) 5.2 (*D*)
Kr	8.4 (*B*) 4.8 (*B*)	8.7 (*B*) 5.0 (*B*)	9.0 (*C*) 5.0 (*C*)	9.1 (*D*) 5.3 (*C*)	9.2 (*D*) 5.4 (*C*)
Xe	13 (*B*) 4.8 (*B*)	12.4 (*A*) 4.9 (*B*)	13 (*D*) 5.2 (*C*)	13 (*D*) 5.3 (*D*)	13 (*D*) 5.4 (*D*)
Hg	110 (*B*) 3.0 (*B*)	55 (*B*) 4.7 (*C*)	52 (*C*) 4.9 (*C*)	49 (*C*) 5.1 (*C*)	50 (*C*) 5.1 (*C*)

Table 12.3. Potential well parameters of diatomic van der Waals molecules: rare gas atom – H, He*, O, F, Ne*, Cl, Br, I (upper line – well depth ε [meV], lower line – position of the potential minimum r_m [Å])

		He	Ne	Ar	Kr	Xe
H		0.52 (*B*) 3.6 (*B*)	2.0 (*B*) 3.2 (*B*)	4.3 (*B*) 3.6 (*B*)	6.0 (*B*) 3.6 (*B*)	7.1 (*B*) 3.8 (*B*)
He ($2\,^3S$)		–	0.4 (*C*) 6.0 (*C*)	3.6 (*C*) 5.7 (*C*)	5.5 (*C*) 5.4 (*C*)	–
He ($2\,^1S$)		–	0.4 (*C*) 6.7 (*C*)	3.7 (*C*) 5.7 (*C*)	6.5 (*C*) 5.6 (*C*)	11 (*C*) 5.7 (*C*)
O (^3P)	Σ	1.0 (*D*) 3.6 (*C*)	–	–	3.5 (*C*) 4.8 (*C*)	3.4 (*C*) 4.8 (*C*)
	Π	2.5 (*D*) 3.1 (*C*)	3.9 (*D*) 3.2 (*C*)	7.6 (*D*) 3.6 (*C*)	7.8 (*C*) 3.8 (*C*)	9.5 (*C*) 3.7 (*C*)
F ($^2P_{3/2}$)	*X* 1/2	–	5.2 (*C*) 2.8 (*C*)	12 (*C*) 2.9 (*C*)	13 (*C*) 3.0 (*C*)	150 (*B*) 2.3 (*C*)
	A 3/2	–	3.9 (*C*) 3.0 (*B*)	6.5 (*C*) 3.4 (*B*)	6.7 (*C*) 3.6 (*B*)	7.0 (*C*) 3.8 (*B*)
Ne (^3P)		–	–	–	8.2 (*D*) 4.9 (*C*)	–
Cl ($^2P_{3/2}$)	*X* 1/2	–	–	–	–	35 (*C*) 3.2 (*C*)
	A 3/2	–	–	–	–	16 (*C*) 4.1 (*C*)
Br ($^2P_{3/2}$)	*X* 1/2	–	–	16 (*C*) 3.7 (*C*)	20 (*C*) 3.9 (*C*)	28 (*D*) 3.8 (*D*)
	A 3/2	–	–	11 (*C*) 3.9 (*C*)	15 (*C*) 4.1 (*C*)	18 (*C*) 4.1 (*C*)
I ($^2P_{3/2}$)	*X* 1/2	–	–	–	24 (*C*) 4.0 (*C*)	30 (*C*) 4.3 (*C*)
	A 3/2	–	–	–	16 (*C*) 4.3 (*C*)	21 (*C*) 4.6 (*C*)

Table 12.4. Spherically symmetric potential well parameters of van der Waals molecules: rare gas atom – diatomic molecule (upper line – well depth ε [meV], lower line – position of the potential minimum r_m [Å])

	He	Ne	Ar	Kr	Xe
H_2	1.3 (*C*) 3.5 (*C*)	3.1 (*C*) 3.3 (*C*)	6.4 (*C*) 3.6 (*C*)	7.4 (*C*) 3.7 (*B*)	8.2 (*C*) 3.9 (*B*)
N_2	2.4 (*C*) 3.8 (*B*)	5.5 (*D*) 3.6 (*C*)	11 (*C*) 3.7 (*C*)	14 (*C*) 3.9 (*C*)	15 (*D*) 3.9 (*C*)
O_2	2.5 (*C*) 3.4 (*B*)	5.8 (*C*) 3.4 (*B*)	12 (*C*) 3.7 (*B*)	15 (*C*) 3.9 (*C*)	16 (*C*) 3.9 (*C*)
CO	2.2 (*C*) 3.8 (*C*)	–	–	–	–
NO	2.3 (*C*) 3.7 (*C*)	6.2 (*C*) 3.1 (*C*)	12 (*C*) 3.8 (*C*)	14 (*C*) 3.9 (*C*)	17 (*C*) 3.8 (*C*)
HCl	–	–	23 (*D*) 4.0 (*B*)	30 (*D*) 4.1 (*B*)	37 (*D*) 4.3 (*B*)

Table 12.5. Spherically symmetric potential well parameters of van der Waals molecules: atomic helium – molecule (upper line – well depth ε [meV], lower line – position of the potential minimum r_m [Å])

	HCl	HBr	I_2	CO_2	H_2O	NH_3	CH_4	SF_6
He	2.3 (*C*) 3.7 (*B*)	1.9 (*D*) 4.0 (*C*)	3.8 (*D*) 4.5 (*C*)	3.0 (*C*) 3.3 (*C*)	2.7 (*C*) 3.4 (*C*)	2.3 (*C*) 3.8 (*B*)	2.3 (*C*) 3.9 (*B*)	5.6 (*D*) 4.2 (*B*)

Table 12.6. Spherically symmetric potential well parameters of van der Waals molecules: rare gas atom – halogen molecule (upper line – well depth ε [meV], lower line – position of the potential minimum r_m [Å])

	Ne	Ar	Kr	Xe
HCl	–	23 (*D*) 4.0 (*B*)	30 (*D*) 4.1 (*B*)	37 (*D*) 4.3 (*B*)
HBr	–	32 (*D*) 4.1 (*C*)	–	–
I_2	11 (*D*) –	29 (*D*) –	–	–

Table 12.7. Spherically symmetric potential well parameters of van der Waals molecules: hydrogen molecule – diatomic, polyatomic molecule (upper line – well depth ε [meV], lower line – position of the potential minimum r_m [Å])

	H_2	CO	N_2	NO	O_2	I_2	CO_2	NH_3	SF_6
H_2	3.0 (*B*) 3.4 (*B*)	5.2 (*D*) 3.8 (*C*)	5.6 (*D*) 3.5 (*C*)	5.9 (*D*) 3.6 (*C*)	5.6 (*C*) 3.7 (*C*)	11 (*C*) –	8.7 (*D*) 3.3 (*C*)	9.8 (*D*) 3.8 (*D*)	10 (*D*) 4.5 (*D*)

Table 12.8. Spherically symmetric potential well parameters of van der Waals molecules: atom, diatomic molecule – diatomic, polyatomic molecule (upper line – well depth ε [meV], lower line – position of the potential minimum r_m [Å])

$H - H_2$	$D_2 - D_2$	$N_2 - N_2$	$N_2 - NO$	$N_2 - O_2$	$O_2 - NO$	$O_2 - O_2$	$H_2O - H_2O$
2.1 (*C*) 3.43 (*B*)	12 (*C*) –	8.7 (*C*) 4.1 (*C*)	3.4 (*C*) 3.6 (*C*)	11 (*C*) 3.7 (*C*)	3.9 (*C*) 3.5 (*C*)	11 (*C*) 3.9 (*C*)	36 (*C*) 2.6 (*C*)

12.2 Potential Well Parameters of Van der Waals Molecular Ions

Tables 12.9, 10 give the values of the well parameters ε and R_{min} characterizing the interaction potential of van der Waals molecular ions with small binding energies in the vicinity of the potential minimum (Sect. 12.1). The key information about the ionic potential minima can be found in [12.2.1–8]. The numerical values were grouped into accuracy classes as usual (see the Introduction).

References

12.2.1 H.-P.Weise: "Elastic scattering of ions", Berichte der Bunsen-Gesellschaft für Physikalische Chemie **77**, 578 (1973)

12.2.2 W.S.Koski: "Scattering of Positive Ions by Molecules", in *Molecular Scattering: Physical and Chemical Applications*, ed. by K.P.Lawley, Adv. Chem. Phys., Vol. 30 (Wiley, New York 1975) p. 333

12.2.3 M.S.Rajan, E.A.Gislason: J. Chem. Phys. **78**, 2428 (1983)

12.2.4 F.E.Budenholzer, E.A.Gislason, A.D.Jorgensen: J. Chem. Phys. **78**, 5279 (1983)

12.2.5 I.R.Gatland: "Ion mobilities and ion-atom interaction potentials", in SASP '82, Symposium At. Surf. Phys., Salzburg, 7–13 February, 1982 (Inst. Atomphysik, Univ. Innsbruck, Innsbruck, Austria 1982) pp. 310–317

12.2.6 L.A.Viehland: Chem. Phys. **78**, 279 (1983)

12.2.7 D.R.Lamm, R.D.Chelf, J.R.Twist, F.B.Holleman, M.G.Thackston, F.L.Eisele, W.M.Pope, I.R.Gatland, E.W.McDaniel: J. Chem. Phys. **79**, 1965 (1983)

12.2.8 M.Waldman, R.G.Gordon: J. Chem. Phys. **71**, 1325 (1979)

Table 12.9. Potential well parameters of diatomic van der Waals molecular ions: (upper line – well depth ε [meV], lower line – position of the potential minimum r_m [Å])

A) Alkali ion – rare gas atom

	He	Ne	Ar	Kr	Xe
Li^+	72 (*B*) 1.97 (*B*)	120 (*B*) 2.1 (*C*)	290 (*C*) 2.3 (*C*)	430 (*C*) 2.4 (*C*)	530 (*C*) 2.6 (*C*)
Na^+	40 (*D*) 2.4 (*C*)	70 (*D*) 2.4 (*C*)	160 (*C*) 2.7 (*C*)	200 (*D*) 2.9 (*C*)	280 (*D*) 3.4 (*C*)
K^+	23 (*C*) 2.7 (*C*)	42 (*C*) 2.8 (*C*)	115 (*B*) 3.0 (*C*)	137 (*B*) 3.2 (*C*)	170 (*C*) 3.4 (*B*)
Rb^+	–	–	90 (*C*) 3.4 (*C*)	120 (*C*) 3.5 (*C*)	130 (*C*) 3.9 (*C*)
Cs^+	15 (*C*) 3.0 (*C*)	25 (*C*) 3.3 (*B*)	75 (*C*) 3.6 (*C*)	110 (*C*) 3.6 (*C*)	110 (*C*) 4.0 (*C*)

B) H^+, He^+, B^+, C^+, N^+, O^+ – rare gas atom

	He	Ne	Ar	Kr	Xe
H^+	2040 (*A*) 0.77 (*A*)	2300 (*D*) 1.0 (*D*)	4100 (*D*) 1.3 (*D*)	4500 (*D*) 1.5 (*D*)	5800 (*D*) 1.7 (*D*)
He^+	2470 (*A*) 1.08 (*A*)	50 (*D*) 2.5 (*D*)	190 (*D*) 2.5 (*D*)	220 (*D*) 2.7 (*D*)	280 (*D*) 2.9 (*D*)
B^+	–	–	300 (*D*) 3.0 (*D*)	500 (*D*) 2.7 (*D*)	820 (*D*) 2.7 (*D*)
C^+	–	–	940 (*C*) 2.0 (*C*)	–	–
N^+	–	400 (*D*) 2.1 (*D*)	2200 (*D*) 2.3 (*D*)	290 (*D*) 2.5 (*D*)	920 (*D*) 2.8 (*D*)
O^+ (^4S)	–	–	670 (*B*) 2.22 (*B*)	1200 (*C*) 2.6 (*C*)	500 (*D*) 2.5 (*C*)
O^+ (^2P)	–	–	–	–	2700 (*C*) 2.3 (*C*)

C) Negative ion – rare gas atom

	He	Ne	Ar	Kr	Xe
H^-	0.4 (*E*) 8 (*E*)	2 (*E*) 5 (*E*)	10 (*E*) 5 (*E*)	–	–
F^-	–	–	–	–	280 (*C*) 2.8 (*C*)
Cl^-	–	–	25 (*E*) 4 (*E*)	–	130 (*C*) 3.8 (*C*)
Br^-	2 (*E*) 5 (*E*)	–	59 (*C*) 3.7 (*C*)	87 (*C*) 3.7 (*C*)	145 (*C*) 3.6 (*C*)

Table 12.10. Potential well parameters of diatomic van der Waals molecular ions: rare gas ion – rare gas atom

	Ne^+–Ne	Ar^+–Ar	Kr^+–Kr	Xe^+–Xe
ε [meV]	1300 (*D*)	1340 (*D*)	1210 (*D*)	980 (*D*)
r_m [Å]	1.7 (*D*)	2.4 (*D*)	2.6 (*D*)	3.2 (*C*)

12.3 Ionization Potentials of Van der Waals Molecules

The numerical values of the ionization potentials (IP) for some van der Waals molecules are collected in Table 12.11. The molecules are arranged mainly in order of increasing number of atoms in the structural element of the van der Waals molecule. We assume that the quantitative data are accurate to one unit in the last significant figure given. The basic information on ionization potentials of van der Waals molecules is presented in [12.3.1–5].

References

12.3.1 Y.Ono, S.H.Linn, H.F.Prest, M.E.Gress, C.Y.Ng: J. Chem. Phys. **73**, 2523 (1980)
12.3.2 S.L.Anderson, T.Hirooka, P.W.Tiedemann, B.H.Mahan, Y.T.Lee: J. Chem. Phys. **73**, 4779 (1980)
12.3.3 S.H.Linn, Y.Ono, C.Y.Ng: J. Chem. Phys. **74**, 3342 (1981)
12.3.4 J.Erickson, C.Y.Ng: J. Chem. Phys. **75**, 1650 (1981)
12.3.5 S.H.Linn, C.Y.Ng: J. Chem. Phys. **75**, 4921 (1981)

Table 12.11. Ionization potentials (IP) of van der Waals molecules [Numbers in parentheses stand for ionization potentials of structural elements of the molecule. For example, in the case of Ar_2 the value of the atomic IP (Ar) is given]

Molecule	IP [eV]	Molecule	IP [eV]
Ar_2	14.54 (15.76)	$(O_2)_2$	11.66 (12.07)
NeAr	15.68	$(CO_2)_2$	13.32 (13.79)
Kr_2	12.87 (14.00)	$(CO_2)_3$	13.24
NeKr	13.95	$(CO_2)_4$	13.18
ArKr	13.42	$(CS_2)_2$	9.36 (10.068)
Kr_3	12.79	$(CS_2)_3$	9.22
Kr_4	12.76	$(CS_2)_4$	9.10
Xe_2	11.13 (12.13)	$(CS_2)_5$	9.04
NeXe	12.09	$(H_2O)_2$	11.21 (12.61)
ArXe	11.98	$(H_2S)_2$	9.74 (10.45)
KrXe	11.76	$(H_2S)_3$	9.63
$(CO)_2$	13.05 (14.013)	$(H_2S)_4$	9.61
$(CO)_3$	12.91	$(H_2S)_5$	9.58
$(HCl)_2$	11.91 (12.74)	$(H_2S)_6$	9.50
$(HBr)_2$	10.83 (11.66)	$(H_2S)_7$	9.63
$(N_2)_2$	14.69 (15.580)	$(NH_3)_2$	9.54 (10.166)
$(N_2)_3$	14.64	$(N_2O)_2$	12.35 (12.89)
$(NO)_2$	8.736 (9.265)	$(OCS)_2$	10.456 (11.174)
$(NO)_3$	8.486	$(OCS)_3$	10.408
$(NO)_4$	8.39	$OCS \cdot CS_2$	9.858
$(NO)_5$	8.32	$(SO_2)_2$	11.72 (12.35)
$(NO)_6$	8.28	$(CH_4)_2$	9.72 (9.74)

13. Polyatomic Molecules

Numerical data are compiled for spectroscopic and geometrical parameters of triatomic molecules, ionization potentials and bond dissociation energies of polyatomic molecules and, also, for transition probabilities and radiative lifetimes of vibrationally excited polyatomic molecules.

13.1 Constants of Triatomic Molecules

Table 13.1 gives quantitative information about geometric, spectroscopic and energetic characteristics of a variety of triatomic molecules that occur most frequently while analyzing the physico-chemical properties of atmospheric gases. Arrangement of molecules in the table is made alphabetically.

We now comment on some of the parameters introduced.

1) The second column of Table 13.1 enumerates the symbols of point symmetry groups characterizing the assembly of coordinate transformations (rotations about some axes and reflections in some plane passing through the given axis) which do not alter the Schrödinger equation for the molecule:

The group C_n contains a single axis of symmetry of the nth order (a rotation through an angle $2\pi/n$ about this axis leaves the molecule unaltered).

The group C_{nh} contains a single axis of symmetry of the nth order and a plane perpendicular to the axis of rotation (the simplest group, C_{1h}, is also denoted by C_s).

The group C_{nv} contains a single axis of symmetry of the nth order and n planes intersecting each other along the axis at angles of π/n ("vertical" planes).

The group D_{nh} contains a single axis of symmetry of the nth order, n axes of the second order perpendicular to it and intersecting at angles π/n, and the horizontal plane passing through the n axes of the second order (hence it also contains the n vertical planes passing through the vertical axis of the nth order and one of the horizontal axes).

2) The third column gives the values of the angle between two straight lines proceeding from the centre of one atom, through which the vertical axis of symmetry passes, towards the other two atoms.

3) The fifth column presents the values of the fundamental vibrational frequencies of the molecule. Classification of these frequencies by symmetry and multiplicity fully depends on the characters of the irreducible representation of a given point symmetry group of the molecule [13.1.1].

4) The sixth column contains the values of the rotational constants of the molecules.

The data incorporated in Table 13.1 are taken from [13.1.2, 3] and give a general idea of the properties of these triatomic molecules.

References

13.1.1 L.D.Landau, E.M.Lifshitz: *Quantum Mechanics (Non-Relativistic Theory)*, 3rd ed. (Pergamon, Oxford 1977)

13.1.2 K.S.Krasnov (ed.): *Molecular Constants for Inorganic Compounds* (Chimia, Moscow 1979) pp. 73–158 (in Russian)

13.1.3 G.Herzberg, L.Herzberg: "Constants of Polyatomic Molecules", in *American Institute of Physics Handbook*, ed. by D.E.Gray, 3rd ed. (McGraw-Hill, New York 1972) Sect. 7h

Table 13.1. Constants of triatomic molecules

Molecule	Point group	Geometrical parameters: bond angle [degrees]	bond length [Å]		Fundamental vibrational frequencies [cm^{-1}] ν_1	ν_2	ν_3	Rotational constants [cm^{-1}] (refer to the lowest vib. level) A_0	B_0	C_0	Bond dissociation energies: energy value [eV]	bond	Ionization potential [eV]	Dipole moment [D]
NCN	$D_{\infty h}$	180	1.23 (NC)	1.23 (CN)	1197	423	1475	–	0.40	–	–	–	–	0
CO_2	$D_{\infty h}$	180	1.160 (OC)	1.160 (CO)	1388	667	2349	–	0.39	–	5.45	(O–CO)	13.8	0
CS_2	$D_{\infty h}$	180	1.55 (SC)	1.55 (CS)	658	397	1532	–	0.109	–	4.5	(S–CS)	10.1	0
FCN	$C_{\infty v}$	180	1.26 (FC)	1.16 (CN)	1076	2290	451	–	–	–	–	–	13.3	2.2
FNO	C_s	110	1.51 (FN)	1.14 (NO)	1844	520	766	3.17	0.39	0.35	–	–	–	1.8
HCN	$C_{\infty v}$	180	1.065 (HC)	1.153 (CN)	3311	2097	712	–	1.48	–	9.6 5.6	(HC–N) (H–CN)	13.7	3.0
HNO	C_s	108.05	1.09 (HN)	1.21 (NO)	2719	1564	1505	18.5	1.41	1.31	2.1	(H–NO)	–	–
H_2O	C_{2v}	104.5	0.957 (HO)	0.957 (OH)	3657	1595	3756	27.88	14.51	9.287	5.12	(H–OH)	12.6	1.85
H_2S	C_{2v}	92.1	1.34 (HS)	1.34 (SH)	2615	1183	2625	10.4	8.99	4.73	3.9	(H–SH)	10.5	0.97
KrF_2	$D_{\infty h}$	180	1.89 (FKr)	1.89 (KrF)	449	233	588	–	–	–	–	–	–	0
NH_2	C_{2v}	103	1.02 (HN)	1.02 (NH)	(3173)	1499	3220	23.7	12.9	8.17	3.9	(H–NH)	10.1	–
N_2O	$C_{\infty v}$	180	1.13 (NN)	1.18 (NO)	2224	1285	589	–	0.419	–	1.67 4.93	(N_2–O) (N–ON)	12.9	0.16
NO_2	C_{2v}	134.25	1.20 (ON)	1.20 (NO)	1320	750	1618	8.0	0.43	0.41	3.11 4.50	(O–NO) (N–O_2)	9.8	0.32
OCS	$C_{\infty v}$	180	1.15 (OC)	1.56 (CS)	2062	859	520	–	0.203	–	–	–	11.2	0.71
O_3	C_{2v}	116.8	1.272 (OO′)	1.272 (O′O)	1103	701	1042	3.55	0.445	0.395	1.04	(O_2–O)	12.5	0.53
S_2O	C_s	118	1.882 (SS)	1.464 (SO)	1165	679	388	1.40	0.169	0.150	3.9	(S_2–O)	–	1.5
SO_2	C_{2v}	119	1.43 (OS)	1.43 (SO)	1151	518	1362	2.03	0.34	0.29	5.66	(OS–O)	12.3	1.63
XeF_2	$D_{\infty h}$	180	1.98 (FXe)	1.98 (XeF)	515	213	555	–	–	–	–	–	12.4	0

13.2 Ionization Potentials of Polyatomic Molecules

The numerical values of the ionization potentials (IP) for some widely spread triatomic, and four-, five-, six-atomic molecules and a few larger polyatomic molecules are shown in Table 13.2 [13.2.1, 2]. The molecules are arranged alphabetically. The values of the IPs are grouped into accuracy classes as usual (see the Introduction).

References

13.2.1 L.V.Gurvitch, G.V.Karachevtsev, V.N.Kondratjev, Y.A.Lebedev, V.A.Medvedev, V.K.Potapov, Y.S.Hodeev: *Bond Dissociation Energies, Ionization Potentials, and Electron Affinities*, 2nd ed. (Nauka, Moscow 1974) (in Russian)

13.2.2 G.Herzberg: *Molecular Spectra and Molecular Structure III. Electronic Spectra and Electronic Structure of Polyatomic Molecules* (Van Nostrand, Princeton 1966)

Table 13.2. Ionization potentials (IP) of polyatomic molecules

Molecule or radical	IP [eV]	Molecule or radical	IP [eV]
Triatomic molecules			
BH_2	9.8 (*C*)	NH_2	10.15 (*A*)
$BaCl_2$	9.2 (*B*)	NO_2	9.78 (*A*)
BaI_2	8.1 (*C*)	N_2O	12.89 (*A*)
BeF_2	14.7 (*C*)	O_3	12.52 (*A*)
BrCN	11.84 (*A*)	PF_2	8.85 (*A*)
CCl_2	13.2 (*B*)	SO_2	12.35 (*A*)
CF_2	11.8 (*B*)	SiC_2	10.2 (*B*)
CH_2	10.396 (*A*)	SiF_2	11.0 (*C*)
CO_2	13.79 (*A*)	SiO_2	11.7 (*C*)
COS	11.18 (*A*)	$SnCl_2$	10.2 (*C*)
CS_2	10.07 (*A*)	$SrCl_2$	9.7 (*B*)
$CaCl_2$	10.3 (*A*)	UO_2	5.5 (*B*)
ClCN	12.34 (*A*)	XeF_2	12.42 (*A*)
ClO_2	11.1 (*C*)		
Cs_2O	4.45 (*A*)	Four-atomic molecules	
CsOH	7.21 (*B*)		
FCN	13.32 (*A*)	BBr_3	10.68 (*A*)
$GeBr_2$	9.5 (*C*)	BCl_3	11.73 (*A*)
$GeCl_2$	10.4 (*C*)	BF_3	15.95 (*A*)
GeF_2	11.8 (*B*)	BH_3	11.4 (*C*)
HCN	13.73 (*A*)	BI_3	9.40 (*A*)
HO_2	11.53 (*A*)	CH_3	9.840 (*A*)
H_2O	12.614 (*A*)	C_2H_2	11.406 (*A*)
H_2S	10.47 (*A*)	H_2O_2	10.9 (*C*)
ICN	10.87 (*A*)	HBO_2	12.6 (*B*)
Li_2O	6.8 (*B*)	NH_3	10.15 (*A*)
NF_2	12.11 (*A*)	PF_3	9.71 (*A*)

Table 13.2 (continued)

Molecule or radical	IP [eV]	Molecule or radical	IP [eV]
Five-atomic molecules		Six-atomic molecules	
CBr_4	11.0 (*B*)	C_2Cl_4	9.34 (*A*)
CCl_4	11.47 (*A*)	C_2F_4	10.12 (*A*)
CF_2Cl_2	12.31 (*A*)	C_2H_4	10.51 (*A*)
$CFCl_3$	11.77 (*A*)	CH_3OH	10.85 (*A*)
CH_4	12.98 (*A*)	N_2F_4	12.0 (*A*)
CH_2Cl_2	11.35 (*A*)	N_2H_4	8.74 (*A*)
CH_3Br	10.53 (*A*)		
CH_3I	9.538 (*A*)		
SiH_4	11.4 (*C*)	Larger polyatomic molecules	
$SnBr_4$	11.0 (*A*)	SF_6	15.7 (*A*)
$SnCl_4$	12.10 (*A*)	UF_6	14.14 (*A*)
XeF_4	12.65 (*A*)	XeF_6	12.19 (*A*)
		B_2H_6	11.41 (*A*)
		C_2H_6	11.50 (*A*)
		C_2H_5Br	10.29 (*A*)
		C_2H_5OH	10.47 (*A*)
		$C_6H_6\,(-h_6)$	9.246 (*A*)
		$(-d_6)$	9.248 (*A*)

13.3 Bond Dissociation Energies of Polyatomic Molecules

Quantitative information about the strengths of chemical bonds in some widely spread polyatomic molecules is presented in Table 13.3. The value of the bond dissociation energy $D_0(\mathrm{X-Y})$ is defined as the heat of the reaction $\mathrm{XY} \rightarrow \mathrm{X} + \mathrm{Y}$ and is given by

$$D_0(\mathrm{X-Y}) = \Delta H^\circ_{\mathrm{f0}}(\mathrm{X}) + \Delta H^\circ_{\mathrm{f0}}(\mathrm{Y}) - \Delta H^\circ_{\mathrm{f0}}(\mathrm{XY}) \ ,$$

where the $\Delta H^\circ_{\mathrm{f0}}$ are the heats of formation of molecules X, Y and XY from elements in their standard states.

The numerical values in Table 13.3 refer to a temperature of 0 K, though chemists often list the bond strengths at 25 °C. The proper conversion factor can be expressed as

$$D_T \cong D_0 + (3/2)\,RT \ ,$$

and so

$$D_{298}\,[\mathrm{kcal/mol}] = D_0 + 0.888 \ ,$$

$$D_{298}\,[\mathrm{eV}] = D_0 + 0.0385 \ ,$$

or one can use the exact relation

$$\varDelta H^\circ_{fT} = \varDelta H^\circ_{f0} + \varDelta (H^\circ_T - H^\circ_0) \,,$$

where $\varDelta (H^\circ_T - H^\circ_0)$ is the enthalpy difference tabulated in thermochemical handbooks.

The key information about the bond dissociation energies of polyatomic molecules can be found in [13.3.1–3]. The numerical values listed in Table 13.3 are grouped into accuracy classes as usual (see the Introduction).

References

13.3.1 L.V.Gurvitch, G.V.Karachevtsev, V.N.Kondratjev, Y.A.Lebedev, V.A.Medvedev, V.K.Potapov, Y.S.Hodeev: *Bond Dissociation Energies, Ionization Potentials, and Electron Affinities*, 2nd ed. (Nauka, Moscow 1974) (in Russian)

13.3.2 G.Herzberg: *Molecular Spectra and Molecular Structure III. Electronic Spectra and Electronic Structure of Polyatomic Molecules* (Van Nostrand, Princeton 1966)

13.3.3 J.A.Kerr, A.F.Trothman-Dickenson: "Strengths of Chemical Bonds", in *CRC Handbook of Chemistry and Physics*, ed. by R.C.Weast, 62nd ed. (CRC, Boca Raton 1981) pp. F180–F200

Table 13.3. Bond dissociation energies of polyatomic molecules

Molecule or radical	Bond breaking	Bond dissociation energy [eV]
Triatomic molecules		
CO_2	CO–O	5.45 (*A*)
CS_2	CS–S	4.51 (*A*)
ClO_2	ClO–O	2.5 (*B*)
HCN	HC–N	9.58 (*A*)
HCO	H–CO	0.75 (*C*)
	HC–O	8.37 (*B*)
H_2O	H–OH	5.12 (*A*)
	H_2–O	5.0 (*B*)
H_2S	H–SH	3.93 (*B*)
HO_2	H–O_2	2.0 (*C*)
NF_2	FN–F	2.85 (*B*)
NH_2	NH–H	3.9 (*C*)
NO_2	NO–O	3.11 (*A*)
N_2O	N–NO	4.93 (*A*)
	N_2–O	1.67 (*A*)
NOH	NO–H	2.04 (*A*)
OCSe	OC–Se	2.68 (*B*)
O_3	O_2–O	1.04 (*B*)
PF_2	FP–F	5.2 (*C*)
SO_2	SO–O	5.66 (*A*)

Table 13.3 (continued)

Molecule or radical	Bond breaking	Bond dissociation energy [eV]
Four-atomic molecules		
BF_3	BF_2–F	7.4 (*B*)
CO_3	CO_2–O	0.5 (*D*)
CH_2O	H–COH	3.77 (*B*)
H_2O_2	HO–OH	2.22 (*B*)
	HO_2–H	3.8 (*C*)
HN_3	H–N_3	4.0 (*C*)
NF_3	F_2N–F	2.47 (*A*)
NH_3	NH_2–H	4.48 (*B*)
N_2H_2	NH–NH	4.9 (*C*)
NO_3	NO_2–O	2.13 (*B*)
PF_3	F_2P–F	5.71 (*A*)
PH_3	PH_2–H	3.2 (*C*)
SO_3	SO_2–O	3.55 (*A*)
Five-atomic molecules		
CBr_4	CBr_3–Br	2.2 (*C*)
CCl_4	CCl_3–Cl	3.18 (*B*)
CF_4	CF_3–F	5.6 (*B*)
CF_2Cl_2	CF_2Cl–Cl	3.26 (*C*)
$CFCl_3$	$CFCl_2$–Cl	3.72 (*C*)
CH_4	CH_3–H	4.51 (*A*)
CH_3Cl	CH_3–Cl	3.6 (*C*)
	CH_2Cl–H	4.4 (*C*)
CH_2Cl_2	CH_2Cl–Cl	3.6 (*C*)
	$CHCl_2$–H	4.3 (*C*)
$CHCl_3$	$CHCl_2$–Cl	3.4 (*C*)
	CCl_3–H	4.1 (*C*)
CH_3I	CH_3–I	2.43 (*B*)
CH_2CO	H–CHCO	4.60 (*B*)
Larger polyatomic molecules		
C_2H_4	C_2H_3–H	4.60 (*B*)
N_2F_4	F_2N–NF_2	0.91 (*B*)
N_2H_4	N_2H_3–H	2.6 (*C*)
P_2F_4	F_2P–PF_2	2.5 (*D*)
C_2Cl_6	CCl_3–CCl_3	3.0 (*C*)
C_2H_6	CH_3–CH_3	3.9 (*C*)
C_2H_5Cl	CH_3–CH_2Cl	4.0 (*C*)
$C_2H_4Cl_2$	CH_3–$CHCl_2$	3.9 (*C*)
	CH_2Cl–CH_2Cl	4.0 (*C*)
$C_2H_3Cl_3$	CH_3–CCl_3	3.8 (*C*)
	CH_2Cl–$CHCl_2$	4.0 (*C*)
$C_2H_2Cl_4$	CH_2Cl–CCl_3	3.7 (*C*)
	$CHCl_2$–$CHCl_2$	3.8 (*C*)
C_2HCl_5	$CHCl_2$–CCl_3	3.4 (*C*)
C_2H_5OH	C_2H_5O–H	4.40 (*A*)

13.4 Lifetimes of Vibrationally Excited Polyatomic Molecules

Table 13.4 presents the values of the Einstein coefficient A^{v}_{ki} for spontaneous radiative transitions between the vibrational states of polyatomic molecules [13.4.1–4]. Separate columns of Table 13.4 give the spectroscopic designations of the relevant vibrational levels of the molecules, transition energies, integrated absorption of the infrared band S_l, transition probabilities A^{v}_{ki} for spontaneous emission and the radiative lifetimes of the upper vibrational levels τ^{v}_{k}.

The absolute (integral) intensity S_l of a vibrational band may be specified by

$$S_l = \frac{1}{N_i} \int k_l(\nu)\, d\nu = \frac{\lambda^2}{4} \frac{g_k}{g_i} A^{v}_{ki} \,,$$

where N_i is the number density of molecules in the lower state i, k_l is the absorption coefficient, ν is the frequency of the radiative transition, λ is the vacuum transition wavelength, A^{v}_{ki} is the Einstein coefficient for spontaneous emission from the upper level k into the lower level i and g_k, g_i are the statistical weights. The S_l entries in Table 13.4 are collected in [13.4.3, 5]. The radiative lifetime τ^{v}_{k} of the upper vibrationally excited k level is defined as usual [13.4.6]:

$$\tau^{v}_{k} = \left(\sum_i A^{v}_{ki} \right)^{-1} .$$

Molecules are arranged in Table 13.4 according to the number of atoms and within such a group they are listed in alphabetical order. Numerical values are rounded off in such a way that any possible corrections will influence only the last significant figure within the range $\pm 1 - \pm 2$.

In Fig. 13.1 the low-lying vibrational energy levels of the CO_2 molecule are shown [13.4.5]. The most interesting laser transitions are marked on the diagram and the values of the transition probability A^{v}_{ki} are indicated.

References

13.4.1 G.Herzberg: *Molecular Spectra and Molecular Structure II. Infrared and Raman Spectra of Polyatomic Molecules* (Van Nostrand, Princeton 1945)

13.4.2 E.B.Wilson, Jr., J.C.Decius, P.C.Cross: *Molecular Vibrations. The Theory of Infrared and Raman Vibrational Spectra* (McGraw-Hill, New York 1955)

13.4.3 L.M.Sverdlov, M.A.Kovner, E.P.Krainov: *Vibrational Spectra of Polyatomic Molecules* (Wiley, New York 1974)

13.4.4 W.B.Person, G.Zerbi (eds.): *Vibrational Intensities* (Elsevier, Amsterdam 1980)

13.4.5 L.S.Rothman, W.S.Benedict: "Infrared energy levels and intensities of carbon dioxide", Appl. Optics **17**, 2605 (1978)

13.4.6 B.M.Smirnov, G.V.Schliapnikov: Usp. Fiz. Nauk **130**, 377 (1980)

Table 13.4. Radiative lifetimes and transition probabilities of vibrationally excited polyatomic molecules

Molecule		Upper and lower vibrational states (or upper only if the lower state is the ground state)	Fundamental vibrational frequencies ν_l [cm^{-1}] or rotationless transition energy [cm^{-1}]	Absolute (integral) intensity of infrared absorption band S_l [10^{-8} cm^2/s]	Transition probability for spontaneous emission $A_{v',v''}$ [s^{-1}]	Radiative lifetime of upper vibrational level $\tau_{v'}$ [ms]
Triatomic molecules						
CO_2		$00^01 \rightarrow 00^00$	2349.15	325	450.4	2.22
	3	02^00	1063.73		0.2	
		10^00	960.96		0.35	
		$01^10 \rightarrow 00^00$	667.38	27.9	1.564	640
		$02^00 \rightarrow 01^10$	618.03		1.209	827
		$02^20 \rightarrow 01^10$	667.75		3.153	317
	2	$03^10 \rightarrow 02^00$	647.06		2.052	383
		02^20	597.34		0.529	
		10^00	544.28		0.0302	
		$03^30 \rightarrow 02^20$	668.11		4.778	209
		$10^00 \rightarrow 01^10$	720.81		2.080	481
		$11^10 \rightarrow 00^00$	2076.86	0.012	$6.6 \cdot 10^{-3}$	265
	1	02^00	791.45		0.0152	
		02^20	741.74		1.202	
		10^00	688.68		2.547	
$^{12}C^{16}O^{18}O$		$00^01 \rightarrow 00^00$	2332.11	1.12	1.526	655
		$01^10 \rightarrow 00^00$	662.37	0.11	$6.16 \cdot 10^{-3}$	$162 \cdot 10^3$
$^{12}C^{16}O^{17}O$		$00^01 \rightarrow 00^00$	2340.0	0.216	0.297	$3.37 \cdot 10^3$
		$01^10 \rightarrow 00^00$	664.73	0.021	$1.17 \cdot 10^{-3}$	$856 \cdot 10^3$
$^{12}C^{18}O_2$			657.3	$1.2 \cdot 10^{-4}$	$6.3 \cdot 10^{-6}$	$1.6 \cdot 10^8$
$^{12}C^{17}O^{18}O$			659.7	$2.1 \cdot 10^{-5}$	$1.2 \cdot 10^{-6}$	$8.5 \cdot 10^8$
$^{13}C^{16}O_2$		$01^10 \rightarrow 00^00$	648.48	0.29	$1.54 \cdot 10^{-2}$	$65 \cdot 10^3$
$^{13}C^{16}O^{18}O$			643.2	$1.2 \cdot 10^{-3}$	$6.3 \cdot 10^{-5}$	$1.6 \cdot 10^7$
$^{13}C^{16}O^{17}O$			645.7	$2.2 \cdot 10^{-4}$	$1.1 \cdot 10^{-5}$	$8.8 \cdot 10^7$
CS_2		$3\,(\Sigma_u^+)$	1532	280	160	6.1
		$2\,(\Pi_u)$	397	2.5	0.049	$20 \cdot 10^3$
HCN		$3\,(\Pi)$	712	23	1.5	69
		$2\,(\Sigma^+)$	2096.8	0.07	0.08	$13 \cdot 10^3$
		$1\,(\Sigma^+)$	3311.5	27	74	14
H_2O		$3\,(B_1)$	3755.8	21	76	13
		$2\,(A_1)$	1594.8	25	16	62
H_2S		$2\,(A_1)$	1182.7	0.24	0.08	$12 \cdot 10^3$
		$1\,(A_1)$	2614.6	0.06	0.1	$10 \cdot 10^3$
N_2O		$3\,(\Pi)$	589	3.4	0.15	$7 \cdot 10^3$
		$2\,(\Sigma^+)$	1285	29	12	80
		$1\,(\Sigma^+)$	2224	160	200	5
NO_2		$3\,(B_1)$	1618	230	150	7

Table 13.4 (continued)

Molecule	Upper and lower vibrational states (or upper only if the lower state is the ground state)	Fundamental vibrational frequencies ν_l [cm^{-1}] or rotationless transition energy [cm^{-1}]	Absolute (integral) intensity of infrared absorption band S_l [10^{-8} cm^2/s]	Transition probability for spontaneous emission $A_{v',v''}$ [s^{-1}]	Radiative lifetime of upper vibrational level $\tau_{v'}$ [ms]
OCS	3 (Π)	520.4	1.3	0.046	$22 \cdot 10^3$
	2 (Σ^+)	859.0	13	2.4	410
	1 (Σ^+)	2062.2	480	510	2
O_3	3 (B_1)	1042.1	29	8	125
SO_2	3 (B_1)	1361.8	94	44	23
	2 (A_1)	517.7	13	0.88	1100
	1 (A_1)	1151.4	10	3.5	290
Four-atomic molecules					
$^{10}BF_3$	4 (E')	482	8.7	0.25	$4 \cdot 10^3$
	3 (E')	1505	300	84	12
	2 (A_1')	718	30	3.9	260
C_2H_2	5 (Π_u)	729	80	5	190
	3 (Σ_u^+)	3287	34	90	11
CH_2O	6 (B_2)	1257	4.9	2.0	510
	5 (B_2)	2847	64	130	8
	4 (B_1)	1180	4.9	1.7	580
	3 (A_1)	1507	4.6	2.6	380
	2 (A_1)	1748	29	22	45
	1 (A_1)	2780	64	124	8
$COCl_2$	6 (B_2)	585	5.0	0.43	$2.3 \cdot 10^3$
	4 (B_1)	850	117	21	47
	2 (A_1)	1827	87	73	14
	1 (A_1)	570	5.0	0.41	$2.5 \cdot 10^3$
NF_3	4 (E)	492.6	0.7	0.02	$48 \cdot 10^3$
	3 (E)	908	200	21	48
	2 (A_1)	647.2	0.8	0.08	$12 \cdot 10^3$
	1 (A_1)	1031.9	15	4.0	250
NH_3	4 (E)	1625	12	4.1	245
	3 (E)	3444	1.5	2.2	460
	2 (A_1)	950	67	15	66
	1 (A_1)	3337	2.2	6.3	160
PH_3	4 (E)	1122	11	1.8	550
	3 (E)	2328	29	20	50
	2 (A_1)	992	9	2.3	440
	1 (A_1)	2323	29	39	25
Five-atomic molecules					
CBr_3H	5 (E)	669	69	1.4	700
	4 (E)	1149	25	4.2	240
CCl_4	4 (F_2)	312	0.1	0.0008	$1.2 \cdot 10^6$
	3 (F_2)	776	194	9.8	100

Table 13.4 (continued)

Molecule	Upper and lower vibrational states (or upper only if the lower state is the ground state)	Fundamental vibrational frequencies ν_l [cm^{-1}] or rotationless transition energy [cm^{-1}]	Absolute (integral) intensity of infrared absorption band S_l [10^{-8} cm^2/s]	Transition probability for spontaneous emission $A_{v',v''}$ [s^{-1}]	Radiative lifetime of upper vibrational level $\tau_{v'}$ [ms]
CCl_3F	4 (E)	847	186	16.8	59
	2 (A_1)	535	0.6	0.04	$25 \cdot 10^3$
	1 (A_1)	1085	87	25.8	39
CCl_2F_2	9 (B_2)	437	0.06	0.003	$350 \cdot 10^3$
	8 (B_2)	902	154.4	31.6	32
	7 (B_1)	446	0.06	0.003	$330 \cdot 10^3$
	6 (B_1)	1159	96.5	32.6	31
	3 (A_1)	458	0.09	0.005	$210 \cdot 10^3$
	2 (A_1)	667	6.2	0.69	$1.4 \cdot 10^3$
	1 (A_1)	1100	149	45.3	22
CF_3Br	5 (E)	547	1.2	0.045	$22 \cdot 10^3$
	4 (E)	1210	232	42.7	23
	2 (A_1)	760	14.6	2.1	470
	1 (A_1)	1089	231	68.9	15
CF_3Cl	5 (E)	563	1.5	0.061	$16 \cdot 10^3$
	4 (E)	1212	336	62.0	16
	2 (A_1)	781	17.5	2.7	370
	1 (A_1)	1105	280	85.9	12
CF_4	4 (F_2)	631	4.7	0.16	$6.4 \cdot 10^3$
	3 (F_2)	1283	510	70	14
CF_3H	6 (E)	508	2.6	0.08	$12 \cdot 10^3$
	5 (E)	1152	430	72	14
	4 (E)	1375	48	12	87
	3 (A_1)	699	7.5	0.92	$1.1 \cdot 10^3$
	2 (A_1)	1137	430	140	7
	1 (A_1)	3035	17	41	25
$CHCl_3$	5 (E)	770	130	10	100
	4 (E)	1220	21	3.9	260
	2 (A_1)	680	3.0	0.35	$3 \cdot 10^3$
	1 (A_1)	3034	0.2	0.5	$2 \cdot 10^3$
CH_2Br_2	9 (B_2)	648	37	3.9	250
	8 (B_2)	1195	23	8.3	120
	7 (B_1)	810	1.8	0.30	$3.4 \cdot 10^3$
	6 (B_1)	3060	0.54	1.28	780
CH_2Cl_2	9 (B_2)	758	60.1	8.7	115
	8 (B_2)	1268	16.2	6.5	150
	7 (B_1)	898	0.6	0.12	$8 \cdot 10^3$
	6 (B_1)	3040	1.9	4.3	230
	4 (A_1)	282	1.4	0.03	$36 \cdot 10^3$
	3 (A_1)	717	3.8	0.49	$2 \cdot 10^3$
	2 (A_1)	1467	1.1	0.61	$1.6 \cdot 10^3$
	1 (A_1)	3000	1.7	3.8	260

Table 13.4 (continued)

Molecule	Upper and lower vibrational states (or upper only if the lower state is the ground state)	Fundamental vibrational frequencies ν_l [cm^{-1}] or rotationless transition energy [cm^{-1}]	Absolute (integral) intensity of infrared absorption band S_l [10^{-8} cm^2/s]	Transition probability for spontaneous emission $A_{v',v''}$ [s^{-1}]	Radiative lifetime of upper vibrational level $\tau_{v'}$ [ms]
CH_2F_2	9 (B_2)	1090	147.2	44.0	23
	8 (B_2)	1435	0.5	0.26	$4 \cdot 10^3$
	7 (B_1)	1178	0.03	0.01	$100 \cdot 10^3$
	6 (B_1)	3014	23.5	53.8	18.6
	4 (A_1)	529	2.6	0.18	$5.6 \cdot 10^3$
	3 (A_1)	1113	33.6	10.5	95
	1 (A_1)	2948	13.0	28.5	35
CH_3Br	6 (E)	954	3.5	0.40	$2.5 \cdot 10^3$
	5 (E)	1444	6.6	1.7	580
	4 (E)	3057	2.3	2.7	380
	3 (A_1)	611	4.6	0.43	$2.3 \cdot 10^3$
	2 (A_1)	1305	6.5	2.8	360
	1 (A_1)	2970	8.4	18.7	53
CH_3Cl	6 (E)	1017	2.3	0.30	$3.3 \cdot 10^3$
	5 (E)	1455	6.6	1.8	570
	4 (E)	3054	5.0	5.8	170
	3 (A_1)	732	11.0	1.5	680
	2 (A_1)	1355	3.8	1.8	570
	1 (A_1)	2968	10.2	22.6	44
CH_3F	6 (E)	1207.5 (ω_6)	0.9	0.15	$7 \cdot 10^3$
	5 (E)	1514.4 (ω_5)	3.0	0.81	$1.2 \cdot 10^3$
	4 (E)	3164.8 (ω_4)	19	21	47
	3 (A_1)	1076.7 (ω_3)	54	15	67
	2 (A_1)	1495.5 (ω_2)	1.7	0.9	1000
	1 (A_1)	3045.7 (ω_1)	19	41	24
CH_4	4 (F_2)	1306	17.5	2.5	400
	3 (F_2)	3019	40	30	33
CH_3I	6 (E)	882	4.5	0.44	$2.3 \cdot 10^3$
	5 (E)	1438	5.8	1.5	670
	4 (E)	3062	1.0	0.2	870
	3 (A_1)	533	0.9	0.07	$15 \cdot 10^3$
	2 (A_1)	1251	10.6	4.2	240
	1 (A_1)	2970	5.8	13	78
CH_2I_2	9 (B_2)	585	18	1.5	660
	8 (B_2)	1113	41	13	78
	7 (B_1)	717	2.9	0.38	$2.7 \cdot 10^3$
	6 (B_1)	3074	1.4	3.3	303
SiH_4	4 (F_2)	913.3	199	13.9	72
	3 (F_2)	2189.1	151	60.7	16

Table 13.4 (continued)

Molecule	Upper and lower vibrational states (or upper only if the lower state is the ground state)	Fundamental vibrational frequencies ν_l [cm^{-1}] or rotationless transition energy [cm^{-1}]	Absolute (integral) intensity of infrared absorption band S_l [10^{-8} cm^2/s]	Transition probability for spontaneous emission $A_{v',v''}$ [s^{-1}]	Radiative lifetime of upper vibrational level $\tau_{v'}$ [ms]
Larger polyatomic molecules					
C_2H_4	12 (B_{3u})	1443	4.8	2.5	390
	11 (B_{3u})	2989	7.0	15.8	63
	10 (B_{2u})	826	0.1	0.02	$49 \cdot 10^3$
	9 (B_{2u})	3105	12.6	30.5	33
	7 (B_{1u})	949	37.5	8.5	118
SF_6	6 (F_{2u})	947	540	40	25
	5 (F_{2g})	615	31	1.0	1000
C_2F_6	12 (E_u)	220	2.1	0.013	$78 \cdot 10^3$
	11 (E_u)	520	3.9	0.13	$7.4 \cdot 10^3$
	10 (E_u)	1250	410	80	12
	6 (A_{2u})	710	16	2.1	480
	5 (A_{2u})	1120	120	37	27
C_2H_6	9 (E_u)	820	28	2.4	420
	8 (E_u)	1472	63	17	58
	7 (E_u)	2995	585	660	1.5
	6 (A_{2u})	1379	20	9.5	105
	5 (A_{2u})	2950	225	490	2.0
C_3H_6	11 (E')	870	14.5	1.4	730
	10 (E')	1030	9.4	1.2	800
	9 (E')	1440	1.1	0.28	$3.5 \cdot 10^3$
	8 (E')	3025	20.7	24	42
	6 (A_2'')	3100	12.2	30	34
C_6H_6	14 (E_u^-)	1040	4.4	0.60	$1.7 \cdot 10^3$
	13 (E_u^-)	1490	6.5	1.8	550
	12 (E_u^-)	3080	30	36	28
	4 (A_{2u})	670	44	5.0	200

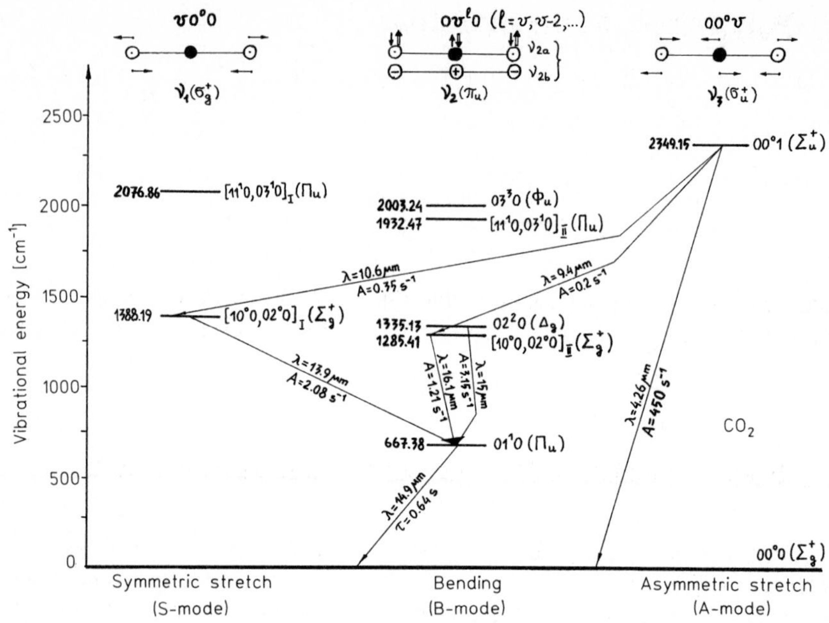

Fig. 13.1. Energy-level diagram of vibrationally excited ground-state CO_2 molecule with transition wavelengths and Einstein coefficients

14. Polyatomic Molecular Ions

Numerical data are presented for bond dissociation energies of positive and negative ionic clusters and for electron and proton affinities of polyatomic molecules.

14.1 Bond Dissociation Energies of Complex Positive Ions

Values of the bond strength for some complex positive ions are given in Tables 14.1–6 (see the definitions in Sect. 13.3). The breaking bond in the ion is indicated by a hyphen. All the values presented are grouped in accordance with the accuracy classes defined in the Introduction. The key information about the energetics of complex positive ions may be found in monographs [14.1.1, 2].

References

14.1.1 B.M.Smirnov: *Complex Ions* (Gordon and Breach, Amsterdam 1985)

14.1.2 H.M.Rosenstock, K.Draxl, B.W.Steiner, J.T.Herron: "Energetics of Gaseous Ions", J. Phys. Chem. Ref. Data **6**, Suppl. 1 (1977)

Table 14.1. Bond dissociation energies of complex positive ions

Clusters with bond indicated	Bond dissociation energy [eV]	Clusters with bond indicated	Bond dissociation energy [eV]
$Ca^+–O_2$	0.56 (*C*)	$O_2^+–N_2$	0.24 (*D*)
$Co^+–CH_2$	3.7 (*C*)	$O_2^+–H_2O$	0.7 (*D*)
$K^+–N_2$	0.31 (*D*)	$O_2^+–N_2O$	0.56 (*D*)
$Li^+–N_2$	0.54 (*D*)	$SO^+–SO_2$	0.6 (*D*)
$Li^+–O_2$	0.33 (*D*)	$CS_2^+–OCS$	0.25 (*D*)
$N^+–N_2$	2.6 (*C*)	$H^+ \cdot N_2–H_2$	0.31 (*D*)
$Na^+–N_2$	0.41 (*D*)	$O_3^+–O_2$	0.26 (*D*)
$Na^+–O_2$	0.25 (*D*)	$O_4^+–O_2$	0.28 (*D*)
$Na^+–O_3$	0.54 (*D*)	$H^+ \cdot N_2 \cdot H_2–H_2$	0.08 (*D*)
$O^+–O_2$	1.86 (*A*)	$UF_4^+–F$	3.1 (*D*)
$Rb^+–N_2$	0.29 (*D*)	$H_2S^+ \cdot H_2S–H_2S$	0.14 (*C*)
$Ar_2^+–Ar$	0.219 (*A*)	$O_6^+–O_2$	0.13 (*D*)
$He_2^+–He$	0.17 (*C*)	$C_3H_5^+–C_2H_4$	0.70 (*C*)
$K_2^+–K$	0.8 (*D*)	$O_8^+–O_2$	0.12 (*D*)
$Li_2^+–Li$	1.5 (*C*)	$C_2H_7^+–C_2H_4$	0.38 (*D*)
$N_2^+–Ar$	1.1 (*C*)	$H_2S^+ \cdot (H_2S)_2–H_2S$	0.05 (*D*)
$Na_2^+–Na$	1.2 (*C*)	$O_{10}^+–O_2$	0.11 (*D*)
$NO^+–N_2$	0.21 (*C*)	$C_2H_4^+ \cdot C_2H_4–C_2H_4$	0.18 (*D*)
$NO^+–H_2O$	0.8 (*D*)	$H_2S^+ \cdot (H_2S)_3–H_2S$	0.06 (*D*)
$NO^+–N_2O$	0.22 (*D*)	$H_2S^+ \cdot (H_2S)_4–H_2S$	0.11 (*D*)
$(NO)_2^+–NO$	0.32 (*C*)		
$O_2^+–O$	0.31 (*B*)		
$O_2^+–H_2$	0.2 (*D*)		

Table 14.2. Bond dissociation energies of positive ionic clusters $X^+–X$

X	Bond dissociation energy [eV]
CO	1.0 (*C*)
HBr	1.0 (*D*)
HCl	0.9 (*D*)
N_2	0.90 (*C*)
NO	0.60 (*C*)
O_2	0.44 (*B*)
CO_2	0.6 (*D*)
CS_2	0.7 (*D*)
H_2O	1.6 (*C*)
H_2S	0.74 (*C*)
N_2O	0.57 (*C*)
OCS	0.76 (*C*)
NH_3	0.79 (*C*)
C_2H_4	0.79 (*C*)

Table 14.3. Bond dissociation energies of ionic clusters $X^+–CO_2$

X^+	Bond dissociation energy [eV]
Cs^+	0.3 (*D*)
K^+	0.3 (*D*)
Na^+	0.5 (*D*)
Rb^+	0.4 (*D*)
NO^+	0.4 (*D*)
O_2^+	0.93 (*C*)
CO_2^+	0.6 (*D*)
H_3O^+	0.62 (*C*)
$H^+ \cdot CO_2$	0.87 (*C*)
$O_2^+ \cdot CO_2$	0.3 (*D*)
$CO_2^+ \cdot CO_2$	0.2 (*D*)
$CO_2^+ \cdot (CO_2)_2$	0.1 (*D*)

Table 14.4. Bond dissociation energies of ionic clusters $X^+ \cdot (H_2O)_n$–H_2O [eV] (typical classes of accuracy – *C* and *D*)

X	*n* 0	1	2	3	4	5
Bi	0.99	0.77	0.61	0.52	0.46	0.42
Cs	0.56	0.52	0.46	0.46	–	–
K	0.78	0.70	0.57	0.51	0.46	0.43
Li	1.50	1.12	0.90	0.71	0.60	0.52
Na	1.04	0.86	0.68	0.60	0.52	0.46
Pb	0.97	0.73	0.53	0.47	0.43	0.42
Sr	1.50	1.32	1.12	0.97	0.89	0.79

Table 14.5. Bond dissociation energies of ionic clusters $X^+ \cdot (NH_3)_n$–NH_3 [eV] (typical classes of accuracy – *C* and *D*)

X	*n* 0	1	2	3	4	5
Bi	1.54	1.01	0.58	0.4	–	–
K	0.87	0.71	0.58	0.50	–	–
Li	1.68	1.44	0.91	0.72	0.48	0.40
Na	1.26	0.99	0.74	0.64	0.47	0.42
Rb	0.81	0.66	0.57	0.49	0.44	–
NH_4	1.09	0.76	0.61	0.53	0.39	–

Table 14.6. Bond dissociation energies of ionic clusters $H^+ \cdot X_n$–X [eV] (typical classes of accuracy – *C* and *D*)

X	*n* 1	2	3	4	5
CO	0.56	0.29	0.27	0.27	0.25
H_2	0.35	0.18	0.16	0.10	–
N_2	0.69	0.17	0.16	0.15	0.14
O_2	0.87	0.29	0.14	–	–
CS_2	0.48	–	–	–	–
H_2O	1.47	0.95	0.72	0.66	0.56

14.2 Bond Dissociation Energies of Complex Negative Ions

The numerical values of bond strength for some complex negative ions are presented in Tables 14.7–11 (see the definitions in Sect. 13.3). The breaking bond in the ion is indicated by a hyphen. All the values given are grouped in accordance with the accuracy classes (see the Introduction). The basic information about the energetics of complex negative ions may be found in monographs [14.2.1, 2] and journal publications [14.2.3–5].

References

14.2.1 B.M.Smirnov: *Complex Ions* (Gordon and Breach, Amsterdam 1985)
14.2.2 H.S.W.Massey: *Negative Ions*, 3rd ed. (Cambridge University Press, Cambridge 1976)
14.2.3 R.N.Compton, P.W.Reinhardt, C.D.Cooper: J. Chem. Phys. **68**, 2023 (1978)
14.2.4 L.N.Sidorov, I.D.Sorokin, M.I.Nikitin, E.V.Skokan: Int. J. Mass. Spectrom. Ion Phys. **39**, 311 (1981)
14.2.5 H.Böhringer, D.W.Fahey, F.C.Fehsenfeld, E.E.Ferguson: J. Chem. Phys. **81**, 2805 (1984)

Table 14.7. Bond dissociation energies of complex negative ions

Clusters with bond indicated	Bond dissociation energy [eV]	Clusters with bond indicated	Bond dissociation energy [eV]
Cl^-–HCl	1.0 (*C*)	HSO_4^-–HCl	0.68 (*C*)
Cl^-–H_2O_2	0.96 (*C*)	HSO_4^-–H_2O_2	0.69 (*C*)
O^-–O_2	1.39 (*B*)	ReF_5^-–F	6.4 (*C*)
O_2^-–O	2.4 (*B*)	SF_5^-–F	1.2 (*D*)
O_2^-–NO	4.3 (*D*)	SeF_5^-–F	1.1 (*D*)
O_2^-–O_2	0.09 (*D*)	$Be_2F_5^-$–KF	3.0 (*C*)
NO_2^-–H_2O_2	0.87 (*C*)	$Be_2F_5^-$–NaF	2.8 (*C*)
NO_2^-–HNO_2	1.4 (*C*)	$Be_2F_5^-$–BeF_2	1.8 (*D*)
NO_3^-–H_2O_2	0.83 (*C*)	$Be_2F_5^-$–$KBeF_3$	2.0 (*D*)
NO_3^-–HNO_3	1.1 (*C*)	$Be_2F_5^-$–$NaBeF_3$	2.0 (*D*)
AlF_4^-–BeF_2	1.8 (*D*)	$NO_2^- \cdot HNO_2$–HNO_2	0.9 (*C*)
AlF_4^-–AlF_3	2.1 (*C*)	$Be_3F_7^-$–KF	3.1 (*C*)
AlF_4^-–$KBeF_3$	2.0 (*C*)	$NO_3^- \cdot HNO_3$–HNO_3	0.77 (*C*)
AlF_4^-–$NaBeF_3$	2.0 (*C*)	$NO_3^- \cdot (HNO_3)_2$–HNO_3	0.70 (*C*)
AlF_4^-–$KAlF_4$	1.6 (*C*)		
AlF_4^-–$NaAlF_4$	1.7 (*C*)		
LaF_4^-–LaF_3	2.6 (*D*)		
YF_4^-–YF_3	2.3 (*D*)		

Table 14.8. Bond dissociation energies of negative ionic clusters X^-–CO_2 [eV]

X =	Cl	I	O	O_2	OH	NO_2	CO_3	SO_3
Bond dissociation energy	0.35 (*D*)	0.24 (*D*)	2.26 (*A*)	0.6 (*C*)	2.5 (*C*)	0.4 (*D*)	0.31 (*D*)	0.28 (*D*)

Table 14.9. Bond dissociation energies of negative ionic clusters $X^- \cdot (H_2O)_n$–H_2O [eV] (typical classes of accuracy – *C* and *D*)

X	*n* 0	1	2	3	4
Br	0.55	0.53	0.50	0.47	–
Cl	0.65	0.55	0.51	0.48	–
F	1.01	0.72	0.59	0.58	0.57
I	0.44	0.42	0.41	–	–
O_2	0.85	0.75	0.67	–	–
OH	0.98	0.71	0.65	0.62	0.61
NO_2	0.66	0.59	0.51	0.50	–
CO_3	0.61	0.59	0.57	–	–
NO_3	0.63	0.62	0.60	–	–
HCO_3	0.68	0.65	0.59	0.58	–
HSO_4	0.52	–	–	–	–

Table 14.10. Bond dissociation energies of negative ionic clusters $X^- \cdot (SO_2)_n$–SO_2 [eV] (class of accuracy – *D*)

X	*n* 0	1	2	3
Cl	0.95	0.53	0.43	0.37
I	0.56	0.44	0.40	–
NO_2	1.05	0.39	0.29	–
NO_3	0.75	–	–	–
SO_2	1.04	–	–	–
SO_3	0.58	–	–	–

Table 14.11. Bond dissociation energies of negative ionic clusters F^-–X [eV]

X	Bond dissociation energy	X	Bond dissociation energy
AlF	4.62 (*B*)	LaF_3	4.5 (*C*)
BF	4.4 (*C*)	MnF_3	4.6 (*C*)
KF	2.3 (*C*)	ScF_3	2.35 (*C*)
BeF_2	4.2 (*C*)	UF_3	5.0 (*C*)
FeF_2	4.6 (*C*)	YF_3	5.0 (*C*)
MnF_2	4.5 (*C*)	HfF_4	4.2 (*C*)
UF_2	4.3 (*D*)	UF_4	4.4 (*C*)
AlF_3	5.16 (*B*)	ZrF_4	4.2 (*C*)
BF_3	4.4 (*C*)	Be_2F_4	4.8 (*C*)
CeF_3	4.8 (*C*)	UF_5	5.1 (*C*)
FeF_3	4.5 (*C*)		

14.3 Electron Affinities of Polyatomic Molecules

Quantitative information about the experimentally determined electron binding energy in molecular ions is given in Tables 14.12–15 for tri-, four-, five-atomic molecules and some larger polyatomic molecules. This energy is called the molecular electron affinity (EA) and is identical to the lowest energy required to remove an electron from the molecular negative ion. The majority of the electron affinities reported here were obtained by the most reliable experimental methods: photoelectron spectroscopy, photodetachment spectroscopy, charge transfer, collisional ionization, etc.

The key information about the values of molecular EAs is given in monographs [14.3.1–3], extensive reviews [14.3.4–6] and some journal publications [14.3.7–9]. The numerical values listed were grouped into the accuracy classes defined in the Introduction.

References

14.3.1 H.S.W.Massey: *Negative Ions*, 3rd ed. (Cambridge University Press, Cambridge 1976)
14.3.2 B.M.Smirnov: *Negative Ions* (McGraw-Hill, New York 1981)
14.3.3 L.V.Gurvitch, G.V.Karachevtsev, V.N.Kondratjev, Y.A.Lebedev, V.A.Medvedev, V.K.Potapov, Y.S.Hodeev: *Bond Dissociation Energies, Ionization Potentials, and Electron Affinities*, 2nd ed. (Nauka, Moscow 1974) (in Russian)
14.3.4 B.K.Janousek, J.I.Brauman: "Electron Affinities", in *Gas Phase Ion Chemistry*, ed. by M.T.Bowers, Vol. 2 (Academic, New York 1979) Chap. 10, pp. 53–86
14.3.5 R.R.Corderman, W.C.Lineberger: "Negative Ion Spectroscopy", in Annu. Rev. Phys. Chem. **30**, 347–378 (1979)
14.3.6 P.S.Drzaic, J.Marks, J.I.Brauman: "Electron Photodetachment from Gas Phase Molecular Anions", in *Gas Phase Ion Chemistry*, ed. by M.T.Bowers, Vol. 3 (Academic, New York 1984) Chap. 21, pp. 167–211

14.3.7 R.G.Keesee, N.Lee, A.W.Castleman: J. Chem. Phys. **73**, 2195 (1980)
14.3.8 R.N.Compton, P.W.Reinhardt, C.D.Cooper: J. Chem. Phys. **68**, 2023 (1978)
14.3.9 L.N.Sidorov, I.D.Sorokin, M.I.Nikitin, E.V.Skokan: Int. J. Mass Spectrom. Ion Phys. **39**, 311 (1981)

Table 14.12. Electron affinities (EA) of triatomic molecules

Molecule or radical	EA [eV]	Molecule or radical	EA [eV]
AlF_2	2.3 (*C*)	LiCN	0.74 (*D*)
AlO_2	4.1 (*C*)	LiNC	0.62 (*D*)
$AsBr_2$	3.5 (*C*)	LiOH	0.22 (*D*)
$AsCl_2$	2.2 (*C*)	NCO	2.6 (*D*)
AsF_2	0.8 (*D*)	NH_2	0.76 (*B*)
AsH_2	1.27 (*B*)	NF_2	1.7 (*D*)
BF_2	2.2 (*C*)	NO_2	2.28 (*B*)
BO_2	4.0 (*C*)	N_2O	0.24 (*D*)
C_3	2.1 (*C*)	N_3	2.7 (*C*)
CF_2	2.1 (*D*)	NiCO	0.80 (*B*)
C_2H	3.73 (*B*)	O_3	2.103 (*A*)
C_2O	1.85 (*B*)	P_3	0.9 (*D*)
CH_2	0.21 (*D*)	PH_2	1.27 (*A*)
CNS	2.0 (*C*)	PF_2	1.4 (*C*)
COS	0.5 (*D*)	SCN	2.2 (*D*)
CS_2	0.8 (*D*)	SH_2	1.1 (*C*)
FCN	4.0 (*D*)	SO_2	1.06 (*B*)
FeCO	1.26 (*B*)	S_3	2.0 (*C*)
GeF_2	1.3 (*D*)	SeCN	2.6 (*C*)
HNO	0.34 (*C*)	SiH_2	1.12 (*B*)
HO_2	1.19 (*A*)		

Table 14.13. Electron affinities (EA) of four-atomic molecules

Molecule or radical	EA [eV]	Molecule or radical	EA [eV]
CCl_3	1.2 (*D*)	PBr_2Cl	1.6 (*D*)
CF_3	1.92 (*C*)	$PBrCl_2$	1.5 (*C*)
CH_3	1.07 (*B*)	PCl_3	0.8 (*D*)
CO_3	2.82 (*B*)	$POCl_2$	3.8 (*C*)
GeF_3	3.0 (*D*)	SF_3	2.9 (*C*)
FeF_3	4.2 (*C*)	SO_3	1.7 (*C*)
HCCO	2.35 (*B*)	SO_2F	2.8 (*C*)
MnF_3	4.4 (*C*)	SiF_3	2.7 (*C*)
NO_3	3.7 (*B*)	SiH_3	1.4 (*D*)
PBr_3	1.6 (*D*)	UF_3	1.5 (*D*)

Table 14.14. Electron affinities (EA) of five-atomic molecules

Molecule or radical	EA [eV]	Molecule or radical	EA [eV]
CCl_3F	1.1 (*D*)	CO_4	1.22 (*C*)
CCl_2F_2	0.4 (*D*)	CeF_4	3.6 (*C*)
CD_3O	1.55 (*B*)	$Fe(CO)_2$	1.22 (*B*)
C_2F_3	2.0 (*D*)	FeF_4	5.4 (*C*)
CF_3Br	0.9 (*D*)	HNO_3	0.6 (*D*)
CF_2CO	2.4 (*D*)	$LiCH_3$	0.24 (*D*)
CF_3I	1.5 (*C*)	MnF_4	5.3 (*C*)
CF_3O	1.4 (*D*)	$Ni(CO)_2$	0.64 (*B*)
CF_3S	1.8 (*D*)	$OH \cdot H_2O$	1.95 (*C*)
CH_3Br	0.4 (*D*)	$POCl_3$	1.4 (*D*)
CH_3O	1.57 (*B*)	PtF_4	5.2 (*C*)
CH_3S	1.86 (*A*)	SF_4	2.3 (*C*)
		UF_4	1.7 (*D*)

Table 14.15. Electron affinities (EA) of larger polyatomic molecules

Molecule or radical	EA [eV]	Molecule or radical	EA [eV]
C_3F_3	4.0 (*D*)	SeF_6	3.0 (*C*)
C_3H_3	2.3 (*D*)	TeF_6	3.3 (*B*)
MoF_5	3.3 (*C*)	UF_6	5.0 (*B*)
PtF_5	6.5 (*C*)	WF_6	4.0 (*C*)
SF_5	3.7 (*C*)	C_3F_5	2.2 (*D*)
UF_5	4.0 (*C*)	C_2H_5O	0.6 (*D*)
CH_3NO_2	0.4 (*D*)	C_2H_5S	1.4 (*D*)
C_2Cl_5	1.6 (*D*)	C_4F_5	1.0 (*D*)
C_2F_5	2.2 (*C*)	C_2H_6N	1.0 (*D*)
C_2H_5	0.89 (*C*)	C_3F_7	2.2 (*C*)
C_4H_2N	1.7 (*D*)	C_3H_7	0.6 (*D*)
$Fe(CO)_3$	1.8 (*D*)	C_5H_5	1.79 (*B*)
IrF_6	4.3 (*C*)	C_6Cl_5	2.8 (*C*)
MoF_6	4.2 (*C*)	C_4F_7	3.1 (*D*)
$Ni(CO)_3$	1.08 (*A*)	C_6F_5	2.74 (*C*)
OsF_6	4.1 (*D*)	C_6H_5	2.3 (*C*)
PtF_6	8.0 (*C*)	C_6H_5NH	1.70 (*B*)
ReF_6	3.9 (*D*)	$C_6H_5CH_2$	0.86 (*B*)
SF_6	0.6 (*D*)		

14.4 Proton Affinities of Molecules

Quantitative data on the experimentally determined proton binding energy in molecular ions are presented in Table 14.16 for some diatomic, triatomic and larger polyatomic molecules. This energy is called the proton affinity (PA) of a molecule X and is defined as the negative of the enthalpy of the reaction $X + H^+ \rightarrow XH^+$, i.e.

$$\mathrm{PA}(\mathrm{X}) = \Delta H^{\circ}_{\mathrm{f0}}(\mathrm{X}) + \Delta H^{\circ}_{\mathrm{f0}}(\mathrm{H}^+) - \Delta H^{\circ}_{\mathrm{f0}}(\mathrm{XH}^+) \,,$$

where the $\Delta H^{\circ}_{\mathrm{f0}}$ are the heats of formation of the species X, H^+ and XH^+ from elements in their standard states.

The main part of the listing of proton affinities in Table 14.16 was determined by the most reliable techniques, namely, pulsed electron beam high-pressure mass spectrometry, and ion cyclotron resonance technique and high-pressure and chemical ionization mass spectrometry, etc.

Basic information about the values of molecular PAs is collected in monograph [14.4.1], extensive reviews [14.4.2–4] and the journal publications [14.4.5, 6]. The numerical values presented were grouped into accuracy classes as usual (see the Introduction).

References

14.4.1 L.V.Gurvitch, G.V.Karachevtsev, V.N.Kondratjev, Y.A.Lebedev, V.A.Medvedev, V.K.Potapov, Y.S.Hodeev: *Bond Dissociation Energies, Ionization Potentials, and Electron Affinities*, 2nd ed. (Nauka, Moscow 1974) (in Russian)

14.4.2 H.M.Rosenstock, K.Draxl, B.W.Steiner, J.T.Herron: "Energetics of Gaseous Ions", J. Phys. Chem. Ref. Data **6**, Suppl. 1 (1977)

14.4.3 R.Walder, J.L.Franklin: Int. J. Mass Spectrom. Ion Phys. **36**, 85 (1980)

14.4.4 S.G.Lias, J.F.Liebman, R.D.Levin: J. Phys. Chem. Ref. Data **13**, 695–808 (1984)

14.4.5 D.K.Bohme, G.I.Mackay, H.I.Schiff: J. Chem. Phys. **73**, 4976 (1980)

14.4.6 A.B.Raksit, D.K.Bohme: Int. J. Mass Spectr. Ion Processes **57**, 211 (1984)

Table 14.16. Proton affinities (PA) of molecules

I) Diatomic molecules

Molecule or radical	CO	CN	D_2	H_2	HBr	HCl	HF	HI	HS	N_2	NH	NO	O_2	OH
PA [eV]	6.15 (*A*)	5 (*D*)	4.56 (*B*)	4.38 (*B*)	6.10 (*A*)	5.86 (*B*)	4.09 (*B*)	6.51 (*B*)	7.2 (*C*)	5.1 (*C*)	6.1 (*C*)	5.50 (*B*)	4.38 (*A*)	6.18 (*B*)

II) Triatomic molecules

Molecule or radical	CH_2	CO_2	HCN	HCO	H_2O	HO_2	H_2S	H_2Se	NH_2	NO_2	N_2O	OCS	PH_2	SO_2
PA [eV]	8.55 (*B*)	5.68 (*B*)	7.43 (*A*)	6.6 (*B*)	7.20 (*B*)	6.9 (*C*)	7.4 (*B*)	7.4 (*B*)	8.1 (*B*)	6.6 (*B*)	5.9 (*B*)	6.6 (*B*)	7.2 (*C*)	7.0 (*B*)

III) Larger polyatomic molecules

Molecule or radical	C_2H_2	CH_3	C_2N_2	F_2CO	H_2CO	HNCO	HNO_2	H_2O_2	$(HF)_2$	NF_3	NH_3	N_2H_2	PF_3	PH_3
PA [eV]	6.7 (*B*)	5.4 (*C*)	7.0 (*B*)	7.0 (*B*)	7.5 (*B*)	7.5 (*B*)	8.1 (*C*)	7.0 (*B*)	5.2 (*C*)	6.6 (*C*)	8.9 (*B*)	7.9 (*C*)	7.1 (*B*)	8.07 (*A*)
Molecule or radical	SO_3	CF_4	CH_3Cl	CH_4	GeH_4	HNO_3	SiH_4	$TiCl_4$	C_2H_4	CH_3CN	CH_3OH	$(HF)_3$	SF_6	C_2H_6
PA [eV]	6.0 (*B*)	5.2 (*C*)	6.9 (*C*)	5.7 (*B*)	7.1 (*B*)	7.5 (*B*)	6.6 (*B*)	7.6 (*B*)	7.1 (*B*)	8.11 (*A*)	7.9 (*B*)	5.8 (*C*)	3.7 (*C*)	6.9 (*B*)

15. Electrical Properties of Molecules

Numerical data are given for dipole and quadrupole moments of molecules and the average dipole polarizabilities of polyatomic molecules.

15.1 Dipole Moments of Molecules

The electric dipole moment of a molecule is defined as

$$\boldsymbol{\mu} = \left\langle \sum_i e_i \boldsymbol{r}_i \right\rangle ,$$

where e_i, $\boldsymbol{r}_i$ are the electron charge and radius vector, respectively. Summation in this expression is carried out with respect to all elementary charges in a molecule, but the choice of coordinate origin for the charges is arbitrary in the case of a neutral molecule. The angled brackets designate averaging over the molecular wavefunction; hence the electric dipole moment μ depends on the electronic-vibration-rotation state of a molecule.

In practice, it appears that the rotational excitation has almost no influence on the μ values; the electric dipole moments for the adjacent vibrational levels may vary in the range 1–2%; the electric dipole moments referred to different molecular electronic states may vary by as much as $\gtrsim 30\%$. The variation of μ values within the range of a few percent may be attributed to molecules with different isotopic compositions.

The numerical values of electric dipole moments of some widely spread molecules are incorporated in Table 15.1. They are grouped into accuracy classes (see the Introduction); we added here one more class S, which covered the molecules with zero electric dipole moment ($\mu \equiv 0$) due to symmetry properties. The measurement unit for the electric dipole moment is called the debye (D) and $1\,\mathrm{D} = 10^{-18}$ e.s.u. · cm.

In a number of cases we give examples of the isotopic variation of μ values, and introduce the values of the electric dipole moment μ_v for different vibrational levels of a diatomic molecule:

$$\mu_\mathrm{v} = \mu_\mathrm{e} + \mu_\mathrm{a}(v + 1/2) + \mu_\mathrm{b}(v + 1/2)^2 + \ldots ,$$

where $v = 0, 1, 2, \ldots$ is the vibrational quantum number, μ_e is referred to the equilibrium position and μ_a, μ_b are the expansion coefficients. Single (*) and double (**) asterisks stand for the values of μ_1 and μ_2, respectively. In all the other cases, the μ's listed are attributed to the molecule with the dominant isotopic composition and a non-excited electronic-vibration state.

Key information about the experimental methods of dipole moment measurement and accuracy estimation is given in [15.1.1–5].

References

15.1.1 A.L.McClellan: *Tables of Experimental Dipole Moments*, Vol. 2 (Rahara Enterprises, El Cerrito 1974)

15.1.2 O.A.Osipov, V.I.Minkin, A.D.Garnovskii: *Handbook of Dipole Moments*, 3rd ed. (High School Publ. House, Moscow 1971) (in Russian)

15.1.3 R.D.Nelson, Jr., D.R.Lide, A.A.Maryott: "Selected Values of Electric Dipole Moments for Molecules in the Gas Phase", Nat. Stand. Ref. Data Ser. Nat. Bur. Stand. **10** (1967)

15.1.4 K.P.Huber, G.Herzberg: *Molecular Spectra and Molecular Structure IV. Constants of Diatomic Molecules* (Van Nostrand Reinhold, New York 1979)

15.1.5 R.Tischer, J.Demaison, B.Starck: "Dipole Moments", in *Molecular Constants*, ed. by K.-H. Hellwege, Landolt-Börnstein, Group II, Vol. 6 (Springer, Berlin, Heidelberg, New York 1974) pp. 2-260 – 2-304

Table 15.1. Dipole moments of molecules

Molecule or radical	Ground state dipole moment μ [debye]	Molecule or radical	Ground state dipole moment μ [debye]
Diatomic molecules			
AgCl	6.1 (*B*)	ClF	0.888 (*A*)
AgF	6.2 (*C*)	ClO	1.24 (*B*)
AlF	1.53 (*B*)	CsBr	10.8 (*A*)
BF	0.5 (*D*)	CsCl	10.387 (*A*)
BH	1.3 (*D*)		10.445* (*A*)
BaO ($X\,^1\Sigma^+$)	7.955 (*A*)		10.503** (*A*)
	7.997* (*A*)	CsF	7.883 (*A*)
	8.039 ** (*A*)		7.953* (*A*)
($A\,^1\Sigma^+$)	2.2 ($v = 7$) (*B*)		8.024** (*A*)
BaS	10.86 (*A*)	CsI	11.7 (*A*)
	10.88* (*A*)	CuF	5.8 (*C*)
	10.91** (*A*)	FO	0.0043 (*C*)
BrCl	0.519 (*A*)		0.027* (*C*)
BrF	1.3 (*D*)	GaF	2.4 (*C*)
BrO	1.76 (*A*)	GeO	3.282 (*A*)
CF	0.65 (*D*)		3.303* (*A*)
CH	1.45 (*C*)		3.324** (*A*)
CN	1.4 (*D*)	GeS	2.00 (*B*)
CO	0.1098 (*A*)	GeSe	1.65 (*A*)
CS	1.97 (*A*)	GeTe	1.06 (*B*)
CSe	2.0 (*B*)	$^1H^2H$	$5.5 \cdot 10^{-4}$ (*B*)
CaBr	4.36 (*A*)	$^1H^2H^+$	0.87 (*B*)
CaCl	4.26 (*A*)	HBr	0.827 (*A*)
CaF	3.1 (*B*)	^2HBr	0.823 (*A*)

Table 15.1 (continued)

Molecule or radical	Ground state dipole moment μ [debye]	Molecule or radical	Ground state dipole moment μ [debye]
HCl	1.108 (A)	^{7}LiRb	4.0 (B)
	1.139* (A)	NH ($a\,^{1}\Delta$)	1.5 (C)
	1.168** (A)	($A\,^{3}\Pi$)	1.3 (C)
^{2}HCl	1.103 (A)	($c\,^{1}\Pi$)	1.7 (C)
	1.126* (A)	NO ($^{2}\Pi_{1/2}$)	0.157 (A)
HF	1.826 (A)		0.142* (A)
^{2}HF	1.819 (A)	NS ($^{2}\Pi_{1/2}$)	1.8 (C)
HI	0.448 (A)	NaBr	9.118 (A)
^{2}HI	0.44 (C)		9.171* (A)
IBr	0.74 (B)	NaCl	9.001 (A)
ICl	1.24 (C)		9.061* (A)
IO	2.45 (C)		9.121** (A)
InCl	3.8 (B)	NaCs	4.7 (B)
InF	3.4 (B)	NaF	8.156 (A)
KBr	10.628 (A)		8.221* (A)
	10.679* (A)		8.287** (A)
	10.729** (A)	NaI	9.236 (A)
KCl	10.269 (A)		9.286* (A)
	10.329* (A)	NaH	6.4 (C)
	10.388** (A)	NaK	2.67 (A)
KF	8.593 (A)	NaRb	3.1 (C)
	8.661* (A)	OH ($X\,^{2}\Pi_{i}$)	1.655 (A)
	8.731** (A)	($A\,^{2}\Sigma^{+}$)	1.98 (B)
KI	10.8 (A)	O^{2}H ($X\,^{2}\Pi_{i}$)	1.653 (A)
^{6}LiBr	7.268 (A)	($A\,^{2}\Sigma^{+}$)	2.16 (B)
	7.352* (A)	PN	2.747 (A)
	7.438** (A)		2.738* (A)
^{7}LiBr	7.265 (A)		2.730** (A)
^{6}LiCl	7.129 (A)	PbO	4.6 (B)
	7.217* (A)	PbS	3.6 (C)
	7.306** (A)	PbSe	3.3 (B)
^{7}LiCl	7.129 (A)	PbTe	2.7 (B)
	7.216* (A)	RbBr	10.9 (A)
	7.305** (A)	RbCl	10.510 (A)
^{7}LiF	6.325 (A)		10.564* (A)
	6.407* (A)		10.618** (A)
	6.491** (A)	RbF	8.546 (A)
^{7}LiH ($X\,^{1}\Sigma^{+}$)	5.882 (A)		8.613* (A)
	5.990* (A)		8.681** (A)
	6.098** (A)	RbI	11.5 (A)
($A\,^{1}\Sigma^{+}$)	1.9* (C)	SF	0.79 (B)
	1.5** (C)	SH	0.758 (A)
^{7}Li^{2}H	5.868 (A)	S^{2}H	0.757 (A)
^{6}LiI	7.428 (A)	SO ($X\,^{3}\Sigma^{-}$)	1.55 (B)
	7.512* (A)	($a\,^{1}\Delta$)	1.32 (B)
^{7}LiK	3.4 (B)	SeF	1.5 (C)
^{7}LiNa	0.46 (B)	SeH	0.50 (C)
^{7}LiO	6.8 (B)	Se^{2}H	0.48 (B)

Table 15.1 (continued)

Molecule or radical	Ground state dipole moment μ [debye]	Molecule or radical	Ground state dipole moment μ [debye]
SeO ($a\,^1\Delta$)	2.01 (*B*)	O_3	0.534 (*B*)
SiH	5.9 (*B*)	SCTe	0.17 (*B*)
SiO	3.098 (*A*)	SF_2	1.0 (*C*)
	3.118* (*A*)	SO_2	1.633 (*A*)
	3.137** (*A*)	S_2F	1.03 (*C*)
SiS	1.7 (*C*)	S_2O	1.47 (*B*)
SnO	4.32 (*B*)	SiF_2	1.23 (*B*)
SnS	3.18 (*A*)		
	3.20* (*A*)	Four-atomic molecules	
SnSe	2.82 (*B*)	BBr_3	0 (*S*)
SnTe	2.19 (*B*)	BCl_3	0 (*S*)
SrF	3.50 (*A*)	BF_3	0 (*S*)
	3.55* (*A*)	C_2H_2	0 (*S*)
SrO	8.90 (*A*)	CHFO	2.35 (*A*)
	8.87* (*A*)	CH_2O	2.33 (*A*)
	8.85** (*A*)	$COCl_2$	1.17 (*A*)
TlBr	4.49 (*B*)	ClF_3	0.6 (*C*)
TlCl	4.543 (*A*)	F_2O_2	1.44 (*B*)
	4.598* (*A*)	$FeCl_3$	1.3 (*C*)
	4.654** (*A*)	HBF_2	0.97 (*B*)
TlF	4.228 (*A*)	HCNO	3.1 (*C*)
	4.297* (*A*)	H_2O_2	2.2 (*C*)
	4.366** (*A*)	HNO_2 (cis-)	1.42 (*A*)
TlI	4.61 (*B*)	(trans-)	1.86 (*A*)
		HNSO	0.911 (*A*)
Triatomic molecules		NF_3	0.24 (*C*)
CF_2	0.46 (*B*)	NFO_2	0.47 (*B*)
Cl_2O	1.7 (*C*)	NHF_2	1.93 (*A*)
ClO_2	1.78 (*A*)	N^2HF_2	1.93 (*B*)
CsOH	7.01 (*A*)	NH_3	1.47 (*A*)
FCN	2.17 (*B*)	$^{15}NH_3$	1.26 (*B*)
F_2O	0.297 (*A*)	N_2F_2	0.16 (*C*)
F_2Si	1.23 (*B*)	N_3H	0.84 (*A*)
HCN	2.98 (*A*)	NO_2Cl	0.53 (*A*)
H_2O	1.8473 (*A*)	PCl_3	0.56 (*C*)
^2H_2O	1.86 (*B*)	PH_3	0.574 (*A*)
H_2S	0.978 (*A*)	$SOCl_2$	1.45 (*B*)
H_2Se	0.24 (*C*)	SOF_2	1.62 (*A*)
HOCl	1.3 (*D*)	SO_2F	0.23 (*B*)
$HgBr_2$	0 (*S*)	SO_3	0 (*S*)
N_2O	0.1608 (*A*)	S_2F_2	1.4 (*C*)
NOF	1.81 (*B*)		
NO_2	0.32 (*B*)	Five-atomic molecules	
$^{15}NO_2$	0.29 (*C*)	B_2O_3	3.5 (*C*)
NSCl	1.87 (*B*)	CCl_4	0 (*S*)
NSF	1.90 (*A*)	CF_3Cl	0.50 (*A*)
OCS	0.715 (*A*)	CH_3Br	1.84 (*B*)
ONBr	1.8 (*C*)	CH_3Cl	1.892 (*A*)
ONCl	1.86 (*B*)	CH_3F	1.847 (*A*)

Table 15.1 (continued)

Molecule or radical	Ground state dipole moment μ [debye]	Molecule or radical	Ground state dipole moment μ [debye]
CH_4	$5.4 \cdot 10^{-6}$ (*B*)	Six-atomic molecules	
$CH_2{}^2H_2$	0.014 (*C*)	BrF_5	1.5 (*D*)
CH_3I	1.62 (*B*)	IF_5	2.3 (*C*)
CH_2O_2	1.41 (*A*)	N_2H_4	1.7 (*C*)
$FClO_3$	0.023 (*D*)	PF_5	0 (*S*)
GeH_3Br	2.00 (*A*)	Larger polyatomic molecules	
HNO_3	2.17 (*A*)	C_2H_4O	2.7 (*B*)
KNO_3	1.6 (*C*)	SF_6	0 (*S*)
N_2O_3	2.12 (*A*)	UF_6	0 (*S*)
NSF_3	1.91 (*B*)	C_2H_6	0 (*S*)
POF_3	1.73 (*B*)	$C_2H_4O_2$	1.70 (*B*)
PSF_3	0.63 (*C*)	C_2H_4OS	3.72 (*B*)
SF_4	0.632 (*A*)	C_3H_6	0.366 (*A*)
SO_2F_2	1.12 (*B*)	C_3H_8	0.084 (*A*)
SeF_4	1.78 (*B*)		
$SrCO_3$	1.9 (*C*)		

15.2 Molecular Polarizabilities

The electric dipole polarizability of a molecule is a tensor quantity according to (5.2). Here we give the average dipole polarizability $\bar{\alpha}_{av}$ for a number of gas-phase molecules. This is derived from the relation

$$\bar{\alpha}_{av} = (\alpha_1 + \alpha_2 + \alpha_3)/3 ,$$

where $\alpha_{1,2,3}$ are the components of the dipole polarizability tensor reduced to the principal axes.

The numerical data contained in Table 15.2 are based mainly on [15.2.1–6], in which the methods of measuring the molecular polarizability are also discussed. The numerical values listed are grouped into accuracy classes (see the Introduction). To convert the measurement units of polarizability, we made use of the conversion factor $1\,a_0^3 = 0.1482\,\text{Å}^3$.

References

15.2.1 J.O.Hirschfelder, C.F.Curtiss, R.B.Bird: *Molecular Theory of Gases and Liquids* (Wiley, New York 1964)

15.2.2 A.A.Maryott, F.Buckley: Nat. Bur. Stand. (U.S.) Circ. **537** (1953)

15.2.3 N.J.Bridge, A.D.Buckingham: Proc. Roy. Soc. **295 A**, 334 (1966)

15.2.4 J.Applequist, J.R.Carl, K.-K.Fung: J. Am. Chem. Soc. **94**, 2952 (1972)

15.2.5 R.W.Molof, T.M.Miller, H.L.Schwartz, B.Bederson, J.T.Park: J. Chem. Phys. **61**, 1816 (1974) (alkali dimers)

15.2.6 J.Rychlewski: Mol. Phys. **41**, 833 (1980) (H_2, HD, D_2)

Table 15.2. Average static polarizabilities $\bar{\alpha}_{av}$ of molecules

Molecule	$\bar{\alpha}_{av}$ [Å^3]	[a_0^3]	Accuracy class
Diatomic molecules			
Br_2	6.5	44	(*D*)
CO	1.95	13.2	(*B*)
Cl_2	4.6	31	(*C*)
Cs_2	13	90	(*C*)
D_2	0.793	5.348	(*A*)
HBr	3.6	24	(*C*)
HCl	2.6	18	(*C*)
HD	0.798	5.387	(*A*)
H_2	0.803	5.417	(*A*)
HF	0.83	5.6	(*C*)
HI	5.3	36	(*C*)
K_2	60	410	(*C*)
Li_2	34	230	(*C*)
LiNa	40	270	(*D*)
N_2	1.75	11.8	(*A*)
NO	1.7	12	(*C*)
Na_2	30	200	(*C*)
O_2	1.59	10.7	(*A*)
Rb_2	70	460	(*D*)
Triatomic molecules			
CO_2	2.6	18	(*C*)
CS_2	8.7	59	(*C*)
HCN	2.6	17	(*C*)
H_2O	1.45	9.8	(*B*)
H_2S	3.7	25	(*C*)
N_2O	3.0	20	(*B*)
NO_2	3.0	20	(*C*)
SO_2	3.8	26	(*C*)
Four-atomic molecules			
C_2H_2	3.5	24	(*D*)
H_2CO	2.5	17	(*D*)
NH_3	2.22	15.0	(*B*)
Five-atomic molecules			
CCl_4	10.4	70	(*B*)
CF_4	2.9	19	(*C*)
$CHBr_3$	11.8	80	(*B*)
$CHCl_3$	8.4	57	(*C*)
CHF_3	2.9	20	(*C*)
CH_2Br_2	8.7	59	(*C*)
CH_2Cl_2	6.7	45	(*B*)
CH_3Br	5.6	38	(*C*)
CH_3Cl	4.56	30.8	(*B*)
CH_3F	2.6	18	(*C*)
CH_4	2.56	17.3	(*B*)
CH_3I	7.6	51	(*C*)
CH_2I_2	12.9	87	(*C*)
CHI_3	18.0	120	(*C*)
Larger polyatomic molecules			
C_2H_4	4.2	28	(*C*)
CH_3OH	3.3	22	(*C*)
$(NO_2)_2$	6.6	45	(*D*)
$(CH_2)_2O$	4.4	30	(*D*)
CH_3NH_2	4.0	27	(*D*)
SF_6	6.55	44.2	(*A*)
C_2H_6	4.5	30	(*C*)
$O(CH_2CH_2)O$	8.6	58	(*D*)
C_3H_6	9.1	62	(*D*)
$C_2H_5 \cdot OH$	5.1	34	(*C*)
$CH_3 \cdot O \cdot CH_3$	5.2	35	(*C*)
C_3H_8	6.28	42.3	(*B*)
C_6H_6	10.4	70	(*B*)
$C_3H_7 \cdot OH$	7.0	47	(*C*)
$C(CH_3)_4$	10.2	69	(*B*)
C_6H_{12}	11.0	74	(*B*)
$C_6H_{11} \cdot OH$	11.5	78	(*B*)

15.3 Quadrupole Moments of Molecules

The electric quadrupole moment of a molecule is defined as the symmetrical tensor

$$Q_{\alpha\beta} = \tfrac{1}{2}\int \varrho\,(3\,r_\alpha r_\beta - r^2\delta_{\alpha\beta})\,d\boldsymbol{r}\ ,$$

where ϱ is the charge density in the molecule, r_α, r_β are the components of the radius vectors of the elementary charges, $\alpha, \beta \equiv x, y, z$, $\delta_{\alpha\beta}$ is the Kronecker symbol, and $d\boldsymbol{r}$ is a volume element.

It has been proved [15.3.1], that if a molecule has an axis of n-fold symmetry, only one independent scalar quantity is required to determine any electric multipole tensor of rank p less than n $(p < n)$. Table 15.3 incorporates the quadrupole moments of molecules which have an axis of at least threefold symmetry (except for SO_2 and C_2H_4 when the tensor component Q_{zz} only is listed); hence their quadrupole moment is a scalar.

The value of the molecular quadrupole moment depends on the location of the coordinate system origin, so that hereafter, the quadrupole moments are given relative to an origin at the centre of mass of the molecule. The orientation of the z axis is chosen so that it coincides with the axis of greatest rotational symmetry, while the x and y axes are orthogonal to z. Taking this into account, we listed in Table 15.3 the values of the molecular quadrupole moment defined as

$$Q = Q_{zz} = -\,2\,Q_{xx} = -\,2\,Q_{yy}\ .$$

The molecular quadrupole moment is also characterized by the sign, which can be determined by several methods [15.3.1]. A number of experimental techniques allow one to derive only the absolute value of the moment Q, which is listed in this case without indicating the sign.

The numerical values of the molecular quadrupole moment are grouped into accuracy classes (see the Introduction). The basic information about the Qs may be found in [15.3.1–4]. The conversion factor for the measurement units of the quadrupole moment is defined as 1 a.u. $Q = 1.345 \cdot 10^{-26}$ e.s.u. $\cdot$ cm^2.

References

15.3.1 D.E.Stogryn, A.P.Stogryn: Mol. Phys. **11**, 371 (1966)

15.3.2 G.Birnbaum: "Microwave Pressure Broadening and Its Application to Intermolecular Forces", in *Intermolecular Forces*, ed. by J.O.Hirschfelder, Adv. Chem. Phys., Vol. 12 (Wiley, New York 1967) pp. 487–548

15.3.3 W.B.Somerville: "Microwave Transitions of Interstellar Atoms and Molecules", Adv. At. Mol. Phys. **13**, 383–436 (1977)

15.3.4 R.Tischer, W.Hüttner: "Magnetic Constants", in *Molecular Constants*, ed. by K.-H. Hellwege, Landolt-Börnstein, Group II, Vol. 6 (Springer, Berlin, Heidelberg, New York 1974) pp. 2-383–2-448

Table 15.3. Scalar quadrupole moments of molecules

Molecule	Quadrupole moment Q [10^{-26} e.s.u. · cm^2]	[a.u.]	Accuracy class
Diatomic molecules			
BF	−4.5	−3.4	(*D*)
Cl_2	+6.1	+4.6	(*D*)
CO	−2.0	−1.5	(*D*)
Cs_2	+1.8	+1.3	(*D*)
D_2	+0.65	+0.48	(*B*)
F_2	+0.9	+0.6	(*D*)
HBr	+4	+3	(*D*)
HCl	+3.8	+2.8	(*D*)
HD	+0.64	+0.48	(*B*)
HF	+2.36	+1.75	(*C*)
H_2	+0.651	+0.484	(*A*)
HI	+6	+4.5	(*D*)
KF	−9.3	−6.9	(*C*)
LiF	+5.8	+4.3	(*D*)
LiH	−5	−3.7	(*D*)
Li_2	+14	+10	(*C*)
NH	−0.3	−0.2	(*D*)
NO	−2.4	−1.8	(*C*)
N_2	−1.4	−1.0	(*C*)
NaF	−2.0	−1.5	(*C*)
O_2	−0.4	−0.3	(*C*)
TlF	−13.5	−10	(*C*)
Triatomic molecules			
BrCN	7	5	(*E*)
ClCN	6.6	4.9	(*D*)
CO_2	−4.3	−3.2	(*C*)
N_2O	−3	−2	(*D*)
OCS	3.1	2.3	(*D*)
SO_2	4.4	3.3	(*D*)
Larger polyatomic molecules			
C_2H_2	3.0	2.2	(*D*)
NH_3	−1	−0.7	(*D*)
C_2H_4	+1.5	+1.1	(*D*)
C_6F	−18.1	−10.0	(*C*)
C_2H_6	−0.65	−0.48	(*D*)
C_6H_6	3.6	2.7	(*D*)

Mathematical Appendices

We give below a brief outline of the coefficients of fractional parentage and Clebsch-Gordan coefficients, which are important for many problems in the theory of atomic spectra, collisions and so on.

A. Coefficients of Fractional Parentage

The wavefunction Ψ of the atomic valence electrons occupying states with the same principal quantum number can be constructed from the wavefunction φ of one valence electron and the wave function Φ of the atomic core as follows:

$$\Psi_{LM_LSM_S}(1, 2, \ldots, N) = \hat{P} \sum_{l, m_l, s, m_s} G_{ls}^{LS}(l_e, N) \cdot (\tfrac{1}{2}s\sigma m_s \,|\, \tfrac{1}{2}sSM_S)$$
$$\times (l_e l \mu m_l \,|\, l_e l L M_L) \cdot \Phi_{lm_l sm_s}(1, 2, \ldots, N-1) \cdot \varphi_{l_e \mu 1/2 \sigma}(N) \,.$$

Here L, S are the orbital and spin angular momenta of the atom, M_L, M_S their projections on a given direction, l, s the orbital and spin angular momenta of the atomic core, m_l, m_s their projections on a given direction, l_e the orbital moment of the valence electron, μ, σ the projections of the electron orbital and spin angular momenta on a given direction and $\hat{P}$ is the operator of electron exchange in the atom. The right subscript, added to the symbol for a wavefunction, characterizes the quantum numbers of the state of a given particle. The argument of a wavefunction includes the electron numbers only and, hence, the latter replace a set of space and spin coordinates of the respective electrons. Finally, $(\tfrac{1}{2}s\sigma m_s \,|\, \tfrac{1}{2}sSM_S)$ and $(l_e l \mu m_l \,|\, l_e l L M_L)$ are Clebsch-Gordan coefficients, which arise in adding the electron and core spins to obtain the spin of the atom and the electron and core orbital momenta to obtain the atomic orbital momentum. The $G_{ls}^{LS}(l_e, N)$ are called the coefficients of fractional parentage.

There is the following relation between the coefficients $G_{ls}^{LS}(l_e, N)$ and $G_{LS}^{ls}(l_e, 4l_e + 3 - N)$ corresponding to the electron configurations l_e^N and $l_e^{4l_e + 3 - N}$:

$$G_{ls}^{LS}(l_e, N) = (-1)^{s - S + l + L - l_e - 1/2} \sqrt{\frac{N(2s+1)(2l+1)}{(4l+1-N)(2S+1)(2L+1)}}$$
$$\times G_{LS}^{ls}(l_e, 4l_e + 3 - N) \,.$$

As a consequence, it is sufficient to calculate the G_{ls}^{LS} for the electron configurations l_e^n with $n \leqslant 2l_e + 1$, i.e. for electron shells less than halffull.

The other property of fractional parentage coefficients arises from the normalization condition for the wavefunction, namely,

$$\sum_{l,s} [G_{ls}^{LS}(l_e, N)]^2 = 1 .$$

Table A.1 gives the fractional parentage coefficients for the electron configurations p^n. Basic information about the role of fractional parentage coefficients may be found in [A.1–3].

References

A.1 E.U.Condon, G.H.Shortley: *The Theory of Atomic Spectra*, 4th ed. (Cambridge University Press, Cambridge 1964)

A.2 G.Racah: Phys. Rev. **62**, 438 (1942); ibid **63**, 367 (1943); ibid **76**, 1352 (1949)

A.3 I.I.Sobelman: *Atomic Spectra and Radiative Transitions*, Springer Ser. Chem. Phys., Vol. 1 (Springer, Berlin, Heidelberg, New York 1979)

Table A.1. Fractional parentage coefficients $G_{S'L'}^{SL}$ for configurations p^n

Electron configuration and state of atomic core: $p^{n-1}(S'L')$	Valence electron configuration and atomic state: $p^n(SL)$	$G_{S'L'}^{SL}$
$p(^2P)$	$p^2(^3P)$	1
	$p^2(^1D)$	1
	$p^2(^1S)$	1
$p^2(^3P)$	$p^3(^4S)$	1
	$p^3(^2D)$	$1/\sqrt{2}$
	$p^3(^2P)$	$-1/\sqrt{2}$
$p^2(^1D)$	$p^3(^4S)$	0
	$p^3(^2D)$	$-1/\sqrt{2}$
	$p^3(^2P)$	$-\sqrt{5/18}$
$p^2(^1S)$	$p^3(^4S)$	0
	$p^3(^2D)$	0
	$p^3(^2P)$	$\sqrt{2/3}$
$p^3(^4S)$	$p^4(^3P)$	$-1/\sqrt{3}$
	$p^4(^1D)$	0
	$p^4(^1S)$	0
$p^3(^2D)$	$p^4(^3P)$	$\sqrt{5/12}$
	$p^4(^1D)$	$\sqrt{3/4}$
	$p^4(^1S)$	0
$p^3(^2P)$	$p^4(^3P)$	$-1/2$
	$p^4(^1D)$	$-1/2$
	$p^4(^1S)$	1
$p^4(^3P)$	$p^5(^2P)$	$\sqrt{3/5}$
$p^4(^1D)$	$p^5(^2P)$	$\sqrt{1/3}$
$p^4(^1S)$	$p^5(^2P)$	$1/\sqrt{15}$

B. Clebsch-Gordan Coefficients

The Clebsch-Gordan coefficients are met in adding together the angular momentum vectors of two particles or angular momenta of different origin (for example, spin and orbital angular momenta) for one particle to obtain the total angular momentum of a system. Let us denote the total angular momentum of a system by j and the angular momenta of its separate parts by j_1 and j_2. Then the wavefunction ψ_{jm} of a system may be represented in the form of an expansion in terms of the wavefunctions for the separate subsystems:

$$\psi_{jm} = \Sigma\,(j_1 j_2 m_1 m_2 | j_1 j_2 jm)\;\psi_{j_1 m_1}\psi_{j_2 m_2}\;.$$

Here $\psi_{j_1 m_1}$, $\psi_{j_2 m_2}$ are the eigenfunctions of the operators of the summed angular momenta, which have definite values of the angular momentum and its projection on a given direction. The wavefunction ψ_{jm} describes the state of the total system with angular momentum j and projection m on the quantization axis. The coefficients of this expansion are called Clebsch-Gordan coefficients[1].

We shall list below some simple relations, which hold for Clebsch-Gordan coefficients, and also give their numerical values (Table B.1) and formulas (Table B.2) for relatively small angular momenta values. Extensive information about the properties of Clebsch-Gordan coefficients is given in monographs [B.1, 2].

1) The angular momentum projections sum algebraically:

$$m = m_1 + m_2\;.$$

2) Orthogonality relations:

$$\sum_{m_1 m_2} (j_1 j_2 m_1 m_2 | j_1 j_2 jm)(j_1 j_2 m_1 m_2 | j_1 j_2 j' m') = \delta_{jj'}\,\delta_{mm'}\;,$$

$$\sum_{jm} (j_1 j_2 m_1 m_2 | j_1 j_2 jm)(j_1 j_2 m_1' m_2' | j_1 j_2 jm) = \delta_{m_1 m_1'}\,\delta_{m_2 m_2'}\;.$$

3) Symmetry properties:

$$(j_1 j_2 m_1 m_2 | j_1 j_2 jm) = (-1)^{j_1 + j_2 - j}\,(j_2 j_1 m_2 m_1 | j_2 j_1 jm)$$

$$= (-1)^{j_1 + j_2 - j}(j_1 j_2 - m_1 - m_2 | j_1 j_2 j - m) = (j_2 j_1 - m_2 - m_1 | j_2 j_1 j - m)\;.$$

---

1 Other designations for Clebsch-Gordan coefficients are often used, namely $(j_1 j_2 m_1 m_2 | jm)$, $C^{jm}_{m_1 m_2}$, $C^{jm}_{j_1 m_1 j_2 m_2}$, or $C^{j_1\ j_2\ j}_{m_1 m_2 m}$.

4) Connection with Wigner $3j$ symbols:

$$(j_1 j_2 m_1 m_2 | j_1 j_2 j m) = (-1)^{j_1 - j_2 + m} \sqrt{2j+1} \begin{pmatrix} j_1 j_2 j \\ m_1 m_2 m \end{pmatrix} .$$

The numerical data on Clebsch-Gordan coefficients for small values of the arguments $j \leqslant 3$, $j_{1,2} \leqslant 5/2$ is presented in Table B.1. It includes the Clebsch-Gordan coefficients with arguments satisfying the conditions $j_1 \geqslant j_2$, $m_1 \geqslant m_2$. The symmetry properties can be used for angular momenta which do not satisfy these conditions. The numerical values listed in Table B.1 were taken from monograph [B.3]. Table B.2 incorporates the formulas for Clebsch-Gordan coefficients with arguments $j_2 = 1/2$, 1, 3/2 and 2 [B.1–4].

References

B.1 E.U.Condon, G.H.Shortley: *The Theory of Atomic Spectra*, 4th ed. (Cambridge University Press, Cambridge 1964)

B.2 A.Edmonds: *Angular Momentum in Quantum Mechanics* (Princeton University Press, Princeton 1968)

B.3 D.A.Varshalovitch, A.N.Moskalev, V.K.Hersonsky: *Quantum Theory of Angular Momentum* (Nauka, Leningrad 1975) (in Russian)

B.4 M.A.Morrison, T.L.Estle, N.F.Lane: *Quantum States of Atoms, Molecules, and Solids* (Prentice-Hall, Englewood Cliffs 1976) Appendix 4, p. 557

Table B.1. Numerical values of Clebsch-Gordan coefficients for $j \leqslant 3$, $j_2 \leqslant j_1 \leqslant 5/2$

j_1	m_1	j_2	m_2	j	m	$(j_1 j_2 m_1 m_2 \| j_1 j_2 j m)$	
0	0	0	0	0	0	1	1
1/2	1/2	0	0	1/2	1/2	1	1
		1/2	1/2	1	1	1	1
			−1/2	0	0	$1/\sqrt{2}$	0.7071
				1		$1/\sqrt{2}$	0.7071
1	0	0	0	1	0	1	1
		1/2	1/2	1/2	1/2	$-1/\sqrt{3}$	−0.5774
				3/2		$\sqrt{2/3}$	0.8165
		1	0	0	0	$-1/\sqrt{3}$	−0.5774
				1		0	0
				2		$\sqrt{2/3}$	0.8165
			−1	1	−1	$1/\sqrt{2}$	0.7071
				2		$1/\sqrt{2}$	0.7071
	1	0	0	1	1	1	1
		1/2	1/2	3/2	3/2	1	1
			−1/2	1/2	1/2	$\sqrt{2/3}$	0.8165
				3/2		$1/\sqrt{3}$	0.5774
		1	0	1	1	$1/\sqrt{2}$	0.7071
				2		$1/\sqrt{2}$	0.7071
			−1	0	0	$1/\sqrt{3}$	0.5774
				1		$1/\sqrt{2}$	0.7071

Table B.1 (continued)

j_1	m_1	j_2	m_2	j	m	$(j_1 j_2 m_1 m_2 \mid j_1 j_2 j m)$	
				2		$1/\sqrt{6}$	0.4082
			1	2	2	1	1
3/2	1/2	0	0	3/2	1/2	1	1
		1/2	1/2	1	1	$-1/2$	-0.5
				2		$\sqrt{3}/2$	0.8660
			$-1/2$	1	0	$1/\sqrt{2}$	0.7071
				2		$1/\sqrt{2}$	0.7071
		1	0	1/2	1/2	$-1/\sqrt{3}$	-0.5774
				3/2		$1/\sqrt{15}$	0.2582
				5/2		$\sqrt{3/5}$	0.7746
		1	-1	1/2	$-1/2$	$1/\sqrt{6}$	0.4082
				3/2		$2\sqrt{2}/\sqrt{15}$	0.7303
				5/2		$\sqrt{3/10}$	0.5477
		1	1	3/2	3/2	$-\sqrt{2/5}$	-0.6324
				5/2		$\sqrt{3/5}$	0.7746
		3/2	1/2	1	1	$-\sqrt{2/5}$	-0.6324
				2		0	0
				3		$\sqrt{3/5}$	0.7746
			$-1/2$	0	0	$-1/2$	-0.5
				1		$-1/2\sqrt{5}$	-0.2236
				2		$1/2$	0.5
				3		$3/2\sqrt{5}$	0.6708
			$-3/2$	1	-1	$\sqrt{3/10}$	0.5477
				2		$-1/\sqrt{2}$	-0.7071
				3		$1/\sqrt{5}$	0.4472
3/2	3/2	0	0	3/2	3/2	1	1
		1/2	1/2	2	2	1	1
			$-1/2$	1	1	$\sqrt{3}/2$	0.8660
				2		$1/\sqrt{2}$	0.7071
		1	0	3/2	3/2	$\sqrt{3/5}$	0.7746
				5/2		$\sqrt{2/5}$	0.6324
		1	-1	1/2	1/2	$1/\sqrt{2}$	0.7071
				3/2		$\sqrt{2/5}$	0.6324
				5/2		$1/\sqrt{10}$	0.3162
		1	1	5/2	5/2	$\sqrt{3/5}$	0.7746
		3/2	1/2	2	2	$1/\sqrt{2}$	0.7071
				3		$1/\sqrt{2}$	0.7071
			3/2	3	3	1	1
			$-3/2$	0	0	$1/2$	0.5
				1		$3/2\sqrt{5}$	0.6708
				2		$1/2$	0.5
				3		$1/2\sqrt{5}$	0.2236
			$-1/2$	1	1	$-1/2\sqrt{5}$	-0.2236
				2		$1/\sqrt{2}$	0.7071
				3		$1/\sqrt{5}$	0.4472
2	0	0	0	2	0	1	1
		1/2	1/2	3/2	1/2	$-\sqrt{2/5}$	-0.6325
				5/2		$\sqrt{3/5}$	0.7746
		1	0	1	0	$-\sqrt{2/5}$	-0.6325
				2		0	0

Table B.1 (continued)

j_1	m_1	j_2	m_2	j	m	$(j_1 j_2 m_1 m_2 \mid j_1 j_2 j m)$	
				3		$\sqrt{3/5}$	0.7746
2	0	1	−1	1	−1	$1/\sqrt{10}$	0.3162
				2		$1/\sqrt{2}$	0.7071
				3		$\sqrt{2/5}$	0.6325
		3/2	1/2	1/2	1/2	$1/\sqrt{5}$	0.4472
				3/2		$-1/\sqrt{5}$	−0.4472
				5/2		$-\sqrt{3/35}$	−0.2928
		3/2	3/2	3/2	3/2	$1/\sqrt{5}$	0.4472
				5/2		$-3\sqrt{2/35}$	−0.7171
		2	0	1	0	0	0
				2		$-\sqrt{2/7}$	−0.5345
				3		0	0
		2	−1	1	−1	$-\sqrt{3/10}$	−0.5477
				2		$-1/\sqrt{14}$	−0.2673
				3		$1/\sqrt{5}$	0.4472
			−2	2	−2	$\sqrt{2/7}$	0.5345
				3		$1/\sqrt{2}$	0.7071
2	1	0	0	2	1	1	1
		1/2	1/2	3/2	3/2	$-1/\sqrt{5}$	−0.4472
				5/2		$2/\sqrt{5}$	0.8944
			−1/2	3/2	1/2	$\sqrt{3/5}$	0.7746
				5/2		$\sqrt{2/5}$	0.6325
		1	0	1	1	$-\sqrt{3/10}$	−0.5477
				2		$1/\sqrt{6}$	0.4082
				3		$2\sqrt{2/15}$	0.7303
		1	1	2	2	$-1/\sqrt{3}$	−0.5774
				3		$1/\sqrt{3}$	0.5774
2	1	1	−1	1	0	$\sqrt{3/5}$	0.7746
				2		$1/\sqrt{2}$	0.7071
				3		$1/\sqrt{5}$	0.4472
		3/2	1/2	3/2	3/2	$-\sqrt{2/5}$	−0.6325
				5/2		$1/\sqrt{35}$	0.1690
			3/2	5/2	5/2	$-\sqrt{3/7}$	0.6546
			−1/2	1/2	1/2	$-\sqrt{3/10}$	−0.5477
				3/2		0	0
				5/2		$\sqrt{5/14}$	0.5976
			−3/2	1/2	−1/2	$1/\sqrt{10}$	0.3162
				3/2		$\sqrt{2/5}$	0.6325
				5/2		$3\sqrt{3}/\sqrt{10}$	0.6211
2	2	0	0	2	2	1	1
		1/2	1/2	5/2	5/2	$\sqrt{3/5}$	0.7746
			−1/2	3/2	3/2	$2/\sqrt{5}$	0.8944
				5/2		$\sqrt{2/5}$	0.6325
		1	0	2	2	$\sqrt{2/3}$	0.8165
				3		$1/\sqrt{3}$	0.5774
			1	3	3	1	1
			−1	1	1	$\sqrt{3/5}$	0.7746
				2		$1/\sqrt{3}$	0.5774
				3		$1/\sqrt{15}$	0.2582
		3/2	1/2	5/2	5/2	$2/\sqrt{7}$	0.7559

Table B.1 (continued)

j_1	m_1	j_2	m_2	j	m	$(j_1j_2m_1m_2\|j_1j_2jm)$	
			$-1/2$	3/2	3/2	$\sqrt{2/5}$	0.6325
				5/2		$4/\sqrt{35}$	0.6761
			$-3/2$	1/2	1/2	$\sqrt{2/5}$	0.6325
				3/2		$\sqrt{2/5}$	0.6325
				5/2		$\sqrt{6/35}$	0.4140
2	2	2	1	3	3	$1/\sqrt{2}$	0.7071
		2	-1	1	1	$1/\sqrt{5}$	0.4472
				2		$\sqrt{3/7}$	0.6547
				3		$\sqrt{3/10}$	0.5477
		2	-2	0	0	$1/\sqrt{5}$	0.4472
				1		$\sqrt{2/5}$	0.6325
				2		$\sqrt{2/7}$	0.5345
				3		$1/\sqrt{10}$	0.3162
5/2	1/2	0	0	5/2	1/2	1	1
		1/2	1/2	2	1	$-1/\sqrt{3}$	-0.5774
				3		$\sqrt{2/3}$	0.8165
		1/2	$-1/2$	2	0	$1/\sqrt{2}$	0.7071
				3		$1/\sqrt{2}$	0.7071
		1	0	3/2	1/2	$-\sqrt{2/5}$	-0.6325
				5/2		$1/\sqrt{35}$	0.1690
		1	1	3/2	3/2	$1/\sqrt{15}$	0.2582
				5/2		$-4/\sqrt{35}$	-0.6761
		1	-1	3/2	$-1/2$	$1/\sqrt{5}$	0.4472
				5/2		$3\sqrt{2}/\sqrt{35}$	0.7171
5/2	1/2	3/2	1/2	1	1	$\sqrt{3/2}\sqrt{5}$	0.3873
				2		$-5/2\sqrt{21}$	-0.5455
				3		$-1/2\sqrt{15}$	-0.1291
			3/2	2	2	$1/\sqrt{7}$	0.3780
				3		$-1/\sqrt{2}$	0.7071
			$-1/2$	1	0	$-\sqrt{3/10}$	-0.5477
				2		$-1/\sqrt{14}$	-0.2673
				3		$1/\sqrt{5}$	0.4472
			$-3/2$	1	-1	$1/2\sqrt{5}$	0.2236
				2		$3/2\sqrt{7}$	0.5669
				3		$3/2\sqrt{5}$	0.6708
		2	0	1/2	1/2	$1/\sqrt{5}$	0.4472
				3/2		$-\sqrt{2/35}$	-0.2391
				5/2		$-2\sqrt{2}/\sqrt{35}$	-0.4781
		2	1	3/2	3/2	$3/\sqrt{35}$	0.5071
				5/2		$-\sqrt{6/35}$	0.4141
		2	2	5/2	5/2	$\sqrt{3/14}$	0.4629
5/2	1/2	2	-1	1/2	$-1/2$	$-\sqrt{2/15}$	-0.3651
				3/2		$-\sqrt{5/21}$	-0.4880
				5/2		0	0
			-2	3/2	$-3/2$	$2/\sqrt{35}$	0.3381
				5/2		$\sqrt{27/70}$	0.6211
		5/2	1/2	1	1	$3/\sqrt{35}$	0.5071
				2		0	0
				3		$-2/\sqrt{15}$	-0.5164
			$-1/2$	0	0	$-1/\sqrt{6}$	0.4082

Table B.1 (continued)

j_1	m_1	j_2	m_2	j	m	$(j_1 j_2 m_1 m_2 \mid j_1 j_2 j m)$	
				1		$1/\sqrt{70}$	0.1195
				2		$-2/\sqrt{35}$	−0.4364
				3		$-2/3\sqrt{5}$	−0.2981
			−3/2	1	−1	$-\sqrt{8/35}$	−0.4781
				2		$1/\sqrt{7}$	0.3780
				3		$1/\sqrt{70}$	0.1826
			−5/2	2	−2	$-\sqrt{5}/2\sqrt{7}$	−0.4226
				3		$-\sqrt{5}/2\sqrt{3}$	−0.6455
5/2	3/2	0	0	5/2	3/2	1	1
		1/2	1/2	2	2	$-1/\sqrt{6}$	−0.4082
				3		$\sqrt{5/6}$	0.9129
			−1/2	2	1	$\sqrt{2/3}$	0.8165
				3		$1/\sqrt{3}$	0.5774
		1	0	3/2	3/2	$-2/\sqrt{15}$	−0.5164
				5/2		$3/\sqrt{35}$	0.5071
			1	5/2	5/2	$-\sqrt{2/7}$	−0.5345
			−1	3/2	1/2	$\sqrt{2/5}$	0.6324
				5/2		$\sqrt{2/7}$	0.5345
5/2	3/2	3/2	1/2	2	2	$-2\sqrt{2}/\sqrt{21}$	−0.5172
				3		$\sqrt{5}/2\sqrt{2}$	0.7906
			3/2	3	3	$-\sqrt{3}/2\sqrt{2}$	−0.6124
			−1/2	1	1	$-\sqrt{3/10}$	−0.5477
				2		$1/\sqrt{42}$	0.1543
				3		$\sqrt{5}/2\sqrt{3}$	0.6455
			−3/2	1	0	$1/\sqrt{5}$	0.4472
				2		$\sqrt{3/7}$	0.6546
				3		$1/2\sqrt{2}$	0.3536
5/2	3/2	2	0	3/2	3/2	$-2\sqrt{3}/\sqrt{35}$	−0.5855
				5/2		$-1/\sqrt{70}$	−0.1195
			1	5/2	5/2	$-\sqrt{3/7}$	−0.6546
			−1	1/2	1/2	$-2/\sqrt{15}$	−0.5164
				3/2		$-\sqrt{2/105}$	−0.1380
				5/2		$\sqrt{6/35}$	0.4140
			−2	1/2	−1/2	$1/\sqrt{15}$	0.2582
				3/2		$4\sqrt{2}/\sqrt{105}$	0.5520
				5/2		$\sqrt{27/70}$	0.6211
		5/2	1/2	2	2	$-3/2\sqrt{7}$	−0.5670
				3		$-1/2\sqrt{3}$	−0.2887
			3/2	3	3	$-2/3$	−0.6667
			−1/2	1	1	$-\sqrt{8/35}$	−0.4781
				2		$-1/\sqrt{7}$	−0.3780
				3		$1/\sqrt{70}$	0.1826
			−3/2	0	0	$-1/\sqrt{6}$	−0.4082
				1		$-3/\sqrt{70}$	−0.3587
				2		$1/2\sqrt{21}$	0.1091
				3		$7/6\sqrt{5}$	0.5218
			−5/2	1	−1	$1/\sqrt{7}$	0.3780
				2		$\sqrt{5/14}$	0.5976
				3		$1/\sqrt{3}$	0.5774
5/2	5/2	0	0	5/2	5/2	1	1

Table B.1 (continued)

j_1	m_1	j_2	m_2	j	m	$(j_1j_2m_1m_2\|j_1j_2jm)$	
		1/2	1/2	3	3	1	1
		1/2	−1/2	2	2	$\sqrt{5/6}$	0.9129
				3		$1/\sqrt{6}$	0.4082
5/2	5/2	1	0	5/2	5/2	$\sqrt{5/7}$	0.8452
			−1	3/2	3/2	$\sqrt{2/3}$	0.8165
				5/2		$\sqrt{2/7}$	0.5345
		3/2	1/2	3	3	$\sqrt{5/8}$	0.7906
			−1/2	2	2	$\sqrt{10/21}$	0.6901
				3		$\sqrt{5/12}$	0.6455
			−3/2	1	1	$1/\sqrt{2}$	0.7071
				2		$\sqrt{5/14}$	0.5976
				3		$1/2\sqrt{2}$	0.3536
		2	0	5/2	5/2	$\sqrt{5/14}$	0.5976
			−1	3/2	3/2	$\sqrt{2/7}$	0.5345
				5/2		$\sqrt{3/7}$	0.6546
			−2	1/2	1/2	$1/\sqrt{3}$	0.5774
				3/2		$2\sqrt{2/21}$	0.6172
				5/2		$\sqrt{3/14}$	0.4629
		5/2	1/2	3	3	$\sqrt{5}/3\sqrt{2}$	0.5270
			−1/2	2	2	$\sqrt{5}/2\sqrt{7}$	0.4226
				3		$\sqrt{5}/2\sqrt{3}$	0.6455
			−3/2	1	1	$1/\sqrt{7}$	0.3780
				2		$\sqrt{5/14}$	0.5976
				3		$1/\sqrt{3}$	0.5774
			−5/2	0	0	$1/\sqrt{6}$	0.4082
				1		$\sqrt{5/14}$	0.5976
				2		$5/2\sqrt{21}$	0.5455
				3		$\sqrt{5}/6$	0.3727

Table B.2. Formulas for Clebsch-Gordan coefficients for $j_2 = 1/2,\ 1,\ 3/2$ and 2

$\left(j_1\ \frac{1}{2}\ m_1 m_2 \middle| j_1\ \frac{1}{2}\ jm\right)$

$j =$	$m_2 = \frac{1}{2}$	$m_2 = -\frac{1}{2}$
$j_1 + \frac{1}{2}$	$\sqrt{\dfrac{j_1 + m + \frac{1}{2}}{2j_1 + 1}}$	$\sqrt{\dfrac{j_1 - m + \frac{1}{2}}{2j_1 + 1}}$
$j_1 - \frac{1}{2}$	$-\sqrt{\dfrac{j_1 - m + \frac{1}{2}}{2j_1 + 1}}$	$\sqrt{\dfrac{j_1 + m + \frac{1}{2}}{2j_1 + 1}}$

$(j_1 1 m_1 m_2 | j_1 1 jm)$

$j =$	$m_2 = 1$	$m_2 = 0$	$m_2 = -1$
$j_1 + 1$	$\sqrt{\dfrac{(j_1 + m)(j_1 + m + 1)}{(2j_1 + 1)(2j_1 + 2)}}$	$\sqrt{\dfrac{(j_1 - m + 1)(j_1 + m + 1)}{(2j_1 + 1)(j_1 + 1)}}$	$\sqrt{\dfrac{(j_1 - m)(j_1 - m + 1)}{(2j_1 + 1)(2j_1 + 2)}}$
j_1	$-\sqrt{\dfrac{(j_1 + m)(j_1 - m + 1)}{2j_1(j_1 + 1)}}$	$\dfrac{m}{\sqrt{j_1(j_1 + 1)}}$	$\sqrt{\dfrac{(j_1 - m)(j_1 + m + 1)}{2j_1(j_1 + 1)}}$
$j_1 - 1$	$\sqrt{\dfrac{(j_1 - m)(j_1 - m + 1)}{2j_1(2j_1 + 1)}}$	$-\sqrt{\dfrac{(j_1 - m)(j_1 + m)}{j_1(2j_1 + 1)}}$	$\sqrt{\dfrac{(j_1 + m + 1)(j_1 + m)}{2j_1(2j_1 + 1)}}$

$\left(j_1 \frac{3}{2} m_1 m_2 \middle| j_1 \frac{3}{2} j m\right)$

$j =$	$m_2 = \frac{3}{2}$	$m_2 = \frac{1}{2}$
$j_1 + \frac{3}{2}$	$\sqrt{\frac{\left(j_1 + m - \frac{1}{2}\right)\left(j_1 + m + \frac{1}{2}\right)\left(j_1 + m + \frac{3}{2}\right)}{(2j_1 + 1)(2j_1 + 2)(2j_1 + 3)}}$	$\sqrt{\frac{3\left(j_1 + m + \frac{1}{2}\right)\left(j_1 + m + \frac{3}{2}\right)\left(j_1 - m + \frac{3}{2}\right)}{(2j_1 + 1)(2j_1 + 2)(2j_1 + 3)}}$
$j_1 + \frac{1}{2}$	$-\sqrt{\frac{3\left(j_1 + m - \frac{1}{2}\right)\left(j_1 + m + \frac{1}{2}\right)\left(j_1 - m + \frac{3}{2}\right)}{2j_1(2j_1 + 1)(2j_1 + 3)}}$	$-\left(j_1 - 3m + \frac{3}{2}\right)\sqrt{\frac{j_1 + m + \frac{1}{2}}{2j_1(2j_1 + 1)(2j_1 + 3)}}$
$j_1 - \frac{1}{2}$	$\sqrt{\frac{3\left(j_1 + m - \frac{1}{2}\right)\left(j_1 - m + \frac{1}{2}\right)\left(j_1 - m + \frac{3}{2}\right)}{(2j_1 - 1)(2j_1 + 1)(2j_1 + 2)}}$	$-\left(j_1 + 3m - \frac{1}{2}\right)\sqrt{\frac{j_1 - m + \frac{1}{2}}{(2j_1 - 1)(2j_1 + 1)(2j_1 + 2)}}$
$j_1 - \frac{3}{2}$	$-\sqrt{\frac{\left(j_1 - m - \frac{1}{2}\right)\left(j_1 - m + \frac{1}{2}\right)\left(j_1 - m + \frac{3}{2}\right)}{2j_1(2j_1 - 1)(2j_1 + 1)}}$	$\sqrt{\frac{3\left(j_1 + m - \frac{1}{2}\right)\left(j_1 - m - \frac{1}{2}\right)\left(j_1 - m + \frac{1}{2}\right)}{2j_1(2j_1 - 1)(2j_1 + 1)}}$
$j =$	$m_2 = -\frac{1}{2}$	$m_2 = -\frac{3}{2}$

Table B.2 (continued)

$j_1 + \frac{3}{2}$	$\sqrt{\frac{3\left(j_1 + m + \frac{3}{2}\right)\left(j_1 - m + \frac{1}{2}\right)\left(j_1 - m + \frac{3}{2}\right)}{(2j_1 + 1)(2j_1 + 2)(2j_1 + 3)}}$	$\sqrt{\frac{\left(j_1 - m - \frac{1}{2}\right)\left(j_1 - m + \frac{1}{2}\right)\left(j_1 - m + \frac{3}{2}\right)}{(2j_1 + 1)(2j_1 + 2)(2j_1 + 3)}}$
$j_1 + \frac{1}{2}$	$\left(j_1 + 3m + \frac{3}{2}\right)\sqrt{\frac{j_1 - m + \frac{1}{2}}{2j_1(2j_1 + 1)(2j_1 + 3)}}$	$\sqrt{\frac{3\left(j_1 + m + \frac{3}{2}\right)\left(j_1 - m - \frac{1}{2}\right)\left(j_1 - m + \frac{1}{2}\right)}{2j_1(2j_1 + 1)(2j_1 + 3)}}$
$j_1 - \frac{1}{2}$	$-\left(j_1 - 3m - \frac{1}{2}\right)\sqrt{\frac{j_1 + m + \frac{1}{2}}{(2j_1 - 1)(2j_1 + 1)(2j_1 + 2)}}$	$\sqrt{\frac{3\left(j_1 + m + \frac{1}{2}\right)\left(j_1 + m + \frac{3}{2}\right)\left(j_1 - m - \frac{1}{2}\right)}{(2j_1 - 1)(2j_1 + 1)(2j_1 + 2)}}$
$j_1 - \frac{3}{2}$	$-\sqrt{\frac{3\left(j_1 + m - \frac{1}{2}\right)\left(j_1 + m + \frac{1}{2}\right)\left(j_1 - m - \frac{1}{2}\right)}{2j_1(2j_1 - 1)(2j_1 + 1)}}$	$\sqrt{\frac{\left(j_1 + m - \frac{1}{2}\right)\left(j_1 + m + \frac{1}{2}\right)\left(j_1 + m + \frac{3}{2}\right)}{2j_1(2j_1 - 1)(2j_1 + 1)}}$

$(j_1 2 m_1 m_2 \mid j_1 2 j m)$ $j =$	$m_2 = 2$	$m_2 = 1$
$j_1 + 2$	$\sqrt{\frac{(j_1 + m - 1)(j_1 + m)(j_1 + m + 1)(j_1 + m + 2)}{(2j_1 + 1)(2j_1 + 2)(2j_1 + 3)(2j_1 + 4)}}$	$\sqrt{\frac{(j_1 - m + 2)(j_1 + m + 2)(j_1 + m + 1)(j_1 + m)}{(2j_1 + 1)(j_1 + 1)(2j_1 + 3)(j_1 + 2)}}$
$j_1 + 1$	$-\sqrt{\frac{(j_1 + m - 1)(j_1 + m)(j_1 + m + 1)(j_1 - m + 2)}{2j_1(j_1 + 1)(j_1 + 2)(2j_1 + 1)}}$	$-(j_1 - 2m + 2)\sqrt{\frac{(j_1 + m + 1)(j_1 + m)}{2j_1(2j_1 + 1)(j_1 + 1)(j_1 + 2)}}$

j_1	$\sqrt{\frac{3(j_1+m-1)(j_1+m)(j_1-m+1)(j_1-m+2)}{(2j_1-1)\,2j_1(j_1+1)(2j_1+3)}}$	$(1-2m)\sqrt{\frac{3(j_1-m+1)(j_1+m)}{(2j_1-1)\,j_1(2j_1+2)(2j_1+3)}}$
j_1-1	$-\sqrt{\frac{(j_1+m-1)(j_1-m)(j_1-m+1)(j_1-m+2)}{2(j_1-1)\,j_1(j_1+1)(2j_1+1)}}$	$(j_1+2m-1)\sqrt{\frac{(j_1-m+1)(j_1-m)}{(j_1-1)\,j_1(2j_1+1)(2j_1+2)}}$
j_1-2	$\sqrt{\frac{(j_1-m-1)(j_1-m)(j_1-m+1)(j_1-m+2)}{(2j_1-2)(2j_1-1)\,2j_1(2j_1+1)}}$	$-\sqrt{\frac{(j_1-m+1)(j_1-m)(j_1-m-1)(j_1+m-1)}{(j_1-1)(2j_1-1)\,j_1(2j_1+1)}}$
$j=$	$m_2=0$	$m_2=-1$
j_1+2	$\sqrt{\frac{3(j_1-m+2)(j_1-m+1)(j_1+m+2)(j_1+m+1)}{(2j_1+1)(2j_1+2)(2j_1+3)(j_1+2)}}$	$\sqrt{\frac{(j_1-m+2)(j_1-m+1)(j_1-m)(j_1+m+2)}{(2j_1+1)(j_1+1)(2j_1+3)(j_1+2)}}$
j_1+1	$m\sqrt{\frac{3(j_1-m+1)(j_1+m+1)}{j_1(2j_1+1)(j_1+1)(j_1+2)}}$	$(j_1+2m+2)\sqrt{\frac{(j_1-m+1)(j_1-m)}{j_1(2j_1+1)(2j_1+2)(j_1+2)}}$
j_1	$\frac{3m^2-j_1(j_1+1)}{\sqrt{(2j_1-1)\,j_1(j_1+1)(2j_1+3)}}$	$(2m+1)\sqrt{\frac{3(j_1-m)(j_1+m+1)}{(2j_1-1)\,j_1(2j_1+2)(2j_1+3)}}$
j_1-1	$-m\sqrt{\frac{3(j_1-m)(j_1+m)}{(j_1-1)\,j_1(2j_1+1)(j_1+1)}}$	$-(j_1-2m-1)\sqrt{\frac{(j_1+m+1)(j_1+m)}{(j_1-1)\,j_1(2j_1+1)(2j_1+2)}}$
j_1-2	$\sqrt{\frac{3(j_1-m)(j_1-m-1)(j_1+m)(j_1+m-1)}{(2j_1-2)(2j_1-1)\,j_1(2j_1+1)}}$	$-\sqrt{\frac{(j_1-m-1)(j_1+m+1)(j_1+m)(j_1+m-1)}{(j_1-1)(2j_1-1)\,j_1(2j_1+1)}}$

Table B.2 (continued)

$j =$	$m_2 = -2$
$j_1 + 2$	$\sqrt{\frac{(j_1 - m - 1)(j_1 - m)(j_1 - m + 1)(j_1 - m + 2)}{(2j_1 + 1)(2j_1 + 2)(2j_1 + 3)(2j_1 + 4)}}$
$j_1 + 1$	$\sqrt{\frac{(j_1 - m - 1)(j_1 - m)(j_1 - m + 1)(j_1 + m + 2)}{j_1(2j_1 + 1)(j_1 + 1)(2j_1 + 4)}}$
j_1	$\sqrt{\frac{3(j_1 - m - 1)(j_1 - m)(j_1 + m + 1)(j_1 + m + 2)}{(2j_1 - 1)\, j_1\, (2j_1 + 2)(2j_1 + 3)}}$
$j_1 - 1$	$\sqrt{\frac{(j_1 - m - 1)(j_1 + m)(j_1 + m + 1)(j_1 + m + 2)}{(j_1 - 1)\, j_1\, (2j_1 + 1)(2j_1 + 2)}}$
$j_1 - 2$	$\sqrt{\frac{(j_1 + m - 1)(j_1 + m)(j_1 + m + 1)(j_1 + m + 2)}{(2j_1 - 2)(2j_1 - 1)\, 2j_1\, (2j_1 + 1)}}$

Subject Index

References to tables are shown by the letter T; references to figures are shown by the letter F.